P9-EDR-808

MACMILLAN ENCYCLOPEDIA OF
ENERGY

EDITORIAL BOARD

MACMILLAN ENCYCLOPEDIA OF ENERGY

JOHN ZUMERCHIK

Editor in Chief

VOLUME 3

Macmillan Reference USA

an imprint of the Gale Group

New York • Detroit • San Francisco • London • Boston • Woodbridge, CT

Macmillan Encyclopedia of Energy

Macmillan Reference USA
1633 Broadway
New York, NY 10019

Macmillan Reference USA
27500 Drake Rd.
Farmington Hills, MI 48331-3535

Library of Congress Catalog Card Number: 00-062498

Printed in the United States of America
Printing number
1 2 3 4 5 6 7 8 9 10

Library of Congress Cataloging-in-Publication Data
Macmillan encyclopedia of energy / John Zumerchik, editor in chief.
p. cm.
Includes bibliographical references and index.
ISBN 0-02-865021-2 (set : hc). — ISBN 0-02-865018-2 (vol. 1). — ISBN 0-02-865019-0 (vol. 2). — ISBN 0-02-865020-4 (vol. 3).
1. Power resources—Encyclopedias. I. Title: Encyclopedia of Energy. II. Zumerchik, John
TJ163.25 .M33 2000
00-062498
621.042'03 dc21

OFFICE EQUIPMENT

TYPES OF EQUIPMENT

Energy use for office equipment ranks behind only lighting, heating, and air conditioning in the commercial sector. And because of growing use of office equipment, the hope is that more efficient equipment will offset the greater use.

Office equipment is an end use that generally is regarded to consist of five major products: personal computers, monitors, copiers, printers, and fax machines. This equipment can be defined by characteristics such as size and speed. Although office equipment is used in homes, this article focuses on office equipment energy use in actual offices.

Personal Computers (PCs)

PCs are desktop and deskside microcomputers that generally include a central processing unit (CPU), storage, and keyboard. PCs first appeared in office spaces during the 1980s. Prior to that time, many offices used simple terminals connected to mainframe computers. PC energy use assessments generally do not include laptops and PC servers—for example, new systems such as webservers—that are left on twenty-four hours a day.

Monitors

Most monitors are display terminals that use cathode-ray tube (CRT) displays, which function by exciting a layer of phosphors with an electron gun. These devices include monitors used with PCs and terminals used with mainframes or minicomputers. Features such as color, resolution, and size influence power requirements. Most PC monitors are sold in screen sizes of fourteen inches, fifteen inches, seventeen inches, and twenty-one inches. In 1995, the typical monitor was fourteen inches and monochrome; by 2000, PC users were purchasing larger, higher-resolution color CRTs. Flat-panel displays such as those used in laptop computers are beginning to replace CRT monitors in desktop computers, though flat-panel models in 2000 cost considerably more than CRTs for the same display size. There are three types of flat-panel displays: liquid crystal (LCD), plasma, and electroluminescent emission (EL). Color LCDs are increasing their share of the monitor market because resolution is improving, costs are dropping, and they can be powered up and down more rapidly than most CRTs.

Copiers

The majority of copiers in the commercial sector use heat and pressure technologies to fix an image to paper. The basic principle consists of forming an image on a photosensitive drum with a laser or a lamp. The drum is then covered with toner, which is transferred to a sheet of paper and fixed to the paper by a fuser roll. The fuser is kept hot while the machine is in a standby mode, ready for the next copy. Some copiers have a low-power "energy-saver" mode in which the drum temperature is reduced; others have an "automatic off" feature, and some have both. Several factors influence energy use per page, such as whether duplexing (two-sided printing) is used and the size of the copy job. Single-page jobs are the most energy-intensive on a per-copy basis.

Copier characteristics vary with the speed of output, with faster machines adding more sophisticated handling of originals, duplexing, finishing of copies as with stapling and collating, and the ability to make more copies each month.

Transmeta Corp. CEO David R. Ditzel demonstrates the benefits of using the new "Crusoe" smart microprocessor and how it will revolutionize the field of mobile computing. (Corbis Corporation)

Printers

Since 1990 there has been rapid change from impact printing, such as daisy wheel and dot-matrix techniques, to nonimpact printing, dominated by energy-intensive laser printing. As is true for copiers, printer energy use is generally linked to print speed. Laser printing is relatively energy-intensive because of the heat and pressure requirements for the fuser. Most of the energy is consumed while the printer sits idle, keeping the fuser warm.

Inkjet printers are suitable in low-volume settings such as personal offices and the home. Inkjets provide low-cost color capability. Power use while the printer is idle is generally low, so that a distinct low-power mode is not needed.

Fax Machines

Facsimile machines, or fax machines, send and receive information from printed documents or electronic files over telephone lines. Three common types of fax machines to consider are direct thermal, laser, and inkjet. Direct thermal faxes apply heat to thermally sensitive paper, while inkjet and laser fax machines are similar to the printers discussed in the previous section. In 2000, laser fax machines had the broadest market share and are of greatest energy use. There were only a few fax machines in 1990 compared to ten years later. It is likely that the technology may change quickly again from 2000 to 2010. Fax cards used with computers suggest the possibility of a future with reduced paper communication by eliminating the need to print a document before sending it. Most contemporary fax machines reduce their power use when not active.

POWER REQUIREMENTS AND ENERGY STAR FEATURES

Office equipment energy use can be characterized by the power requirements in each "mode" of operation and by the number of hours per year for each mode. In 1993 the U.S. Environmental Protection Agency launched the Energy Star program, beginning with power management in PCs and monitors. This program now promotes the use of power management technology in all office equipment (see Table 1). The power modes in PCs and monitors can be described as follows:

Active mode: This is the power of the device when in operation.
Standby mode: This mode represents an intermediate state that attempts to conserve power with instant recovery. The system is idle.
Suspend mode: This mode, also known as sleep mode, has the lowest power level (without being off) but has a longer recovery time than for standby.
Off mode: The power in this mode is that drawn when the device is switched off (essentially zero), or for copiers when the device is unplugged.

Many office equipment models consume some electricity when nominally "off." This results in "leaking" or "standby" energy use. Usually this amounts to just 2 or 3 W, but in some copiers can be up to considerably more. Power management modes vary most widely in PCs, with the number of modes and sophistication of operation increasing. Power management is most transparent to the user with fax machines.

EQUIPMENT POWER LEVELS, OPERATING PATTERNS, AND ANNUAL ENERGY USE

Energy use of office equipment is primarily a function of three factors: the power characteristics of the device, the operation of power management, and manual turn-off patterns. Table 2 shows typical power levels for office equipment, though within each category there is a significant range. Printers and copiers are subdivided into categories based on their maximum speed of making images (in pages or copies per minute), with power levels and usage patterns varying across these categories.

In addition to the power levels above, extra energy is required for each image made by a device. While this varies considerably, a value of 1 Wh/image is common. For copiers, about 10 percent of annual energy use is used for making images, with the other 90 percent primarily used to maintain the copier in the "ready to copy" mode.

Power management capability has reached most office equipment, but on many devices the feature is not enabled or (for some PCs or monitors) is prevented from functioning. Equipment with functioning power management or that is turned off manually at night and on weekends use considerably less, and devices left fully on constantly use much more.

The Energy Star program has affected the office equipment market in several ways: increasing the portion of equipment capable of automatic power management; reducing low-power power levels; adding auto-off capability to many copiers; and increasing enabling rates of power management. All of this has saved energy for consumers with little or no additional manufacturing cost. Monitor power management has risen with larger screens becoming more common, though as LCD monitors replace CRTs the average should drop. The number of PCs and monitors that are properly enabled is likely to increase.

Another aspect of the Energy Star program is that "efficient" copy machines not only have the potential to lower direct use of electricity, they also should

Equipment Category	Default Time to Low-Power State	Max. Power in Low-Power State	Date in Force
PC (without monitor)[1]	**N.A.[2]**	**30 W**	**mid-1993**
Monitors[1]	**N.A.[2]**	**30 W**	**mid-1993**
Printers and printer/fax combos:			
1-7 pages per min.	15 min.	15 W	Oct 1, 1995
8-14 pages per min.	30 min.	30 W	Oct 1, 1995
Color and/or >14 pages per min.	60 min.	45 W	Oct 1, 1995
Fax machines:			
1-7 pages per min.	5 min.	15 W	July 1, 1995
8-14 pages per min.	5 min.	30 W	July 1, 1995
>14 pages per min.	15 min.	45 W	July 1, 1995
	Default time to Low/Off	*Max Power Low/Off*	
Copiers-Tier 1[3]:			
1-20 copies per min.	N.A./30 min.	N.A./5W	July 1, 1995
21-44 copies per min.	N.A./60 min.	N.A./40W	July 1, 1995
>44 copies per min.[4]	N.A./90 min.	N.A./40W	July 1, 1995
Copiers-Tier 2[5]:			
1-20 copies per min.	N.A./30 min.	N.A./5W	July 1, 1997
21-44 copies per min.	15 min./60 min.	(3.85xcpm + 5W)/10W	July 1, 1997
>44 copies per min.	15 min./90 min.	(3.85xcpm + 5W)/15W	July 1, 1997

Table 1.
EPA Energy Star Office Equipment Characteristics
NOTES: (1) PCs and monitors are to be shipped with the power saving features enabled and those features must be tested in a networked environment. Some PCs qualify for a second standard of 15 percent of the rating of the power supply, usually at least 30 W. For monitors, an initial low-power mode of 15 W is also required. (2) "NA" means "not applicable," which means that no requirement exists. (3) Additional Tier 2 requirement for copiers includes a required recovery time of 30 seconds for midspeed copiers (recommended for high speed copiers).

reduce paper use by encouraging double-sided copying, or duplexing. Paper costs in 2000 are about $0.005/sheet. An average 40-cpm copier produces about 140,000 images per year. Approximately 124,600 pages are simplexed (copied on one side only) and 7,700 pages are duplexed (15,400 images). Doubling the duplex rate to handle 30,000 images per year saves about 15,000 pages per year or about $75 per year, which is nearly twice the value of the energy savings.

In addition to the direct savings from reduced energy and paper, there are indirect savings from the energy embodied in the paper. This is about 16 Wh per sheet. Reducing annual paper use by 15,000 pages cuts embodied energy use by 300 kWh per year. By comparison, an Energy Star copier saves about 260 kWh per year directly.

Energy Star office equipment typically does not cost any more than the non–Energy Star devices, and power-management technologies have greatly improved since 1990. The first Energy Star PCs had slow recovery times from the low-power modes, which was one of the reasons power management was disabled by users. Some network technologies have not been compatible with PC power management. The PC operating system and hardware characteristics also influence the success of the power management. Monitor power management is gener-

Year	Active (W)	Standby (W)	Suspend (W)
Pre-Energy Star	70	70	70
Energy Star	44	25	25
Low Power	44	15	15

Monitor Equipment Power

Year	Active (W)	Standby (W)	Suspend (W)
Pre-Energy Star	60	60	60
Typical Energy Star	65	50	15
Low Power	65	30	3

Laser Printer Equipment Power

Year	Active (W)	Standby (W)	Suspend (W)
Pre-Energy Star	250	80	80
Typical Energy Star	250	80	25
Low Power	250	33	10

Copier Equipment Power

Year	Active (W)	Standby (W)	Suspend (W)	Plug (W)
Pre-Energy Star	220	190	190	10
Typical Energy Star	220	190	150	10
Low Power	220	190	150	2

Fax Equipment Power

Year	Active (W)	Standby (W)
Pre-Energy Star	175	35
Typical Energy Star	175	15
Low Power	175	7

Table 2.
Typical Power Requirements by Product Category

ally simpler to operate and saves more energy than PC power management.

Overall, office equipment currently uses about 7 percent of all commercial sector electricity, with that fraction projected to grow to about 8 percent by 2010. Total electricity used by office equipment would grow from 58 TWh in 1990 to 78 TWh in 2010 in the absence of Energy Star or any other government policies. While total energy use for office equipment has grown rapidly in recent years, this growth is likely to slow from 2000 to 2010 because the market is becoming saturated and because mainframe and minicomputer energy use is declining. Home office energy use is on the rise, but is still well below the energy use of office equipment in commercial buildings.

The likely energy and dollar savings in the commercial sector from the Energy Star program are significant on a national scale. Total electricity savings are estimated to range from 10 to 23 TWh/year in 2010, most likely about 17 TWh/year. Energy bill savings will exceed $1 billion per year after the year 2000. The cost of achieving Energy Star efficiency levels is estimated to be negligible, while the cumulative direct cost of funding the Energy Star program is on the order of a few million dollars. This policy therefore saves U.S. society large amounts of money with minimal expenditure of public funds.

Mary Ann Piette
Bruce Nordman

See also: Conservation of Energy; Consumption; Economically Efficient Energy Choices; Efficiency of Energy Use.

BIBLIOGRAPHY

Kawamoto, K.; Koomey, J. G.; Nordman, B.; Brown, R. E.; Piette, M. A.; and Meier, A. K. (2000). "A Simplified Analysis of Electricity Used by Office Equipment and Network Equipment in the U.S." In *Proceedings of the ACEEE 2000 Summer Study on Energy Efficiency in Buildings*. Washington, DC: American Council for an Energy-Efficient Economy.

Koomey, J. G.; Cramer, M.; Piette, M. A.; and Eto, J. H. (1995). "Efficiency Improvements in U.S. Office Equipment: Expected Policy Impacts and Uncertainties." Report No. LBNL-41332. Berkeley, CA: Lawrence Berkely National Laboratory.

Nordman, B.; Piette, M. A.; Kinney, K.; and Webber, C. (1997). "User Guide to Power Management for PCs and Monitors." Report No. LBNL-39466. Berkeley, CA: Lawrence Berkeley National Laboratory.

Nordman, B.; Meier, A. K.; and Piette, M. A. (2000). "PC and Monitor Night Status: Power Management Enabling and Manual Turn-Off." In *Proceedings of the ACEEE 2000 Summer Study on Energy Efficiency in Buildings*. Washington, DC: American Council for an Energy-Efficient Economy.

U.S. Environmental Protection Agency. (1999). *Energy Star Office Equipment Program*. <http://www.epa.gov/office/>.

OFF-SHORE DRILLING RIGHTS

See: Oil and Gas, Drilling for; Property Rights

OIL AND GAS, DRILLING FOR

After all the exploratory analyses, drilling determines whether the exploration geophysicist has accurately located the reservoir (exploratory drilling) and whether the sites chosen for drilling into the same reservoir are optimal for efficient production (developmental drilling). When an exploratory hole produces neither oil nor gas, it is capped and abandoned. But if it does yield oil or gas, it is readied for production and becomes a completed well. To extract oil and gas requires drilling a well.

Drilling a well involves much more than making a hole. It entails the integration of complex technologies, requiring the driller to make individual decisions related to unexpected pressure regimes, practices, and rock formations. The resulting well is the sole conduit to move fluids from a reservoir to the surface—a conduit that must last at least fifty years and be flexible enough in design to allow for the application of future production technologies.

Drilling operators must confront and solve extremely difficult technical, safety, and control problems as they bore through layers of subsurface rock to access oil- or gas-bearing strata. Furthermore, drilling must protect the geologic formation, the ultimate productive capacity of the well, and the surface environment. Drilling problems must first be diagnosed using the information or data that is transmitted from the bottom of the well to the surface, where the information is collected on the rig floor.

Depending on the depth of the well, this time lag can consume valuable time needed to address the problem—either technical or geological—before it becomes worse and/or causes drilling operations to stop. Drilling a well involves all types of technical, geological, and economic risks. The greatest economic risk occurs when drilling operations must be halted after time and money have been invested. This is the nature of the challenge faced during the drilling process.

Oil rig workers begin drilling. (Corbis Corporation)

When a well has been drilled and lined with pipe, the connection between the geological formation and the well must be established. Well "completion" includes installing suitable metal pipe or casing, cementing this casing using rock section isolation devices, and perforating the casing to access the producing zones. In some reservoirs, the geological conditions dictate that "stimulation" processes be applied to improve reservoir permeability or fluid conductivity of the rock, thereby facilitating production through the well bore.

With an understanding of the nature of the rocks to be penetrated, the stresses in these rocks, and what it will take to drill into and control the underground reservoirs, the drilling engineer is entrusted to design the earth-penetration techniques and select the equipment. Substantial investments are at stake. Drilling and completion costs can exceed $400,000 per well, and offshore operations can easily escalate to more than $5 million per well. However, overall per-well costs (adjusted for inflation) have been dropping, falling more than 20 percent between 1980 and 2000. This technology-driven gain in extraction efficiency has more than made up for the higher costs associated with the increasingly more geologically complex operations and deeper depths faced in trying to extract the remaining oil and gas resources.

In the United States, 85 percent of wells are drilled by independent oil and gas operators, more than 90 percent of whom employ fewer than twenty people. Therefore, U.S. oil and gas production is dependent on the economic health of independent producers to offset the rising tide of imported crude oil.

Drilling is indeed a high-risk investment. Even with modern technology, drilling successful ex-ploratory wells can be elusive. In the United States, for exploratory wells drilled more than a mile from production, the chance of striking hydrocarbons are about one in ten, and exceeds one in forty for drilling in unproven frontier areas. Because of these daunting odds, the ever-increasing complexity of recovery, and the dwindling resource base, most energy experts in the late 1970s and early 1980s projected that the price of crude oil would double, triple, or quadruple by 2000. This did not happen. Profound advances in drilling technology as well as in other exploration and production technologies have allowed the inflation-adjusted price of crude oil to remain stable through the twentieth century except for short-term price distortions caused by geopolitical turmoil or misguided policies.

THE EARLY METHODS

On August 27, 1859, at a depth of only 69.5 feet, Francis Drake drilled and completed the first well at Titusville, Pennsylvania. By the end of the nineteenth century there were about 600,000 oil wells in more than 100 countries, with the United States and Russia dominating world production. In the United States, drilling was concentrated in the Appalachian oil fields until the Beaumont Texas, Spindletop Hill discovery on January 10, 1901, which then shifted exploration and production to the Southwest.

The world's largest petroleum reserves, in Turkey and the Middle East, were discovered just prior to World War I. Exploration and production continued to expand throughout the region in the 1920s, and culminated in the discovery of the vast oil resources of Saudi Arabia in 1938.

Early methods entailed cable-tool drilling. This crude impact-type drilling involved dropping a weighted bit by a cable and lever system to chisel away at the rock at the bottom of the hole. Periodically drilling had to be stopped to sharpen the bit and remove rock fragments and liquids from the bottom of the hole with a collecting device attached to the cable. Removal of liquids was necessary so the bit could more effectively chip away at rock. This dry method created the well-recognized "gusher," since a dry borehole allowed oil and gas to flow to the surface from the pressure of natural gas and water once the bit penetrated a producing reservoir. Usually considerable oil and "reservoir energy" were wasted until the well could be capped and controlled.

MODERN DRILLING

Rotary drilling became the preferred method of drilling oil and gas wells in the middle of the twentieth century (Figure 1). Using this method, the drill bit rotates as it pushes down at the bottom of the hole much like a hand-held power drill. Unlike in cable-tool drilling, the borehole is kept full of liquid during rotary drilling for two important reasons: The drilling mud circulates through the borehole to carry crushed rock to the surface to keep drilling continuous, and once the bit penetrates the reservoir, it does not create a gusher.

To drill a well, a site is selected and prepared. A drilling rig is transported to the site and set up. A surface hole is drilled, followed by drilling to the total planned depth of the well. The well is then tested, evaluated, and completed. Finally, production equipment is installed, and the well is put on production.

Drill-Site Selection and Preparation

Selection of the drill site is based largely on geological evidence indicating the possible accumulation of petroleum. The exploration drilling company wants to drill the well at the most advantageous location for the discovery of oil or gas. However, surface conditions also must be taken into consideration when selecting the drill site. There must be a nearly level area of sufficient size on which to set up the drilling rig, excavate reserve pits, and provide storage for all the materials and equipment that will be needed for the drilling program. All required legal matters must have been attended to, such as for acquiring a drilling permit and surveying of the drill site. When all of these matters have been resolved, work on site preparation will begin. Once the drill site has been selected and surveyed, a contractor or contractors will move in with equipment to prepare the location. If necessary, the site will be cleared and leveled. A large pit will be constructed to contain water for drilling operations and for disposal of drill cuttings and other waste. Many environmental regulations guide these practices. A small drilling rig, referred to as a dry-hole digger, will be used to start the main hole. A large-diameter hole will be drilled to a shallow depth and lined with conductor pipe. Sometimes a large, rectangular cellar is excavated around the main borehole and lined with wood. A smaller-diameter hole called a "rat hole" is drilled near the main borehole. The rat hole is lined with pipe and is used for temporary storage of a piece of drilling equipment called the "kelly." When all of this work has been completed, the drilling contractor will move in with the large drilling rig and all the equipment required to drill the well.

The Rig

The components of the drilling rig and all necessary equipment are moved onto the location with large, specially equipped trucks. The substructure of the rig is located and leveled over the main borehole. The mast or derrick is raised over the substructure, and other equipment such as engines, pumps, and rotating and hoisting equipment, are aligned and connected. The drill pipe and drill collars are laid out on racks convenient to the rig floor so that they may be hoisted when needed and connected to the drill bit or added to the drill strings. Water and fuel tanks are filled. Additives for the drilling fluid (drilling mud) are stored on location. When all these matters have been attended to, the drilling contractor is ready to begin drilling operations ("spud the well").

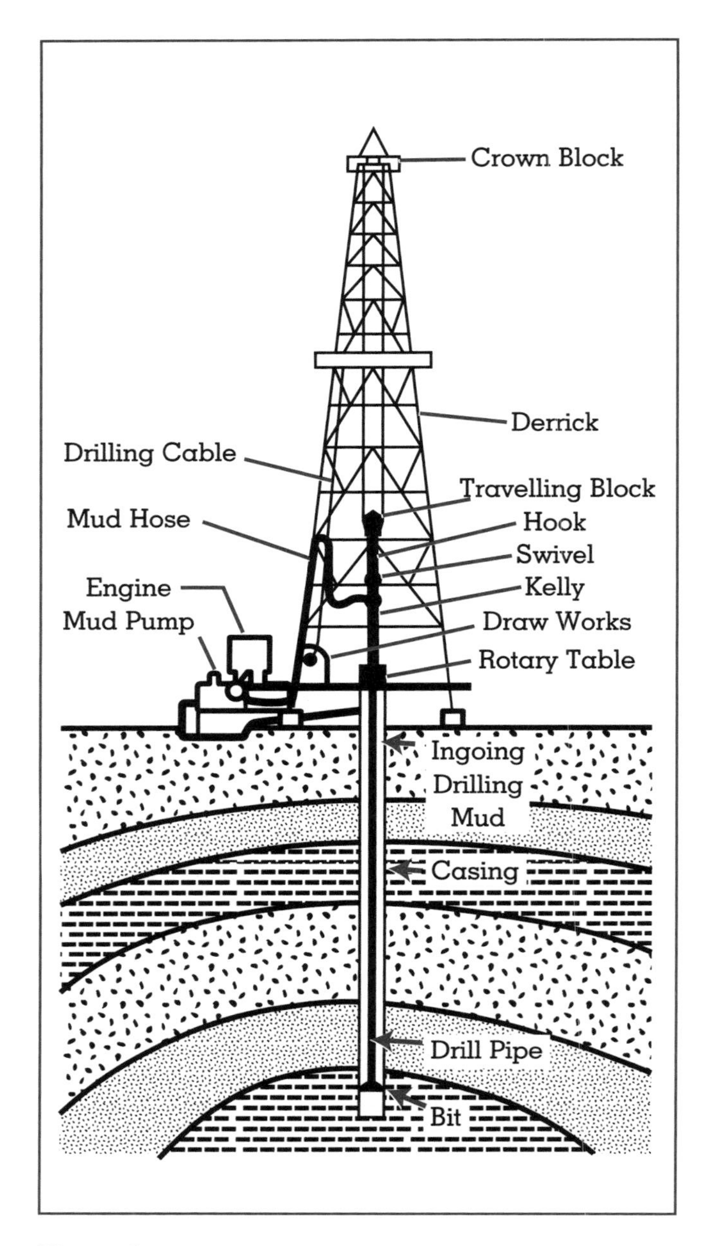

Figure 1.
A typical drilling rig operation. The string of pipe and drill bit rotate, with their weight adding to the cutting effectiveness of the bit. Borehole pressure is maintained by drilling mud, which also brings up the drill cuttings. Casing is used for weak formations or when there is risk of groundwater contamination.

The drill string, consisting of a drill bit, drill collars, drill pipe, and kelly, is assembled and lowered into the conductor pipe. Drilling fluid, better known as drilling mud, is circulated through the kelly and the drill string by means of pipes and flexible hose connecting the drilling fluid or mud pumps and a swivel device attached to the upper end of the kelly. The swivel device enables drilling mud to be circulated while the kelly and the drill string are rotated. The mud pump draws fluid from mud tanks or pits located nearby. The drilling mud passes through the kelly, drill pipe, drill collars, and drill bit. The drilling mud is returned to the surface by means of the well bore and the conductor pipe where it is directed to a shale shaker, which separates the drill cuttings and solids from the drilling mud, which is returned to the mud tanks to be recirculated. As the drill string is rotated in the well bore, the drill bit cut into the rock. The drilling mud lubricates and cools the drill bit and the drill string and carries the drill cuttings to the surface.

Drilling the Surface Hole

When a well is spudded in, a large-diameter drill bit is used to drill to a predetermined depth to drill the surface hole, which is lined with casing. The casing protects aquifers that may contain freshwater, provides a mounting place for the blowout preventer, and serves as the support for the production casing that will be placed in the well bore if the drilling program is successful. The surface hole may be several hundred or several thousand feet deep. When the predetermined depth is reached, the drill string will be removed from the well bore. Steel casing of the proper diameter is inserted. Sufficient cement is pumped down the surface casing to fill the space between the outside of the casing and the well bore all the way to the surface. This is to ensure protection of freshwater aquifers and security of the surface casing. The casing and the cement are tested under pressure for twelve hours before drilling operations may be resumed. The blowout preventer is attached at the top of the surface casing. This device is required to control the well in the event that abnormal pressures are encountered in the borehole that cannot be controlled with drilling fluid. If high-pressure gas or liquid blows the drilling fluid out of the well bore, the blowout preventer can be closed to confine the gas and the fluid to the well bore.

Drilling to Total Depth

After the surface casing has been tested and the blowout preventer installed, drilling operations are resumed. They will continue until the well has been drilled to the total depth decided upon. Usually the only interruptions to drilling operations will be to remove the drill string from the well bore for the replacement of the drill bit (a procedure known as tripping) and for testing of formations for possible occurrences of oil or gas (known as drill-stem testing). Other interruptions may be due to problems incurred while drilling, such as the shearing off the drill string (known as "twisting off") and loss of drill-bit parts in the well bore (known as "junk in the hole").

Deep drilling oil rig at Naval Petroleum Reserve's Elk Hills site near Bakersfield, California. (U.S. Department of Energy)

As drilling operations continue, a geologist constantly examines drill cuttings for signs of oil and gas. Sometimes special equipment known as a mud logger is used to detect the presence of oil or gas in the drill cuttings or the drilling fluid. By examining the drill cuttings, a geologist determines the type of rock that the drill bit is penetrating and the geologic formation from which the cuttings are originating.

Today's conventional drill bit utilizes three revolving cones containing teeth or hardened inserts that cut into the rock as the bit is revolved. The teeth or inserts chip off fragments of the rock which are carried to the surface with the drilling fluid. The fragments or chips, while they are representative of the rock being drilled, do not present a clear and total picture of the formation being drilled or the characteristics of the rock being penetrated as to porosity and permeability. For this purpose a larger sample of the rock is required, and a special type of drill bit is used to collect the sample, called a core. The core is usually sent to a laboratory for analysis and testing.

If the geologist detects the presence of oil or gas in the drill cuttings, a drill-stem test is frequently performed to evaluate the formation or zone from which the oil show was observed. Drill-stem tests may also be performed when the driller observes a decrease in the time required to drill a foot of rock, known as a "drilling break." Since porous rock may be drilled easier than nonporous or less porous rock, a drilling break indicates the presence of the type of rock that usually contains oil or gas. A drill-stem test enables the exploration company to obtain a sample of the fluids and gases contained in the formation or interval being tested as well as pressure information. By testing the well during the drilling phase, a decision as to whether to complete it can be made. When the well has been drilled to the predetermined depth, the drill string will be removed from the well bore to allow insertion of tools that will further test the formation and particularly the well. These tools are suspended on a special cable. Specific properties of the formation are measured as the tools are retrieved. Signals detected by the tools are recorded in a truck at the surface by means of the electrical circuits contained in the cable.

SlimDril drill bits are manufactured from polycrystalline with natural diamonds on the face of the bit for oil and natural gas drilling. (Corbis-Bettmann)

Completing the Well

When drill-stem testing and well logging operations have been completed and the results have been analyzed, company management must decide whether to complete the well as a producing well or to plug it as a dry hole. If the evidence indicates that no oil or gas is present, or that they are not present in sufficient quantity to allow for the recovery of drilling, completion, and production costs and provide a profit on investment, the well probably will be plugged and abandoned as a dry hole. On the other hand, if evidence indicates the presence of oil or gas in sufficient quantity to allow for the recovery of the cost and provide a profit to the company, an attempt will be made to complete the well as a producing well. The casing is delivered and a cementing company is called in.

The well bore is filled with drilling fluid that contains additives to prevent corrosion of the casing and to prevent movement of the fluid from the well bore into the surrounding rock. The casing is threaded together and inserted into the well bore in much the same manner as the drill string. Casing may be inserted to the total depth of the hole, or a cement plug may have been set at a specific depth and the casing set on top of it. Cement is mixed at the surface, just as if the well were to be plugged. The cement is then pumped down the casing and displaced out of the bottom with drilling fluid or water. The cement then flows up and around the casing, filling the space between the casing and the well bore to a predetermined height. After cementing of the casing has been completed, the drilling rig, equipment, and materials are removed from the drill site. A smaller rig, known as a workover rig or completion rig, is moved over the well bore. The smaller rig is used for the remaining completion operations. A well-perforating company is then called to the well site. It is necessary to perforate holes in the casing at the proper depth to allow the oil and the gas to enter the well bore. The perforating tool is inserted into the casing and lowered to the desired depth on the end of a cable. Shaped charges are remotely fired from the control truck at the surface, and jets of high-temperature and high-velocity gas perforate the casing, the cement, and the surrounding rock for some distance away from the well bore.

A smaller-diameter pipe, called tubing, is then threaded together and inserted into the casing. If it is expected that oil or gas will flow to the surface via the pressure differential between the well bore and the formation, a wellhead is installed; it is equipped with valves to control the flow of oil or gas from the well. The wellhead is known as a "Christmas tree." If there is not sufficient pressure differential to cause the oil and the gas to flow naturally, pumping equipment is installed at the lower end of the tubing.

During well completion it is sometimes desirable or necessary to treat or "stimulate" the producing zone to improve permeability of the rock and to increase the flow of oil or gas into the casing. This may be accomplished by use of acid or by injection of fluid and sand under high pressure to fracture the rock. Such a treatment usually improves the ability of the rock to allow fluid to flow through it into the well bore. At this point the drilling and completion phases have ended.

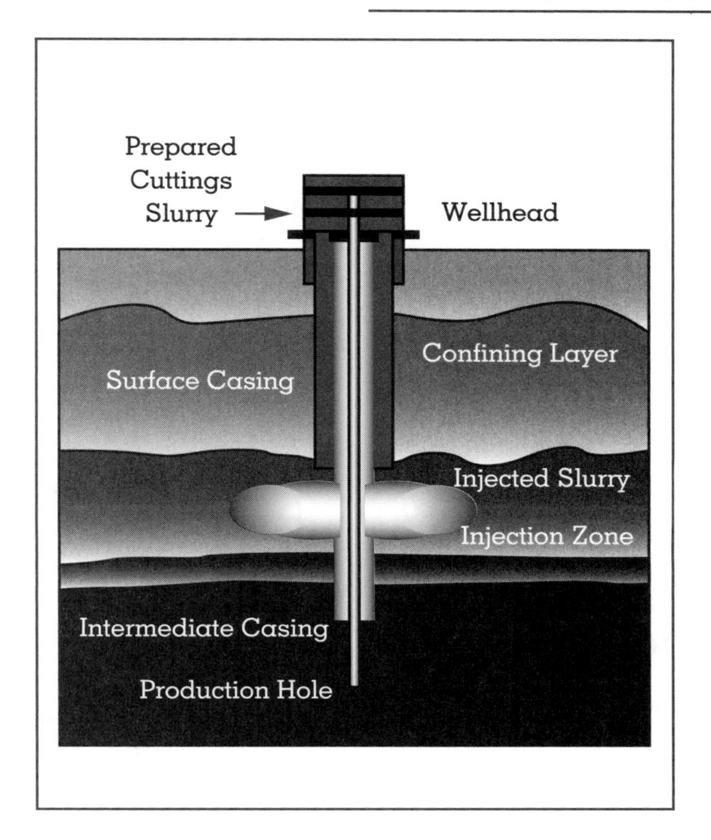

Figure 2.
To reduce the amount of waste and disposal costs, drill cuttings are increasingly being reinjected into the formations from which they came.

The Drill Bits and Well Design

Designing more effective and durable drill bits is important because drilling depths of reservoirs can reach as deep as 5 mi (8,052 m). Worldwide use of the more expensive polycrystalline diamond compact (PDC) drill bit has slowly been gaining on the conventional roller cone bit because of faster rates of penetration and longer bit life. The 1998 PDC bit rotated 150 to 200 percent faster than similar bits a decade earlier, and averaged more than 4,200 ft (1,281 m) per bit as opposed to only 1,600 ft (488 m) per bit in 1988. This 260 percent improvement dramatically cuts time on site since the less often the drill bit needs changing, the less time and energy must be devoted to raising and lowering drill pipe. It also means that the 15,000-ft well in the 1970s that took around eighty days could be completed in less than forty days in 2000. Because wells can be drilled more quickly, more profitably and with less of an environmental impact, the total footage drilled by diamond bits went from about 1 percent in 1978 to 10 percent in 1985 and to 25 percent in 1997.

To reduce drilling and development costs, slimhole drilling is increasingly being used. Slimhole wells are considered wells in which at least 90 percent of the hole has been drilled with a bit fewer than six inches in diameter. For example, a typical rig uses a 8.5-in bit and a 5-in drill pipe, whereas a slimhole rig may use a 4-in bit and a 3.7-in drill pipe. Slimhole drilling is especially valuable in economically marginal fields and in environmentally sensitive areas, since the fuel consumption can be 75 percent less (mud pumps, drill power), the mud costs 80 percent less, the rig weight is 80 percent less, and the drill site is 75 percent smaller.

Another important advance for developmental drilling is coiled tubing technology. Often used in combination with slimhole drilling technology, coiled tubing is a continuous pipe and thus requires only about half the working space as a conventional drilling pipe operation. Moreover, coiled tubing

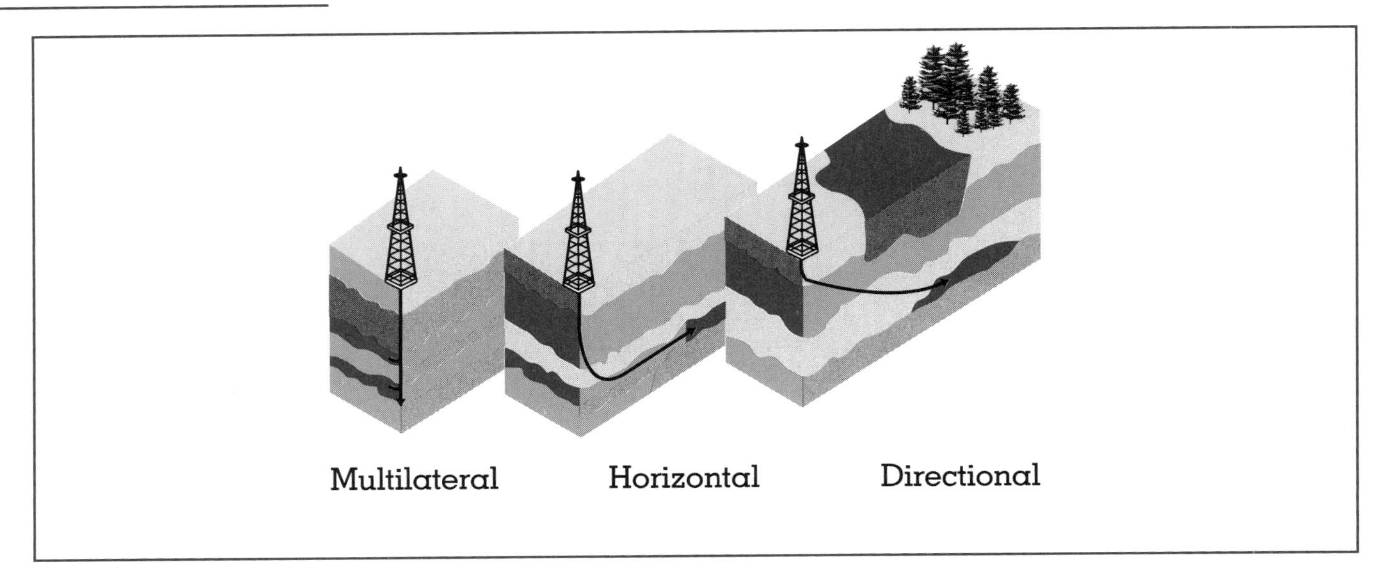

Figure 3.
Horizontal, Directional and Multilateral Drilling. Prior to 1988, fewer than 100 horizontal wells were drilled, but this total grew to more than 2,700, since horizontal drilling makes it possible to reach tight and thin reservoirs, and reservoirs inaccessible by vertical drilling. And because horizontal wells penetrate a greater cross section of any formation, more of the reserve can be extracted. Horizontal wells costs are 80 percent greater, yet production averages 400 percent more.

eliminates the costs of continuous jointing, reinstallation and removal of drill pipe. Since the diameter of coiled tubing is smaller than that of conventional pipe, coiled tubing also reduces operational energy use, noise, and the quantity of mud generated. The material of choice is currently steel, yet titantium is slowly replacing steel because titanium's greater strength promises a life cycle five to ten times longer than steel.

In weak formations or where groundwater needs to be protected, the borehole is lined with casing to prevent any transfer of fluids between the borehole and its surroundings. Strings of casing are progressively smaller than the casing strings above it, and the greater the depth of the well, the more casing sizes used. For very deep wells, up to five sizes of ever smaller diameter casings are used during the drilling process. This tiered casing approach has not only reduced the volume of wastes but also has been refined so that drilling cuttings can be disposed of by reinjecting them into the area around the borehole between the surface casing and the intermediate casing (Figure 2). In this way, wastes are returned to the geologic formations far below the surface, eliminating the need for surface disposal, including waste management facilities, drilling waste reserve pits, and off-site transit transportation.

Drilling is not a continuous operation. Periodically drilling must stop to check bottomhole conditions. To reduce the time lag between occurrence and surface assessment, measurement-while-drilling (MWD) systems are increasingly being used to measure downhole conditions for more efficient, safer, and accurate drilling. These systems transmit real-time data from the drill bit to the surface by encoding data as a series of pressure pulses in the mud column that is then decoded by surface sensors and computer systems. The accelerated feedback afforded by MWD systems helps keep the drill bit on course, speeds reaction to challenging drilling operations (high-pressure, high-temperature and underbalanced drilling), and provides valuable and continual clues for zeroing in on the reservoir's most productive zones. Since MWD systems provide a faster and more accurate picture of formation pore pressure and fracture pressure while the well is being drilled, it also substantially reduces the risks of life-threatening blowouts and fires.

HORIZONTAL, DIRECTIONAL AND MULTILATERAL DRILLING

Horizontal and directional drilling are non-vertical drilling that allow wells to deviate from strictly vertical by a few degrees, to horizontal, or even to invert toward the surface (Figure 3). Horizontal drilling is not new. It was first tried in the 1930s, but was abandoned in favor of investing in hydraulic fracturing technology (first introduced in 1947) that made vertical wells more productive. Rebirth of horizontal drillings began in the mid-1970s due to the combination of steerable downhole motor assemblies, MWD systems, and advances in radial drilling technology that made horizontal drilling investments more cost-effective. In 1996 more than 2,700 horizontal wells were dug, up from a very minimal number only a decade earlier.

In the simplest application, the borehole begins in a vertical direction and then is angled toward the intended target. The drill pipe still needs to move and rotate through the entire depth, so the deviation angle of the borehole needs to be gradual. A large deviation is made up of many smaller deviations, arcing to reach the intended target. The redirection of drill pipe is accomplished by use of an inclined plane at the bottom of the borehole. More recently, greater precision in directional drilling has been accomplished by using a mud-powered turbine at the bottom to drill the first few feet of the newly angled hole.

Horizontal and directional drilling serves four purposes. First, it is advantageous in situations where the derrick cannot be located directly above the reservoir, such as when the reservoir resides under cities, lakes, harbors, or environmentally sensitive areas such as wetlands and wildlife habitats. Second, it permits contact with more of the reservoir, which in turn means more of the resource can be recovered from the single well. Third, it makes possible multilateral drilling—multiple offshoots from a single borehole to contact reservoirs at different depths and directions. Finally, horizontal drilling improves the energy and environmental picture, since there is less of an environmental impact when fewer wells need to be drilled, and the energy future is brighter since more oil can be expected from any given well. Horizontal drilling has been credited with increasing the United States oil reserves by close to 2 percent or 10 billion barrels. Although horizontal drilling is more expensive than vertical drilling, the benefits of increased production usually outweigh the added costs.

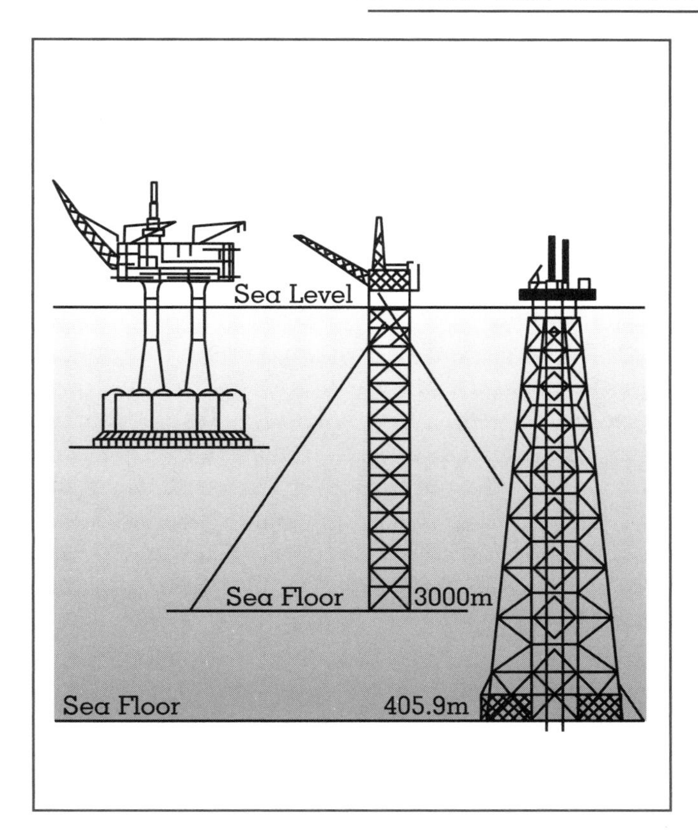

Figure 4.
Submersible base rigs. Flotational tanks built into the base are flooded so that the base rests on the ocean floor. The legs are constructed to the correct height for each site.

OFFSHORE DRILLING

Since oceans and seas cover more than two-thirds of the Earth, it came as no surprise when oil was discovered ten miles off the Louisiana coast in 1947. Fifty years later, offshore production had grown to approximately a third of all world output, most coming from the North Sea, the Persian Gulf, and the Gulf of Mexico. Technology has evolved from shallow-water drilling barges into fixed platforms and jack-up rigs, and finally into semisubmersible and floating drill ships (Figures 4–7). Some of these rigs are among the largest and most massive structures of modern civilization. And because drilling often takes place in waters as deep as 10,000 ft (more than 3,080 m) deep, as far as 200 mi (more than 300 km) from shore, and in unpredictable weather (it must be able to withstand hurricanes and uncontrollable releases of oil), the challenges facing designers, builders, and users of these rigs are daunting.

Offshore drilling operations entail much of the same technology as land-based drilling. As with onshore drilling, the offshore drill pipe must transmit rotary power and drilling mud to the bit for fixed rigs, floating rigs, and drill ships. Since mud also must be returned for recirculation, a riser (a flexible outer casing) extends from the sea floor to the platform. As a safety measure—to contend with the stress created from the motion and movement of the boat in relation to the sea bottom—risers are designed with material components of considerable elasticity.

Besides the much grander scale of offshore rigs, there are numerous additional technologies employed to contend with deep seas, winds, waves, and tides. The technical challenges include stability, buoyancy, positioning, mooring and anchoring, and rig designs and materials that can withstand extremely adverse conditions.

Shallow Water and Arctic Areas

For shallow-water drilling, usually water depths of 50 ft (15.4 m) or less, a drilling platform and other equipment is mounted on a barge, floated into position, and sunk to rest on the bottom. If conditions dictate, the drilling platform can be raised above the water on masts. Either way, drilling and production operations commence through an opening in the barge hull.

For drilling in shallow water Arctic environments, there is an added risk from the hazard of drifting ice. Artificial islands need to be constructed to protect the platform and equipment. Because of permafrost, drilling onshore in Arctic regions also necessitates building artificial islands. These usually are built from rock and gravel so that the ground does not become too unstable from permafrost melting around the drill site. To leave less of an environmental footprint, these artificial islands are increasingly being built from ice pads rather than gravel. Because the ice pads are insulated to prevent thawing, the drilling season in some Arctic areas has been extended to 205 days, and well operations to as long as 160 days. This heightens the chances of single-season completions that dramatically reduce costs, since equipment does not need to be removed to nontundra areas and brought back again.

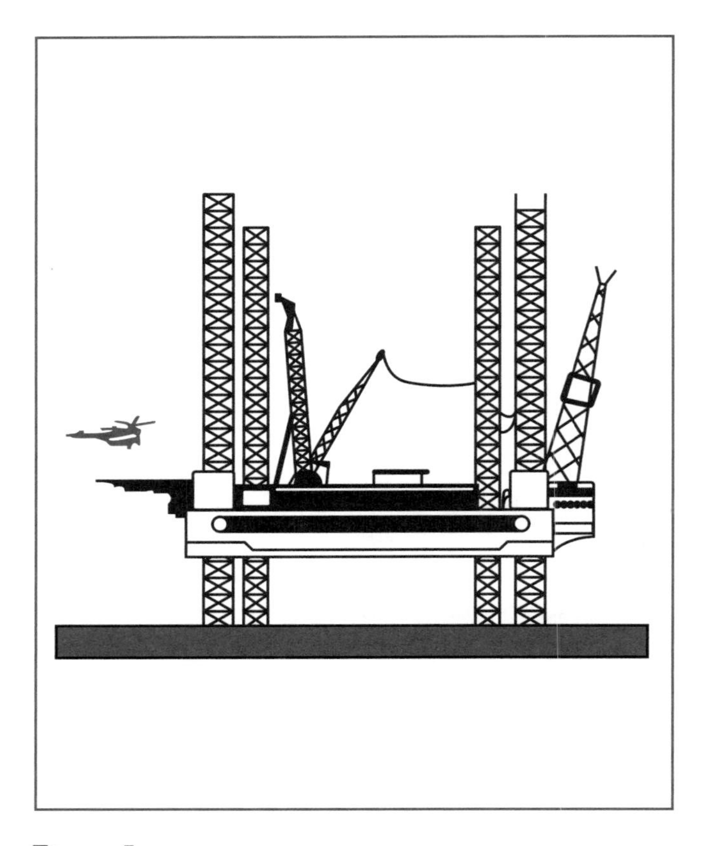

Figure 5.
Jack-up rigs. For drilling in water depths of up to 400 ft (122 m), the legs are cranked downward until they reach the sea floor and then jacked up so that the platform is 15 to 30 ft (4.6 m to 9.1 m) above the surface. Workers often reach the rig by helicopter (left).

Deep Water

For deeper drilling in open waters over the continental shelves, drilling takes place from drill ships, floating rigs and rigs that rest on the bottom. Bottom-resting rigs are mainly used for established developmental fields (up to 3,000 ft or 924 m); the semisubmersible and free-floating drill ships (up to 10,000 ft or 3,080 m) are the primary choice for exploratory drilling.

Where drilling and production are expected to last a long time, fixed rigs have been preferred. These rigs stand on stilt-like legs imbedded in the sea bottom and usually carry the drilling equipment, production

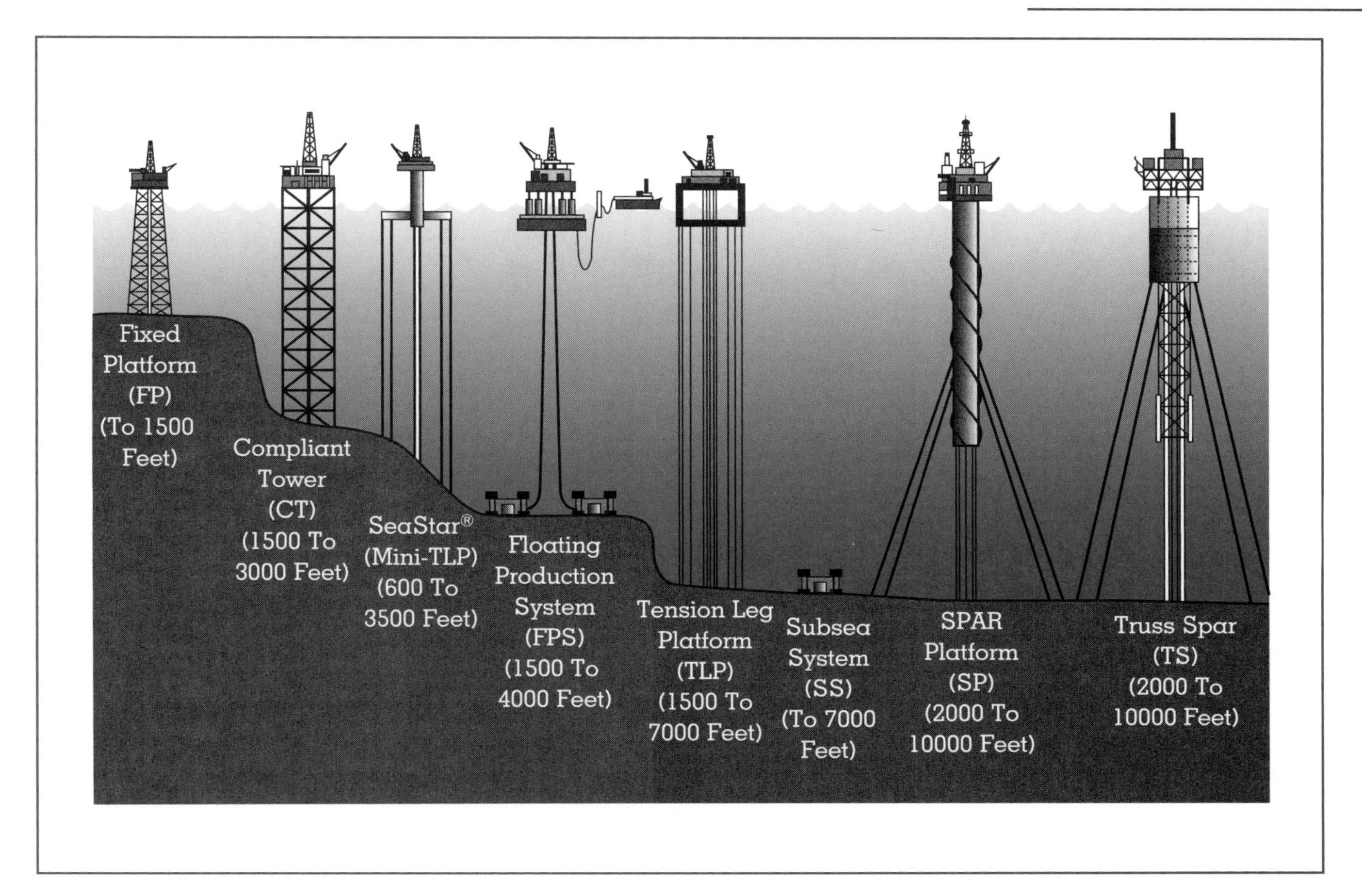

Figure 6.
Fixed rigs. Used in depths up to 1,500 ft (457 m), these rigs are imbedded in the ocean floor and tapered from bottom to top for stability. Compliant tower. Used in water depths from 1,500 to 3,000 ft (457 m to 915 m), these towers rely on mooring lines for stability.

equipment, and housing and storage areas for the work crews.

The jack-up platform, which is most prevalent, is towed to the site and the rack-and-pinion geared legs are cranked downward toward the bottom. Once the legs are stabilized on the bottom with pilings, the platform is raised 30 to 60 ft (9.2 to 18.5 m) above the surface. Another common fixed rig rests on rigid steel or concrete bases that are constructed onshore to the correct height. The rig is then towed to the drilling site, where the flotation tanks, built into the base, are flooded to sink the base to the ocean floor. For drilling in mild seas, a popular choice is a tender-assisted rig, where an anchored barge or tender is positioned alongside the rig to help with support. The hoisting equipment sits on the rig, with all the other associated support equipment (pumps, generators, cement units, and quarters) on the barge.

As more offshore exploratory operations move into deeper water—in 1997 there were thirty-one deep water rigs drilling in water depths greater than 1,000 ft (305 m) as opposed to only nine in 1990—there is sure to be more growth in the use of semi-submersible rigs and drill ships. In rough seas, semi-submersible drilling rigs are the preferred option since the hull is entirely underwater, giving the rig much more stability. The operational platform is held above the surface by masts extended upward from the deck. Cables are moored to the sea floor, with the buoyancy of the platform creating a tension in the cables that holds it in place.

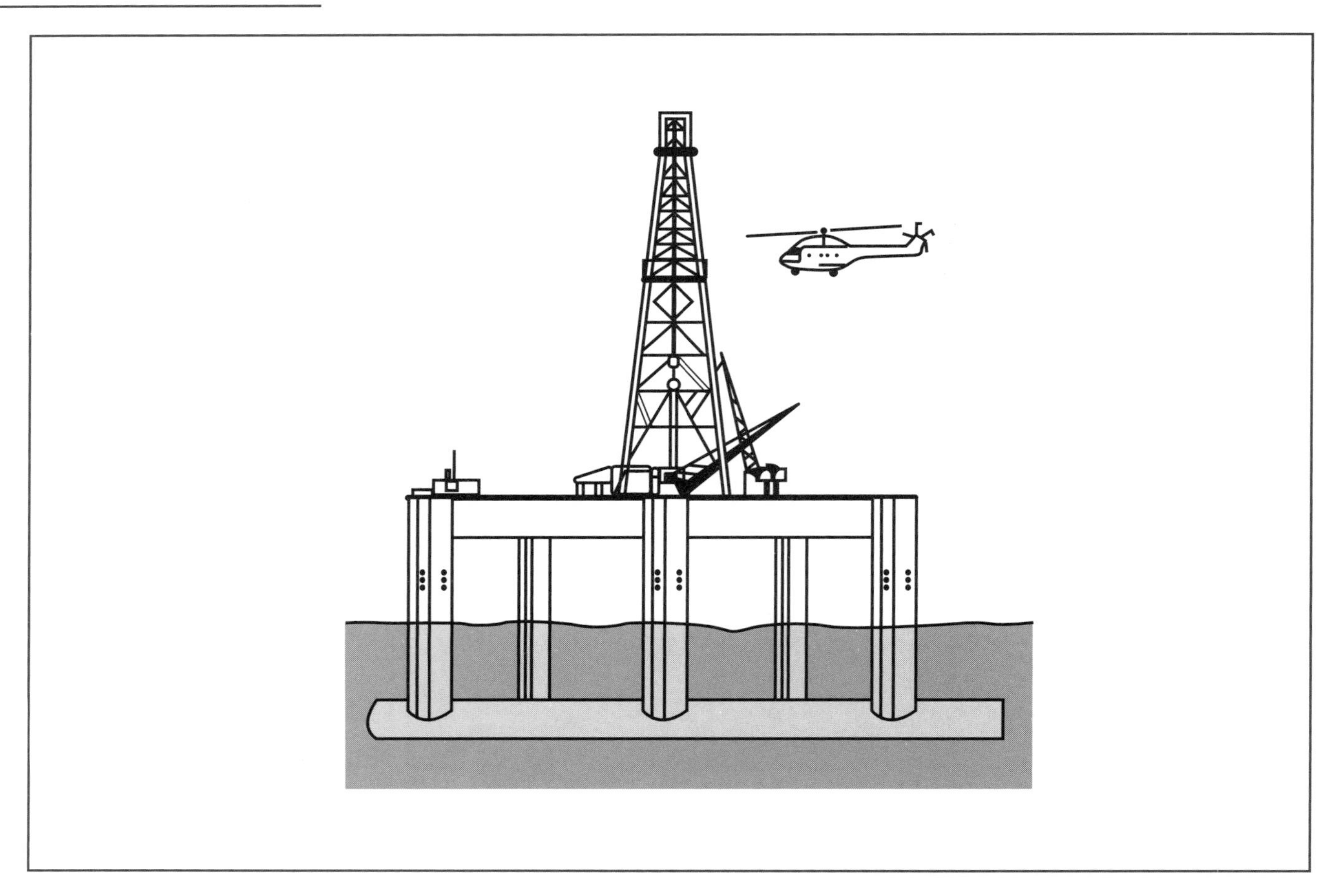

Figure 7.
Semisubmersible rigs. Huge cables and moorings anchor these floating rigs to the ocean bottom. The buoyancy of the rig creates the tension in the cables that keep the rig in place. When setting and removing mooring lines is excessively difficult or expensive, dynamic positioning systems are used instead.

When seas are less treacherous and there is greater need for mobility and flexibility, the drill ship is preferred over the semisubmersible. The drill ship has a derrick mounted in the middle over an opening for drilling operations, with several moorings used to hold the ship in position.

The use of dynamic positioning systems, instead of using moorings and cables, is becoming more popular in situations where it is expensive or extremely difficult to set and remove mooring lines. With the aid of advanced monitoring systems—gyrocompass wind sensors, global positioning systems, underwater sonar beacons, and hydroacoustic beacons—precise directional thrust propellers are used to automatically hold the vessel's orientation and position over the borehole. Despite the forces of wind, waves and ocean currents, at a water depth of 5,000 ft (1,526 m), a dynamic positioning system can reliably keep a drill ship within 50 ft (15.2 m) of the spot directly over the borehole.

Future Trends

As larger offshore discoveries become harder to find, more attention will be directed toward maximizing development from existing known reserves (workovers and recompletions), which seems to ensure a future for self-contained and jack-up rigs. There are certain to be reductions in size and weight of these new rigs. Self-elevating telescoping masts, which have improved dramatically in the 1990s, will become lighter and install more quickly. There also will be efficiency gains in equipment, and setup will be quicker, since modular units are becoming smaller, lighter, and less bulky.

The frontier remains deep water. Since the trend is toward longer-term operations, the design and materials of rigs that rest on the sea bottom will need to improve to handle the increased loads the extended reach needed in greater water depths. Productivity also will be increasing, since some of the latest ships can conduct simultaneous drilling operations using two drilling locations within a single derrick. Furthermore, because the costs of building and transporting massive drilling rigs can easily exceed several billion dollars, material research will continue to look for ways to prolong rig life cycles, and ways to better test structures regularly to avoid a catastrophic collapse. In particular, better welds, with fewer defects, will greatly extend the life cycle of offshore platforms from the continual forces of wind, waves, and ocean currents.

ANOTHER EQUIPMENT USE: CARBON DIOXIDE AND METHANE SEQUESTRATION

Since methane (CH_4) and carbon dioxide (CO_2) are greenhouse gases that governments of the world want reduced, the oil and gas industry—a major emitter of both—is looking for ways to effect such reductions. During drilling and production for oil and gas, significant amounts of CO_2 and CH_4 are produced. Where it is uneconomical to produce natural gas, this "stranded gas" is either vented as methane, flared (about 95% conversion to CO_2), converted to liquids (liquefied natural gas or methanol), or reinjected into the well. Over half of the oil exploration and production CO_2 emissions come from flaring, and almost all the methane emissions come from venting. The industry is trying to reduce emissions of both by better flare reduction practices, focusing on ways to convert more stranded gas to liquid fuels, and developing vapor recovery systems.

Another problem is when the carbon dioxide content of natural gas is too high and must be lowered to produce pipeline-quality gas. Although the current practice is to vent this CO_2, sequestration of CO_2 in underground geologic formations is being considered. Already, in the Norwegian sector of the North Sea, CO_2 has been injected into saline aquifers at a rate of 1 million tons a year to avoid paying the Norwegian carbon tax of $50 per ton of CO_2.

It may turn out that the same drilling technology that is being used to extract oil and gas, and that has been adapted for mining, geothermal, and water supply applications, will someday be equally useful in sequestering CO_2 in appropriate subsurface geologic formations.

John Zumerchik
Elena Melchert

See also: Climatic Effects; Fossil Fuels; Gasoline and Additives; Governmental Intervention in Energy Markets; Liquefied Petroleum Gas; Methane; Natural Gas, Processing and Conversion of; Natural Gas, Transportation, Distribution, and Storage of; Oil and Gas, Exploration for; Oil and Gas, Production of; Risk Assesment and Management.

BIBLIOGRAPHY

Backer, R. (1998). *A Primer of Offshore Operations*, 3rd ed. Austin, TX: University of Texas Press.

Benefits of Advanced Oil and Gas Exploration and Production Technology. (1999). Washington DC: United States Department of Energy.

Carter, G. (2000). "Offshore Platform Rigs Adapting to Weight-Space Restrictions for Floaters." *Offshore* 60(4).

Devereux, S. (2000). *Drilling Technology.* Tulsa, OK: PennWell.

Drilling for Oil, Unit 4: Activity 4-4/Teacher, pp. 4–48 to 4–53. Washington, DC: Teacher Training Program, National Petroleum Technology Office, Office of Fossil Energy, U.S. Department of Energy.

Hyne, N. J. (1995). *Nontechnical Guide to Petroleum Geology, Exploration, Drilling and Production.* Tulsa, OK: PennWell.

Pratt, J. A.; Priest, T.; and Castaneda, C. J. (1997). *Offshore Pioneers: Brown & Root and the History of Offshore Oil and Gas.* Houston, TX: Gulf Publishing Company.

U.S. Department of Energy. (1999). *Securing the U.S. Energy Environmental, and Economic Future.* Washington, DC: Author.

OIL AND GAS, EXPLORATION FOR

Exploration for oil and gas has evolved from a basic trial-and-error drilling to the application of sophisticated geophysical techniques to predict the best locations for drilling.

BASICS OF OIL AND GAS TRAPS

Oil and gas are usually associated with sedimentary rocks. The three basic types of sedimentary rocks are shales, sands, and carbonates. The shales are the sources of the hydrocarbons while the sands and carbonates act as the conduits and/or the containers.

Source Rocks

Shales often are deposited with some organic matter, for example, microscopic plant or animal matter, that become part of the rock. When the shales are squeezed under the pressure of overlying rocks and heated by the natural heat flowing from the earth, oil and gas may be formed depending on the type of organic matter involved and the temperature to which the source rock is raised. If a rock is capable of producing oil or gas, it is referred to as a source rock because it is a source for the hydrocarbons. An important element of modern exploration is making certain that a new area being explored has source rocks capable of generating hydrocarbons.

The petroleum industry relies on organic geochemists to analyze the source rocks in new exploration environments. The geochemists evaluate the potential of a source rock for producing oil and gas by determining the amount and type of organic matter as well as the amount of heating and pressure applied to the rock. The whole process of making hydrocarbons can be compared to cooking in the kitchen. The final product depends upon the ingredients and the temperature of the oven (or sedimentary basin in the case of a source rock). Geochemists perform studies of potential source rocks in order to evaluate both the thermal history and the basic ingredients.

The important ingredient in a source rock is the organic matter or kerogen (Figure 1). Kerogen originally meant "mother of oil" but is used to today to describe the organic component of sediment not solvable in common solvents. Kerogens consist of basic biochemicals including carbohydrates, proteins, polyphenols, and lipids. Oil and gas are derived from kerogens rich in lipids. These types of kerogens are often deposited in lakes or marine environments. Other types of kerogens derived primarily from land plants are more prone to producing only gas. The geochemist has to identify the type of kerogen in a source rock in order to predict the types of hydrocarbons generated.

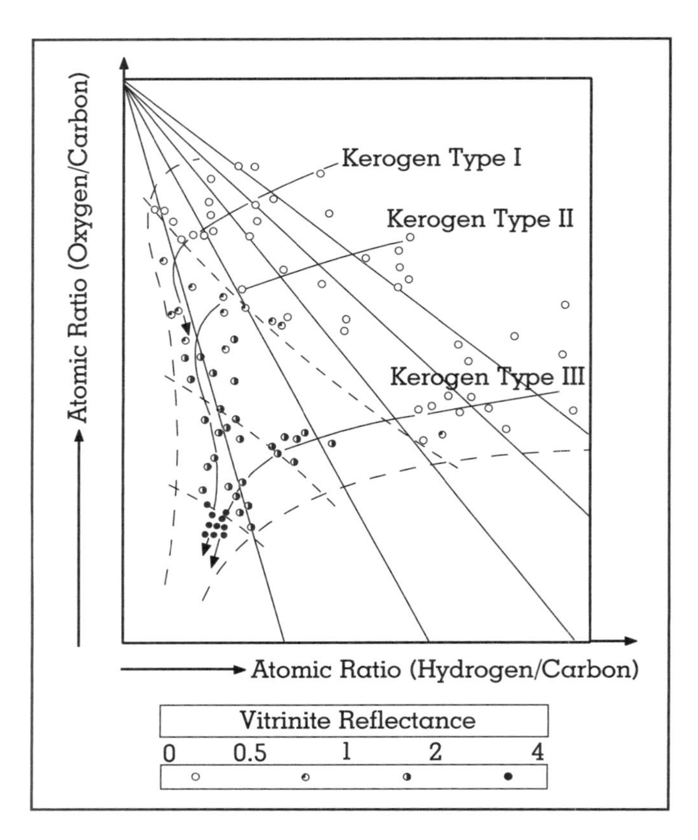

Figure 1.
General scheme of kerogen evolution from diagenesis to metagenesis in the van Krevelen diagram.

As the source rocks are buried and their temperature increases, the individual kerogens undergo chemical changes that are referred to as maturation. When the temperature reaches around 60–70° C and the rocks are buried approximately 1000 m below the surface, hydrocarbons begin to be released from the kerogen. At first oil is produced from the rock if the kerogen is the right type. Next, "wet" gas (or a gas with many different components) and finally "dry" gas (or a gas made up primarily of methane) is produced. As the rock continues to cook beyond the dry gas stage, the residual kerogen is of little use for producing oil and gas. Just as in the kitchen, too much heating can spoil even the best of ingredients.

Geochemists try to determine where hydrocarbons begin to be liberated and how their quantity and composition may vary with increasing maturity. This is equivalent to evaluating the amount and type of kerogen present in the rocks as well as the maturity (or amount of cooking in the analogy used above) for the potential source rocks in a region being explored. In addition, geochemists try to estimate when and in what directions the hydrocarbons migrated. This helps with the exploration process because some traps may have been formed too late in order to hold hydrocarbons.

One important measurement made by geochemists is the Total Organic Carbon (TOC) measurement. The results are expressed as a weight percentage. When the TOC is less than 0.5 percent, the rock is considered unlikely to have enough kerogen to produce oil or gas.

Other measurements made by geochemists try to determine the state or level of maturity of a rock. Just as with cooking, the color and appearance of kerogen changes as the source rock matures. As a result, there are several schemes for using color changes to measure the maturity of a source rock. One example is called the vitrinite reflectivity. Vitrinite is a main component of coal and is also found dispersed in source rocks. The reflectance of light from the vitrinite is measured electronically under a microscope and is denoted R_0 (the vitrinite reflectance). When R_0 is less than 0.5 to 0.7 percent, the rock is immature and still in the diagenesis stage of maturing. For R_0 greater than 0.5 to 0.7 percent to around 1.3 percent, the rock is mature and in the oil window of development. When R_0 is greater than around 1.3 percent and less than 2 percent, this stage of maturity implies the source rock is producing condensate and wet gas. When R_0 is greater than 2 percent and less than 4 percent methane is the primary hydrocarbon generated (dry gas zone). When the reflectance is greater than 4 percent, the source rock has been overcooked. Vitrinite reflectance R_0 combined with the measurement of the TOC and a determination of the type of kerogen can then be used to predict the amount and type of hydrocarbons expected from a source rock. The type or kerogen in a source rock is determined using laboratory techniques to determine the relative quantities of hydrogen and oxygen in the kerogen. For example, if the atomic ratio of hydrodren to carbon of the kerogen is plotted on a graph against the atomic ratio of oxygen to carbon, different types of kerogens can be identified. Three categories of kerogen can be identified using this type of plot. Type I kerogen is rich in lipids and is an excellent source of oil. Type II is still a good source rock but is not as rich in lipids and not as oil prone at the Type I kerogen. Type III kerogen is primarily a gas prone kerogen. Armed with results of this type of study, the geochemist is able to make predictions regarding the quality of the source rocks in an exploration area and the expected products (oil, gas, or both) expected from them. In addition, geochemists work with geologists to predict the timing and the direction of flow of hydrocarbons as they are expelled from the source rocks. This information is then used in exploration to evaluate the relative timing between trap formation and the creation of the oil. A potential oil trap might be capable of holding a great deal of oil and still not have one drop of oil because the trap was formed after the oil had migrated past the trap.

A piece of oil shale formed in the bed of Uintah Lake has been polished to reveal strata. (Corbis Corporation)

Reservoir Rocks

Once the potential for hydrocarbons has been verified in a region, explorationists look for reservoir rocks that potentially contain the hydrocarbons. Reservoir rocks are typically comprised of sands or carbonates that have space, called "porosity," for holding the oil. Sands are present at most beaches, so it is not surprising that a great deal of petroleum has been found in sands that were once part of an ancient beach. Other environments responsible for sandstones include rivers, river deltas, and submarine fans. An example of a carbonate environment is a coral reef, much like the reef located off the coast of Florida. The sediment deposited on reefs consists of the skeletons of tiny sea creatures made up primarily of calcium carbonate. The sandstone and carbonate rocks become a type of underground pipeline for transporting and holding the oil and gas expelled from the source rocks. These reservoir rocks have pore spaces (porosity) and the ability to transmit fluids (permeability). A rock with a high porosity and high permeability is considered a desirable reservoir rock because it can not only store more petroleum, but the petroleum will flow easily when being produced from the reservoir. Finding reservoir rocks is not always easy and much of the effort of exploration is devoted to locating them. One approach to looking for reservoir rocks is to make maps over a region based on data from wells (Figure 2). Geologists can use these maps to predict the range of distribution for the reservoir rocks. For example, if the reservoir rocks were deposited by a river system, the map can show a narrow patch of sand following the meander of an ancient river channel. Because the speed of sound in sands and carbonates (reservoir rocks) is usually faster than in the surrounding shales (usually the source rock), the difference can be used to identify reservoir rocks by the reflection character and strength on the seismic data. One popular seismic method for examining the strength of the reflection coefficient as a function of source-receiver separation (referred to as "offset") is called "amplitude-versus-offset" or "AVO." Another way to identify reservoir rocks is their anisotropy when compared to the anisotropy of shales. Anisotropy is used here to describe the directional dependence of a material property (the velocity in this case). Shales have a natural anisotropy that causes the speed of sound to travel faster horizontally than vertically. Reservoir rocks tend to have a different type of anisotropy caused by fractures or no anisotropy at all (they are said to be "isotropic" in this case). When searching for reservoir rocks, the key is to separate the type of anisotropy caused by shales from the anisotropy (or lack of it) due to the reservoir rocks that are present. This approach can be used to seismically evaluate the potential for reservoir rocks in a region. Still another seismic approach to predicting the presence of reservoir rocks is based on a geological interpretation of the seismic data called "seismic stratigraphy." All of the above methods are used to identify the presence of reservoir rocks in an area being explored.

Traps

Once the petroleum has been generated and squeezed from a source rock into a neighboring sand or carbonate, the petroleum is free to move through the porous and permeable formations until it reaches the surface or meets a place beyond which the petroleum can no longer flow. When the petroleum flows to the surface, the hydrocarbons are referred to as seeps. When the hydrocarbons are prevented from reaching the surface, they are "trapped" (see Figure 3). Traps are divided into two basic categories: structural and stratigraphic. Structural traps are caused by a deformation of the earth that shapes the strata of reservoir rocks into a geometry that allows the hydrocarbons to flow into the structure but prevents their easy escape. The deformation can be caused by thrust faults such as those responsible for mountain building or normal faults such as those found in subsiding sedimentary basins such as the Gulf of Mexico. Stratigraphic traps are the result of lateral variations in layered sedimentary rocks so that porous rocks grade into nonporous rocks, that will not allow the petroleum to move any further. The easiest traps to find have been structural traps, because deformation usually affects all adjacent layers or strata. This means that the effects are easier to see on seismic data. The stratigraphic traps are the most difficult to find because they occur without affecting much of the surrounding geology. Many reservoirs are constrained by both stratigraphic and structural trapping.

GEOLOGICAL MAPPING FOR OIL AND GAS

Geologists can be thought of as the historians of the earth. History is important to exploration success. When a well is drilled, a number of devices are lowered down the well to log or identify the different formations that have been penetrated. The geologist uses the logs from many different wells to put together an interpretation of the geology between wells. A

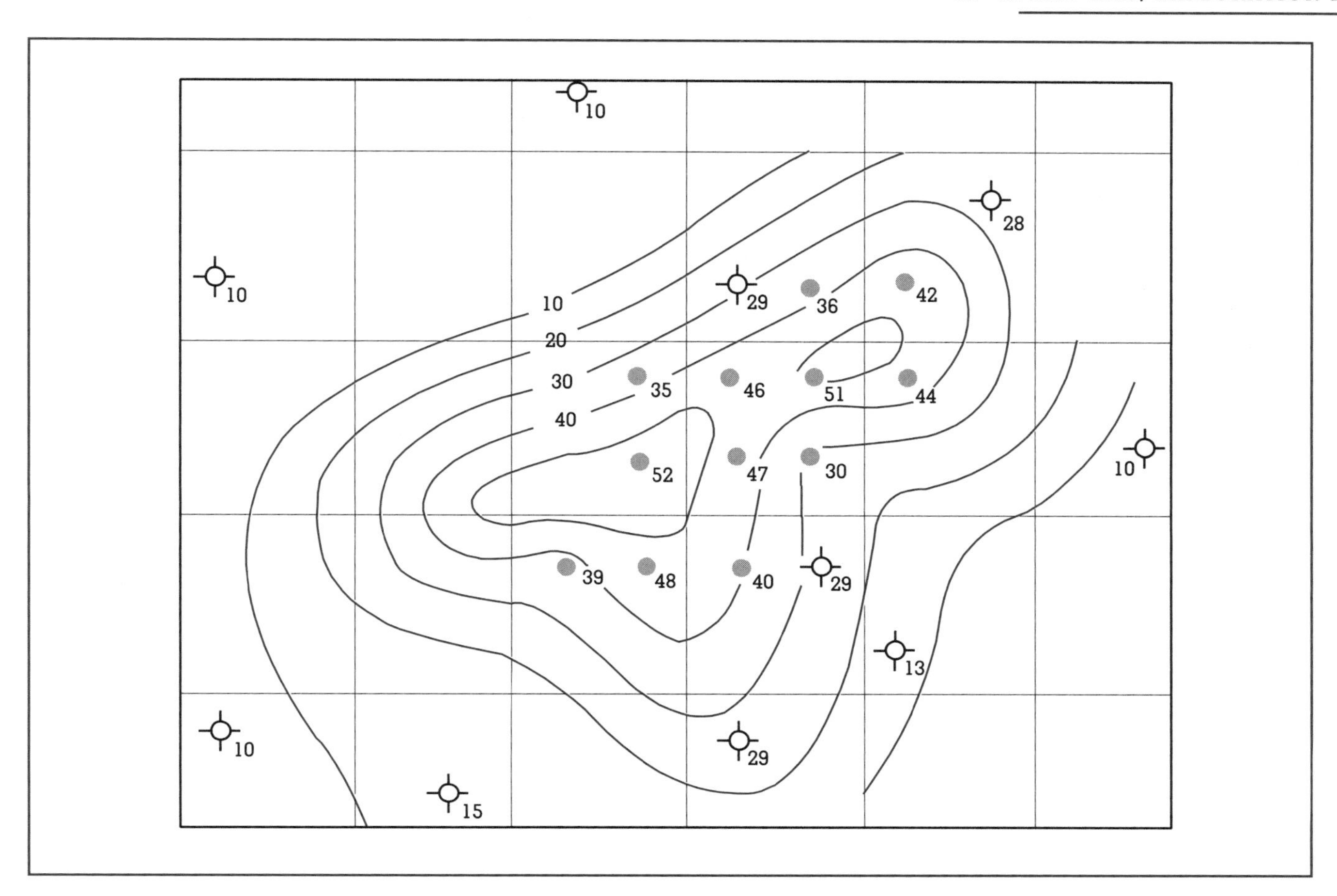

Figure 2.
Isopach maps depict thickness of a given interval.

part of the interpretation process involves making structure maps of the depth of sedimentary rock formations beneath the surface of Earth. In addition, geologists try to predict stratigraphic variations between wells using other mapping techniques. Besides being historians, geologists must solve geometric puzzles because they use the bits and pieces of information about rock layers observed from the well logs in order to form a complete picture of the earth. They can, for example compare well logs from two adjacent wells and determine if a fault cuts through one of the bore holes. The interpretation for modern geologists has been more challenging because many oil and gas wells purposely deviate from the vertical. Modem geologists employ numerous geometric tricks of the trade to assemble complicated 3-D interpretations of the earth and to identify where the reservoir rocks are located.

EARLY GEOPHYSICAL METHODS (1900s TO 1950s)

Geophysical methods are used to obtain information on the geology away from the wells where the location of oil- and gas-producing formations is known. In the early 1900s, the use of geophysics to find structural traps was accomplished by employing gravity and seismic methods. Gravity methods were the earliest geophysical method used for oil and gas exploration. The torsion balance gravimeter that was originally invented by Roland von Eotvos of Hungary (1890) later became an exploration tool. The first discovery by any geophysical method in the United States was made using a torsion balance in January 1924. Nash Dome was discovered in Brazoria County, Texas, by Rycade Oil Company, a subsidiary of Amerada. Sensitive gravity instruments measure the changes in the pull of Earth's

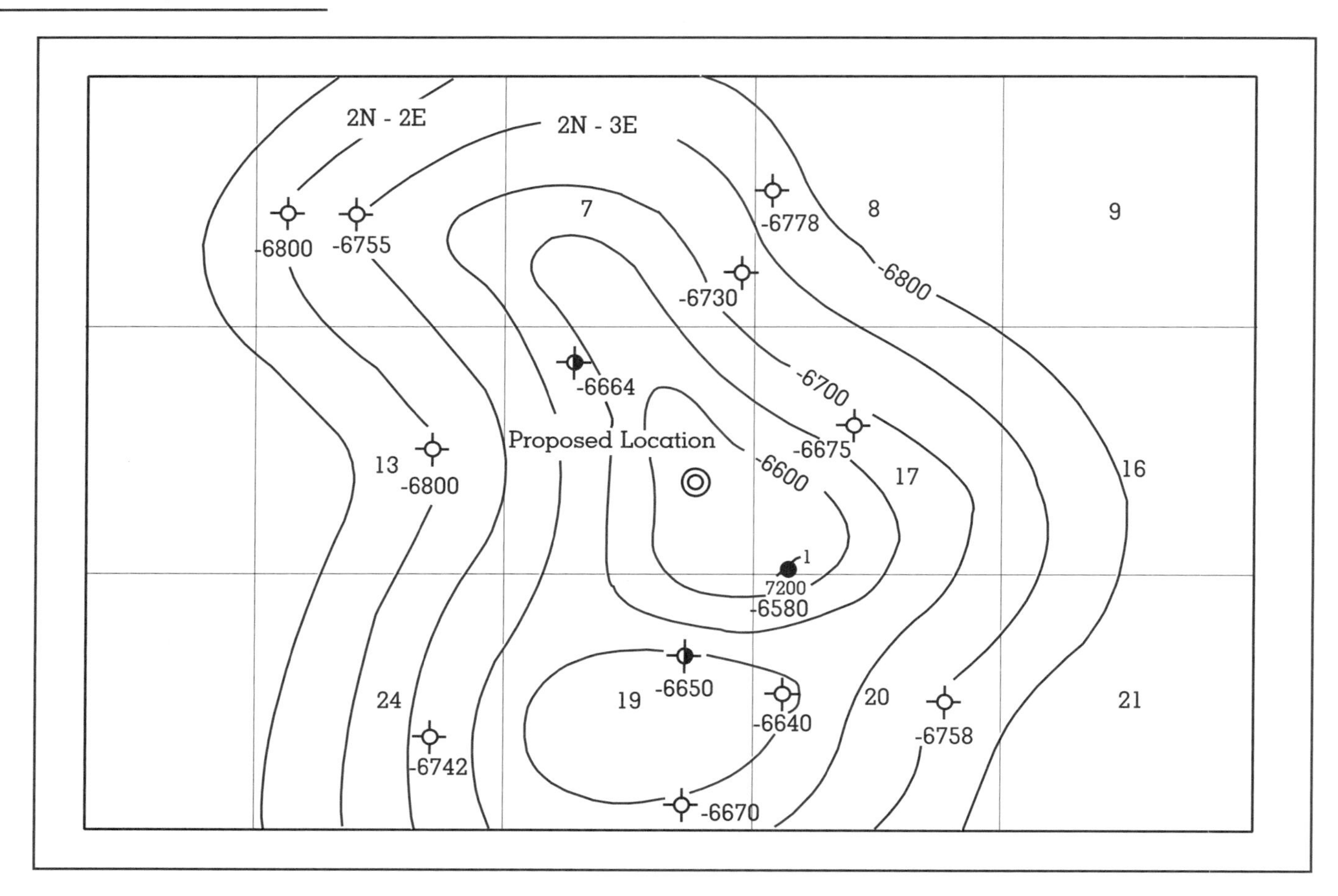

Figure 3.
Example of a structure map used to determine where oil is likely to be trapped.

gravity at different locations. The gravitational pull varies due to density variations in the different types of rocks. Because salt is lighter than most sedimentary rocks, gravity measurements were a popular approach to finding reservoirs associated with salt domes.

Sounding methods for spotting enemy artillery that were developed during World War I led to the major seismic exploration methods. On the German side, Ludger Mintrop developed the refraction exploration method that was most useful for early exploration of salt domes. Although Robert Mallet shot the first refraction survey in 1845, Mintrop recognized the commercial value of refraction surveying. He was able to verify the refraction method as an exploration tool by finding two salt domes in Germany in the period 1920–1921. In June 1924, shortly after the first torsion balance discovery, a Mintrop refraction crew (a company named "Seismos") discovered Orchard Dome in Fort Bend County, Texas. A patent filed in the United States in January 1917 by Reginald Fessenden, a Canadian by birth, laid down the fundamental ideas of using sound waves for exploration. The title of Fessenden's patent was "Methods and Apparatus for Locating Ore Bodies." In addition, the French, British, and U.S. wartime sound ranging efforts (for artillery) laid some of the foundation for the development of early seismic instrumentation and thinking developed by John Clarence Karcher. In 1921, Karcher used seismic reflections to map the depth to the top of the Viola limestone in Oklahoma. In 1927, Geophysical Research Corporation made the first seismic reflection discovery near Potawattamie County, Oklahoma. The successful well was drilled September 13, 1928. John Clarence Karcher and Everette Lee DeGolyer were the major contributors to this new exploration tool. Today the seismic

This is a team of trucks that create and record seismic vibrations, in an attempt to detect the presence of oil. (Corbis Corporation)

reflection method is the dominant form of exploration. The depth to the top of the oil and gas formations can be mapped with this technique, so the seismic reflection method can be used to directly identify structural traps.

Many other geophysical methods were attempted, but the gravity and seismic methods have survived the test of time. Gravity and seismic methods are often used in tandem because they give complementary information. This is helpful when exploring beneath salt or volcanic lava flows where seismic methods suffer difficulties. Some early work on magnetic measurements were accomplished, but magnetic surveys did not become popular until more sophisticated magnetometers were developed after World War II. When Conrad and Marcel Schlumberger joined forces with Henri Doll in the 1920s, they achieved the first geophysical logging of wells. The electric logs and other logs they developed played a major role in the early history of oil and gas exploration. The company they formed, Schlumberger Limited, has been an active contributor to numerous technological advances.

EXPLOSION OF TECHNOLOGY

In the 1950s the petroleum industry experienced an explosion of technology, a trend that continues today. One of the contributors in this effort was Harry Mayne, who developed common depth-point stacking, a method of adding seismic signals. The resulting stacked signals were then plotted in the form of a picture called a seismic section. The seismic sections, that appear as cross-sections of the earth, were initially used to find structural traps. The creation of the stacked sections benefited from digital recording that could be used to accurately record the amplitudes of seismic signals. True amplitude recording led to the discovery that the presence of hydrocarbons, especially gas, caused high-amplitude seismic reflections. These reflections, that appeared as bright spots on the seismic sections, enabled researchers to distinguish hydrocarbons from the surrounding rocks. Thus, in the 1970s bright-spot technology was developed by the petroleum industry to directly detect hydrocarbons from the surface. Shell Oil Company was the first company to use this new technology in the offshore region of the Gulf of Mexico, lowering the risk of falsely identifying oil and gas reservoirs. Unfortunately, other geological circumstances can create bright reflectors and some of these locations have been drilled.

As a result of these limitations in identifying stratigraphic traps, the industry sought a better understanding of stratigraphy. An important contribution to this understanding was the development of improved seismic methods to map stratigraphy. This improved method shaped the waveform leaving a seismic source into a shorter duration signal (the wave form is said to have been deconvolved). Enders Robinson, who had been a part of radar research efforts at MIT during World War II, introduced the deconvolution methods to the industry. Robinson applied some of the same radar technology to compress the seismic wavelets so that small changes in the stratigraphy could be mapped.

Seismic stratigraphy, originally developed by Peter Vail and his associates at Exxon, was another important contribution to the industry's ability to unravel the stratigraphy of the earth. As the industry pushed to find even more information about stratigraphic variations, methods were developed to use information about the shear-wave properties of rock as well as the compressional-wave velocities that had been the primary tool for seismic exploration. Shear waves or shear-related methods of exploration, such as AVO are used by the industry to identify changes in lithology (e.g., from sandstones to shales). These methods can also be used to evaluate bright spots or other types of reflection amplitude anomalies. For example, AVO methods are used to find rocks that have been fractured, thus making them into better reservoir rocks. In the early 1970s Jon Claerbout at Stanford University introduced an important seismic imaging principle that is still used today migrate seismic data accurately. "Migration" is a process used by geophysicists to plot seismic reflections in their true spatial positions. Today, many companies are using more sophisticated seismic imaging and analysis techniques called "inversion".

An important part of interpreting surface seismic data is the identification of the oil- and gas-bearing zones on the seismic data. Vertical seismic profile (VSPs) data are often recorded with a source on the surface and with receivers down the well to accomplish this task. In this way, the travel time to the reservoir can be measured along with other information relating the well data to the surface seismic data.

Another type of technology that has influenced oil and gas exploration is the acquisition of 3-D surface seismic data. This technology produces so much information that modern interpreters have been forced to give up looking at paper sections and use computers just to view their data. The 3-D seismic surveys are very much like a solid section of the earth. An interpreter can sit at a workstation and literally slice through the data viewing the seismic picture of the earth in almost every way possible. Some companies have assembled virtual reality rooms where the interpreter actually feels as if he or she is walking inside the earth to make the interpretation. Maps that might have taken months to produce in the early days of seismic mapping can now be accomplished in minutes. Computers are used not only for seismic processing and interpretation but for geological modeling and more accurate reservoir modeling. Integrated teams of scientists consisting of geologists, geophysicists, and petroleum engineers are pooling their talents to construct detailed 3-D models of reservoirs.

EXPLORATION TODAY

If the area being explored is a new one, the company has to make an assessment of the source rock potential. Next, they look for the reservoir rock and the trap. Sophisticated seismic methods including 3-D seismic surveys and VSPs are used to create a picture of where the oil and gas traps are located. They can also be used to determine when the traps were formed as well as the amount of oil and/or gas they contain. Seismic methods are often integrated with other geophysical measurements such as gravity, and magnetic or magnetotelluric measurements. In this way, the complementary data can be used to give a more complete picture of the geology.

EVALUATING RISK

Because financial risk is a major element of exploration, putting together a map that indicates where to drill is only part of the exploration process. Someone has to be convinced to put money into the effort to drill. Large companies go through multiple computer runs to evaluate the risks of drilling at a prospective site. Smaller companies tend to use a formula in which they multiply the probability of success times the current value of the oil that will be obtained if the well is successful. Next, they subtract from this product the probability of failure times the cost of drilling a dry well. The difference is called the "Expected Monetary Value" (EMV). If the EMV is a positive number, the well is judged to be a potentially profitable investment. Geophysical methods act to raise the probability of success for finding an economically successful well. However, the geophysical methods raise the cost of a dry well. Modern methods of geophysical exploration have raised the probability of success by as much as 30 percent in some instances over the original methods implemented in the 1920s. However, the question remains as to when the modern technology, such as 3-D seismic data, should be employed because it is more expensive. In offshore areas such as the North Sea, the deepwater Gulf of Mexico, and offshore Africa, there is the potential of finding large reserves. There is no question that 3-D data should be applied in these areas. The probability of drilling a successful well in these deep ocean areas is approximately the same as the shallow offshore areas, but what is driving deep ocean basin exploration is the potentially greater size of the reservoirs that remain to be found.

Raymon L. Brown

See also: Fossil Fuels; Oil and Gas, Drilling for; Oil and Gas, Production of.

BIBLIOGRAPHY

Allaud, L. A. (1977). *Schlumberger: The History of a Technique.* New York: John Wiley & Sons.

Curtis, D.; Dickerson, P. W.; Gray, D. M.; Klein, H. M.; and Moody, F. W. (1989). *How to Try Find an Oil Field.* Tulsa, OK: PennWell Books.

Mayne, H. W. (1989). *Fifty Years of Geophysical Ideas.* Tulsa, OK: Society for Exploration Geophysicists.

McCaslin, J. C. (1983). *Petroleum Exploration Worldwide: A History of Advances Since 1950 and a Look at Future Targets.* Tulsa, OK: PennWell Books.

Nettleton, L. L. (1940). *Geophysical Prospecting for Oil.* New York: McGraw-Hill.

Schlumberger, A. G. (1982). *The Schlumberger Adventure.* New York: Arco.

Sweet, G. E. (1966). *The History of Geophysical Prospecting.* Los Angeles: Science Press.

Tissot, B. P., and Welte, D. H. (1978). *Petroleum Formation and Occurrence.* New York: SpringerVerlag.

OIL AND GAS, PRODUCTION OF

The terms oil production and gas production refer to rates of extraction of liquid and gaseous hydrocarbon materials from natural underground deposits. Reserves and resources, on the other hand, refer to amounts of oil and gas that are present in the deposits, the difference between reserves and resources being whether or not the amounts can be economically recovered under current conditions. Supply refers to the amount of a product that becomes available for use in a given time, such as a year.

Worldwide production of crude oil increased through 1998 in response to increasing demand, although U.S. production was declining. The annual world supply of oil, which includes the annual production of crude oil and natural gas liquids grew from eighteen billion barrels in 1970 to twenty-seven billion barrels in 1998. A barrel is forty-two U.S. gallons. At the same time, the U.S. domestic supply declined from 4.3 billion barrels in 1970 to 3.5 billion barrels in 1998. So far, producing regions of the world have been able to meet the world demand without constantly escalating prices. However, the supply has become increasingly dependent on production in a few regions, notably the Middle East, where the remaining resources are concentrated. The rest of the world is becoming more dependent on these producers. Declining domestic production and growing demand have led to increasing imports of foreign oil into the United States. In 1997, one-half of U.S. consumption came from foreign imports.

Production of natural gas also has been increasing. World production of dry gas rose from sixty-six trillion cubic feet in 1987 to eighty-two trillion cubic feet in 1996. U.S. domestic dry gas production rose from seventeen trillion cubic feet in 1987 to nineteen trillion cubic feet in 1996, and the nation imported an additional three trillion cubic feet in 1996 to meet demand. Dry natural gas is produced from wellhead gas by removing most of the hydrocarbons heavier than methane. These heavy components, which tend to liquefy from the wellhead gas, are added as natural gas liquids to the oil supply and appear in the crude oil statistics.

Production of crude oil and natural gas involves technologies that have become increasingly complex as the remaining resources have become more difficult to locate and remove from their subsurface locations. Many new discoveries are made in sediments below the ocean floor in deep-water, and thus require removal of the oil and gas through long water columns. Other situations now require directional drilling of wells so that production involves transfers along wells that are far from vertical.

Most petroleum scientists believe that crude oil and natural gas formed over millions to tens of millions of years through the decomposition of organic matter buried by sediments. Generally, marine sediments have led to oil and gas, while freshwater

Gas production platform under construction in the Gulf of Thailand. (Corbis-Bettmann)

(swamp) sediments have produced coal and gas, although there are some exceptions. The liquid and gaseous hydrocarbons, in many cases, migrated through permeable subsurface rock, or along fractures, until they met an impermeable rock barrier, called a trap. There they collected to form an oil or gas reservoir. Oil and gas exploration teams using traditional and sophisticated geological techniques locate new reservoirs by studying the surface and subsurface geology in promising regions.

A reservoir is not a subterranean lake of pure oil or a cavity filled with gas. It is a porous and possibly fractured rock matrix whose pores contain oil, gas, and some water, or else, more rarely, it is a highly fractured rock, whose fractures contain the fluids. Such a reservoir is usually located in sandstone or carbonate rock. The rock matrix of an exploitable reservoir must be porous or fractured sufficiently to provide room for the hydrocarbons and water, and the pores and fractures must be connected to permit fluids to flow through the reservoir to the wells. In order to evaluate the production potential of a reservoir, a crew takes cores for analysis from appraisal wells drilled into the reservoir, and then examines properties of downhole formations using instruments lowered into the holes, a process called well logging.

When the exploration and appraisal phases at a reservoir are complete, a drilling crew drills production wells to extract the valuable hydrocarbons. Generally, a well is not left unlined. A pipe, called the casing, lines the well to protect the hole and the interior well structures, especially the tubing that transfers the oil or gas to the surface. In order to begin production, the well crew perforates the casing in the formations that contain the oil or gas and installs appropriate packing, plugs, valves, tubing, and any other necessary equipment required to remove the product.

A major factor in the production process is the driving force that moves the oil or gas to the borehole and lifts it to the surface. The fluid is, of course, under high pressure in the reservoir. If the pressure at the borehold exceeds the pressure of the rising column of oil or gas inside the tubing, it will flow out by itself. There are natural mechanisms that can provide the drive (energy) required to raise the product to the wellhead. One is the potential energy of the compressed oil or gas in the formation. The magnitude of this energy depends on the compressibility of the fluid or fluids involved. For gas, this mechanism is very effective because gas is quite compressible. As the formation gas expands into the borehole under the reduction of pressure at the borehole, it enters the tubing of the well and flows up to the surface. A gas well generally does not need to be pumped.

The situation is more complicated for oil because oil is relatively incompressible and is much denser. The natural driving pressure from the oil decompression is quickly reduced as oil flows into the borehole and, furthermore, the back-pressure produced by the heavy column of oil in the tubing is much greater than is the case for gas. Thus, one could expect this drive mechanism to be quickly lost. However, if the oil reservoir is in contact with either a large reservoir of water below it or a reservoir of gas above it, this contiguous fluid reservoir can supply the energy to maintain the drive for a much longer period. One says in such a case that the reservoir is water-driven or gas-cap-driven. There is, in fact, a third possibility called solution gas drive that originates from gas that is dissolved in the oil. When the formation pressure declines to a certain point, called the bubble point, the dissolved gas will begin to come out of solution. When this happens, the greater compressibility of the gas phase thus formed can contribute to an extended drive. However, this condition can be a mixed blessing because the gas can become the major product from the well, and it is less valuable than the oil.

The contiguous reservoirs of gas or water that contribute the drive pressures for an oil well can cause a problem. Gas and, in some cases, water, are more mobile than oil in an oil reservoir. As a result, during production the oil-gas or oil-water surface can move toward the region of reduced pressure around the borehole. If the interface reaches the borehole, the driver fluid will enter the well and be produced along with the oil. Since gas is not as valuable as oil, and water is unwanted, several measures are taken to limit ground water extraction. For example, in a gas-cap-driven reservoir, the production crew can locate the perforated section, where the oil enters the well, near the bottom of the pay zone so as to keep the collection point as far as possible from the oil-gas boundary. On the other hand, a water-driven well can be completed near the top of the zone to accomplish the same purpose. Reducing the rate of production from the well, and hence the drop in pressure around the borehole, is another way to avoid this problem.

The initial pressure in a reservoir is usually high enough to raise the oil to the surface. However, as production proceeds and the reservoir becomes depleted, the formation pressure falls and a new source of energy is required to maintain the flow of oil. A common practice is to inject either gas or water into the producing formation through special injection wells in order to maintain the pressure and move the oil toward the collection points in the producing wells. This procedure is called pressure maintenance or secondary recovery. Its success depends on maintaining a stable front between the oil and the injected fluid so that the oil continues to arrive at the collecting wells before the injected fluid. To the extent that the gas or water passes around the oil, arrives at the producing wells, and becomes part of the product coming up to the surface, the process fails to achieve its full objective. This undesirable outcome can occur, as noted above, when the mobility of the injected fluid exceeds that of the oil, as is the case for injected gas and for injected water with heavy oils. Gravity also can play a role in the success of secondary recovery depending on the geometry of the reservoir and the location of the injection and production wells.

In addition to pressure maintenance in a reservoir, there are other ways to maintain or improve well productivity as the formation pressure falls. A pump can be used to raise the oil, a process called artificial lift. One type is the familiar beam pump with its surface power unit driving up and down one end of a center-mounted beam while the other end executes the opposite down-up motion. The second end is attached to a string of "sucker rods" that extend down to the bottom of the well and operate a pump arrangement, consisting of a cylinder, plunger, and one-way valves. Other types of pumps in use eliminate the sucker rod and may have the power unit at the bottom of the well. A method for extending the period of natural flow is gas lift. In this procedure the well operators inject gas bubbles into the tubing to reduce the weight of the column of liquid that must be lifted to the surface. A third way to improve productivity is to "workover" an existing well. Mud, sand, or corrosion products may be plugging the perforations, or tubing, packing, pumps, or casing may be damaged and need repair. A workover crew will carry out repairs remotely from the surface or else remove interior components of the well for service.

Infill drilling is another possible way to extend production. For a variety of reasons, some oil may not be available to the original wells in the reservoir. Some wells may be spaced too far apart to capture the oil between them. Gas or water flooding may have bypassed some oil, or fractures or faults may block off certain parts of the reservoir from the rest so that they cannot be drained from existing wells. In these cases drilling new wells between existing ones can be an effective way to capture more of the resource.

Secondary recovery, infill drilling, various pumping techniques, and workover actions may still leave oil, sometimes the majority of the oil, in the reservoir. There are further applications of technology to extract the oil that can be utilized if the economics justifies them. These more elaborate procedures are called enhanced oil recovery. They fall into three general categories: thermal recovery, chemical processes, and miscible methods. All involve injections of some substance into the reservoir. Thermal recovery methods inject steam or hot water in order to improve the mobility of the oil. They work best for heavy oils. In one version the production crew maintains steam or hot water injection continuously in order to displace the oil toward the production wells. In another version, called steam soak or "huff and puff," the crew injects steam for a time into a production well and then lets it "soak" while the heat from the steam transfers to the reservoir. After a period of a week or more, the crew reopens the well and produces the heated oil. This sequence can be repeated as long as it is effective.

Chemical processes work either to change the mobility of a displacing fluid like water, or to reduce the capillary trapping of oil in the rock matrix pores. Reducing the mobility of water, for example by adding polymers, helps to prevent "fingering," in which the less viscous water bypasses the oil and

Oil well pumps in Midway-Sunset Oil Field. (Corbis-Bettmann)

reaches the production well. The alternate chemical approach is to inject surfactant solutions that change the capillary forces on the oil and release it from pores in which it is trapped. Another way to mobilize the oil is to inject a fluid with which the oil is miscible at reservoir conditions so that the oil trapped in the pores can dissolve in the fluid and thus be released. Carbon dioxide, nitrogen, and methane have been used for this purpose.

When oil or gas from a reservoir reaches the surface at the wellhead, it is processed to meet specifications for transportation and delivery to a customer such as a refinery or pipeline. The first step is usually to collect the output of a group of producing wells through a pipeline network into a gathering station where field processing takes place. In the case of crude oil, the objective is to produce a product relatively free of volatile gases, water, sediments, salt, and hydrogen sulfide, and fluid enough to transport to the refinery. In the case of wellhead gas, the objective is to produce a field-processed gas relatively free of condensable hydrocarbons, moisture, and various gas contaminants such as hydrogen sulfide and carbon dioxide.

The field processing facilities separate the components of the wellhead liquids and gases using their differing physical and chemical properties. Tank and baffle arrangements that utilize density differences are effective for initial separation of the water, oil, and gas. Drying agents to take water out of the separated hydrocarbons, screens, or other means to remove mists of small droplets in the gas, and heat, de-emulsifying agents, or electrostatic means to eliminate emulsions can further refine the components. For wellhead gas, refrigeration of the raw gas will remove heavier molecular weight hydrocarbons, such as propane and butane, improving the quality of the purified gas and coincidentally producing valuable hydrocarbon liquids.

After field processing of wellhead products is complete, the oil and gas production phase of the industry passes into the refining stage, where the crude oil and field gas are further processed to make the products that ultimate users will purchase. The prices that refiners will pay determine the maximum allowable costs for oil and gas exploration and production. An oil or gas producer must be able to provide the product at a competitive price. For a vertically integrated oil company, no actual sale may take place, but the economics are much the same.

Oil and gas development in new fields is a somewhat risky business for several reasons. The future trend in the price of crude oil or natural gas is uncertain, especially since existing low-cost producers in the world can exert considerable monopoly power in the industry. Furthermore, the amount of product that a reservoir will produce at a given cost is uncertain, and there is a long delay between the time when the expenses of developing a new field are incurred and the time when the revenues come in from the sale of the products. These uncertainties affect the likelihood that an investment in a new prospective development will be worthwhile. Even the continuation of production in an existing field can pose uncertainty. For example, when should operators shut down operations in a declining oil field? Shutting down marginally profitable wells because the price of crude oil has dropped may be an irreversible action because of governmental decommissioning requirements, or even just the cost of restarting. Yet the price may go up again shortly so that profitable operations could resume. Thus, the business aspects of oil and gas production play a major role in decision-making in this extractive industry, and the players must be attuned to the risks involved.

George W. Hinman
Nancy W. Hinman

See also: Oil and Gas, Exploration for; Oil and Gas, Drilling for.

BIBLIOGRAPHY

Energy Information Administration. (1998). *Historical Natural Gas Annual 1930 Through 1997.* Washington, DC: U.S. Department of Energy.

Energy Information Administration. (1999). *International Petroleum Statistics Report.* Washington, DC: U.S. Department of Energy.

Guiliano, F. (1989). *Introduction to Oil and Gas Technology*, 3rd ed. Englewood Cliffs, NJ: Prentice-Hall.

Jahn, F.; Cook, M.; and Graham, M. (1998). *Hydrocarbon Exploration and Production.* Amsterdam, The Netherlands: Elsevier Science Publishers.

Royal Dutch Shell Group. (1983). *The Petroleum Handbook.* Amsterdam, The Netherlands: Elsevier Science Publishers B.V.

Tissot, B., and Welte, D. (1984). *Petroleum Formation and Occurrence*, 2nd ed. Berlin, Germany: Springer-Verlag.

OIL SHALE

See: Synthetic Fuel

ONSAGER, LARS (1903–1976)

Lars Onsager was a Norwegian-American chemist and physicist who received the 1968 Nobel Prize in Chemistry for "the discovery of the reciprocal relations bearing his name which are fundamental for the thermodynamics of irreversible processes."

Onsager was born on November 27, 1903, to Ingrid and Erling Onsager. The family lived in Oslo (then Kristiania), Norway, where his father was a barrister. He grew up with two younger brothers, Per and Knut. Onsager's mathematical acumen was apparent early. At the age of fifteen he discovered for himself how to solve cubic equations. His school performance enabled him to skip one year at school, and at the age of sixteen he was ready for university studies. He entered the Norwegian Institute of Technology at Trondheim, where he graduated in 1925 with a chemical engineering degree.

As a freshman in Trondheim, he had studied the most recent theory, attributed to Peter Debye and Erich Hückel, of electrolytic solutions, a topic to which he returned continually throughout his life. His first contact with the scientific world was when he, at the age of twenty-two, entered Debye's office in Zürich, saying without introducing himself: "Good morning, Herr Professor, your theory of electrolytic conduction is incorrect" (Hemmer et al., 1996). Debye, impressed by Onsager's arguments, hired him as his assistant. Onsager's revision of the Debye-Hückel theory resulted in the so-called Onsager limiting law for electrolytic conduction, an important result in good agreement with experiments.

In 1928 Onsager emigrated to the United States, first to Johns Hopkins University, then in the same year to Brown University. Here he struggled with an attempt to base symmetry relations found in his work on electrolytes on general physical laws. Thus he considered more general irreversible processes than conduction and diffusion, the basic ingredients in electrolytic conduction. Irreversible transport processes are centered around fluxes of matter, energy, and electric charge and their dependence upon the forces that give rise to them. The processes are called irreversible because the fluxes are unidirectional and cannot run in the reverse direction. The main effects are well-known: diffusion (transport of matter) due to a concentration gradient or concentration difference (e.g., across a membrane), heat conduction (transport of energy) due to a temperature gradient or temperature difference, and electrical conduction (transport of electric charge) due to a potential difference.

However, a potential may give rise to more than one type of flux. There are cross-effects: A temperature difference can also result in diffusion, called thermal diffusion, and a concentration difference can result in a heat current. The general relation between fluxes J_i and the driving potentials X_i is of the form of linear relations

$$J_i = \Sigma_j L_{ij} X_j, \qquad (1)$$

with coefficients L_{ij}, when the fluxes are not too great. Onsager showed that the coefficients of the cross-effects were equal:

$$L_{ij} = L_{ji}. \qquad (2)$$

These are the famous Onsager reciprocal relations. Thus there is symmetry in the ability of a potential X_i to create a flux J_j, and of the ability of a potential X_j to create a flux J_i. The reciprocal relations are experimentally verifiable connections between effects which superficially might appear to be independent.

Such symmetries had been noticed in special cases already in the nineteenth century. One example was heat conduction in a crystal, where the X_i are the temperature gradients in the three spatial directions. Cross-effects between electric conduction and diffusion, and between heat conduction and diffusion, were also studied. Reciprocal relations of the form $L_{ij} = L_{ji}$ were confirmed experimentally for these cases, but a general principle, from which these would follow, was lacking. This situation changed dramatically in 1931 when Onsager, then at Brown University, derived the reciprocal relations. In the derivation, Onsager used the fact that the microscopic dynamics is symmetric in time, and he assumed that microscopic fluctuations on the average follow macroscopic laws when they relax towards equilibrium. The Onsager formulation was so elegant and so general that it has been referred to as the Fourth Law of Thermodynamics. It is surprising to note that the immediate impact of Onsager's work, which later earned him the Nobel Prize, was very minor, and that it was virtually ignored during the following decade.

Onsager spent the summer of 1933 in Europe. During a visit to the physical chemist Hans Falkenhagen, he met Falkenhagen's sister-in-law, Margrethe (Gretl) Arledter. They fell in love and married in September the same year. Another important event in 1933 was Onsager's move to Yale University, where he remained until his retirement. In the course of time the Onsager family grew to include one daughter and three sons. At Yale he was first a Stirling and Gibbs fellow, then assistant professor, associate professor, and finally in 1945 J. Willard Gibbs Professor of Theoretical Chemistry.

Lars Onsager (right), receiving the 1968 Nobel Prize in Chemistry from King Gustav Adolf of Sweden (left). (Archive Photos, Inc.)

Although Onsager's first appointment at Yale was a postdoctoral fellowship, he had no doctorate. It disturbed him that everybody called him "Dr. Onsager," and he decided to seek the Ph.D. from Yale. He was told that any of his published works would do for the thesis, but he felt he should write something new, and he quickly submitted a lengthy dissertation on Mathieu functions. Both the Department of Chemistry and the Department of Physics, found it difficult. The Department of Mathematics, however, was enthusiastic and was prepared to award the degree, whereupon the Department of Chemistry did not hesitate in accepting the thesis.

If Onsager's great achievement with the thermodynamics of irreversible processes met with initial indifference, Onsager's next feat created a sensation in the scientific world. In a discussion remark in 1942, he disclosed that he had solved exactly the two-dimensional Ising model, a model of a ferromagnet, and showed that it had a phase transition with a specific heat that rose to infinity at the transition point. The full paper appeared in 1944. For the first time the exact statistical mechanics of a realistic model of an interacting system became available. His solution was a tour de force, using mathematics almost unheard of in the theoretical physics of the day, and it initiated the modern developments in the theory of critical phenomena. Basing their decision especially on this work, the faculty at Cornell University nominated Onsager for the Nobel Prize in both physics and chemistry.

Although the reciprocal relations, electrolyte theory, and the solution of the Ising model were highpoints in Onsager's career, he had broader interests. From about 1940 Onsager became active in low-temperature physics. He suggested the existence of quantized vortices in superfluid helium, and provided the microscopic interpretation of the oscillatory diamagnetism in metals. In 1949, four years after he became an American citizen, he laid the foundation for a theory of liquid crystals. In his later years he studied the electrical properties of ice and took much interest in biophysics.

The importance of Onsager's work on irreversible processes was not recognized until the end of World War II, more than a decade after its publication. From then on irreversible thermodynamics gained momentum steadily, and the reciprocal relations have been applied to many transport processes in physics, chemistry, technology, and biology. The law of reciprocal relations was eventually recognized as an enormous advance in theoretical chemistry, earning Onsager the Nobel Prize in 1968. Thus more than a third of a century passed before Onsager's work was suitably recognized. In his presentation speech at the Nobel ceremonies S. Claesson said: "Here we thus have a case to which a special rule of the Nobel Foundation is of more than usual applicability. It reads: 'Work done in the past may be selected for the award only on the supposition that its significance has until recently not been fully appreciated.'"

Onsager was reluctant to publish. Many of his original discoveries appeared first in the discussion periods at scientific meetings and were not published for years, if at all. He was very much an individualist. Although he had students and coworkers, especially in his later years, he preferred to work alone. Therefore, he never created a school around him. As a person, Onsager was modest and self-effacing, with a wry sense of humor. He had an awesome memory and was at ease with history, language, and the literature of the West. Games of all kinds appealed to him.

After retirement he went to the Centre for Theoretical Studies at Coral Gables, Florida, where one of his main interests was the question of the origin of life. He died at Coral Gables on October 5, 1976.

P. C. Hemmer

See also: Thermodynamics.

BIBLIOGRAPHY

Hemmer, P.C., Holden, H., and Ratkje, S. Kjelstrup, eds. (1996) *The Collected Works of Lars Onsager*. Singapore: World Scientific.

Onsager, L. (1931) "Reciprocal Relations in Irreversible Processes." *Physical Review* 37:405–426 and 38:2265–2279.

OPTOELECTRONICS

See: Lighting

OTTO, NIKOLAUS AUGUST (1832–1891)

Otto was born in the village of Holzhausen, near Frankfurt. His father, an innkeeper, died when Otto was a small boy. This was a period of declining prosperity in Germany, although Otto's mother had hoped her son might receive a technical education, Otto was forced to leave school and take up employment in a Frankfurt grocery store. His brother Wilhelm, who had a textile business in Cologne, then secured him a position as a salesman in tea, sugar, and kitchen ware to grocery stores. Otto's travels took him along the German border with France and Belgium as far as Cologne where in 1859 he met Anna Gossi and began a nine-year courtship.

In 1860 Jean Lenoir built the world's first mass production internal combustion engine, a noncompression double-acting gas engine. Otto saw the potential of the Lenoir engine and perceived that the future development of the internal combustion engine would be dependant on its fuel source. In an attempt to make the engine mobile he devised an alcohol-air carburetor, and with his brother filed for a patent in 1861. Hampered by a lack of funds and a technical education, Otto spent all his leisure time in developing an engine that would surpass that of Lenoir. In 1861 Otto commissioned a Cologne instrument maker, Michael Zons, to build an experimental model Lenoir engine. Otto carried out a series of experiments with the purpose of lessening piston shock loading at the start of combustion. In one series of experiments he turned the engine over drawing in the explosive mixture, the piston was then returned to top center compressing the mixture that was then ignited. Otto found that ignition took place with great violence. This undoubtedly set the scene for the development of Otto's compression engine. In 1862

Zons built for Otto a flat four-cylinder engine. Each cylinder contained a free "cushion" piston, this engine, however, was a failure. In 1863 Otto had Zons built a one-half horsepower atmospheric engine.

In 1864 Otto's luck took a turn for the better when his engine came to the attention of Eugen Langen. Langen was a partner in a sugar refining business that he had expanded into refining sugar beets as well as cane sugar, using processing equipment of his own design. Although Langen could see the deficiencies in Otto's design he quickly made up his mind to invest in the engine. With the help of his father, Langen raised sufficient capital to form a company. In March 1864 Otto and Langen signed an agreement forming N.A. Otto & Cie, the world's first company to be set up to manufacture internal combustion engines. By 1867 the engine design was perfected and a patent applied for. The improved engine was exhibited at the 1867 Paris Exposition. After a series of tests that examined the engine efficiency and performance, the Otto and Langen engine was found to consume less than half the gas of the best noncompression engine and was awarded a gold medal. Although orders now flooded in, the partners still lacked sufficient funds. New partners were sought and Otto, whose stock holding, fell to zero, accepted a long-term employment contract. In 1872 Gasmotoren-Fabric Deutz AG came into being with Langen in control and Otto as Technical Director. Gottlieb Daimler joined the firm as production manager bringing with him Wilhelm Maybach to head the design department. Improvements were made to the engine probably by Maybach, but Daimler demanded they be patented in his name, an action that led to disputes between him and Langen. Langen also had to arbitrate numerous disagreements between the autocratic Daimler and the oversensitive Otto. Through the use of licenses and subsidiaries, Otto and Langen established an international industry with engines being built in most industrialized nations.

Around 1872 Otto returned to the problem of shockless combustion. It occurred to him, if only air was brought into the cylinder first and then the gas/air mixture, the charge would become progressively richer towards the source of ignition. Otto built a small hand-cranked model with glass cylinder to study his stratification idea by using smoke. A prototype four-stroke cycle engine was constructed in 1876. Although, at the time, Daimler thought Otto's new ideas would prove a waste of time, Langen assigned Otto an engineer and let him continue his research. Otto's ideas concerning stratification were patented in 1877. The new engine had a single horizontal cylinder and bore some similarity to the Lenoir engine. The important difference was in the admission of gas and air, also gas flame ignition. Otto seemingly had more faith in ignition by flame rather than electrical spark. The basic elements of the modern four-stroke engine are to be found in this experimental engine. A slide valve controlled air intake with a second cam operated slide valve controlling gas admission. As the piston began its outward stroke, the air valve opened and the gas valve remained closed until the piston had completed half its travel. At the end of the stroke, the valves closed and the piston began the return stoke compressing the air gas mixture. At the end of the return stroke, the gas air mixture is ignited and the piston commences its second outward stroke, the power stroke. At the end of this stroke the exhaust valve opens, and the piston returns exhausting the burnt gasses. In Otto's new design there was one power stroke for every four strokes of the piston or two revolutions of the flywheel.

Nikolaus August Otto. (Granger Collection, Ltd.)

Wilhelm Maybach redesigned Otto's prototype engine for mass production. The "Otto Silent" engine rapidly established the gas engine as a practical power source and Deutz jealously protected their position as the world's sole supplier of Otto's new engine. Through litigation and threats of action Deutz pursued any manufactures who attempted to infringe on the Otto patent. These actions, while effectively emasculating other engine builders, acted as a stimulus for basic research into the correctness of Otto's ideas on stratified charge and in turn the validity of his patent. Deutz's manufacturing competitors in endeavoring to discredit the stratified charge theories discovered an obscure patent filed by Frenchman A. B. de Rochas in 1862. Although de Rochas's patent had never been published, through a failure to pay the required fees, nor was there any evidence he had produced a working engine, his patent was used to challenge Deutz's monopoly. During litigation Otto strove to establish prior invention, but suffered, as have many other inventors, from inadequately documenting his research work. Although Otto must be considered the true inventor of the four-stroke cycle, patent law is more a matter of establishing priority of ideas than production of working machines. The Otto patents were overturned in Germany in 1886. In Britain, Otto's designs were licensed by Crossley who successfully defended their position by having the de Rochas patent ruled inadmissible.

Although his ideas were vindicated in Britain, the loss of his German patents were a blow to Otto's pride from which he apparently never recovered. Otto died five years later in 1891. Otto's only son Gustov became an engineer and aircraft designer in Munich, and in 1901 founded an airframe company that produced aircraft for World War I.

Robert Sier

See also: Engines; Gasoline Engines.

BIBLIOGRAPHY

Barlow, K. A. (1994–1995). "Nikolaus August Otto and the Four-Stroke Engine." *Transactions of the Newcomen Society* 66: suppl. no. 1.

Clerk, D. (1896). *The Gas and Oil Engine*, 6th ed. London: Longman, Green and Co.

Cummins, C. L. (1976). *Internal Fire*. Oregon: Carnot Press.

Donkin, B. (1896). *Gas, Oil and Air engines*, 2nd ed. London: Griffin and Co.

PARSONS, CHARLES ALGERNON (1854–1931)

Charles Algernon Parsons was born in London on June 13, 1854, the son of a wealthy, aristocratic Anglo-Irish family that was scientifically very distinguished. His father, William Parsons (third Earl of Rosse), a member of parliament, was an engineer, astronomer, and telescope-maker who had built the largest telescope in the world. His mother, Mary Countess of Rosse, best remembered as a photographer, was adept at architectural design and cast-iron foundry work.

The greater part of Parsons' childhood was spent at the family's castle and estate at Parsonstown (now Birr), County Offaly, Ireland. The workshops there were extensive, and he grew up surrounded by tools and machinery. He was privately educated by, among others, Sir Robert Ball, a man who encouraged him to spend time not just in the classroom, but also in the workshops. With these facilities and the guidance of his parents and tutor, he developed an exceptional talent for engineering.

The key to Parsons' success lies in the nature of his early training in science and engineering. In a letter concerning the design of yachts, he wrote that "they seem to design by rule of thumb in England—they should rely on model experiments, which would put them right." This is illustrative of an approach to engineering that was highly practical. He solved problems by building experiments and then meticulously observing and analyzing the results. It was an aspect of his character that was formed by the time he left Birr for his university education.

In July 1871, Parsons entered Trinity College Dublin, where he studied mathematics. In 1873, he entered St. John's College Cambridge and took his B.A. degree in mathematics in 1877. Thus, at the age of 23 he graduated from the university armed with a battery of practical and theoretical skills that allowed him to gain unique insights into the engineering problems of the time.

Having spent five years at the university honing his theoretical skills, Parsons then returned to the practical by entering the engineering firm of W. G. Armstrong & Co. in Newcastle-upon-Tyne. As a premium apprentice he was allowed to spend some of his time constructing his own experiments, provided that he paid the material and labor costs himself. Using this time, he developed a high-speed, four-cylinder epicycloidal steam engine.

The Armstrong company was not interested in producing the engine; therefore in 1881, Parsons moved to Kitson's of Leeds who took it into production. At Kitson's he occupied himself with experiments in rocket powered torpedoes that, although unsuccessful, provided useful background for his next project, the steam turbine.

Parsons left Kitson's and joined Clarke, Chapman & Co. of Gateshead as a junior partner. This was the turning point in his career, his previous working life having been devoted to learning and developing his skills. In 1884, he decided "to attack the problem of the steam turbine and of a very high speed dynamo." During the next ten years of his life, he made huge strides in the construction of dynamos and steam turbines.

In the late 1880s the Industrial Revolution was demanding ever-increasing amounts of energy. Electrical power was an obvious solution to this demand; however, it was not yet possible to generate very high power. Parsons realized that high-speed

Charles Algernon Parsons. (Library of Congress)

dynamos were the answer, and that the best source of rotational energy would be a steam turbine, whose nonreciprocating nature makes it capable of far higher speeds than conventional reciprocating engines.

Turbine designers had been experimenting with machines in which the blades were driven by a high-velocity steam jet that expanded through them in a single stage. It had been shown that steam-jet velocities of the order of 1000 m/s were required.

For a steam turbine to work efficiently, it is necessary that the velocity of the tips of the turbine blades be proportional to the velocity of the input steam jet. If the velocity of the input steam jet is high, it is necessary to increase the radius of the blades so that their tip velocity becomes correspondingly high. These long rotor blades subject the entire assembly to unmanageable mechanical stresses at high speed.

Parsons realized that he could, by using multiple rings of turbine blades on the same shaft, allow the steam to expand in stages through the turbine, and therefore extract energy more efficiently from the steam jet. This allowed him to utilize much lower jet velocities, of the order of 100 m/s, with correspondingly shorter turbine blades.

Parsons attacked the problem with characteristic gusto, so much so that by 1885 his first prototype was running successfully, giving 4.5 kW at 18,000 rpm. In 1889, frustrated by what he perceived as a lack of progress, he broke his partnership with Clarke Chapman and set up on his own in Heaton. He lost all his patent rights to the steam turbine; however, he was not daunted by this and simply developed a radial-flow machine that circumvented his own patents. In a radial-flow turbine, the steam expands outwards at right-angles to the rotational axis in contrast to the axial-flow machine where the steam expands along the length of the rotational axis.

Parsons realized that his invention was ideal for powering ships. Having received an unenthusiastic response from the admiralty, he set about proving his point in a characteristically irreverent manner. He built the *Turbinia*, a vessel powered by steam turbine and screw propellers. On the occasion of her diamond jubilee in 1897, Queen Victoria was reviewing the royal navy fleet at Spithead when Parsons appeared in the *Turbinia* and raced through the fleet at the then unbelievable speed of 30 knots. The point was well made, and on November 21, 1899, HMS *Viper* had her trials, reaching a speed of 32 knots.

Parsons was, first and foremost, a compulsive inventor. He spent his days inventing everything from children's toys to the Auxetophone, a mechanical amplifier for stringed musical instruments. His success as an inventor lies in his inquisitive nature and the fact that he was equally comfortable with theory and practice.

Parsons loved travel and the sea. It was on a cruise to the West Indies that he died aboard ship on February 11, 1931, in Kingston Harbor, Jamaica.

Bob Strunz

BIBLIOGRAPHY

Appleyard, R. (1933). *Charles Parsons*. London: Constable & Company Ltd.

Parsons, C. A. (1911). *The Steam Turbine*, Cambridge, England: Cambridge University Press.

Richardson, A. (1911). *The Evolution of the Parsons Turbine*. London: Offices of Engineering.

Scaife, W. G. (1992). "Charles Parsons—Manufacturer." *Journal of Materials Processing Technology* 33:323–331.1

PARTICLE ACCELERATORS

A particle accelerator is a device that accelerates charged particles—such as atomic nuclei and electrons—to high speeds, giving them a large amount of kinetic energy. The most powerful accelerators give particles speeds almost as great as the speed of light in a vacuum (300,000 kilometers per second), which is a limiting speed that electrons and nuclei can approach but cannot reach.

If a particle is moving much slower than the speed of light, for example, one-tenth the speed of light, then its kinetic energy is proportional to the square of its speed, in accordance with Newtonian mechanics. For example, if an accelerator quadruples the kinetic energy of a slowly moving particle, its speed doubles. However, if the particle is already traveling fast, for example, at 90 percent of the speed of light, and the accelerator quadruples the kinetic energy of the particle, then the speed cannot double, or the particle would go faster than the speed of light. In this case, one has to use Einstein's more general formula for the relation between speed and kinetic energy, according to which the speed goes from 90 percent of the speed of light to a little under 99 percent. No matter how much the kinetic energy of the particle is increased by the accelerator, the particle still will travel slower than light. If the particle is traveling at 99 percent of the speed of light, no matter how much energy it is given, its speed can only increase by about 1 percent. For this reason, powerful accelerators are really particle energy boosters and not speed boosters.

Fermilab designed and built the proton therapy accelerator as the first proton accelerator specifically for the treatment of cancer. (U.S. Department of Energy)

ACCELERATOR PHYSICS

The basic unit of energy used in accelerator physics is the electron volt (eV), which is the energy acquired by an electron when accelerated through a potential difference of one volt. An electron volt is a very small unit compared to an energy unit such as a food calorie (kilocalorie). A kilocalorie is about 26 billion trillion times as large as an eV. Common multiples of eV are MeV (million eV), GeV (billion eV), and TeV (trillion eV).

It is instructive to compare the kinetic energy acquired by a particle to its rest energy. If an airplane travels at 1000 kilometers per hour, its rest energy is more than 2 million million ($2 (10^{12})$) times its kinetic energy. In contrast, the highest energy electron accelerator gives an electron a kinetic energy almost 200,000 times its rest energy. The speed of such an electron is very nearly equal to the speed of light. If a burst of light and a 100 GeV electron leave a distant star at the same time, and it takes the light 2,500 years to reach the Earth, the electron will reach the earth about one second later.

The main reason particle accelerators were invented and developed is so that scientists can observe what happens when beams of particles strike ordinary matter. However, particle accelerators are also used in a wide variety of applications, such as irradiating cancers in people.

Ernest Lawrence with cyclotron. (Corbis Corporation)

HISTORY

The development of particle accelerators grew out of the discovery of radioactivity in uranium by Henri Becquerel in Paris in 1896. Some years later, due to the work of Ernest Rutherford and others, it was found that the radioactivity discovered by Becquerel was the emission of particles with kinetic energies of several MeV from uranium nuclei. Research using the emitted particles began shortly thereafter. It was soon realized that if scientists were to learn more about the properties of subatomic particles, they had to be accelerated to energies greater than those attained in natural radioactivity.

The first particle accelerator was built in 1930 by John Cockroft and Ernest Walton. With it, they succeeded in accelerating protons (hydrogen nuclei) by means of a high-voltage source. The protons reached an energy of 300,000 eV. The following year Robert Van de Graaff succeeded in obtaining a high voltage by means of a charged moving belt. Subsequently, accelerators of the Van de Graaff type accelerated particles to more than 10 MeV.

The Cockroft–Walton and Van de Graaff accelerators are linear; that is, they accelerate particles in a straight line. A short time later Ernest Lawrence got the idea to build a circular accelerator, called a cyclotron, and with the help of M. Stanley Livingston he constructed it in 1932. This first cyclotron accelerated protons to about 4 MeV. Since then, many other cyclotrons have been built, and they have been used to accelerate particles to more than 50 times as much kinetic energy as the original one. Also, other kinds of circular accelerators, such as synchrotrons, have been constructed.

OPERATION

The principle of a circular accelerator is that forces from properly arranged electromagnets cause the charged particles of the beam to move in circles, while properly arranged electrical forces boost the energy of the particles each time they go around. The radius of the circle depends on the mass and speed of

1. The first accelerator, built by J.D. Cockroft and E.T. Walton in 1930 at the Cavendish Laboratory of the University of Cambridge, England, accelerated protons to an energy of 300,000 electron volts (eV). At this energy, the protons have a speed of about 7,600 kilometers per second, and are traveling about 2.5% of the speed of light. The first circular accelerator, or cyclotron, was built by Ernest Lawrence with the help of M. Stanley Livingston in 1932 at the University of California in Berkeley.
2. The highest energy accelerator in existence today, the Tevatron, is at the Fermi National Accelerator Laboratory, located west of Chicago. It accelerates both protons and antiprotons to an energy of about 1 trillion electron volts (TeV). At this energy, the protons and antiprotons have a speed almost as great as the speed of light, which is 300,000 kilometers per second. The difference in speed is only about one part in 2 million.
3. The Large Hadron Collider is an accelerator scheduled for completion in the year 2005 at the CERN laboratory near Geneva Switzerland. It will accelerate protons to an energy of 8 TeV. At that energy, the protons will have a speed only about one part in 130 million slower than the speed of light.
4. Accelerators are responsible for many fundamental discoveries as well as many practical applications. An example of a fundamental discovery was the observation of the W and Z particles that carry the weak force that is responsible, along with the strong and electromagnetic forces, for the fact that the sun shines. An example of a practical application is the use of accelerator beams to kill cancerous tumors in patients.
5. In the coming decades, by colliding particles at ever greater energies, physicists hope to discover what causes mass to exist.

the particle and on the strength of the magnetic forces. In a cyclotron, the magnetic forces are kept constant, and as the particles gain energy they spiral outward into larger and larger circles. In a synchrotron, as the particles gain energy the magnetic forces are increased enough to keep the particles going around in the same circle.

After a beam of particles is accelerated to its final energy, it can be made to strike atoms in a target fixed in space or else collide with particles in a high–energy beam traveling in the opposite direction. Accelerators with two beams going in opposite directions are called colliders. In all particle accelerators, the particles move in an enclosed region from which nearly all the air has been removed. Otherwise, too many of the particles in the beam would collide with the molecules of air rather than with the target.

If subatomic particles moving at speeds close to the speed of light collide with nuclei and electrons, new phenomena take place that do not occur in collisions of these particles at slow speeds. For example, in a collision some of the kinetic energy of the moving particles can create new particles that are not contained in ordinary matter. Some of these created particles, such as antiparticles of the proton and electron, can themselves be accelerated and made to collide with ordinary matter.

If one wants to achieve the highest possible energy, the accelerator of choice is a collider. If two particles moving in opposite directions with the same kinetic energy collide with each other, the total kinetic energy of both particles is available to create new particles. On the other hand, when a beam strikes a stationary target, the law of conservation of momentum requires that the struck particle and any created particles carry off the same total momentum as the particle in the beam initially had. This fact requires the final particles to have significant kinetic energy—energy that cannot be used to create additional particles.

The highest energy accelerators in existence are synchrotron colliders. The one with the highest energy is called the Tevatron, and is located at the Fermi Accelerator Laboratory near Chicago. Each beam has an energy of about a TeV (hence the name), about 250,000 times the energy achieved in Lawrence's first cyclotron. The Tevatron accelerates protons in one beam and antiprotons (antiparticles of the proton) in the other beam. The antiprotons are accelerated after being created in collisions of protons with a target. The Tevatron is about six kilometers in circumfer-

The Collider Detector at Fermi National Accelerator Laboratory, home of the Tevatron. (Corbis Corporation)

ence and gives both protons and antiprotons kinetic energies of about 1,000 times their rest energies.

When an electrically charged particle is accelerated, it radiates electromagnetic energy. The higher the energy of the particle, the more energy it radiates, and, for a given energy, the smaller the mass of the particle and the smaller the radius of the circle in which it goes, the more energy it radiates. In a synchrotron, the particles go around the accelerator thousands of times, each time radiating electromagnetic energy, appropriately called synchrotron radiation.

Synchrotron radiation is not a serious problem for proton synchrotrons, but it is for high energy electron synchrotons because the mass of the electron is only about one two-thousandth as large as the mass of the proton. For this reason, a high-energy electron synchrotron has to have a large circumference or it wil radiate too much power to be economically useful. The largest electron synchrotron is located at the European Laboratory for Nuclear Research (CERN) near Geneva, Switzerland, and partly in France. Sited in an underground tunnel, it has a circumference of 27 kilometers. It gives electrons kinetic energies of about 100 GeV or 200,000 times their rest energy. After attaining this energy, the beam of electrons is made to collide with an equally energetic beam of positrons (antiparticles of electrons) going in the opposite direction.

SYNCHROTON RADIATION

If the object of a synchrotron is to accelerate electrons to the highest possible energy, synchrotron radiation is a serious obstacle that limits the energy attainable. On the other hand, the electromagnetic radiation from a synchrotron can be useful for experiments on the properties of solids and for other purposes. For this reason, some electron synchrotrons are built primarily for the synchrotron radiation they emit.

To avoid the problem of synchrotron radiation, linear electron accelerators have been built. The principle of acceleration is not a static voltage, as in the Cockroft–Walton and Van de Graaff accelerators, but a traveling electromagnetic wave that has the right phase to accelerate the electrons as they travel inside a long straight tube. The longest linear accelerator is located in California at the Stanford Linear Accelerator Center. It is two miles long and accelerates electrons and positrons to an energy of about 50 GeV, after which they are made to collide.

The 27-kilometer tunnel of the CERN laboratory also houses a proton collider that is scheduled for operation beginning in the year 2005. This collider will be the highest energy accelerator in the world, with the protons in each beam attaining energies of about 8 TeV, which is about 8000 times the rest energies of the protons. Although so far, electron synchrotrons have achieved electron kinetic energies that are the highest multiple of their rest energies, proton synchrotons, which are less limited by synchroton radiation, have achieved the highest actual particle energies.

Don Lichtenberg

See also: Matter and Energy.

BIBLIOGRAPHY

Lederman, L. (1993). *The God Particle.* New York: Houghton Mifflin.

Newton, D. (1989). *Particle Accelerators.* New York: Franklin Watts.

PEAK LOAD MANAGEMENT

See: Electric Power, Generation of

PERPETUAL MOTION

The idea of perpetual motion, which has been around for centuries, is to make a device that will produce more work output than the energy input—in short, to get something for nothing. Robert Fludd in 1618 was one of the first to discover it is impossible to get something for nothing. He designed a pump to drive some of the output from a water wheel to recirculate water upstream, which would then run over the wheel again. The remaining portion of the output could then be used to operate a flour mill. The only problem with this device is that it took more energy to pump the water than the entire energy output of the water wheel. Friction will always cause such a water wheel to "grind" to a halt even in the absence of doing useful work.

Would-be inventors frequently employ magnetic or electrostatic interactions because these forces are less understood (by them). For instance, a weight can be lifted by a magnetic field, perhaps produced by a superconducting magnet, with no power loss. Then, when the magnetic field is turned off or taken away, the falling weight can be harnessed to do useful work. The magnet can continually lift and drop the weight. The details can be quite complicated, but when any design is correctly analyzed it is always found that the energy required to turn the magnetic or electric field on and off, or to move the field from place, to place exceeds the work obtained by the falling weight. The first perpetual motion device using magnetic forces

John W. Keely (1827–1898) seated next to what he claimed was the first perpetual motion machine. It was proved to be a fake, but not until after he died. (Corbis Corporation)

was proposed by John Wilkins in the 1670s. Wilkins proposed using the magnetic material lodestone, not a superconducting magnet.

To prove that a particular design of a perpetual motion machine will not work can be very time consuming, and the predictable negative result has never been worth the effort. Therefore, the U.S. Patent Office has a policy to not examine applications covering perpetual motion machines unless the applicant furnishes a working model.

The nonexistence of perpetual motion machines, despite centuries of effort to design them, has been used to support the law of conservation of energy. This law is based, however, not on this negative result, but on all the experiments performed to date in which energy is carefully accounted for. It has never been observed to fail. This law is, therefore, a good basis from which to analyze perpetual motion machines. It clearly states that the goal of getting more energy output than the energy input is impossible. It also gives a basis for considering another lesser goal of perpetual motion, which is to produce a device that will run forever with no further external inputs.

There are many systems in nature that for practical purposes are perpetual. The rotation of the Earth does not change perceptibly in any person's lifetime. Very careful measurements can detect such changes, and they are predictable based on the tidal interaction with the sun and the moon. At the current rate of decrease, the Earth's rotation relative to the Sun would stop in 5.4 billion years. Systems that are unchanging for practical purposes, but which are running down ever so slowly, do not count as perpetual motion.

There are clocks that do not need winding or battery replacement. They run by taking advantage of small changes in atmospheric pressure or the movement, or by utilizing solar cells. These do not count as perpetual motion either because they use the flow of energy from other sources to maintain their motion.

The electrons of atoms in their ground state are in perpetual motion. Also, in a gas, each atom has an average kinetic energy (motion) depending on the temperature of the gas. These motions don't count either because humans demand something on their own scale that they can see or take advantage of before they consider it as true perpetual motion.

Perpetual motion itself would not defy the conservation of energy. A pendulum swinging forever at the same amplitude does not change its energy. However, energy comes in several different forms and it is impossible to keep a system in a single energy mode because of all the possible interactions with the environment.

The pendulum, for instance, will encounter air resistance, transforming its kinetic energy into random motion or heat in the air. If the pendulum were mounted in a vacuum, it would swing for a longer period of time, but any friction in the support would eventually bring it to rest. Suppose that the support could be made absolutely frictionless, would that do it? Again, no, because as the pendulum moves back and forth in the light of the room, the light pressure is greater as the pendulum moves toward the light than it is when it moves away. The light itself will slow and eventually stop the pendulum. Suppose the frictionless pendulum were mounted in a dark vacuum? That still would not do it, because the heat (infrared) radiation from the walls of the chamber will slow the pendulum the same way as the light does. The frictionless pendulum mounted in a dark vacuum chamber at absolute zero might do it if the small tidal effects gravitational interaction of the moving pendulum with the surroundings (including the Earth) could be neglected, which they cannot. Without gravitational interaction, the pendulum would not swing downward. If true perpetual motion is pursued, no interaction, no matter how small, can be neglected.

The consideration of the simple pendulum illustrates the basic problem behind devising a perpetual motion machine. The problem is the fact that energy exists in several forms and is transformed from one form to the other, especially when motion is involved. Even if friction is eliminated, there are still the electromagnetic radiation and gravitational interactions which, given the present understanding of physics, would be impossible to eliminate.

Since transformation of energy from form to form is inevitable, what about devising a system that will just pass the energy around and thus keep going forever? The problem with such a device is more subtle. It involves the second law of thermodynamics. Once the energy of a pendulum swinging in air, for example, is transferred as heat to the air, there is no way to transfer all of the heat energy back to the kinetic energy of the pendulum. According to the second law, some energy could be returned to the pendulum, but not all of it. Gradually, even with a perfect heat engine (Carnot cycle or Stirling cycle) operating to restore energy to the motion of the pendulum, its swinging would eventually die away.

There would remain some very small residual motion of the pendulum due to the air molecules striking it at random (Brownian motion), but that does not count in the "game" of perpetual motion. In the condition of residual motion, the pendulum is just another (big) molecule sharing equally in the average kinetic energy of all the individual air molecules. In other words, the pendulum eventually comes to thermal equilibrium with the air.

One statement of the second law of thermodynamics is that there is a tendency in nature for an isolated system (no external inputs) to move toward conditions that are statistically more probable (higher entropy). For the swinging pendulum, given a certain energy, there is only one amplitude at which it can swing. That is, there is only one way the pendulum can have this energy. When this energy is transformed as heat to the air, there are almost an infinite number of ways for the air molecules to contain the same energy. The swinging pendulum mode is just one mode out of many for the energy to exist in the system. Thermodynamics does not predict that all the energy can never return to the swinging mode; it just states that in the foreseeable duration of the universe it is not likely to happen because there are so many other possibilities.

Similar considerations apply to all other schemes to produce perpetual motion. No matter how it is designed, an isolated system will always tend toward a condition of thermal equilibrium, which never involves motion of objects on a human scale, and is thus not "perpetual motion."

Don C. Hopkins

BIBLIOGRAPHY

Feynman, R. P.; Leighton, R. B.; and Sands, M. (1963). *Lectures on Physics, Vol. I*, Chapters 39–46. Reading, MA: Addison-Wesley.

Ord-Hume, A. (1977). *Perpetual Motion: The History of an Obsession*. New York: St. Martin's Press.

PETROLEUM CONSUMPTION

NATURE OF PETROLEUM

Crude oil generally appears as a dark-colored liquid. It is found underground under pressure and therefore wells must be drilled in order to bring it to the surface. Part of the crude oil is in the form of a gas. The latter separates out from the oil upon reaching the lower surface pressures, and is commonly referred to as natural gas.

Structurally, oil is made up of hydrocarbon molecules containing various combinations of carbon and hydrogen. The configurations of these molecules depend on the number and arrangement of the carbon atoms. These may be linked in straight chains, branched chains, and circular, or ring formation. With respect to molecular weight, the lighter molecules (fewer than 4 carbon atoms) tend to be gases; the medium weight molecules compose liquids; and the heavier (15 or more carbon atoms) are heavy liquids and solids.

Hydrocarbons are segmented into a variety of categories. Each category possesses a distinct molecular profile and, in turn, set of chemical and physical properties. Each class of hydrocarbons therefore has historically served different markets. Crude petroleum is composed of four major hydrocarbon groups: paraffins, olefins, naphthenes, and aromatics.

The paraffins are saturated hydrocarbons with formula $C^nH^{(zn+z)}$. All possible positions on the carbon atoms that can combine with other atoms are occupied by hydrogen, and paraffins are chemically inactive under a wide range of conditions.

There are two major types of paraffin molecules found in crude oil: "normal" paraffins are those with straight or linear carbon chains; "iso-" paraffins are branched (i.e., portions of the molecule that are attached at angles to the main, linear chain). The former tend to lower the octane number of gasoline resulting in engine knock. Because of their superior burning properties, they are desirable components of diesel fuel, kerosene, and fuel oils. Compared to the normal paraffins, the iso-paraffins are selected for use in automotive and aviation fuel due to the high octane numbers they impart to gasolines.

In contrast to the paraffin group, the olefins are unsaturated compounds. Simple olefins (mono-olefins) contain a single double bond [-c=c-] with formula, C^nH^{zn}. Thus, all available positions around the carbon atoms are not occupied by hydrogen, and olefins are far more reactive than the paraffins. The "di-" olefins $C^nH^{(zn-z)}$ contain two double bonds between carbon atoms. These bonds break apart readily resulting in a highly unstable intermediate that can combine easily with other molecular types. Diolefins have a low octane number and tend to produce gummy materials that affect combustion and foul surfaces. In contrast, the mono-olefins have a high octane number and, with only a single double bond in a molecule, are stable enough to prevent the rapid buildup of gum-like residues.

The naphthenes and aromatics both have cyclic (or ring-like) molecular structures and both possess high octane numbers. Napthenes are saturated and aromatics contain alternate double bonds on their ring. They are typically found in gasoline. The naphthenes also are an important part of kerosene.

In addition to the above hydrocarbon groups, crude oil also contains a number of inorganics, including sulfur, nitrogen, oxygen, metals, and mineral salts. Not all crudes are the same. They differ in average molecular weight (boiling point, viscosity) and composition according to the geographical locations at which they are found.

PRODUCTS AND MARKETS OF CRUDE OIL

Crude oil is the source for over 3,000 petroleum-based products for both industrial and consumer applications. The technique of distillation, the first stage processing of petroleum, exploits the different boiling points of the various petroleum fractions to separate out and isolate for use the different portions of the crude. The type and proportions of hydrocarbons present in each fraction depends upon the type of crude oil used and the range of temperatures employed. The major products produced directly from crude oil from distillation processes include gasoline, kerosene, lubricants, and waxes.

These various fractions are processed further into additional products. These value-added operations generally involve chemical transformations often using catalysts. They include cracking, hydrogenation, reforming, isomerization, and polymerization. The main output from these processes is fuels and petrochemicals.

Gasoline has been the primary product extracted from petroleum since the 1920s. The major portion of the gasoline group used for automotive and aviation purposes is cracked gasoline obtained through the thermal or catalytic cracking of the heavier oil fractions (e.g., gas oil). A wide variety of gasoline types are made by mixing straight run gasoline (gasoline obtained through distillation of crude oil), cracked gasoline, reformed and polymerized gasolines with additives, such as tetraethyl lead, alcohols, and ethers.

Kerosene is heavier than gasoline and lighter than gas oil. The lighter portion of kerosene is most suitable as an illuminant for lamps. The heavier portions of kerosene traditionally have been used as stove oil. Since the 1950s, kerosene has been used as a major component in jet fuel.

Gas oil, which is heavier than kerosene, is the raw material of choice in cracking and other refinery operations. Cracking of gas oil produces a variety of fuels for automotive, industrial, and domestic (furnace) use.

The heaviest products obtained directly from oil are lubricants, waxes, asphalt, and coke. These products have both domestic and industrial uses. Lubricants, for example, are applied in the operation and maintenance of industrial equipment and machinery. Asphalt, because it is not reactive to chemicals in the environment, is a superb material of construction in the building of roads and in roofing. It is also used in the waterproofing of concrete, the manufacture of black paints, and as a material for tire threads, battery housing, electrical insulation, and other applications. The heaviest of all the petroleum products, coke, is used extensively as a major component of industrial electrodes and as a commercial fuel.

PRODUCTION PATTERNS

United States

Between the 1860s and the mid-1880s, the bulk of petroleum crude came form Pennsylvania, particularly the area near Pittsburgh. This oil, which had a high paraffin content and low level of impurities, could be readily distilled into illuminant products and lubricants.

After 1885, oil production moved into the Midwest, chiefly Ohio and Indiana. While these fields held large amounts of oil, compared to Pennsylvania, the crude was of an inferior grade. It had both a high asphalt and sulfur content. The former reduced its use as an illuminant; and the high sulfur content created a foul smell that made it unsuitable for domestic consumption. At the turn of the century, the Standard Oil Trust, adopting a new type of separations technology, was able to remove the sulfur from "sour" crude and produce commercially usable "sweet" fuel products, including kerosene.

By 1900, other products from petroleum began to take on importance. Lubricants especially became prominent. This was due to the growth of industrialization in the United States, a shortage of naturally occurring lubricants (e.g., vegetable and whale oils), and an intense and creative marketing effort on the part of Rockefeller and his Standard Oil Trust. By 1910, Standard Oil Trust was also marketing coke and asphalt to a variety of manufacturers as well as the construction industry.

During the period 1900 and 1919, there was tremendous growth in U.S. oil production. Geographically, oil production also underwent important changes. In 1900, total annual U.S. oil production reached 63.6 million barrels. By this time, major production had shifted from the Northeast to the Appalachian region and the Midwest, and in particular the famous Lima-Indiana fields. In 1900, these areas together produced over 91 percent of total U.S. oil.

By 1920, total U.S. oil production had grown seven-fold, to 442.9 million barrels. The Appalachian and Lima-Indiana fields no longer dominated oil production: in 1920, in aggregate, they accounted for less than 8 percent of total production. Now the mid-continental fields of Louisiana and Oklahoma represented over 56 percent of U.S. output. California also had grown rapidly as an oil producing region accounting for nearly a quarter of all U.S. oil production. With the opening of the Spindle Top fields in Texas, this period also saw the beginning of oil producing dominance in the Gulf states. In 1920, this region already drilled more than 6 percent of all U.S. oil. The Rocky Mountain region and Illinois also increased their status as oil producing areas; together they also accounted for over 6 percent of total U.S. oil output.

Over the next two decades, oil production within the United States continued a generally upward path as producers discovered new fields and worked established fields more intensely. By 1929, total U.S. oil output exceeded the one billion barrel per year level. Despite the Depression, and in part due to the still growing demand for automotive gasoline, oil production grew over 39 percent over the following twelve years, topping off at over 1.4 billion barrels in 1941.

At this time, the four important producing areas were the mid-continental states, The Gulf region, California, and Illinois. The Mid-Continental states continued to generate over 50 percent of the nation's oil. But its share had fallen from 1920 levels, from over 56 percent to about 51 percent. California too had seen a decline in its share, from about 24 percent to 16 percent. The fastest growing area was the Gulf region: between 1920 and 1941, its share of U.S. oil production had climbed from 6 percent to 16 percent.

Annual domestic crude oil production expanded over 50 percent between 1945 and 1959, from 1.7 billion barrels annually to almost 2.6 billion, induced by a rise in domestic oil prices. During this period, Texas was the largest single producing state followed by California, and Louisiana. These states combined accounted for 60 to 80 percent of total domestic production through the period. There was also a resurgence of production in Illinois and Kansas, and the emergence of new fields in Colorado, New Mexico, and Wyoming.

Overall, U.S. oil production continued to climb through the 1960s as the Gulf, Mid-Continental, and West Coast fields were more actively exploited. In particular, drilling went deeper and off-shore sites were mined to a greater extent than previously. By 1973, U.S. producers were shipping about 3.4 billion barrels of oil annually. A major shift in oil production came in the 1970s with the construction of the Alaskan pipeline. By the late 1980s Alaska accounted for well over 10 percent of U.S.-produced crude. By the end of the 1990s, this figure had risen to close to 19 percent, or about one-fifth of total U.S. oil production.

But overall there was a steady contraction in domestic production during the 1970s and 1980s. This unprecedented decline was the result of increasing imports of foreign oil, especially from the Middle East, Venezuela, and Mexico, and fluctuations in the price of crude worldwide, largely due to OPEC policies. By 1983, U.S. production had declined to 3.2 billion barrels per year.

The tail end of the century has seen a continued decline in U.S. oil production as a result of the precipitous fall in prices. In 1997, U.S. production stood at only 2.3 billion barrels, or less than 1959 production volume. This declining petroleum activity has continued from 1997 to early 1999, during which period the number of rotary oil drillingrigs fell from 371 to 111.

The fact remains however that the United States still has considerable untapped supplies of petroleum. It is estimated that total remaining recoverable U.S. oil from all sources may exceed 200 billion barrels, or about 70 years worth of energy, assuming the current rate of U.S. consumption. During the first decade of the twenty-first century, producers will be extracting more oil from such regions as the Gulf of Mexico and other redundant off-shore sites. Employing new types of alloys in their equipment, they will also be drilling deeper into the ground to reach more corrosive high sulfur crudes and other previously untouched reserves.

International

As the automobile came to control the market for fossil fuel, the United States increasingly dominated world oil production. From the end of the nineteenth century to the World War I period, U.S. share of world oil production grew from around 40 percent to over 70 percent. By the early 1920s, there were other oil-producing powers, notably Mexico and Russia, although they could not compete with the United States in the production and processing of petroleum.

Over the next twenty years, world production rose from nearly 1.4 billion barrels per year to over 2 billion barrels. Foreign countries greatly increased their production and refining capability. Between 1929 and 1941, U.S. exports of petroleum abroad declined (although Canada remained the leading foreign consumer of American oil). During this period, foreign production increased its world share from 34 percent to 41 percent.

This growth resulted from additional countries entering as major producers of petroleum. By 1941, Venezuela was producing over one-half of the crude oil extracted in the Western Hemisphere. During this period as well, the Middle and Near East first began to flex its muscles as an oil producing region. Iran,

Iraq, the Bahrain Islands, and Saudi Arabia were the major oil-producing countries in the region. As a whole, by the end of World War II, the region supplied about 13 percent of the total world production. Middle East oil wells tend to be much more economical than those in North America. There are fewer dry holes, and each mideast well produces on average ten times the volume of a U.S. well.

Following the war, there was a steady increase in imports into the United States relative to total supply. During this period, the U.S. share of world oil production declined from 60 percent to 40 percent (1959). Crude oil imports as a percentage of total crude oil going to refineries increased from about 5 percent to 12 percent and the U.S. industry for the first time shifted its position from a net exporter to a net importer of mineral oils. At the same time, U.S. companies were establishing a strong presence abroad. Indeed, much of the oil produced abroad and imported into the United States at that time was owned by American-based companies.

Petroleum imports into the United States have continued to increase through the 1970s, 1980s, and 1990s. In terms of energy content, between 1970 and 1996, the energy contained in petroleum imported into the United States almost tripled, from 7.47 to 20.1 quadrillion Btus. Between 1970 and 1997, the Middle East countries in particular have greatly expanded their imports into the United States, from 222 million barrels to 1,349 million barrels. By the latter half of the 1990s, imports as a percentage of total domestic petroleum deliveries had reached the 50 to 60 percent level.

It is believed that world reserves may total 2 trillion barrels. This represents sufficient energy to last at the least to the end of the twenty-first century. Oil production worldwide is on the rise. Both Canada and Mexico have been expanding production. Brazil is becoming an important oil producing center. In Eastern Europe, and especially Romania, new oil is being drilled in and around the Black Sea. The former Soviet Union, especially in and around the Caspian Sea, has access to one of the world's richest oil reserves. Southeast Asia, especially Indonesia and Malaysia, are extending their petroleum production and processing capabilities. Western Europe is working regions of the North Sea. Other promising oil-producing countries include Australia, India, Pakistan, China (off-shore), and parts of the West Coast of Africa.

U.S. CONSUMPTION PATTERNS

The Nineteenth Century

From the late 1850s to the turn of the century, the major interest in petroleum was as a source of kerosene, as both a fuel and illuminant. As early as the 1820s, it was generally known from chemical experiments that illuminants, heating fuels, and lubricants could be obtained relatively cheaply from the distillation of crude oil.

In the 1840s, America's first petroleum reservoir was discovered in Tarentum, Pennsylvania. By the late 1850s, lubricants and kerosene were being extracted commercially from crude oil.

At this time, the petroleum industry was still a small scale affair and largely decentralized: production, transportation, refining, and distribution of refined products were undertaken by separate companies. Periods of overcapacity caused price declines that cut deep into the profits of producers. In the last two decades of the nineteenth century, John D. Rockefeller imposed order on the industry by building a fully integrated corporation: The Standard Oil Trust.

The Early Twentieth Century: 1900–1920

The period 1900 to 1919 set the stage for the take-off of oil as a major energy source. The four main products for consumption extracted from petroleum were naphtha and gasoline, illuminating oil, lubricating oil, and fuel oil. During these years demand for these products increased in the wake of significant increases in the auto population, especially in the urban areas, growing industrialization, and an expanding national income.

Thus we find that between 1900 and 1919, total energy consumption more than doubled from 7,529 trillion Btus to 18,709 Btus and the percentage of total energy consumed in the United States in the form of oil increased from less than 5 percent to about 12 percent. On the other hand, the trend for coal was downward as its percentage of total energy dropped from over 89 percent to 78 percent.

Use for the various petroleum products grew at different rates. The expansion in consumption of fuel oil production was an important trend. Increasingly, fuel oil, due to its lower price, more complete consumption, and relative ease of transport, became the fuel of choice over coal in industrial, commercial, and residential facilities as well as a fuel in rail and ship transport. During these years, fuel oil output showed the greatest increase, nearly 2,387 percent, from 7.3 million barrels to 180.3 million barrels.

Even more important was the growing importance of gasoline. At the turn of the century, virtually all of the gasoline produced (6.2 million barrels) was used as a solvent by industry (including chemical and metallurgical plants and dry cleaning establishments). But by 1919, 85 percent of the 87.5 million barrels of gasoline produced went into the internal combustion engine to power automobiles, trucks, tractors, and motor boats. During this period, production of gasolines grew from 6.7 million barrels to 97.8 million barrels.

At the same time gasoline was finding its first large markets, kerosene was in the fight of its life with the burgeoning electrical power industry. (The temporary exception occurred in the rural areas without access to electricity.) Not surprisingly, gasoline and kerosene exhibited a widening price differential: between 1899 and 1919, the per gallon price for gasoline exceeded that for kerosene and the price differential grew over 86 percent, from 1.3 cents to 12.3 cents. Overall, production of Illuminating oil increased from 29.5 million barrels to 56.7 million barrels.

With respect to the various byproducts of petroleum, including paraffin wax and candles, lubricating jellies, asphalt, and coke, total value of this group grew 543 percent, from $12.3 million to $79.1 million. But this good performance would be the last major growth for the group. The one exception was asphalt which would continue in demand as a material of choice for the highway and roofing construction industry. By 1920, more than 1.3 million tons of the substance entered the market, up from the 44,000 tons produced annually at the turn of the century.

The 1920s and 1930s

Over the next ten years (1919–1929), U.S. annual energy demand grew from nearly 18.7 to 25.4 quadrillions Btus. The percentage of energy represented by petroleum increased from a little over 12 percent to 25 percent. Natural gas also nearly doubled its share, from 4.3 perent to 8 percent. Coal continued to lose ground as a source of energy, from 78 percent to 63 percent of total energy consumption.

With respect to petroleum products, during the 1920s, the amount of kerosene placed on the market remained unchanged at close to 56 million barrels annually for the period, and lubricant output increased somewhat, from 20.2 to 35.6 million barrels. Production of both gasoline and fuel oil outpaced all other petroleum products in growth. For example, output of fuel oil increased two and one half times, from near 182 million barrels to close to 449 million barrels.

But these years put the spotlight on gasoline above all. The years following World War I saw a rapid growth in the number of automobiles on the road and in miles of highway constructed. The number of registered vehicles increased from 7.6 million to over 26.5 million and the length of surface highways expanded from 350,000 to over 694,000 mi. As a result, the average distance traveled per car grew from 4,500 mi to 7,500 mi and the average annual consumption of gasoline per car rose from 400 gal to over 590 gal. In response, during the 1920s, gasoline output expanded fourfold from 94.2 million barrels to over 435 million barrels. By 1930, gasoline accounted for almost 48 percent of all petroleum products produced in the United States.

From the onset of the Depression (1929) to the beginning of World War II (1941), petroleum's share of total U.S. energy consumption continued to expand, from 24 percent to more than 34 percent (while natural gas increased its share from 8 to 11 percent). Coal continued to lose share in the nation's energy output, from 62 percent to 54 percent. Over the same period, output of motor fuel, while not matching the growth of the previous decade, nevertheless continued its upward trend, from 256.7 million barrels to 291.5 million barrels.

It was during the 1930s that aviation solidified its position as a major and expanding gasoline market. Through the 1930s, consumption of aviation fuel increased over eightfold, from only 753 thousand barrels in 1929 to over 6.4 million barrels at the start of World War II. The other petroleum products—kerosene, lubricating oils, asphalt—saw growth as well, but behind gasoline and refinery products as a whole.

The Rise of Petrochemicals and World War II

By the start of World War II, fossil fuel products were being used increasingly to make advanced organic synthetics. The petrochemicals industry began in the years immediately following World War I. In the 1920s and 1930s, petrochemicals—chemicals made from either petroleum or natural gas—focused

The San Francisco metropolitan area had the nation's highest average price per gallon of gas as consumer costs escalated during 2000. (Corbis Corporation)

on making and utilizing for chemical synthesis the basic olefin compounds: ethylene, propylene, and the butylenes. These were generated in increasing volumes from the refinery plants as off-gases or byproducts of the cracking process. From these building blocks issued a range of chemical products and intermediates. Compared to coal, the traditional raw material, petroleum offered companies such as Union Carbide and Dow Chemical, the efficient, large-scale production of a great variety of synthetics.

Even with direct access to the basic raw materials, the refiners were slow to enter the field. Eventually, in the years leading up to World War II, the refiners began to perceive how their refining operations could be supplemented by petrochemical manufacture. By the start of world War II, they were beginning to compete in earnest with the chemical industry in petrochemical synthetics markets. Between 1929 and 1941, the byproduct refinery gases consumed by both the chemical and petroleum industries for the purpose of manufacturing chemicals more than doubled, from 38.6 million barrels to 83.4 million barrels.

This activity expanded greatly during World War II. Refiners, such as Jersey Standard, Sun Oil, Shell, and Sacony-Vacuum, pioneered the mass production of advanced and strategically critical petrochemicals including butadiene for use as raw-material for synthetic rubber, toluene for in advanced explosives, as well as high-octane motor fuel and aviation gasoline.

The Postwar Period: 1945–2000

The postwar period witnessed increased use of petroleum both for chemicals and energy. Petrochemicals enabled the production of synthetic fibers and high-performance plastics as well as synthetic resins and rubber. By the early 1950s, 60 percent of organics were produced from petroleum. By the early 1970s, over 2,000 chemical products, having a combined weight of over 25 million tons, were derived from petroleum within the United States By the 1990s, petrochemicals represented over 90 percent of all man-made commercial organics.

Petroleum made further gains as a source of energy in the United States in the form of gasoline and liquid gas. The portion of total energy consumption in the United States represented by petroleum and petroleum liquid gas rose from over 34 percent to nearly 46 percent. During these years, the production of gasoline increased from 695 million barrels to 1.5 billion barrels.

If chemicals and gasoline represent the high-growth areas for petroleum, the other petroleum products, for the most part, did not perform quite as well. The exception was kerosene. Output of kerosene rose from 100 million barrels in 1949 to over 200 million barrels in 1959, a result of increased demand in jet fuel. At the same time, demand for fuel oil declined as customers switched to natural gas.

From the 1970s through the 1990s, consumption of petroleum fluctuated closely with prices that to a large extent depended on OPEC supply policies. The OPEC oil crisis of 1973, in particular, forced oil prices upward and, in turn, caused a sharp reduction in petroleum consumption. Between 1973 to 1975, for example, in the face of high oil prices, U.S. consumption of petroleum declined to a level of 16 million barrels per day (bpd). In the 1975 to 1979 period, prices declined and consumption increased to a level near 19 million bpd. The period 1984 to 1994 was one of price reduction and growing consumption. During these years, crude oil prices declined regularly from $36/barrel to a low of $12/barrel in 1994. Then, between 1994 to 1996, with tightening supplies worldwide, prices increased to $18.50/barrel. There was another reversal during the late 1990s as prices steeply contracted, in some areas of the United States to below $10/barrel. By 1997, consumption of petroleum reached 18.6 million bpd, or the level close to that achieved by the 1978 record consumption. Overall, petroleum consumption in absolute terms in the United States has increased at an average of 0.5 percent between 1972 and 1997.

Even so, the last three decades of the twentieth century have been a period of a declining presence for petroleum in the U.S. energy picture. Between 1970 to 1996, energy consumption represented by petroleum has declined from 44.4 percent to 38.1 percent. This contraction is the result of increasing efficiency in the use of energy and growth of such alternate energy sources as nuclear, natural gas, and, more recently, coal. However, a fall in petroleum prices and the decommissioning of nuclear power plants will result in a resurgence of petroleum usage. By the year 2020 government predictions are that petroleum will comprise over 40 percent of the nation's total consumed energy.

THE TRANSPORT AND STORAGE OF PETROLEUM

The petroleum transportation and distribution network constitutes a major portion of the industry's total infrastructure and represents a large capital expenditure for producers. This means that production often outpaces demand.

Economies of scale dictate full-capacity use as companies must keep the infrastructure working at full throttle in order to maintain their investment. This is especially true of the pipeline that must be run at a throughput for which it was designed because the costs per barrel involved in operating pipelines varies inversely with level of throughput. Over the last century, the transportation and distribution network has expanded greatly all the while undergoing significant innovation.

Pipelines

The U.S. pipeline system began in the 1870s and is today composed of a maze of trunk lines and side pipelines that split off in numerous directions and over many miles. The pipeline grew over the years with the volume of oil production. Since the turn of the century, the U.S. petroleum pipeline has expanded from less than 7,000 mi to nearly 170,000 mi. The greatest growth of the pipeline system occurred in the period between the wars in the Mid-Continent and Gulf regions. This network expansion was required not simply to carry larger volumes but also because fields increasingly were further removed from population centers (and thus markets) as well as refining and distribution points.

The carrying capacity of the pipeline did not depend only on the laying down of more pipe. A number of technical improvements expanded the capacity of a given mile of pipe. Pipe making technology has extended pipe size from only 5 in. at the turn of the century up to 12 in. by the 1960s. Larger pipes carrying oil longer distances required greater pumping force. At first, this was provided by steam-powered engines at pumping stations. By the 1920s, these were replaced, at first by diesel engines, and soon thereafter electric power.

From the 1930s and well into the postwar period, pipeline technology advanced through the application of numerous improvements that increased the economies of laying down pipe and facilitated oil transport over long distances and in a variety of climatic conditions. These included new methods of welding pipe joints, the introduction of seamless tubing and pipe, new types of pipe linings and materials of construction to fight against corrosion and advance streamline oil flow, and development of automatically controlled central pipeline pumping stations and remote controlled line stations.

Shipping

Prior to 1900, water transport of oil was carried out in makeshift vessels, such as barges and converted freighters and ore-carrying ships. These were often contracted out (in contrast to pipelines that more often were owned by the producers). There was an increase in water transport of oil during the first two decades of the century.

The larger companies led the way in acquiring water transport fleets. Increasingly after World War I, producers, to be sure of sufficient capacity when required and to avoid charter prices, invested in their own tankers. In this period, about one-third of all U.S. oil was shipped by barge or tanker at some point in its trip to the refiner or market. This figure has increased to over 60 percent through the post-World-War-II period. In the years leading up to World War II, approximately 400 U.S. vessels were involved in oil transport. It is estimated that in the postwar period, the number of oil carrying vessels has fluctuated between 200 and 600 vessels.

Oil tankers today can be either single- or double-hulled. The capacity of tankers usually lies between 1 to 2 million barrels. They are often designed for special purposes. For example, in the 1970s, specially constructed tankers were used in the transport of Alaskan oil. A new generation of oil tankers is being designed as the fleet built during the 1970s and 1980s ages. Designs of the newer double-hulled tankers are most concerned with preventing oil spills, minimizing corrosion, improving maneuverability, advancing engine power and efficiency, strengthening hull structures, and installing vapor-control systems.

Storage

Oil has been stored in tanks since the late nineteenth century. Stored oil is essentially inventory waiting to be distributed to markets. Storage tanks are used to hold both crude oil and the variety of distilled and refined products.

There are two types of storage tanks: underground and aboveground. The former are typically found at service stations. Aboveground tanks are usually used to store crude oil and various refined products. They are often installed at marketing terminals, at various points along a pipeline, and at refineries. In addition, both types of storage tanks are utilized by a wide range of industrial facilities to hold oil used in fueling heir operations. It is estimated that there are 3 to 4 million tanks store oil in the United States.

Technology has improved storage facilities so that oil can be held for longer periods of time with minimal degradation of the liquid, internal corrosion and structural damage, evaporation loss, and leakage. Storage of crude is an expensive operation. Evaporation, corrosion, and leakage increase the costs of storing oil, as do insurance expenses as well as opportunity costs involved in keeping inactive large amounts of petroleum. Accordingly, producers attempt to reduce their inventories of crude and so minimize their reliance on storage. This means the maintaining of continuous flow conditions from well to refinery.

PETROLEUM AND THE REGULATORY CLIMATE

Since the 1970s, the petroleum industry has been increasingly affected by the regulatory push at the international as well as federal and state levels. Oil spills, such as the one involving the Exxon Valdez in Alaska in the 1980s, put the regulatory spotlight on the environmental dangers inherent in moving oil by marine tanker. Since the 1970s, The International Maritime Organization (IMO), in conjunction with the U.S. Department of Transportation (DoT) and the U.S. Environmental Protection Agency (EPA), has stepped up its efforts to implement measures to increase protection of the environment on U.S. and international waters.

The pipelines and storage facilities also present environmental hazards, mainly from leakage and subsequent seepage of oil into the soil and groundwater. Pipeline regulations are the responsibility of the EPA and the DoT. DoT's Office of Pipeline Safety in particular is responsible for regulating interstate hazardous liquid pipeline systems. With respect to storage tanks, both the under- and above-ground types are regulated by the EPA under the Resource Conservation and Recovery Act (RCRA). The major regulation affecting oil-bearing underground storage tanks (USTs), effective on December 22, 1998, required the upgrade of tanks including improved leak detection systems as well as measures to prevent future corrosion of tanks.

Sanford L. Moskowitz

See also: Diesel Fuel; Fossil Fuels; Gasoline and Additives; Kerosene; Liquified Petroleum Gas; Oil and Gas, Drilling for; Oil and Gas, Exploration for; Refining, History of; Residual Fuels.

BIBLIOGRAPHY

Amuzegar, J. (1999). *Managing the Oil Wealth: OPEC's Windfalls and Pitfalls.* New York: I. B. Tauris and Co. Ltd.

Bromley, S. (1991). *American Hegemony and World Oil: The Industry, the State System, and the World Economy.* University Park: Pennsylvania State University Press.

Business Trend Analysts, Inc. (1998). "The Consumption of Specialty Steel in U.S. and International Petroleum and Petrochemical Markets." In *The U.S. Stainless Steel Industry.* Commack, NY: Business Trend Analysts, Inc.

Department of Commerce. (1998). *Statistical Abstracts of the United States.* Washington, DC: U.S. Government Printing Office.

Economides, M., et al. (2000). *The Color of Oil: The History, the Money and the Politics of the World's Biggest Business.* Houston: Round Oak Publishing Company.

Enos, J. (1962). *Petroleum, Progress and Profits: A History of Process Innovation.* Cambridge, MA: MIT Press.

Giebelhaus, A. W. (1980). *Business and Government in the Oil Industry: A Case Study of Sun Oil, 1876–1945.* Greenwich, CT: JAI Press.

Hartshorn, J. (1993) *Oil Trade: Politics and Prospects.* New York: Cambridge University Press.

International Energy Association. (1996). *Global Offshore Oil Production: Prospects to 2000.* Paris: Author.

Larson, H. M., et al. (1971). *History of Standard Oil Company (New Jersey): New Horizons, 1927–1950.* New York: Harper and Row.

Moskowitz, S. L. (1999). "Science, Engineering and the American Technological Climate: The Extent and Limits of Technological Momentum in the Development of the U.S. Vapor-Phase Catalytic Reactor, 1916–1950." Diss. Columbia University.

Spitz, P. (1988). *Petrochemicals: The Rise of an Industry.* New York: John Wiley and Sons.

Waddams, A. L. (1980). *Chemicals from Petroleum*, 4th ed. Houston: Gulf Publishing Co.

Williamson, H. F., et al. (1959). *The American Petroleum Industry, Vol. I: The Age of Illumination, 1859–1899.* Evanston, IL: Northwestern University Press.

Williamson, H. F., et al. (1963). *The American Petroleum Industry, Vol. II: The Age of Energy, 1899–1959.* Evanston, IL: Northwestern University Press.

PHANTOM LOAD

See: Electricity

PHOTOSYNTHESIS

See: Biological Energy Use, Cellular Processes of

PIEZOELECTRIC ENERGY

Piezoelectric energy is a form of electric energy produced by certain solid materials when they are deformed. (The word *piezo* has its roots in the Greek word *piezein* meaning "to press.") Discovery of the piezoelectric effect is credited to Pierre and Jacques Curie who observed in 1880 that certain quartz crystals produced electricity when put under pressure.

Quartz is a crystalline piezoelectric material composed of silicon and oxygen atoms and is illustrative of a piezoelectric material. If no forces are applied to a quartz crystal the distribution of charges is symmetric at the sites of the atoms and there is no net internal electric field. Squeezing or stretching a quartz crystal displaces atoms from their normal positions producing a separation of positive charges and negative charges (electrons) that give rise to a net internal electric field. A voltage develops across the whole crystal making it, in a sense, a battery. As with a battery, an electric current is produced by the voltage when something is connected to the crystal.

The piezoelectric effect will not be used to produce energy to energize a light bulb, for example, yet there are many applications. One common usage is as an igniter in an outdoor gas-fired grill used for cooking. A spring-loaded plunger released by a button on the front of the grill gives a sharp blow to a piezoelectric crystal. The voltage produced generates a spark between two separated metal contacts near the incoming gas, causing the gas to ignite.

Another important but little-known piezoelectric effect is found in some electronic systems. Speaking produces pressure variations that propagate through the air. Forces are produced on anything in contact with this vibrating air so that when contact is with a piezoelectric crystal, tiny voltage variations are produced. A crystal microphone is designed to make use of this piezoelectric effect.

Pressure gauges used to monitor pressures in the fluids pumped throughout an automobile engine are also often based on the piezoelectric effect. An experimenter in a laboratory can apply a known pressure to a piezoelectric crystal and measure the voltage produced. Once the relationship between voltage and pressure is known, a pressure can be inferred from a measure of the voltage. In this manner, the piezoelectric material functions as a pressure gauge. A variety of gases and oils that rely on piezoelectric gauges are pumped throughout the working components of an automobile engine.

Interestingly, a reverse piezoelectric effect is the basis for a number of applications. In this reverse process, a piezoelectric crystal vibrates when an electric current is provided from an external source such as a battery. The precise frequency of these vibrations is used to great advantage in watches and other timepieces. The most common piezoelectric crystal used is quartz. Quartz occurs naturally, but quartz crystals can also be grown in the laboratory. The most common crystals in timepieces are tiny U-shaped tuning forks that vibrate 32,768 times per second. An electronic circuit in the timepiece records time by counting the vibrations. If you peek inside a "quartz" watch you will not see wheels and a spring as in a mechanical watch. Rather, you will see a battery that provides electric current for the crystal and an electronic circuit. You might also see the U-shaped tuning fork.

Since discovering and making use of the piezoelectric effect in naturally occurring crystals such as quartz and Rochelle salts, scientists have produced a wide range of piezoelectric materials in the laboratory. An early example is barium titanate, used in an electrical component called a capacitor. Currently, most piezoelectric materials are oxide materials based on lead oxide, zirconate oxide, and titanium. These very hard piezoelectric materials are termed piezoceramics.

Piezoelectricity is also a natural by occurring phenomenon in the human body. Studies have shown that piezoelectricity causes electric potentials in dry bones of humans and animals. Whether or not this is the cause of the electric potentials that occur in wet, living bone is debatable. More than likely, any piezoelectric effect occurring in moist bone is overshadowed by other sources of electric potentials.

Joseph Priest

See also: Capacitors and Ultracapacitors.

BIBLIOGRAPHY

Ikeda, T. (1990). *Fundamentals of Piezoelectricity.* New York: Oxford University Press.

PISTONS

See: Engines

PLUTONIUM

See: Nuclear Energy; Nuclear Fission

POTENTIAL ENERGY

Potential energy is the energy that something has because of its position or because of the arrangement of its parts. A baseball in flight has potential energy because of its position above the ground. A carbohydrate molecule has potential energy because of the arrangement of the atoms in the molecule.

A diver standing on a platform above the water has potential energy because of a capacity for doing work by jumping off. The higher the diver is above the water and the greater the diver's mass, the greater the potential energy—that is, the capacity for doing work. It is the gravitational force that pulls the diver downward and into the water. For this reason, the potential energy of the diver is called "gravitational potential energy." As an equation, gravitational

The bow stores elastic potential energy as the hunter pulls on it. The energy is transferred to the arrow as kinetic energy upon release. (Corbis Corporation)

potential energy (U) measured in joules (J) is given by $U = mgh$, where m is the mass in kilograms (kg), g is the acceleration due to the gravitational force (9.80 meters/second^2, and h is the height in meters (m). An 80-kg diver 3 m above the water would have 2,350 J of gravitational potential energy.

Any type of potential energy is associated with a force and involves some favorable position. A compressed spring in a toy gun does work on a projectile when the spring is released. Before release, the spring had potential energy that we label "elastic potential energy." The force involved is that of the spring, and the favorable position is the amount of compression. Other examples of elastic potential energy are involved in flexing the bow of a bow and arrow, flexing a vaulter's pole, compressing the suspension springs of an automobile, and stretching a rubber band.

A loss in potential energy by a system is accompanied by a gain of energy in some other form. All the potential energy of a mass held above a spring is converted to kinetic energy just before the mass hits the spring. The mass loses kinetic energy as it pushes against the spring, but the spring gains potential energy as a result of being compressed. Water atop a dam in a hydroelectric power plant has gravitational potential energy. In falling, it loses potential energy and gains kinetic energy. When the water impinges on the blades of a water turbine, the rotor of the turbine gains rotational kinetic energy at the expense of the water losing kinetic energy.

Molecules have potential energy associated with electric forces that bind the atoms together. In a chemical reaction liberating energy—heat, for example—the atoms are rearranged into lower potential energy configurations. The loss of potential energy is accompanied by an increase in other forms of energy.

Protons and neutrons in the nucleus of an atom have potential energy associated with nuclear forces. In a nuclear reaction liberating energy, the nuclei are rearranged into lower potential energy structures. The loss of potential energy is accompanied by an increase in other forms of energy. A single nuclear reaction in a nuclear power plant liberates nearly 10 million times the energy of a single chemical reaction in a coal-burning power plant.

Joseph Priest

See also: Conservation of Energy; Gravitational Energy; Kinetic Energy; Nuclear Energy.

BIBLIOGRAPHY

Hobson, A. (1995). *Physics: Concepts and Connections.* Englewood Cliffs, NJ: Prentice-Hall.

Priest, J. (2000). *Energy: Principles, Problems, Alternatives*, 5th ed. Dubuque, IA: Kendall/Hunt Publishing Co.

Serway, R. A. (1998). *Principles of Physics*, 2nd ed. Forth Worth, TX: Saunders College Publishing.

POWER

Power is the rate of doing work or the rate of converting energy. A unit of work or energy is called a joule. A person doing 4,000 joules of work in four seconds would be doing work at a rate of 1,000 joules per second. A joule per second is called a watt (W) in honor of James Watt, inventor of the steam engine. Hence, the person developed a power of 1,000 watts. When a hundred watt light bulb is lit, electric energy is converted to heat and light at a rate of one hundred joules per second. In the United States, the power developed by engines and motors is usually expressed in horsepower (hp). The horsepower unit was coined by Watt who estimated the rate at which a typical work horse could do work: One horsepower equals 746 watts. In an effort to use metric units for all physical quantities, there is a trend in Europe to rate engines and motors in watts.

In terms of the rate at which work is done or energy is converted, consider that

- the human heart in a person at rest does work at a rate of about 0.01 watt,
- a flashlight converts energy at a rate of about four watts,
- a student in daily activity does work at a rate of about ninety watts or about 0.1 hp,
- an automobile engine develops about 40,000 watts, a large electric power plant produces about 1,000,000,000 watts of electric power, and
- the solar input to the earth (at the top of the atmosphere) amounts to about hundred million billion watts.

When the power is a large number, as in the case of an electric power plant, it is convenient to express the power in megawatts (MW) where one megawatt equals one million watts. An electric power of 1,000,000,000 watts would be expressed as 1,000 MW. A large coal-burning or nuclear power plant produces about 1,000 MW of electric power. The sum total of the electric power produced by all electric power plants is expressed in units of gigawatts (GW). One gigawatt equals one billion watts. An electric power of 1,000,000,000,000 watts would be expressed as 1,000 GW.

A unit of energy divided by a unit of time is a unit of power. If you are burning wood at a rate of 10,000 Btu per hour then Btu per hour denotes power. Similarly, a unit of power multiplied by a unit of time is a unit of energy. Electric power in a home is usually measured in kilowatts (kW) (i.e., thousands of watts, and time in hours [h]). Therefore, kilowatts multiplied by hours (kWh) is a measure of energy. A light bulb rated at one hundred watts (0.1 kW) converts electric energy at a rate of one hundred joules per second. If the bulb is on for two hours then the electric energy provided by the power company is $0.1\ \text{kW} \times 2\ \text{h} = 0.2\ \text{kWh}$. The power company would be paid for 0.2 kWh of electric energy. Typically, the cost is about 10¢/kWh, so the amount due the company is two cents. It is interesting that electric utilities are referred to as power companies when we pay them for energy, not power.

A bicyclist has to expend energy to cycle up a steep hill. The energy the cyclist expends is determined by the product of the power the cyclist develops and the time the cyclist is pumping (energy = power × time). The cyclist may choose to develop a relatively large amount of power and pump for a short period of time. Or he may prefer to take it easy developing a relatively small amount of power and pump for a longer period of time. The energy would be the same in both cases, but the effort would be different. Even after shifting to the lowest gear, cyclists often find that it is still easier to make their way up a hill by traversing back and forth across the road. The traversing takes less effort (less power), but requires more time. This is also why roads up a mountain take winding

Each cyclist must convert the energy stored in his or her muscles into torque to move the pedals. To go uphill, the necessary rate of work (power) increases as does the required torque. (Corbis Corporation)

paths rather than trying to go directly up—vehicles with less reserve capacity power can still climb the mountain.

By definition, a force (F) acting on some object as it moves some distance (x) along a line does work (W):

$$W = Fx.$$

This idea says nothing about the time involved in doing the work. Asking for the rate at which work was done (i.e. the power) and how fast was the object moving, we find power (P) equals force times velocity (v):

$$P = Fv.$$

The power developed by a person pushing a box on a floor with a force of 300 newton (N) at a speed of 1 m/s is 300 W or about 0.4 hp.

A force is required to rotate an object. The response to the force depends not only the size of the force, but also on the manner in which the force is applied. A bicyclist must push down on a pedal to cause the sprocket to rotate. But if the shaft to which the pedal is attached is vertical, no rotation results. The greatest response occurs when the bicyclist pushes down on the pedal when the shaft is horizontal. The concept of torque is used to describe rotational motion. The bicyclist pushing down on the pedal when the shaft is vertical produces zero torque. The maximum torque is produced when the shaft is horizontal. In this case the torque is the product of the force and the length of the shaft. For any other position between vertical and horizontal, the torque is the product of the force, shaft length, and sine of the angle made by the shaft and direction of the force. A force develops power when linear motion is involved and a torque develops power when rotational motion is involved. The power developed by a force is the product of force and linear velocity ($P = Fv$) and the power developed by a torque is the product of torque and angular velocity ($P = T\omega$).

Power and torque are directly related. An automobile engine develops power when producing a torque on the drivetrain. Generally, the owner of an automobile is more interested in power than torque. However, technicians in the automobile industry usually evaluate the performance in terms of torque rather than power. The power developed in the drive chain of an automobile is essentially constant. But the rotational speed of the drivetrain depends on the torque developed. The larger the torque, the smaller the rotational speed.

Joseph Priest

See also: Watt, James; Work and Energy

BIBLIOGRAPHY

Hobson, A. (1999). *Physics: Concepts and Connections.* Englewood Cliffs, NJ: Prentice-Hall.

Priest, J. (2000). *Energy: Principles, Problems, Alternatives,* 5th ed. Dubuque, IA: Kendall/Hunt Publishing.

Serway, R. A. (1997). *Physics for Scientists and Engineers,* 4th ed. Fort Worth, TX: Saunders College Publishing.

Serway, R. A., and Faughn, J. (1999). *College Physics,* 5th ed. Fort Worth, TX: Saunders College Publishing.

POWER PLANTS

See: Electric Power, Generation of

PRESSURE

Pressure is force divided by the area over which the force acts:

$$\text{pressure} = \text{force}/(\text{area of contact})$$

or

$$P = F/A.$$

In the metric system, pressure has a unit of newtons per square meter, which is called a pascal (Pa). Although the pascal is the scientific unit and is preferred, pounds per square inch (lbs/in^2) is common in the United States. For example, in most of Europe, tire pressure is recorded in pascals (typically 220,000 Pa), whereas tire pressure in American cars is measured in pounds per square inch (typically 32 lbs/in^2). As a point of reference, the pressure that the earth's atmosphere exerts on anything at the earth's surface is roughly 100,000 Pa or 15 lbs/in^2.

A large pressure does not necessarily mean a large force because a large pressure can be accomplished by making the contact area small. For example, a 160-pound man having fifty square inches in his shoe soles produces a pressure of 3.2 lbs/in^2 on the ground when he stands flatfooted. But the same person wearing ice skates having 2.5 square inches in the runners of the blades produces a pressure of 64 lbs/in^2 when standing on ice. When a person pushes down on the head of a thumbtack, the pressure at the point is quite large because the contact area of the sharp thumbtack is small.

Blaise Pascal, the French scientist for whom the pascal pressure unit is named, was the first to discover that a pressure applied to an enclosed fluid is transmitted undiminished to every point of the fluid and the walls of the containing vessel. The earth's atmosphere exerts a pressure on the surface of water in a swimming pool (a container) and this pressure is felt equally throughout the pool including the walls. This principle is employed in a hydraulic press that has many applications. The pressure due to the force f on the piston of area a is $P = f/a$. The piston of area A feels a force F creating a pressure $P = F/A$. According to Pascal's principle, the two pressures are equal so that $F/A = f/a$ or $F = (A/a)f$. If the area A is larger than a, then the force F is larger than f. Therefore, a small force on the left piston causes a larger force on the right piston. Lifting an object requires an upward force larger than the weight of the object. This is why rather massive objects such as automobiles, trucks, and elevators can be lifted with a hydraulic press. For example, a car weighing 10,000 N (about 2,000 pounds) can be raised with a hundred N force if the ratio of the areas of the pistons (A/a) is one hundred.

Work is done by both forces when the pistons move and the larger force cannot do more work (force × displacement) than the smaller force. Accordingly, the larger piston moves a much smaller distance than the smaller piston. Theoretically, if the ratio of the areas is one hundred then the smaller piston must move one hundred times further than the larger piston.

The mechanism for rotating the wheels in a car or truck has its origin in pistons in the engine that are forced to move by expanding gases in the cylinders. The expanding gas is a result of igniting a mixture of gasoline vapor and air. An expanding gas exerts a force on a piston doing work as the piston moves. In principle, it is no different than a person doing work by pushing a box on the floor of a hallway. The energy for the work comes from the energy of the gas produced by the ignition of the gasoline vapor and air mixture. As in the calculation of work done by a constant force the work done by the gas is the product of the force (F) and displacement (d, the distance the piston moves). Because the force on the piston is equal to the pressure (P) times the area (A) of the piston the work (W) can be written $W = PAd$. The quantity Ad is just the change in volume (ΔV) of the gas so that the work can also be written as $W = P\Delta V$. This is the basic idea in determining the performance of engines, which are used to power cars, trucks, and all sorts of other equipment.

The air pressure created by the clown's lungs on the inner walls of the balloon force it to expand. (Corbis Corporation)

The expansion of a balloon when it is inflated is evidence of the pressure exerted on the interior walls by molecules blown into the balloon. If you tie off the open end of the balloon, you can squeeze the balloon with your hands and actually feel the result of the pressure in the balloon. If you were to cool the balloon you find that the pressure goes down. Warming causes the pressure to increase. This happens also with the pressure in an automobile tire after it has warmed by running on a road. When there is no change in the pressure, temperature, and volume, we say the gas is in equilibrium. The pressure, volume, and temperature of a gas in equilibrium and how they are related is an important feature for describing the behavior of a gas when the gas does work. The relationship for all possible variations of P, V, and T is very complicated, but for moderate temperatures (around 300 K or 70°F) the equilibrium state of all gases is given by PV = NkT where N is the number of atoms (or molecules) and k is the Boltzmann constant, 1.38×10^{-23} joules/kelvin. A gas is said to be ideal if it subscribes to the relationship PV = NkT and the equation is called the ideal gas equation.

Joseph Priest

BIBLIOGRAPHY

Hobson, A. (1999). *Physics: Concepts and Connections.* Englewood Cliffs, NJ: Prentice-Hall.

Priest, J. (2000). *Energy: Priciples, Problems, Alternatives,* 5th ed. Dubuque, IA: Kendal/Hunt Publishing.

Serway, R. A. (1997). *Physics for Scientists and Engineers,* 4th ed. Fort Worth, TX: Saunders College Publishing.

Serway, R. A., and Faughn, J. (1999). *College Physics,* 5th ed. Fort Worth: Saunders College Publishing.

PROPANE

See: Liquified Petroleum Gas

PROPELLERS

A propeller converts through helical motion the energy supplied by a power source into thrust, a force that moves a vehicle forward in a fluid medium. They are used primarily for marine and aerial propulsion, but they are found on other technologies such as hovercraft and wind turbines as well. Propellers, which are essentially a series of twisted wings, or blades, connected to a central hub, are efficient energy transmission devices for those applications. The blades strike the air or water at a certain angle, called the pitch, and create an area of low pressure in front of the propeller. As a result, the blades generate thrust through either fluid or aerodynamic means by pushing forward through the low-pressure area. Slip, or the energy lost as the propeller rotates, offsets the full output of thrust. The effectiveness of the propeller is measured by propulsive efficiency, the ratio of engine power to the actual thrust produced minus slip, during one complete revolution of the propeller. Because this process resembles the twisting movement of a carpenter's screw as it advances through wood, marine propellers are often called "screws" and aerial propellers are called "airscrews."

There exists a wide range of propellers for different applications. Fixed pitch propellers, most commonly found on ships and small aircraft, are simple in operation and efficient for one operating regime. Transport aircraft use variable-pitch propellers that can alter their pitch (sometimes automatically) to perform efficiently over a variety of conditions. Counter-rotating propellers are two propellers placed in tandem and operating from the same power source. They produce thrust efficiently, but they are also very complicated in design. Feathering propellers can set their blades parallel to the movement of either water or air in case of engine failure, which increases passenger safety. Reversible pitch propellers decrease landing distances for aircraft and aid in the maneuvering of ships by changing the direction of thrust. Depending on their application and operating regime, propellers can be made from wood, metal, or composite materials.

The concept of the propeller originates from three important antecedents. First, the Greek mathematician Archimedes developed a method of transporting water uphill through helical motion during the third century B.C.E. Second, the emergence of windmills in western Europe in the twelfth century C.E. indicated an inverse understanding of the principles and capabilities of thrust. Finally, Leonardo da Vinci's adaptation of the Archimedean helical screw to the idea of aerial propulsion during the fifteenth century inspired later propeller designers. These ideas influenced the development of both marine and aerial propellers but were mainly conceptual. It was not until the nineteenth and twentieth centuries that the availability of adequate power sources such as steam and internal combustion engines would make propellers feasible power transmission devices.

MARINE PROPELLERS

Propellers for the marine environment appeared first in the eighteenth century. The French mathematician and founder of hydrodynamics, Daniel Bernoulli, proposed steam propulsion with screw propellers as early as 1752. However, the first application of the marine propeller was the hand-cranked screw on American inventor David Bushnell's submarine, *Turtle* in 1776. Also, many experimenters, such as steamboat inventor Robert Fulton, incorporated marine propellers into their designs.

The nineteenth century was a period of innovation for marine propellers. Reflecting the growing importance of the steam engine and the replacement of sails and side paddle wheels for propellers on ocean-going vessels, many individuals in Europe and America patented screw propeller designs. Two individuals are credited with inventing the modern marine propeller: English engineer Francis P. Smith and American engineer John Ericcson. Both placed important patents in 1836 in England that would be the point of departure for future designs. Smith's aptly named ship *Archimedes* featured a large Archimedean screw and demonstrated the merits of the new system. Swedish-born Ericcson employed a six-bladed design that resembled the sails on a windmill. He immigrated to the United States in 1839 and pioneered the use of screw propellers in the United States navy. The first American propeller-driven warship, the U.S.S. *Princeton*, began operations in 1843. Other notable screw propulsion designs included British engineer Isambard Kingdom Brunel's 1843 iron-hull steamship, the *Great Britain*, which was the first large vessel with screw propellers to cross the Atlantic. In 1845, the British Admiralty sponsored a historic "tug-of-war" between the paddle wheel-driven H.M.S. *Alecto* and screw propeller-driven H.M.S. *Rattler*. The victory of the *Rattler* indicated the superiority of propellers to both sail and paddle wheel technology. However, naval vessels did not rely exclusively on screw propulsion until the 1870s. The early twentieth century witnessed the universal adoption of screws for oceangoing ships, which include the majority of vessels from the largest battleships to the smallest merchant marine vessels.

A worker walks underneath a ship's propeller in a dry dock in Canada. (Corbis Corporation)

Typical marine propellers are fixed pitch and small in diameter with very thin, but broad, blade sections. They are made from either cast metal, corrosion-resistant metal alloys such as copper, or composite materials. Marine propellers normally operate at 60 percent efficiency due to the proximity of the ship's hull, which limits the overall diameter of the propeller and disturbs the efficient flow of water through the blades. As a result, the blades have to be very wide to produce adequate thrust. Marine propeller designers use innovations such as overlapping blades and wheel vanes to offset those problems and improve efficiency.

Another important consideration for marine propeller design is cavitation, the rapid formation and then collapse of vacuum pockets on the blade surface at high speed, and its contributions to losses in propulsive efficiency. The phenomenon can cause serious damage to the propeller by eroding the blade surface and creating high frequency underwater noise. Cavitation first became a serious problem in the late nineteenth and early twentieth centuries when innovations in steam and diesel propulsion technology drove propellers at unprecedented speeds. The introduction of gearing to make propellers rotate more efficiently at high revolutions and blades designed to resist cavitation alleviated the problem. However, research in the 1950s found that cavitation was a desirable trait for many high-speed marine propeller designs such as hydrofoils.

AERIAL PROPELLERS

As early as the eighteenth century, flight enthusiasts gradually began to consider the aerial propeller a practical form of transmitting propulsion. French mathematician J. P. Paucton revived the idea of the aerial propeller in Europe in his 1768 text, "Theorie de la vis d'Archimede." His *Pterophore* design used propellers for propulsion as well as for overall lift. Pioneers of lighter-than-air flight such as the French aeronauts Jean-Pierre Blanchard and Jean Baptiste Meusnier, used propellers on their balloons and airships in the 1780s. The increased attention given to the development of a practical heavier-than-air flying machine during the late nineteenth century ensured that experimenters rejected power transmission devices such as flapping wings, oars, sails, and paddle wheels, and incorporated propellers into their designs. Even though the propeller became a common component of proto-aeroplanes such as wings, aeronautical enthusiasts did not recognize the importance of propeller efficiency.

Wilbur and Orville Wright first addressed the aerial propeller from a theoretical and overall original standpoint during the successful development of their 1903 *Flyer.* They conceptualized the aerial propeller as a rotary wing, or airfoil, that generated aerodynamic thrust to achieve propulsion. They determined that the same physics that allowed an airfoil to create an upward lifting force when placed in an airstream would produce horizontal aerodynamic thrust when the airfoil was positioned vertically and rotated to create airflow. As a result, the Wrights created the world's first efficient aerial propeller and the aerodynamic theory to calculate its performance that would be the basis for all propeller research and development that followed. Used in conjunction with the reciprocating internal combustion piston engine, the aerial propeller was the main form of propulsion for the first fifty years of heavier-than-air flight.

Others built upon the achievement of the Wrights by improving the overall efficiency of the propeller. American engineer William F. Durand developed the standard table of propeller design coefficients in his landmark 1917 National Advisory Committee for Aeronautics study, "Experimental Research on Air Propellers." Besides the need for an aerodynamically efficient blade shape, propellers needed to be efficient over a variety of flight regimes. Developed simultaneously by Frank W. Caldwell in America, H. S. Hele-Shaw and T. E. Beacham in Great Britain, and Archibald Turnbull in Canada in the 1920s and becoming widely adopted in the 1930s, the variable-pitch mechanism dramatically improved the performance of the airplane. It linked aeronautical innovations such as streamline design, cantilever monoplane wings, and retractable landing gear with the increase in power brought by sophisticated engines, fuels, and supercharging. The properly designed variable-pitch propeller has been critical to the economic success of commercial and military air transport operations since the 1930s.

After World War II, the variable-pitch popeller was combined with a gas turbine to create the turbine propeller, or turboprop. The use of a gas turbine to drive the propeller increased propulsive efficiency, fuel economy, and generated less noise than conventional piston engine propeller aircraft. Turboprop airliners, first developed in Great Britain, began commercial operations in the early 1950s and were considered an economical alternative to turbojet airliners. Propellers are the most efficient form of aerial propulsion because they move a larger mass of air at a lower velocity (i.e., less waste) than turbojets and rockets. That efficiency is only present at speeds up to 500 mph (800 km/h). Beyond that, the tips of the rotating propeller suffer from near-sonic shockwaves that degrade its aerodynamic efficiency to the point where it loses power and cannot go any faster. Given that limitation and the higher efficiencies of turbojets at speeds above 500 mph for long distance flights, the propeller appeared to be obsolete as a viable energy conversion device for air transport. What occurred was that each propulsion technology proved the most efficient in different applications. The high efficiency of the aerial propeller, between 85 to 90 percent, and its lower operating costs created a strong economic impetus for air carriers to encourage the improvement of propeller-driven aircraft for the rapidly

Turboprop engine. (Allied Signal Corporation)

expanding short haul commuter market in the 1970s.

Further attempts to improve the efficiency of airplanes resulted in advances in propulsion technology based on the aerial propeller. Introduced in the 1960s, the turbofan, a large, enclosed multiblade fan driven by a turbojet core, harnessed the efficiency of the propeller while developing the thrust of the turbojet. NASA's long-range aircraft energy efficiency program of the 1980s developed the propfan, which is an advanced turboprop that employs multiple scimitar-shaped blades with swept-back leading edges designed for high speed. Propfans are capable of operating at speeds comparable to those generated by turbofan and turbojet-powered aircraft at a 25 percent savings in fuel.

The extreme climatic and physical environments in which propellers operate test the limits of aerodynamics, mechanical engineering, and structural theory. Depending upon the size of the power source, aerial propellers can be made from wood, metal, or composite materials and feature from two to six long and slender blades. The blades must be able to withstand exposure to harsh weather and stand up to aerodynamic loading, engine vibration, and structural bending while efficiently creating thrust.

A significant characteristic of the aerial propeller since its invention has been noise. Aircraft propellers, especially those made of metal, often produce an extremely loud "slapping, beating" sound when operating at high speeds. The increased use of propeller-driven commuter aircraft in the 1970s contributed to the growing noise pollution around busy urban airports. Measures to quiet aircraft included propeller synchronization to prevent high frequency vibration on multiengine aircraft and the use of elliptical blades, thinner airfoil sections, smaller propeller diameters, and lower rotational speeds.

Technologies other than conventional aircraft use aerial propellers for propulsion. Helicopters and vertical and short takeoff and landing aircraft (V/STOL) use rotor blades shaped to produce aerodynamic lift much like a propeller produces thrust. Hovercraft, or air-cushion vehicles, use propellers designed to be efficient at slower speeds for both movement and maneuverability when mounted on swiveling pylons. The search for an alternative energy source to fossil fuels in the 1970s encouraged many experimenters to use propellers to catch the wind and generate electricity through a turbine. Wind turbines benefit from advances in propeller design to create energy rather than converting the energy of a power source into forward movement.

Jeremy R. Kinney

See also: Aircraft; Ships; Turbines, Wind.

BIBLIOGRAPHY

Anderson, J. D., Jr. (1989). *Introduction to Flight.* New York: McGraw-Hill.

Anderson, J. D., Jr. (1997). *A History of Aerodynamics and Its Impact on Flying Machines.* New York: Cambridge University Press.

Anderson, J. D., Jr. (1999). *Aircraft Design and Performance.* New York: McGraw-Hill.

Breslin, J. P., and Andersen, Poul. (1994). *Hydrodynamics of Ship Propellers.* New York: Cambridge University Press.

Durand, W. F. (1917). Technical Report No. 14, "Experimental Research on Air Propellers." *Annual Report of the National Advisory Committee for Aeronautics, 1917.* Washington, DC: U.S. Government Printing Office.

Gilfillan, S. C. (1935). *Inventing the Ship.* Chicago: Follett Publishing.
Gillmer, T. C. (1986). *Modern Ship Design.* Annapolis, MD: Naval Institute Press.
Gillmer, T. C., and Johnson, B. (1982). *Introduction to Naval Architecture.* Annapolis, MD: Naval Institute Press.
Jakab, P. L. (1990). *Visions of a Flying Machine: The Wright Brothers and the Process of Invention.* Shrewsbury, England: Airlife.
Kerwin, J. E. (1986). "Marine Propellers." *Annual Review of Fluid Dynamics* 18:367–403.
Metzger, F. B. (1995). *An Assessment of Propeller Aircraft Noise Reduction Technology.* Hampton, VA: Langley Research Center, National Aeronautics and Space Administration.
Miller, R., and Sawers, D. (1968). *The Technical Development of Modern Aviation.* London: Routledge and Keegan Paul.
Rosen, G. (1984). *Thrusting Forward: A History of the Propeller.* Windsor Locks, CT: United Technologies Corporation.
Smith, E. C. (1937). *A Short History of Naval and Marine Engineering.* Cambridge, England: Babcock and Wilcox.
Sparenberg, J. A. (1984). *Elements of Hydrodynamic Propulsion.* Boston: Martinus Nijhoff.
Weick, F. E. (1930). *Aircraft Propeller Design.* New York: McGraw-Hill.

PROPERTY RIGHTS

In many situations, property rights are easily determined. Boundaries to stationary, observable assets such as land can be defined and enforced clearly. Property rights to crude oil and natural gas are far more troublesome, for two reasons. Both resources lie below the surface, making it difficult to locate the exact limits of the deposits, and they migrate within subsurface reservoirs. Some reservoirs cover many square miles, including hundreds of property owners, each with a competing claim to the oil and gas. Moreover, other reservoirs straddle national boundaries, encouraging international competition for the oil and gas. For these reasons energy production from crude oil and natural gas can involve serious common-pool losses. Although there is much to be gained from cooperation among property owners to avoid these losses, reaching a cooperative solution can be very difficult.

The problems of the common pool were most dramatically illustrated in Garrett Hardin's famous 1968 article "The Tragedy of the Commons." Exploitation of valuable resources when property rights are absent or poorly defined can be extremely wasteful. Because the parties must capture the resource before it is appropriated by another, time horizons become distorted, with undue emphasis on short-term production and neglect of long-term investment. Even if the trend of market prices indicates that resource values would be greater from future, rather than current, use, production will not be postponed. The losses are compounded by excessive capital investment in extraction and storage.

Common-pool problems characterize many resources where it is difficult to define property rights to restrain access and use. In North America, common-pool conditions in oil and natural gas production are created when firms compete for hydrocarbons in the same formation under the common-law rule of capture (which also governs fisheries and many other natural resources).

Under the rule of capture, ownership is secured only upon extraction. In the United States, production rights are granted to firms through leases from those who hold the mineral rights, typically surface land owners. Each of the producing firms has an incentive to maximize the economic value of its leases rather than that of the reservoir as a whole. Firms competitively drill and drain, including the oil of their neighbors, to increase their private returns, even though these actions can involve excessive investment in wells, pipelines, and surface storage, higher production costs, and lower overall oil recovery.

Oil production requires pressure from compressed gas or water to expel oil to the surface. There are three main types of reservoir drives to flush oil to wells: dissolved gas drive, gas-cap drive, and water drive. With a gas drive, the oil in the reservoir is saturated with dissolved gas. As pressures fall with oil production, the gas escapes from solution, expands, and propels oil to the surface. Hence it is important to control gas production so it remains available to remove the oil. With a gas-cap drive, the upper part of the reservoir is filled with gas, and oil lies beneath it. As oil is withdrawn, the compressed gas expands downward, pushing oil to the well bore. As with a dissolved gas drive, gas production from the gas cap should be restricted to maintain reservoir pressure to expel the oil. Finally, with a water drive, the oil lies above a layer of water. The compressed water migrates into the oil zone as the oil moves to the well and helps to push it from the rock formation. High oil recovery rates require that subterranean water pressures be maintained.

Maximizing the value of the reservoir requires that full reservoir dynamics be considered in drilling wells and in extracting oil. Gas and water must be recycled through the strategic placement of injection wells; wells with high gas-oil or water-oil ratios must be closed or not drilled; and the rate of oil production must be controlled to maintain underground pressure.

Each reservoir generally has a dominant drive, an optimal pattern of well locations, and a maximum efficient rate of production (MER), which, if exceeded, would lead to an avoidable loss of ultimate oil recovery. Unfortunately, oil, gas, and water are not evenly distributed within the reservoir. With multiple leases above the reservoir, some lease owners will have more oil, gas, or water than will others, and coordination among competing firms in well placement and in controlling production rates is difficult. Efficient production of the reservoir suggests that some leases not be produced at all. Further, since each firm's production inflicts external costs on the other firms on the formation, some mechanism must be found to internalize those costs in production decisions.

Maintaining subsurface pressures; effectively placing wells; coordinating timely investment; and controlling oil production across leases while protecting the interests of the various lease owners under conditions of geologic and market uncertainty are formidable challenges. Yet, if they are not adopted, common-pool losses are likely.

Common-pool problems in oil and gas have been observed since commercial production began in the United States in 1859. The solution, unitization, also has been understood for a long time. With unitization, a single firm, often with the largest leased area, is designated as the unit operator to develop the reservoir, ignoring lease boundaries. With unitization, optimal well placement and production are possible as is coordination in other activities to increase overall recovery from the reservoir; These activities include pressure maintenance; secondary recovery; and enhanced oil recovery (EOR), whereby heat, carbon dioxide, or other chemicals are injected into the reservoir.

All lease owners share in the net returns of unitized production, based on negotiated formulas. As residual profit claimants, the lease owners now have incentives to develop the reservoir (rather than the lease) in a manner that maximizes its economic value over time. When producers expect unitization to occur, exploration is encouraged because greater recovery rates and reduced costs are anticipated. Bonuses and royalties to landowners are higher because the present value of the oil and gas resource is greater with unitization.

Despite its attractions for reducing common-pool losses, use of unitization has been more limited than one would expect. Joe Bain (1947, p. 29) noted: "It is difficult to understand why in the United States, even admitting all obstacles of law and tradition, not more than a dozen pools are 100 percent unitized (out of some 3,000) and only 185 have even partial unitization." Similarly, in 1985 Wiggins and Libecap reported that as late as 1975, only 38 percent of Oklahoma production and 20 percent of Texas production came from field-wide units.

To be successful, a unit agreement must align the incentives of the oil-producing firms over the life of the contract to maximize the economic value of the reservoir without repeated renegotiation. Unit contracts involve a number of difficult issues that have to be addressed by negotiators. Because remaining production often lasts twenty years or more, unit agreements must be long-term and be responsive to considerable uncertainty over future market and geological conditions. They must allocate unit production and costs among the many firms that otherwise would be producing from the reservoir. Additionally, they must authorize investments that may be made later to expand reservoir production, and distribute the ensuing costs among the individual parties.

These are formidable tasks for unit negotiators. The negotiating parties must agree to a sharing rule or participation formula for the allocation of costs and revenues from production. There may be different sharing rules for different phases of unit production, such as primary and secondary production, and the rules should apply to all firms on the reservoir. This arrangement is termed a single participating area. There should not be separate participating areas for oil and gas; otherwise, different incentives for oil and gas production will emerge. Development, capital, and operating costs must be allocated in proportion to revenue shares in an effort to align all interests to maximize the economic value of the reservoir. In

STAKING A CLAIM: THE HOMESTEADING ALTERNATIVE

Property rights assignment by homesteading is based on a theory of philosopher John Locke. Locke believed "as much land as a man tills, plants, improves, cultivated and can use the product of, so much is his property." Such labor, Locke continues, "enclose(s) it from the common."

In a Lockean world, mineral rights do not accompany surface rights in either original or transferred ownership. Minerals would not be owned until homesteaded by the acts of discovery and intent to possess. In the case of oil and gas, initial ownership would occur when the oil or gas entered the well bore and was legally claimed by the driller. The reservoir would then turn from a "state of nature" into owned property. The homesteader (discoverer) could be the property owner directly overhead, another land owner, or a lessee of either.

The oil and gas lease under homestead law would simply be the right to conduct drilling operations on a person's surface property. The lease would not constitute a claim to minerals found under a particular surface area as under the rule of capture rule. Since the reservoir could be reached from different surface locations with slant drilling, the economic rent of a homestead lease would be far less than the value of a capture-rule lease. The difference in value would accrue to the driller-finder, thereby encouraging production by making drilling more efficient and profitable.

Under a homestead theory of subsurface rights, the first finder of a mineral area would have claim to the entire recognized deposit. In the case of oil and gas, the geologic unit is the entire reservoir, however shaped, as long as it can be proven to be contiguous. Separate and distinct reservoirs in the same general area, whether vertical or horizontal, would require separate and distinct discovery and claim. Alien wells draining a claimed reservoir would be liable for trespass if the homesteader can reasonably prove invasion (whether by well distance, well depth, crude type, geological formation, reservoir pressure, etc.). The pool owner would also be able to use his discovery as chosen. The pool could be left untapped, depleted, or used for storage. The reservoir space, in addition to the virgin contents of the reservoir, is newfound property.

The operation of homestead property rights system can be facilitated by pragmatic rules established either by law or judicial opinion. When a new oil or gas field is discovered, the surrounding acreage could be declared off-limits for other producers until the discoverer has a "reasonable" opportunity to drill frontier wells to delineate the field. An exception would be made if another operator had begun to drill before the nearby discovery was made so long as opportunistic slant drilling did not occur. If the discoverer does not drill development wells to delineate the field, then other operators may drill and become co-owners of the same contiguous reservoir upon discovery, requiring cooperation to minimize drilling costs.

While the homestead theory was never seriously considered in the United States experience, it has been recommended as part of a subsoil privatization program for Latin America. The major attraction is increased economic efficiency: Production and management of a contiguous oil and gas reservoir is most efficiently done by one operator, and homesteading creates one owner in place of multiple owners under the rule of capture.

Robert L. Bradley

BIBLIOGRAPHY

Bradley, R. L., Jr. (1996). *Oil, Gas, and Government: The U.S. Experience.* Lanhan, MD: Rowman & Littlefield.

Locke, J. (1947). "An Essay Concerning the True Original Extend and End of the Civil Government." In *Social Contract,* ed. E. Barker. London: Oxford University Press.

that case, each party will share in the net returns from production and hence will have an incentive to minimize development and production costs and to maximize revenues. Further, a single unit operator must be selected to develop the field. Multiple unit operators lead to conflicting objectives and hinder the coordinated production practices necessary to maximize the value of the reservoir. Supervision of the unit operator by the other interests must be determined, with voting based on unit participation.

All of these issues can be contentious, but reaching agreement on the revenue sharing or participation formula is the most difficult. Shares are based on estimates of each firm's contribution to the unit. Those firms with leases that reflect a natural structural advantage (e.g., more resources or easier access to resources) seek to retain this advantage in the unitization formula. Such firms are unlikely to agree to a unitization agreement that does not give them at least as much oil or gas as they would receive by not unitizing. Even if the increase in ultimate recovery from unitization is so great that these parties will receive more from unit operations than from individual development, they have a much stronger bargaining position in negotiations than less-favored tract owners. The former can hold out for the most favorable allocation formula, secure in the knowledge that the regional migration of oil will continue toward their tracts during any delay in negotiations. Indeed, holding out may increase the value of a structurally advantageous location. If the firms form a subunit without the participation of the owners of better-located tracts, the pressure maintenance operations of the unit may increase the amount of oil migration toward the unsigned tracts. The holdouts, then, benefit from the unit without incurring any costs of the pressure maintenance activity.

The information available to the negotiating parties for determining lease values depends upon the stage of production. During exploration, all leases are relatively homogeneous, and unitization agreements can be comparatively easy to reach using simple allocation formulas, often based on surface acreage. Information problems and distributional concerns, however, arise with development as lease differences emerge. Because reservoirs are not uniform, the information released from a well is descriptive of only the immediate vicinity. There will be disagreements over how to evaluate this information in setting lease values. As a result of disagreements over subsurface parameters, unit negotiations often must focus on a small set of variables that can be objectively measured, such as cumulative output or wells per acre. These objective measures, however, may be poor indicators of lease value.

Conflicts over lease values and unit shares generally continue until late in the life of a reservoir. As primary production nears zero, accumulated information causes public and private lease value estimates to converge, so that a consensus on share values is possible. Without unit agreement, secondary production from all leases may be quite limited. The timing of a consensus on lease values, however, suggests that voluntary unit agreements are more likely to be reached only after most of the common-pool losses have been inflicted on the reservoir.

Reaching agreement on these issues is a complex process. In a detailed examination of seven units in Texas, Wiggins and Libecap showed in 1985 that negotiations took four to nine years before agreements could be reached. Moreover, in five of the seven cases, the area in the final unit was less than that involved in early negotiations. As some firms became frustrated, they dropped out to form subunits, which led to a partitioning of the field and the drilling of additional wells but generally did not minimize common-pool losses. For example, 28 subunits, ranging from 80 to 4,918 acres, were established on the 71,000-acre Slaughter field in West Texas. To prevent migration of oil across subunit boundaries, some 427 offsetting water injection wells were sunk along each subunit boundary, adding capital costs of $156 million.

Other costs from incomplete unitization are shown on Prudhoe Bay, North America's largest oil and gas field, which was first unitized in 1977. Two unit operators, separate sharing formulas for oil and gas interests, and a disparity between revenue sharing and cost sharing among the working interests have resulted in protracted and costly conflicts. The parties on the field do not have joint incentives to maximize the economic value of the reservoir, but rather seek to maximize the values of their individual oil or gas holdings. Because ownership of oil and gas in Prudhoe Bay is very skewed, with some firms holding mostly oil in the oil rim and others holding mostly natural gas in the gas cap, the firms have been unable to agree to a complete, reservoirwide agreement. Disputes have centered on investment and whether natural gas should be reinjected to raise oil production, or liquefied and sold as natural-gas liquids. In 1999 Libecap and Smith showed that skewed ownership in the presence of gas

Prudhoe Bay Oil Field. (Corbis Corporation)

caps characterizes other incomplete unit agreements elsewhere in North America.

Most states, as well as the U.S. government, have some type of compulsory unitization rule to limit the ability of a minority of holdouts to block a unit. Mandatory unitization began in the United States in Louisiana, in 1940. Due to political opposition by small firms that receive regulatory-related benefits, Texas, surprisingly, has not had a compulsory unitization law. Texas court rulings have tended to restrict unitization.

Although unitization is most problematical in North America, where there generally are more operators on a given oil field, the problems of coordinated production to avoid common-pool losses exist elsewhere. In the North Sea, Argentina, Ecuador, and the Caspian Sea, lease and national boundaries often do not coincide with reservoir boundaries. Under these conditions a nation may engage in competitive drilling to drain the oil and gas before its neighbor does. This incentive appears to explain the initial pattern of Norwegian drilling only along its boundaries with the United Kingdom (Hollick and Cooper, 1997).

Gary D. Libecap

See also: Governmental Approaches to Energy Regulation; Government Intervention in Energy Markets; Oil and Gas, Drilling for; Oil and Gas, Exploration for; Oil and Gas, Production of; Pipelines.

BIBLIOGRAPHY

Bain, J. S. (1947). *The Economics of the Pacific Coast Petroleum Industry, Part III: Public Policy Toward Competition and Prices.* Berkeley: University of California Press.

Hardin, G. (1968). "The Tragedy of the Commons." *Science* 162:1243–1248.

Hollick, A. L., and Cooper, R. N. (1997). "Global Commons: Can They Be Managed?" In *The Economics of Transnational Commons*, ed. P. Dasgupta, K-G. Maäler, and A. Vercelli. Oxford: Clarendon Press.

Libecap, G. D. (1989). *Contracting for Property Rights.* New York: Cambridge University Press.

Libecap, G. D., and Smith, J. L. (1999). "The Self-Enforcing Provisions of Oil and Gas Unit Operating Agreements: Theory and Evidence." *Journal of Law, Economics, and Organization* 15(2):526–48.

Libecap, G. D., and Wiggins, S. N. (1984). "Contractual Responses to the Common Pool: Prorationing of Crude Oil Production." *American Economic Review* 74:87–98.

Libecap, G. D., and Wiggins, S. N. (1985). "The Influence of Private Contractual Failure on Regulation: The Case of Oil Field Unitization." *Journal of Political Economy* 93:690–714.

McDonald, S. L. (1971). *Petroleum Conservation in the United States.* Baltimore: Johns Hopkins University Press.

Smith, J. L. (1987). "The Common Pool, Bargaining, and the Rule of Capture." *Economic Inquiry* 25:631–644.

Sullivan, R. E., ed. (1960). *Conservation of Oil and Gas: A Legal History, 1948–1958.* Chicago: American Bar Association.

Weaver, J. L. (1986). *Unitization of Oil and Gas Fields in Texas: A Study of Legislative, Administrative, and Judicial Policies.* Washington, DC: Resources for the Future.

Wiggins, S. N., and Libecap, G. D. (1985). "Oil Field Unitization: Contractual Failure in the Presence of Imperfect Information." *American Economic Review* 75:376–385.

PROPULSION

Whenever anything is set in motion, there must be some type of propulsive force that moves it. Propulsion is a key element of many activities, including athletic events, recreation, transportation, weapons, and space exploration. This article explains the basic principles involved in any propulsion system, differentiates the types of propulsion systems, and discusses some practical aspects of propulsion.

BASIC PHYSICS

Momentum/Impulse

To change an object's speed it is necessary to exert a net force on it for some duration of time. The product of the force and the length of time it is applied is called the impulse and is the same as the change of momentum. Any increase in speed is directly proportional to the impulse and inversely proportional to its mass:

$$\text{Change of Speed} = \text{Force} \times \text{Time/Mass} = \text{Impulse/Mass}$$

A 1,000 kg (metric ton) car accelerating from 0–30 m/s (68 mph) in 6 seconds, thus requires an average force of 5,000 newtons. This is a force approximately half the weight of the car. It is likely that the car would "burn rubber" as the drive wheels spun on the road during this acceleration. To accelerate to the same speed in 12 seconds would require a force of only 2,500 newtons.

Energy/Work

An increase of speed corresponds to an increase of kinetic energy. This increase of kinetic energy is a result of the net force acting in the direction of motion through some distance. Net force times distance is the work done on the object:

$$\text{Change of Kinetic Energy} = \text{Force} \times \text{Distance} = \text{Work}$$

Kinetic energy equals 1/2 mass × speed^2. Therefore, a formula for the change of the square of the speed is

$$\text{Change of speed}^2 = 2 \times \text{Force} \times \text{Distance/Mass} = 2 \times \text{Work/Mass}$$

The car used in the example above would travel 90 meters during its acceleration (assumed constant) to 30 m/s in 6 seconds. The increase in kinetic energy would be 450,000 joules, which exactly equals the work input of 5,000 newtons × 90 meters.

In the case of maintaining the speed of an object against opposing forces such as friction or air resistance, the energy analysis is more appropriate than a momentum analysis. For friction forces, the conservation of energy requires that work by propelling force is equal to the energy dissipated by friction, or, in the case of lifting an object at constant speed through a gravitational field, that the work by propelling force is equal to the gain in potential energy.

In designing a rocket with sufficient propulsive power to launch a satellite into orbit, all three energy considerations are involved. The satellite gains speed (more kinetic energy), it is lifted upward against gravity (more potential energy), and it encounters air resistance (energy dissipated by friction).

Whatever the type of propulsion, it is necessary to exert a force on the object for a period of time or through a distance.

Newton's Third Law

The force of propulsion must be exerted by some object other than the one being propelled. Forces always appear as pairs. As expressed by Newton's third law, the object being propelled and the propelling agent interact by equal, opposite and simultaneous forces. For instance a runner pushes against the starting blocks and the starting blocks push against the runner, impelling him forward. A ship's screw pushes against the water propelling it backward, and the water pushes against the ship's screw, propelling the ship forward. The rocket exhaust thrusts against the rocket, propelling it forward as the rocket thrusts the exhaust backward. And the wind pushes against the sail, propelling the boat forward as the sail pushes against the wind, slowing it down or deflecting its motion.

Efficiency

There are numerous ways to look at the efficiency of propulsion systems, depending on the goal of the system. One way might be to define it in terms of an energy ratio—the ratio of the energy required to achieve the desired motion to the energy consumed by the propelling agent. For instance, consider accelerating the one-ton car to 30 m/s, giving it a kinetic energy of 450,000 joules. Suppose during the acceleration the engine uses 1/2 cup of gasoline with a heat of combustion of 4,200,000 joules, then the energy efficiency would be 450,000/4,200,000 = 10.7 percent.

Other definitions of efficiency arise where energy considerations are irrelevant. In rocket propulsion, for instance, the goal is to give a certain mass of payload a very large speed. A more useful measure here might be the ratio of the speed boost given to the payload mass compared to the total mass of fuel, including its container and the engine.

The most well-known propulsion efficiency is the amount of energy (fuel) required to move a vehicle a certain distance against opposing forces. Efficiencies of road vehicles, for instance, are usually measured in miles per gallon.

TYPES OF PROPULSION

Propulsion can be classified according to the type of object with which the propulsive device interacts. It might interact with (A) a massive object, (B) a fluid medium in which it is immersed, (C) material that it carries along to be ejected, or (D) a flowing stream.

Thrust Against a Massive Object

The object interacts directly with another relatively massive object. For example, with every fall of a foot, a runner interacts ultimately with the earth, as do automobiles, trains and bicycles. A pitcher pushes against a mound of dirt as he throws a ball, and bullets interact with the gun.

Thrust Against the Ambient Medium

The object interacts with a relatively lightweight medium such as air or water that it encounters along its path. A ship takes some of the water in which it floats, propels it backward, and the ship is in turn propelled forward. Other examples of this type are propeller airplanes and jet planes. The jet plane takes in air (23 percent oxygen by weight) through which it is flying, adds some fuel to it, and the chemical reaction with the oxygen (burning) results in a high-velocity exhaust. The mutual interaction between the exhaust (which, in the case of the fan jet, mixes in even more air) and the airplane propel it forward. Unlike a thrust against a massive object, which barely moves, the fluid carries away significant amounts of kinetic energy during the interaction.

Rockets

The object to be propelled carries the material to be pushed against, and usually carries the energy source as well (fuel and oxidant). Some examples of this type are: rockets, inflated balloons released to exhaust their air, and ion engines. The inflated balloon would be propelled by interaction with its released air even if it were released in space. It does not push against the air around itself. In fact, the ambient air only tends to impede its motion.

Flowing Streams

The object interacts with a flowing medium. Sails interact with the wind, rafts float downstream, a spacecraft is propelled by the stream of photons (light) from the sun. This is more of a "channeling" type of propulsion in that the propelled object deflects the flowing stream in such a way that it is forced to move in desired direction.

COMPARISON

Theoretically, the first type of propulsion, where the interaction is with a relatively immovable object, is the most energy efficient. In the absence of friction, it is theoretically possible to convert 100 percent of the available energy into kinetic energy, the energy of motion. This is because the more massive immovable object takes away a very small amount of kinetic energy even though it shares equally (but oppositely) in the impulse or change of momentum.

A common energy source for this type of propulsion is an internal combustion engine, in which fuel is burned and expanding gasses convert the energy into mechanical motion. This process is governed by the laws of thermodynamics, which limit the efficiency to much less than 100 percent. Further losses are encountered through transmission and through rolling friction. The efficiencies are in the 20 percent range for automobiles.

Human propulsion, which depends on muscle power, obtains its energy through the metabolism of food. This process has an efficiency of about 30 percent for a long-distance runner during a race. A cyclist who pedals at a tenth horsepower (climbing a 5 percent grade at 5 miles per hour) for eight hours, and who eats a diet of 3,000 calories, has an efficiency of 17 percent for the day, assuming no additional work is done.

The efficiency of propulsion for a given energy source will always be less when the interaction is with a fluid medium such as air or water. For example, the kinetic energy of the wind generated by the propellers of an airplane is energy that is dissipated and does not end up as part of the airplane's kinetic energy.

In this second type of propulsion the efficiency depends on the speed of the object relative to the speed at which the medium is being pushed backward. The greatest theoretical efficiency occurs when the propeller of the ship, for example, pushes a large quantity of water backward at a relative speed just slightly greater than the speed of the ship moving forward. Practical limits on the size of the screw determine that the water is usually pushed backward much faster than the ship's speed in order to get adequate thrust. The efficiency turns out to be very small when the screw is pushing water backwards at high rates of speed.

The jet engine also interacts with the ambient medium as it goes along, so it is basically the same type of propulsion as with the propeller. The decision of whether to use a propeller or a jet engine for an airplane depends on the design speed. The dividing line is about 300 miles per hour. The propeller is more efficient at pushing large quantities of air at relatively low speeds, so it is the choice for low-speed flight. Since it is always necessary to push the air backward at a relative speed greater than that at which it is desired to fly, a jet engine is necessary for high speed flight.

When the efficiency of transportation, by whatever propulsion method, is measured on a distance per amount of energy basis (miles per gallon), it is found that two very important factors are speed and aerodynamic design. At normal road speeds, and for subsonic (most commercial) aircraft, the wind resistance for a given size and shape depends on the square of the speed. The fuel consumption for a given vehicle is influenced directly by the friction forces, including wind resistance. An aircraft flying the same distance will need four times the energy to overcome wind resistance flying at 400 miles per hour as it uses flying at 200 miles per hour. The actual fuel consumption difference is less than this because of constants such as the amount of fuel needed to take off and to gain altitude.

The same rule is true of automobile travel. An automobile traveling at 70 miles per hour will need nearly twice the energy to overcome wind resistance as if it made the same trip at 50 miles per hour. Again, the difference in fuel consumption will be less due to the effect of rolling friction (nearly independent of speed), the fuel used to accelerate up to speed after every stop, and the energy used to operate lights, air conditioner, etc.

The third type of propulsion is best typified by a rocket engine. The rocket must carry its energy supply and ejecta material as well as the object to be launched. The energy supply is usually two chemicals that, when they combine, give off a large amount of heat, vaporizing the reaction products. This hot gas is ejected at high velocity, propelling the rocket. The two chemicals may be hydrogen and oxygen, which combine to produce very hot water (steam). The fuel might also be a solid mixture of chemicals that are designed to react at a controlled rate.

To achieve the goal of the greatest speed with a given mass of fuel, it is necessary that the ejecta come out of the rocket as fast as possible. Therefore, the efficiency (or effectiveness) of rocket fuels is better measured in terms of the speed of the ejected material than by energy efficiency.

One advantage of the rocket is that the payload can reach speeds in excess of the relative speed of the ejecta, which is not the case with a jet engine. The thrust of the jet engine goes to zero when its speed through the air reaches the relative speed of the ejecta, whereas the thrust of the rocket stays the same no matter what the speed. The jet engine has the advantage that it uses the air it encounters along the way as both the oxidant for the fuel and most of the ejecta. It, therefore, does not need to carry nearly as much mass as the rocket to accomplish the same impulse.

An ion rocket engine does not need to carry its energy supply like the chemical rocket does. It can use solar cells to collect energy from the sun to provide an electric field to accelerate ions (charged xenon atoms, for instance) that provide a thrust when they are ejected at high speed. Since the energy is "free," the energy efficiency is not relevant. The relative speed of the ejected ions from NASA's Deep Space 1 ion propulsion system is about 62,000 miles per hour, which is ten times that from conventional chemical rockets. Therefore, for a tenth of the "fuel," it can achieve the same boost in speed as a conventional rocket. By this measure the ion rocket engine is ten times more efficient than the chemical rocket.

The ion engine produces a very tiny thrust, about the weight of a sheet of typing paper, but exerts it continuously for a long period of time (several months). This very long time multiplied by the relatively small thrust (force) still results in large impulse to the space probe. Because it is not possible to push more xenon ions through the engine in a given time only low thrust is possible. Greater thrust would require a much larger engine and a more powerful source of energy.

Nuclear energy could be a compact source for application in an ion engine, but no practical devices have yet been constructed. Nuclear energy would be necessary for the ion drive of an interstellar mission for which solar energy would not be available.

Launching a satellite from earth could not be done with the low thrusts of currently conceived ion rockets. Chemical rockets, which can generate large amounts of thrust, will be needed for the foreseeable future.

The fourth type of propulsion makes use of a flowing stream. In this case, as in the ion engine, the energy efficiency is irrelevant because the supply of energy is unlimited for practical purposes.

The sailboat being propelled downwind is easy to understand. Also it is not too hard to see that turning the sail at an angle to bounce the wind off to the side propels the boat at some angle to the wind. What is amazing is that it is possible for the sailboat to be propelled faster than the wind, and to make headway upwind. These feats are possible because it is easier to move a boat in the direction of its keel (lengthwise) than it is to move it abeam (crosswise).

Another flowing stream that might be used for propulsion in the future consists of photons (light particles) from the sun. These photons have momentum and thus exert a small pressure on everything they strike. If reflected from a mirror, the mirror would act exactly like a sail in the wind. The pressure of the light would always be at right angles to the plane of the mirror. Therefore, it would be possible to use a large, lightweight reflecting surface as a sail to navigate anywhere in the solar system. The solar sail, like the ion engine, would generate a very small thrust. At the earth it would be only 7 newtons per square kilometer (4 pounds per square mile) of reflecting sail.

Don C. Hopkins

BIBLIOGRAPHY

Reynolds, W. C., and Perkins, H. C. (1977). *Engineering Thermodynamics.* New York: McGraw-Hill.

Harrison, J. P. (1996). "The Jet Age." In *Mastering the Sky.* New York: Sarpedon.

Zumerchik, J. (1997). "Sailing." In *Encyclopedia of Sports Science, Vol. 1.* New York: Macmillan.

Leary, W. E. (1998). "Ion Propulsion of Science Fiction Comes to Life on New Spececraft." *The New York Times,* Oct. 6.

PROSPECTING, GEOPHYSICAL

See: Oil and Gas, Exploration for

PULLEY

See: Mechanical Transmission of Energy

Q–R

QUANTUM MECHANICS

See: Molecular Energy

RADIANT ENERGY

See: Heat Transfer

RADIOACTIVE WASTES

See: Nuclear Waste

RAILWAY FREIGHT SERVICE

See: Freight Movement

RAILWAY PASSENGER SERVICE

Travelers usually have a choice among several transportation alternatives, most prominent being automobile, bus, airplane and railway. Rail was the preferred means of travel in the mid-nineteenth century and continuing until the 1920s, but began to decline afterward because it could not compete with the greater mobility of the automobile and the greater speed of air travel. Despite this decline, rail passenger service may yet find a new role where its potential to be more fuel-flexible, environmentally friendly, and energy efficient can be realized. Moreover it has the flexibility to avoid the congestion delays of air and roadway traffic that is likely to only worsen.

About 60 percent of transport energy in the United States is used for passenger transport, almost entirely by autos, light trucks and aviation. In fact, rail passenger services in the U.S. carry only about 2 percent of total passenger-kilometers (a passenger-kilometer is one passenger moved one kilometer), with the remainder carried by auto, air and bus. In other countries the rail role is larger, ranging from 6 to 10 percent of passenger-kilometers in many European countries to about 20 percent in India and as high as 34 percent of passenger-kilometers in Japan. Depending on the country, energy in transport could be saved by making rail passenger services more energy-efficient (not so important in the United States but much more so in Europe, India and Japan) or, probably more important, in getting passengers to switch from less-energy-efficient modes to potentially more efficient modes of rail passenger service.

Rail passenger service offers many alternatives, with different characteristics and customers. In urban areas, slow (50 km/hr), frequent-stop trolleys ("streetcars") have long operated in many cities, often in the same right-of-way where autos and buses drive. There are "light rail trains" (LRT), midway in speed (up to about 80 km/hr) and capacity between trolleys and traditional subways, operating in a number of large metropolises. "Heavy rail" metros (up to 110 km/hr) form the heart of the transport network in many of the world's largest cities. Many cities supplement their metros with longer-range, higher-speed (up to 140 km/hr) suburban rail services, and some have mixtures of all of these (e.g., Cairo and or Moscow).

In the intercity market there are glamorous high-speed trains (200 to 300 km/hr) such as the Japanese Shinkansen, the French TGV, and Amtrak's Metroliner. Such trains carry enormous numbers of passengers in Japan (almost 300 million per year), France (about 65 million), and Germany (more than 20 million). These trains provide high-quality, city-center-to-city-center passenger service in competition with air (below 500 km or so) and auto (over 150 km or so). Though they are important, only rarely do these high-speed trains carry a sizeable percentage of the country's total rail passengers.

The workhorses of most rail passenger systems are ordinary passenger trains operating at 70 to 160 km/hr. Table 1 shows the rail passenger traffic carried by most of the world's railways. The top three railway systems account for more than half of the world's passenger services, and the top six account for over two-thirds. Table 1 demonstrates another important point: the *entire* developed world accounts for only 30 percent of rail passenger transport. North American, European, and Japanese experiences are not representative of the vast bulk of the world's rail passengers.

	Passenger-Kilometers (000,000)	*Percent of Passenger-Kilometers*
India	357,013	20.1
China	354,261	19.9
Japan	248,993	14.0
Russia	168,679	9.5
Rest of W. Europe	68,107	3.8
Ukraine	63,752	3.6
Germany	60,514	3.4
France	55,311	3.1
Egypt	52,406	2.9
Italy	49,700	2.8
Rest of dev. Asia	42,849	2.4
Republic of Korea	29,292	1.6
United Kingdom	28,656	1.6
Rest of CEE	27,963	1.6
Poland	20,960	1.2
Kazakhstan	20,507	1.2
Pakistan	19,100	1.1
Romania	18,355	1.0
Rest of Middle East	17,803	1.0
Latin America	17,204	1.0
Rest of CIS	16,079	0.9
All of Africa	14,242	0.8
US Commuter	11,135	0.6
Amtrak	8,314	0.5
Australia	4,904	0.3
Canada VIA	1,341	0.1
World Total	**1,777,440**	**100.0**
Developing Countries		69.8
Developed Countries		30.2

Table 1.
The World's Rail Passenger Traffic

ENERGY USE BY RAIL PASSENGER TRAINS

Rail passenger energy efficiency depends on many variables, and a complete analysis would consume volumes of argument. In large part, though, an analysis of energy consumption in passenger transport begins with four basic factors: surface contact friction (rolling wheels), air drag (wind resistance), mass and weight (which influence acceleration/deceleration requirements and contact friction), and load factor (the percentage of occupied seats or space).

Surface friction is a source of rail's advantages in transport energy efficiency. Under similar conditions, steel wheels on steel rail generate only about 20 to 30 percent of the rolling friction that rubber wheels on pavement generate, both because rails are much smoother than pavement and because steel wheels are much more rigid than rubber tires, so steel wheels deform much less at the point of contact with the ground. Each rail wheel has only about 0.3 sq in of surface in contact with the rail, whereas an automobile tire can have 20 to 30 sq in of rubber in contact with the pavement. The greater rubber tire deformation takes energy that is wasted as heat in the tread, and the lower the pressure, the greater the deformation. Slightly offsetting this difference is the friction of the rail wheel flange, which keeps the train on the track.

Air drag is minimal at very low speeds, but it rises rapidly with increasing speed. In fact, drag is related to speed squared, and the power required to overcome drag is related to speed cubed, and it affects all vehicles. With good design of rolling bearings, rolling friction does not increase as rapidly with speed as does air drag, so in most cases air drag begins to exceed rolling friction at speeds above 60 to 100 km/hr in rail passenger vehicles, and it dominates energy requirements at speeds above about 150 Km/Hr. Streamlining can reduce air drag for any vehicle, but rail again has an inherent advantage because, for the same number of passengers, trains can be longer and thinner than buses or airplanes, and the thinness of the form affects air drag significantly.

Mass affects energy consumption because it takes energy to accelerate the vehicle and its passengers, and most of this energy is subsequently wasted as heat in the braking system when the vehicle slows. Weight also increases rolling friction. Surprisingly, rail passenger vehicles are relatively heavy. A fully loaded five-passenger automobile will have a gross weight of no more than 800 pounds per occupant and a bus slightly less, whereas a fully loaded rail passenger coach will have a gross weight of between 2,000 and 3,000 pounds per occupant. The average train, including locomotive, diner, and sleeping cars, can average 4,000 pounds or more per passenger.

Engineering models of tractive effort calculation derive from a basic model, generally called the Davis formula (after its initial formulator, W. J. Davis). The Davis formula calculates the resistance (Rt, in pounds per ton of weight being pulled) as:

$$Rt = (1.3 + 29/w + bV + CAV^2/wn)/wn$$

In this formula, w is the weight in tons per axle; b is a coefficient of flange friction (0.03 for passenger cars); V is the speed in miles per hour; C is the air drag coefficient (0.0017 for locomotives, 0.00034 for trailing passenger cars); and A is the cross-sectional area of locomotives and cars (120 sq ft for locomotives, 110 sq ft for passenger cars).

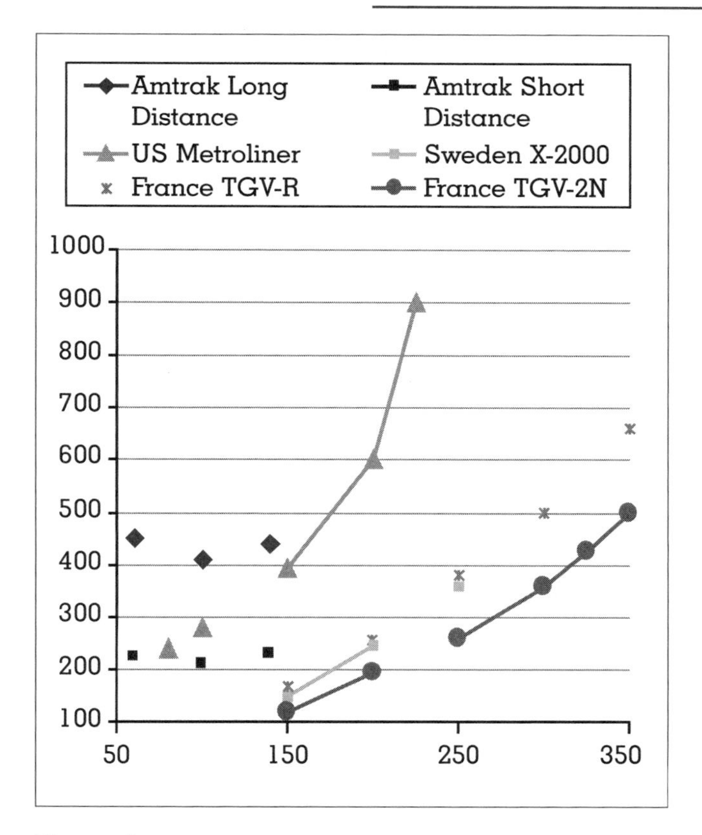

Figure 1.
Rail Passenger Fuel Consumption at Various Speeds (KJ/P-km for full trains).

The coefficients shown for the formula are examples; they would differ with changing designs of freight and passenger rolling stock and are specific to each train. The purpose here is to show the form of the relationship, not the exact values. Figure 1 displays energy consumption on various full-passenger trains over a range of speeds. Energy consumption rises rapidly with speed—and the energy consumed by the highly streamlined TGV or Swedish X-2000 is far less than the energy consumed by the boxy Metroliner.

The figure leaves out another important determinant of energy efficiency: *load factor*. Roughly speaking, a half-empty vehicle consumes twice as much energy per passenger-kilometer as a full vehicle, and a one-fourth-full vehicle four times as much. No matter what the potential energy advantages might be, empty trains waste energy, and full autos can be highly efficient. Unless modes are compared while operating at the appropriate load factor, the energy efficiency conclusions reached can be seriously flawed—and it is not necessarily valid to assume the same load factor for all modes. Indeed, the annual average load factor for Amtrak in the United States is only 46 percent, while the average load factor for U.S. airlines is 65 percent. Commuter trains do run full (or more than full) in the loaded direction during rush hour, but they often run nearly empty in the outbound direction and during midday, and actually average less than 35 percent load factor.

Even after the readily quantifiable engineering and operating issues are argued, the picture is still incomplete. Energy consumption in passenger transport is also affected by a large number of less predictable, real-world factors. Mountainous countries, for example, cause reduced railway efficiency because railways cannot climb steep grades (the downside of reduced rolling friction), so railway lines usually have to be 30 to 50 percent longer than highways between the same two end points in mountainous terrain. Bad maintenance practices and older equipment can severely reduce energy efficiency and increase pollution. These factors, and others such as driver expertise, can easily dominate the engineering and operating factors that can be readily measured.

Taking all these factors into account, well-calibrated computer models show that U.S. freight trains of various types could theoretically operate at energy consumption rates of 75 to 140 kJ per ton-km (kJ/t-km) of cargo depending on the type of train, with low-speed, heavy coal trains being the most efficient and high-speed, high-drag, double-stack container trains being the most energy-consuming. In 1997, the actual rail industry average of 250 kJ/t-km was nearly twice the theoretical upper bound of 140 kJ/t-km, reflecting the running of partially empty trains, wagon switching, and locomotive idling practices, among other factors. Though the technology is effectively the same, average energy consumption in Kenya Railways freight traffic is more than 1,000 kJ/t-km as a result of far less than optimum operating and maintenance practices.

Similar variations occur in reported cases of rail passenger service where engineering models show trains potentially operating at 200 to 400 kJ per passenger-kilometer. In practice, though, passenger trains are rarely full, and some trains, such as many long haul Amtrak trains, carry sleepers and diners which add significantly to the weight of the train but do not add many passenger seats. Commuter trains do not usually carry diners, of course, but they often operate essentially empty on their midday trips or on trips against the flow of rush-hour traffic. Thus the theoretically high rail passenger energy efficiency gets transmuted into an Amtrak system-wide average of 1,610 kJ per passenger-kilometer and an average for all United States commuter railways of 2,131 kJ per passenger-kilometer. Because of the effect of actual practices, the optimal efficiency from the engineering model is, in practice, a factor of four to eight better than what is experienced.

In general, the range of reported experience supports the conventional wisdom. Rail is potentially the most efficient method of motorized passenger travel; but bus is almost the equal of rail. Motorcycles are also quite efficient (and this assumes only the driver—motorcycles with passengers are the most energy-efficient vehicles of all). Automobiles and airlines do typically consume more energy per passenger-kilometer than rail.

More important, though, is the fact that there are significant overlaps in the energy consumption ranges, depending on the factors discussed above. While rail and bus are generally a toss-up, there are conditions in which a fully loaded Boeing 747-400 can be more energy-fficient than a partially loaded Amtrak Metroliner. A fully loaded automobile can well be a better energy and environmental choice than an empty bus or an empty train. There is no single answer, and actual conditions are very important.

Though only partly energy-related, passenger trains have additional features that deserve highlighting, such as the potential for electrification, space efficiency, and impact on urban form.

Electrification

Because trains travel path is well defined, where train traffic is dense it can be economically feasible to construct either overhead electric power supply wires (the "catenary" system) or ground-level electric supply systems (usually called the "third rail"), which permit trains to be powered directly by electric traction motors. Though a few cities have electric trolley buses, the cost of building and operating a system to supply electricity to buses, autos, and trucks is normally so high as to restrict road travel to use of fossil fuels, though advances in battery technology and hydrogen fuels may eventually reduce this dependence.

Electric traction power in railways offers three significant advantages: diversity of fuels, easier control of pollution, and high tractive effort per unit of train weight. Because electricity can be generated by a wide range of fossil fuels such as coal or natural gas, electric trains are not solely dependent on petroleum products for their operation. In fact, hydroelectric or nuclear power uses no fossil fuels at all (though each has its own environmental implications), and some countries, such as Switzerland, can use hydropower to drive almost all of their passenger and freight trains. Even when the electricity used by trains is generated by fossil fuels, the pollution is far more easily controlled from a single electric power station than from thousands of cars and buses, and the pollution and engine noise will be emitted at the power station and not in the city centers.

Because it takes time for an electric motor to overheat when exposed to high power currents, properly designed electric motors can operate for limited periods of time at power ratings as much as twice the level at which they can operate continuously. This means that electric motors can deliver short bursts of acceleration that require power well beyond the power needed for cruising. Diesels, by contrast, are limited by the power of the engine, and cannot exceed this rating even for short periods of time. In addition, in the diesel-electric system, which almost all diesel trains use, the diesel engine drives a generator, which in turn feeds electric traction motors, which drive the train; this involves extra weight, so that diesel systems are heavier than all-electric systems. As a result, electric traction has advantages where high acceleration is required, or where (as in high-speed trains) the weight of the train must be controlled to reduce the forces the train exerts on the track. Offsetting the performance advantages of electric traction is the added cost of the catenary or third rail and transformers needed to feed the train with electricity.

Most urban rail passenger systems are electrically operated because electric power permits high acceleration and thus closer spacing of trains. Most high-speed trains are electrically driven because diesel engines of the high power required are too heavy (and damage the track), and gas turbines do not handle acceleration well because they are not energy-efficient at speeds other than the optimum speed for which the turbine is designed. Ordinary, longer-haul passenger trains are a mix of diesel and electric traction: Diesel traction is far cheaper on light-density lines, and high acceleration and light weight are not so important on slower trains.

Because railway electric motors can be made to slow a train as well as drive it, the energy of braking can be regenerated onboard and put back into the catenary or third rail for use by other trains. Regeneration has not been prevalent in railways because of the complexity and weight of the onboard equipment involved, and regeneration has generally not been economically feasible where the catenary is providing alternating current (ac) traction current because of the difficulty of matching the frequency and phase of the regenerated power to the power in the catenary. Technology is changing in this respect, and ac regeneration is emerging as a possibility where train traffic is dense enough to support the added investment.

Railway electric traction systems use either ac or direct current (dc). Third rail systems operate on dc at 600 to 700 V. Overhead systems can be found with either 1,500 or 3,000 V dc, or using ac at a range of voltages (from 11,000 to 50,000) and frequencies (16 2/3 to 60 Hz). The modern standard for overhead catenary systems is usually 25,000 V and either 50 or 60 Hz depending on the industrial standard for the country (50 Hz for most of Europe, 60 Hz for the United States and Japan).

The total energy efficiency of a diesel system is not much different from an electrically driven system—a point that many energy comparisons neglect. The percentage of the energy in diesel fuel that is actually translated into tractive effort seems low, about 26 percent, in comparison with electric drive trains, which convert 90 percent of the power received from the wire or third rail into traction. However, the percentage of the fuel consumed in a power plant that is converted into electric energy is only about 40 percent (slightly higher in modern, compound cycle plants), and electric systems lose energy due to resistance in the transmission lines and transformers. Measured by the percentage of the initial fuel's energy that is actually translated to tractive effort on the train, there is not a great deal of difference between diesel at about 26 percent efficiency and electric traction systems that deliver about 28 percent of the initial fuel energy into tractive effort at the rail.

Space Efficiency

The land area required for a double-track railway (a minimum right-of-way approximately 44 ft wide) is less than that required for a four-lane highway (at least 50 ft wide at an absolute minimum, and usually much more). Depending on assumptions about load factors, a double-track railway can carry 20,000 to 50,000 passengers in rush hour in each direction, about twice to four times the peak loading of a four-lane highway (2,200 vehicles per lane per hour). Putting a double-track railway underground is also much cheaper than the equivalent capacity by highway because rail tunnels are smaller and require much less ventilation. Overall, high-density railways can carry three to eight times the amount of traffic a highway can carry per unit of land area used, and can do so with limited and controlled environmental impact.

Urban Form

Because of their ability to handle massive numbers of people effectively, using little space and emitting little or no pollution (in the urban area, at least), railways can support the functioning of much denser urban areas than auto-based urban sprawl; in fact, the term "subway" was coined to describe the underground railway that consumes almost no surface space. Where urban congestion is important (in Bangkok there have been estimates that traffic congestion lowers the gross domestic product of Thailand by several percentage points), various rail options can have a high payoff in moving people effectively. In densely populated countries, as in most of Europe and Japan, longer-haul rail passenger service plays a similar function of moving people with minimum burden on land space and overloaded highway and air transport facilities.

Technology in transport is also on the move. Automobiles have seen advances in engine design (reducing energy and pollution), body design (reducing air drag), radial tires and tread design (reducing rolling friction), and replacement of steel with plastics and aluminum (reducing weight). As a result, auto energy mileage per gallon of fuel, which was about 13.5 miles per gallon in the United States in 1970, has improved to about 21.5 miles per gallon in 1997 (a 37% improvement), and there are vehicles now available that can produce 50 to 70 miles per gallon. Diesel engine technology has improved in parallel, as has vehicle design for both rail and bus vehicles, yielding roughly a 46 percent improvement in rail efficiency since 1970. Advances in aircraft design and size, along with improved engines, have kept pace with the surface modes. Since the basic technology in engine, air drag, and vehicle frame weight management is similar and is available to all modes, there is reason to believe that all modes will (or could) improve, but this is no convincing basis for arguing that any of the modes will dramatically improve its energy efficiency relative to the others.

CONCLUSIONS

The energy/transport relationship is complex and resists easy generalization. Against this backdrop, what useful conclusions can be drawn about the role of passenger rail services in saving energy or reducing environmental impacts from transport?

In the urban arena, energy-efficient rail passenger services can reduce air pollution and help manage urban congestion. Rail passenger services can be vital to the form and function of large cities by putting large movements of people underground or overhead without consuming undue space. This potential cannot be reached, however, unless rail services are carefully planned and managed so that they operate where needed and at high load factors. Rail's effects are likely to be localized, though, and the potential will be limited if competitive modes are not fully charged for the congestion and pollution they cause. Urban rail probably does not offer much to the effort to control greenhouse gas emissions, because of the relatively small amounts of energy involved and because of urban rail's inherently low load factors.

Rail's contribution in the intercity passenger area is less clear. Where population density is high and travel distances short, and especially where fuel prices and tolls are high and airline travel expensive, there is a need for rail passenger service that can operate efficiently. This describes conditions in Japan, and parts of Western Europe and possibly the northeast corridor in the United States, where high-density and high-speed rail services do exist. But with rail passenger services carrying less than 2 percent of the intercity traffic in the United States, the contribution of this traffic to energy-efficiency objectives is probably minimal. Where populations are high and extremely poor, and where rail tariffs are kept low, there is also a significant demand for rail passenger services, as in India, China, the CIS countries, and Egypt. It is possible that rail passenger service in these countries is making a contribution to the reduction of energy consumption and thus to control of CO_2 emissions. In the "middle" countries, including the United States and Canada (outside a few dense urban corridors), which have long distances and low population densities, it seems doubtful that rail passenger services can make a measurable contribution to energy efficiency and CO_2 reduction. In all countries it is unlikely that intercity rail passenger services will be useful in reducing localized urban air pollution.

Louis S. Thompson

See also: Air Travel; Automobile Performance; Diesel Cycle Engines; Diesel Fuel; Electric Motor Systems; Electric Vehicles; Engines; Freight Movement; Hybrid Vehicles; Locomotive Technology; Magnetic Levitation; Mass Transit; Power; Steam Engines; Tires; Traffic Flow Management; Transportation, Evolution of Energy Use and.

BIBLIOGRAPHY

Association of American Railroads. (1997). *Railroad Facts.* Washington, DC: Author.

Archondo-Callao, R., and Faiz, A. (1992). "Free Flow Vehicle Operating Costs." Washington, DC: World Bank.

Anderson, D., and Mattoon, D. (1993). *Train Operation and Energy Simulator (TOES) Technical Manual.* Chicago: Association of American Railroads.

Anderson, D.; Mattoon, D.; and Singh, S. (1992). *Revenue Service Validation of Train Operation and Energy Simulator (TOES)*, Version 2.0. Chicago: Association of American Railroads.

Association of American Railroads. (1990). *Locomotive Improvement Program, Eleventh Research Phase Final Report.* Washington, DC: Author.

Chan, Lit-Mian, and Weaver, C. (1994). "Motorcycle Emission Standards and Emission Control Technology." Washington, DC: World Bank.

Chester, A., and Harrison, R. (1987). *Vehicle Operating Costs: Evidence from Developing Countries.* Baltimore: Johns Hopkins University Press.

Commission of the European Communities. (1992). *Green Paper on the Impact of Transport on the Environment: A Community Strategy for Sustainable Mobility.* Brussels: Author.

Commission of the European Communities. (1996). *White Paper: A Strategy for Revitalizing the Community's Railways.* Brussels: Author.

Davis, S. C. (1999). *Transportation Energy Data Book,* Edition 19 of ORNL-6958. Washington, DC: U.S. Department of Energy.

DeCicco, J., and Ross, M. (1994). "Improving Automotive Efficiency." *Scientific American* 271(6):52-57.

Drish, W. F., and Singh, S. (1992). *Train Energy Model Validation Using Revenue Service Mixed Intermodal Train Data.* Chicago: Association of American Railroads.

Faiz, A.; Weaver, C. S.; and Walsh, M. P. (1996), *Air Pollution from Motor Vehicles.* Washington, DC: World Bank.

Gordon, D. (1991). *Steering a New Course: Transportation, Energy and the Environment.* Washington, DC: Island Press.

Hay, W. W. (1961). *An Introduction to Transportation Engineering.* New York: John Wiley & Sons.

Heierli, U. (1993). *Environmental Limits to Motorization Non-Motorized Transport in Developed and Developing Countries.* St. Gallen, Switzerland: SKAT, Swiss Centre for Development Cooperation in Technology and Management.

Imran, M., and Barnes, P. (1990). "Energy Demand in the Developing Countries: Prospects for the Future." Washington, DC: World Bank.

Ishiguro, M., and Akiyama, T. (1995). *Energy Demand in Five Major Asian Developing Countries.* Washington, DC: World Bank.

Levine, M. D., et al. (1991). *Energy Efficiency, Developing Nations, and Eastern Europe.* A Report to the U.S. Working Group on Global Energy Efficiency.

Martin, D. J., and Michaelis. (1992). *Research and Technology Strategy to Help Overcome the Environmental Problems in Relation to Transport: Global Pollution Study.* Luxembourg: EEC.

Noll, S. A. (1982). *Transportation and Energy Conservation in Developing Countries.* Washington, DC: Resources for the Future.

Organisation for Economic Cooperation and Development. (1985). *Energy Savings and Road Traffic Management.* Paris: Author.

Redsell, M.; Lucas, G.; and Ashford, N. (1988). *Comparison of on-road Fuel Consumption for Diesel and Petrol Cars.* Berkshire, Eng.: Transport and Road Research Laboratory.

Rose, A. B. (1979). *Energy Intensity and Related Parameters of Selected Transportation Modes: Passenger Movement.* Oak Ridge, TN: Oak Ridge National Laboratory.

Schipper, L., and Meyers, S. (1992). *Energy Efficiency and Human Activity: Past Trends, Future Prospects.* New York: Cambridge University Press.

Schipper, L., and Marie-Lilliu, C. (1998), *Transportation and CO2 Emissions: Flexing the Link.* Washington, DC: World Bank.

Thompson, L. S., and Fraser, J. (1995). *Energy Use in the Transport Sector.* Washington, DC: World Bank.

United Nations. Economic Commission for Europe. (1983). *An Efficient Energy Future: Prospect for Europe and North America.* Boston: Author.

U.S. Congress, Office of Technology Assessment. (1991). *Improving Automobile Fuel Economy: New Standards and Approaches.* OTA-E-504. Washington, DC: U.S. Government Printing Office.

U.S. Department of Transportation. (1990). *Compendium of National Urban Mass Transportation Statistics from the 1987 Section 15 Report.* Washington, DC: Urban Mass Transportation Administration.

U.S. Department of Transportation. (1973). *The Effect of Speed on Automobile Gasoline Consumption Rates.* Washington, DC: Federal Highway Administration.

U.S. Department of Transportation. (1993). *Transportation Implications of Telecommuting.* Washington, DC: Author.

U.S. Department of Transportation. (1997). *National Transport Statistics.* Washington, DC: Bureau of Transportation Statistics.

U.S. Department of Transportation. (1991). *Rail vs. Truck Efficiency: The Relative Fuel Efficiency of Truck Competitive Rail Freight and Truck Operations Compared in a Range of Corridors.* Washington, DC: Federal Railway Administration.

Wagner, F. A., and Institute of Transportation Engineers. (1980). *Energy Impacts of Urban Transportation Improvements.* Washington, DC: Institute of Transportation Engineers.

World Bank. (1988). *Road Deterioration in Developing Countries: Causes and Remedies.* Washington, DC: World Bank.

World Bank. (1993). *Energy Efficiency and Conservation in the Developing World.* Washington, DC: World Bank.

World Bank. (1996). *Sustainable Transport: Priorities for Policy Reform.* Washington, DC: World Bank.

RANKINE, WILLIAM JOHN MACQUORN (1820–1872)

William Rankine has been credited with many things derived from his brilliant career, with perhaps the most unique being the transition of his empirical work into scientific theories published for the benefit of engineering students. He is considered the author of the modern philosophy of the steam engine and also the greatest among all founders of and contributors to the science of thermodynamics.

Born in Edinburgh, Scotland, on July 5, 1820, Rankine received most of his education from his father, David Rankine, a civil engineer, plus various private tutors. The elder Rankine worked as superintendent for the Edinburgh and Dalkeith Railway, imparting to his son a love of steam engines. On top of this fine education came two inspiring years at the University of Edinburgh, which helped launch his career in civil engineering, even though he left without earning a degree. He spent a year assisting his father, then worked in Ireland for four years on projects including railroads and hydraulics. Returning to Scotland, he apprenticed under Sir John Benjamin MacNeil, a respected civil engineer of his time. Rankine remained in this profession until the late 1840s before switching to the practice of mathematical physics.

At the age of thirty-five, Rankine was appointed by the Queen's Commission to the Chair of Civil Engineering and Mechanics at the University of Glasgow in Scotland. This Regius Chair was established in 1855 with great distinction at the fourth oldest university in Great Britain. Despite the great achievements of British inventors, most self–educated, they were often deprived of an appropriate claim to true professional recognition in the areas of architecture and engineering. As a professor, Rankine argued passionately with his University commissioners for the establishment of a diploma in engineering science, later adding a Bsc in science. These were just a few of the achievements that earned him accolades as a pioneer in engineering education.

Rankine's most famous textbook, *A Manual of the Steam Engine and Other Prime Movers* (1859), set an engineering standard for years to come, delivering systematic instruction to upcoming engineers. This publication was credited with affording inventors a new basis beyond their experimentation to advance the development of the steam engine. Other books Rankine authored include *A Manual of Applied Mechanics* (1858); *A Manual of Civil Engineering* (1862); and *A Manual of Machinery and Millwork* (1869). He also published numerous practical manuals on civil and mechanical engineering and scientific tables for construction, carpentry, architecture and surveying. Somehow in his spare time, he found time to write for entertainment, including *Songs and Fables* (1874) and *A Memoir of John Elder* (1871). His research led to such renowned legacies as the Rankine cycle, for ideal operation of a steam engine, as well as his Rankine Tables for cycle efficiencies (comparing performances of steam engines and steam turbines), the construction strength of columns, and his Rankine absolute temperature scale. Many of his tables remain in use today.

Moving into other fields of practical applications, Rankine developed improved strength for iron rails by conducting fatigue testing of existing railroad ties. He was a pioneer in promoting a theory of open water sailing vessels that debunked a simple but common belief of his time—that metal ships could not be made to float. Through his successful research, the size, length, and safety of ships was significantly improved. Through his knowledge of thermodynamics, he played a key role in the advancements in refrigeration equipment. His study of soil mechanics (earth pressures) led to improved retaining walls.

Rankine's works placed him among the greatest of many famous scholars at the University of Glasgow—from Adam Smith in modern economic science, to James Watt with his invention of the double acting steam engine, to Rankine's peer in thermodynamics, William Thomson (Lord Kelvin), after whom the absolute scale of temperature was named. Before passing away on Christmas Eve, 1872, in Glasgow, Rankine served as president of the Institute of Engineers in Scotland and honorary member of the Literary and Philosophical Society of Manchester.

Making a statement on the importance of recording and teaching the history of technology, Rankine began his *A Manual of the Steam Engine and Other Prime Movers* as follows: "Nations are wrongly accused of having, in the most ancient of times, honored and remembered their conquerors and tyrants only, and of having neglected and forgotten their benefactors, the inventors of the useful arts. On the contrary, the want of authentic records of those benefactors of mankind has arisen from the blind excess of admiration, which led the heathen nations of remote antiquity to treat their memory with divine honors, so that their real history has been lost amongst the fables of mythology."

Dennis R. Diehl

See also: Steam Engines; Thermodynamics.

BIBLIOGRAPHY

Cardwell, D. S. L (1971). *From Watt to Clausius: The Rise of Thermodynamics in the Early Industrial Age*. Ithaca, NY: Cornell University Press.

Rankine, W. J. M. (1859). *Steam Engine and Other Prime Movers*. London and Glasgow: Richard Griffin & Company

Thurston, R. H. (1939). *A History of the Growth of the Steam Engine*. Ithaca, NY: Cornell University Press

RECYCLING

See: Materials

REFINERIES

The process of refining begins with the delivery of crude oil—a mixture of thousands of different hydrocarbon compounds that must be separated into fractions of varying boiling ranges in order to be used or converted into products such as gasolines, diesel fuels, and so on. Crude oils from different sources vary widely in terms of density (gravity), impurities (sulfur, nitrogen, oxygen, and trace metals), and kinds of hydrocarbon compounds (paraffins and aromatics). In general, the heavier the crude oils, the less transportation fuel (gasoline, diesel fuel and jet fuel) boiling-range components are contained in the crudes. The average crude oil charged to U.S. refineries in 1998 contained about 20 percent material in the gasoline boiling range and about 25 percent in the middle distillate (diesel and jet fuel) boiling range. The U.S. market requirements are about 50 percent gasoline and 35 percent middle distillate fuels per barrel of crude oil charged to U.S. refineries.

CRUDE OILS

Crude oils are classified by major hydrocarbon type (paraffinic, intermediate, or naphthenic), gravity (light or heavy), and major impurities—high sulfur (sour), low sulfur (sweet), high metals, etc. Everything else being equal, the heavier the oil the lower the value of the oil, because a smaller percentage of that particular crude is in the transportation fuel boiling range. The boiling ranges of the heavier fractions can be reduced, but the additional processing is costly, adding anywhere from fractions of a cent to several cents per gallon. The same is true of the sulfur, nitrogen, oxygen, and metal contents of crude oils. The impurities have to be removed by reaction with hydrogen, and hydrogen processing is very expensive.

Paraffinic crude oils are defined as those crude oils containing waxes, naphthenic crude oils as those containing asphalts, and intermediate crude oils those containing both waxes and asphalts. Usually the residues—material boiling above 1050°F (565°C)—of naphthenic crude oils make good asphalts, but the intermediate crude oils can contain enough wax so that the asphaltic part may not harden properly, and therefore they cannot be used to produce specification asphalts.

Refinery processes are available to reduce the boiling points of components of the crude oils that boil higher than the transportation fuels by cracking the large high-boiling-point molecules into smaller molecules that boil in the desired transportation fuel boiling range. Lighter materials can also be converted to transportation fuels (usually gasoline) by processes such as alkylation or polymerization. There are also processes to improve the quality of the transportation fuel products because the crude oil components with the proper boiling range do not have the other characteristics required. These include octane and cetane numbers for gasoline and diesel fuel, respectively, and low sulfur and nitrogen contents for gasoline, diesel, and jet fuels.

Over 50 percent of the crude oils used in the United States are imported, because U.S. crude oil production has passed its peak and has been declining in recent years. Also, other countries (Mexico, Saudi Arabia, and Venezuela) have been buying full or part ownership into U.S. refineries in order to ensure there will be a market for their heavy, high-sulfur crude oils. These efforts to ensure a market for their imported crude oils have resulted in more equipment being installed to crack the very heavy portions of the crude oils into as much transportation fuel boiling-range material as is economically feasible. As of January 1, 1999, U.S. refineries had the ability to crack more than 50 percent of the crude oil charged to the refineries. As the demands for heavier products decreases, additional cracking equipment will be added to some U.S. refineries to increase the cracking ability to 60 percent or higher. Also, as the demand for cleaner burning transportation fuels with lower concentrations of aromatics and olefins increases, additional process equipment will be added to produce larger volumes of alkylates.

Most of the U.S. refineries are concentrated along the Gulf Coast, because most imported crude oils come from Venezuela, Mexico and Saudi Arabia and a large amount of U.S. domestic crude oil is produced near the Gulf Coast in Texas, Louisiana, Mississippi and Oklahoma. California also has a concentration of refineries that process crude oil from deposits in California and Alaska, in order to avoid the costs of transferring crude oils and products over or around the Rocky Mountains while also satisfying the large demands for transportation fuels in the West. These logistics explain in large part why over 54 percent of U.S. refining capacity is in the states of Texas, Louisiana, and California. Texas contains the most with more than 26 percent, Louisiana is second with almost 16 percent, and California is third with more than 12 percent. Environmental regulations, resistance to "having a refinery in my neighborhood," and economics have stopped the construction of completely new refineries in the U.S. since the 1970s. It is improbable that any will be built in the foreseeable future. Since the mid-1970s, all new complete refineries have been built overseas.

Because transportation fuels vary in boiling range and volatility depending upon the type of engine for which they are suitable, refineries are designed to be flexible to meet changes in feedstocks and seasonal variations in transportation demands. Typically, spark ignition engines require gasolines and compression ignition engines (diesels) require diesel fuels. Jet fuel (similar to kerosene) is used for most commercial and military aircraft gas turbines. Jet fuel demand fluctuates widely depending on the economy and whether the United States is involved in military operations. Most of the refineries in the United States are fuel-type refineries; that is, more than 90 percent of the crude oil feedstocks leave the refineries as transportation or heating fuels. Although there are other profitable products, they are small in volume compared to the fuels and it is necessary for the refinery to make profits on its transportation fuels in order to stay in operation. In 1999, there were 161 refineries in the United States, owned by eighty-three companies, and having a capacity of over 16,000,000 barrels per day (672,000,000 gallons per day). Worldwide, there are over 700 refineries with a total crude oil feed rate of more than 76,000,000 barrels per day. In the United States, for refinery crude oil feedstock capacities and crude oil purchases throughout the world, the standard unit describing amount is the barrel. The term *barrel* is outmoded and refers to the 42 U.S. gallon capacity British

Starter Material	Boiling Point or Range °F	°C	Product	Uses
Methane & Ethane	-259°F -132°F	161°C - 91°C	Fuel gas	Refinery fuel
Propane	- 44°F	- 42°C	LPG	Heating
i-Butane	+11°F	-12.2°C		Feed to alkylation unit
n-Butane	31°F	-0.5°C	LPG, lighter fluid, gasoline	Gasoline blending or to LPG
Propylene and butylenes				Feed to alkylation or polymerization units
Light naphthas	80-180°F	27-82°C	Light gasoline	Gasoline blending or isomerization unit
Heavy naphthas	180-380°F	82-193°C	Heavy gasoline	Catalytic unit feed
Kerosine	380-520°F	193-271°C	Kerosine	Jet fuel
Atmospheric gas oil	520-650°F	271-343°C	Light gas oil	Blending into diesel fuels and home heating oils
Vacuum gas oils	650-1050°F	343-566°C		Feeds to FCCU and hydrocracker
Vacuum resid	1050°F+	566°C+	Vacuum tower bottoms (VRC)	Heavy and bunker fuel oils, asphalts

Table 1.
Characteristics of Some Typical Refineries

wooden salted-fish barrel used to haul crude oil to the refineries in horsedrawn wagons in the later part of the nineteenth century. Most of the world uses either metric tons or cubic meters as the unit of measurement.

REFINING SYSTEMS

All refineries are similar and contain many of the same basic types of equipment, but there are very few that contain exactly the same configuration, due to differences in the crude oils used as feedstocks and the products and product distributions needed. In addition, most refinery processing is very expensive, and many of the smaller companies do not have the necessary capital to install the latest highly capital-intensive technology such as alkylation units and hydrocrackers. Instead they use less costly units such as fluid catalytic cracking units (FCC) and polymerization units. However, except for the least complicated refineries, topping and hydroskimming refineries, they all have processes to change boiling range and improve quality.

A small number of refineries produce lubricating oils, but the total volumes produced are only about 3 percent of crude oil charged to U.S. refineries. Processes to make lubricating oil blending stocks are expensive from both capital and operating costs viewpoints. Nevertheless, the business is still profitable because lubes demand high retail prices. Older processes use solvent extraction to improve the temperature-viscosity characteristics (viscosity index) and low-temperature refrigeration tech-nologies, such as dewaxing, to lower the pour points of the lube oil blending stocks. Newer plants use selective hydrocracking and high-pressure hydrotreating to produce high-quality lube oil blending stocks. The newer equipment is so expensive that many of the new facilities are joint projects of two or more companies in order to gain enough volume throughout to justify the high capital investments.

Some refineries also install processes to produce petrochemical feedstocks. However, from a volume viewpoint, it is not a large part of refinery income. Even though profit margins can be high, most refineries restrict their capital investments to those

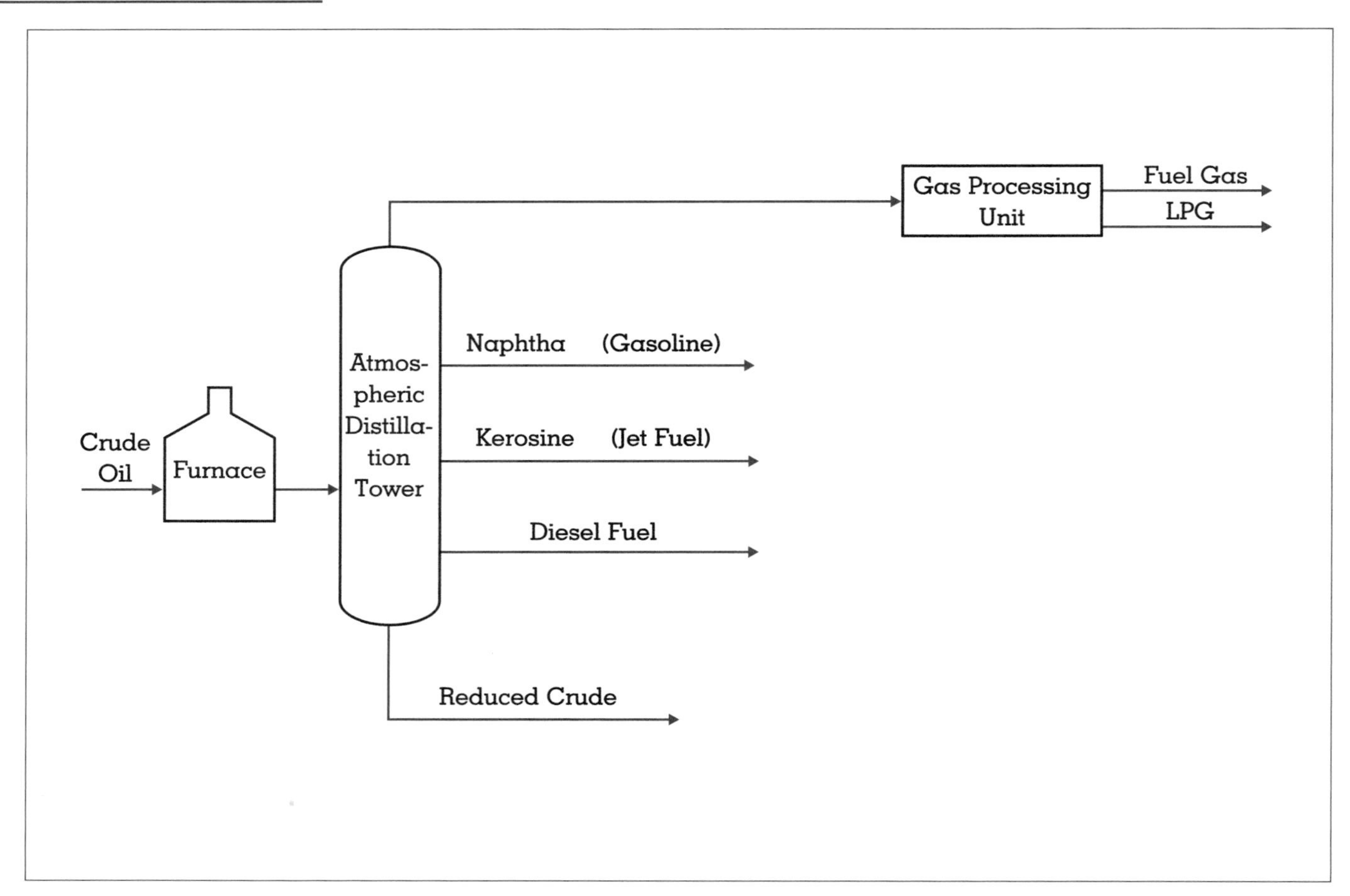

Figure 1.
Diagram of a Topping Refinery.

needed to produce transportation fuels. The most common petrochemical feedstocks produced are olefins (chiefly ethylene and propylene) and aromatics (benzene, methylbenzene [toluene], ethylbenzene, and xylenes [BTX compounds]). BTX compounds are produced in a catalytic reformer by operating at high reactor temperatures to produce 102 to 104 research octane products containing over 75 percent aromatics. Using solvent extraction, fractionation, and adsorption technologies the products are separated into individual aromatic compounds for sale as petrochemical feedstocks.

Topping and Hydroskimmimg Refineries

The *topping* refinery (see Figure 1) is the simplest type of refinery, and its only processing unit is an atmospheric distillation unit which *tops* the crude by removing the lower–boiling components in the gasoline, diesel fuel, and diesel fuel boiling ranges (those boiling up to 650°F [343°C]). The bottoms product from the atmospheric distillation unit (atmospheric resid, atmospheric reduced crude, or ARC) is sold as a heavy fuel oil or as a feedstock to more complex refineries containing conversion equipment that can convert the high–molecular–weight components into transportation fuel boiling–range products. Until recently the topped components could easily be converted into retail products by blending additives (such as lead compounds) into them. When lead compounds were banned from gasolines in the United States, it was necessary to add equipment to improve the octanes of the gasoline boiling–range components. Catalytic reforming units increased the octanes of the naphthas by converting paraffin molecules into aromatic compounds plus hydrogen. As a result, a topping refinery producing unleaded gasoline is a hydroskimming refinery.

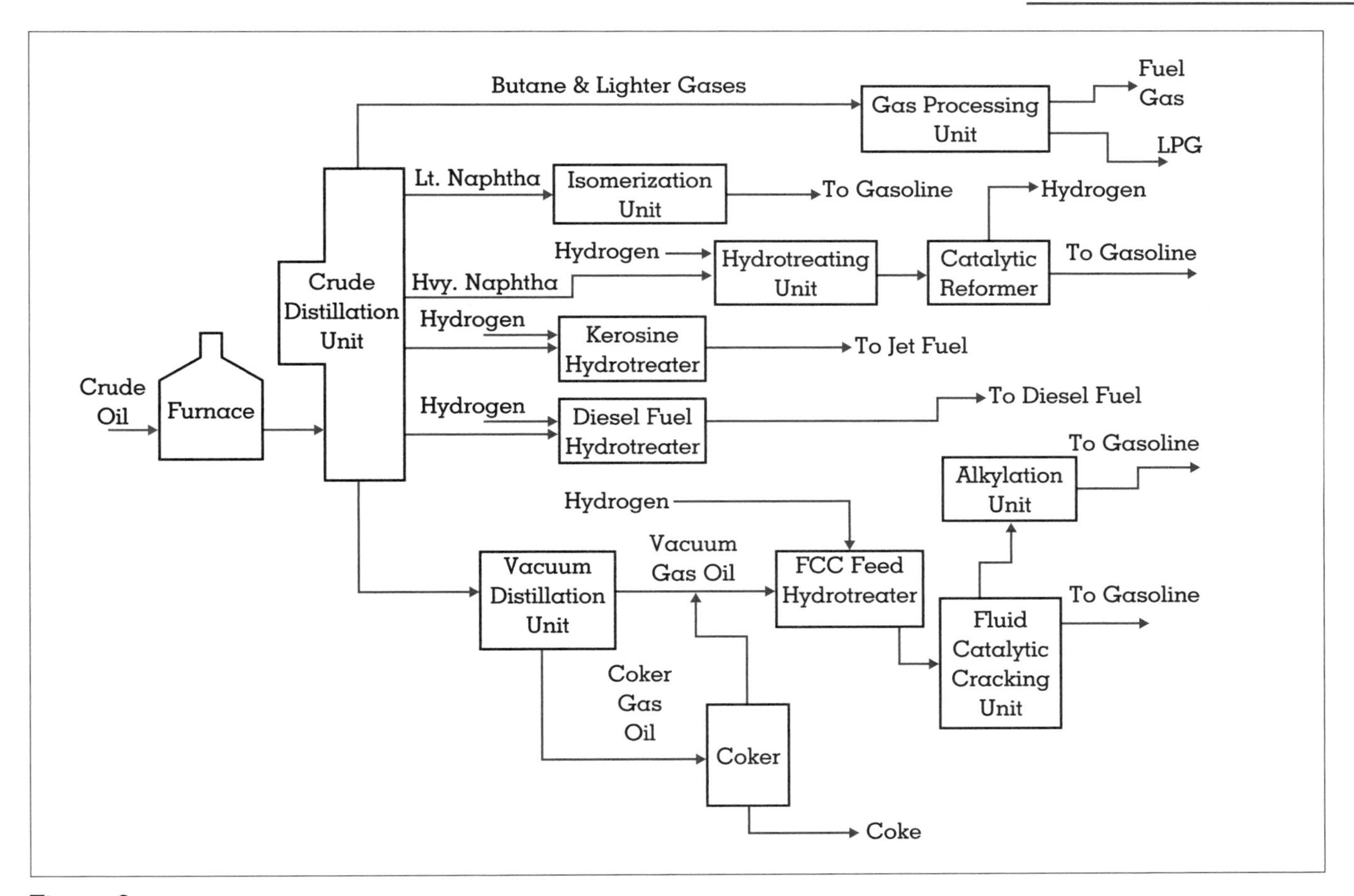

Figure 2.
Integrated Refinery.

Conversion and Integrated Fuels Refineries

Although a small amount of crude oils charged (less than 5 percent) to U.S. refineries is converted to specialty products, the remainder is either used in the refinery as a fuel or is processed into asphalt or fuel products including motor fuels and heating oils. The market for heavy fuel oils—those with boiling ranges above 650°F (343°C)—in the United States is less than 7 percent of the crude oil charged to U.S. refineries. It is therefore necessary to change the boiling ranges of the heavier portions of the crude to meet the product demands of the transportation fuels. Refineries with the necessary boiling-range conversion equipment are integrated or conversion refineries (see figure 2). The majority of the conversion equipment is to reduce boiling point—these are cracking units. When the molecules are cracked to reduce boiling range, secondary reactions produce some light hydrocarbons with boiling points too low for typical transportation fuels. Conversion equipment is also included to increase the boiling points of these components to permit blending into gasolines. As a result, about 70–80 percent of the crude oil charged to U.S. refineries is converted into gasolines, diesel and jet fuels, and home heating oils.

REFINERY PRODUCTS

Refineries produce more than 2,000 products, but most of these are very similar and differ in only a few specifications. The main products, with respect to volume and income, are liquefied petroleum gases (LPG), gasolines, diesel fuels, jet fuels, home heating oils (No. 1 and No. 2), and heavy heating oils (No. 4, No. 5, No. 6, and bunker fuel oil). Some refineries also produce asphalts and petroleum coke.

LPGs are gases that are liquefied at high pressures—up to 250 psig (17 bars)—and consist mainly

Generic name	*Characteristics*
Naphtha	Any liquid hydrocarbon material boiling in the gasoline boiling range
Middle distillate	Liquid hydrocarbons boiling in the jet fuel, diesel fuel, and home heating oil ranges
Gas oil	Hydrocarbons boiling at temperatures above 400°F (205°C) that have been vaporized and reliquefied
Coke	A solid residue from cracking operations, mostly carbon but with a small amount of high-molecular-weight hydrocarbons

Table 2.
Generic Names of Selected Refinery Products.

of mixtures of propane and butanes. These are sold in pressurized cylinders and are used for heating. Their selling price is related to the heating value.

Gasolines are the largest volume refinery products. In 1998, for instance, more than 330,000,000 gallons were consumed each day in the United States. This amounts to more than one million tons per day. The best known specifications for gasoline are octane numbers and Reid vapor pressure (RVP). The octanes are numbers determined by laboratory tests and can be related to performance of gasolines at normal road operating conditions. Octane numbers measure the resistance of the gasoline to self–ignition (knocking) and are determined by comparison with reference fuels. Self-ignition causes a decrease in engine efficiency and power, and when it occurs there are noticeable reductions in mileage and acceleration. Research octane number can be correlated with city driving performance (RON) and motor octane number (MON) with highway driving performance. Federal regulations require that the arithmetic average of these octanes ([RON + MON]/2) be posted on the pumps dispensing the gasolines. In 1998, about 70 percent of the gasolines sold in the United States were regular grade, and 30 percent premium grades. Regular gasoline posted octanes are 87 for lower elevations and 85 or 86 at higher elevations (above 3,000 feet). The octane requirement of an engine decreases with elevation. Refiners try to produce gasolines with the RON and MON as close together as possible, so their gasolines perform equally well for both city and highway driving but it is not practical. Refinery processing equipment has major effects on the sensitivity to driving conditions of the gasolines produced. Typically the RON and MON of gasolines will differ by six to twelve octane numbers (RON–MON). For gasolines the RON is always higher than the MON.

Reid vapor pressure determines ease of starting, engine warmup time, and tendency for vapor lock to occur. Vapor pressure is regulated by the addition of n–butane to the gasoline mixture. During summer months, EPA regulations limit the maximum vapor pressure of gasolines to as low as 7.2 psi in the southern U.S. and Rocky Mountain areas in order to reduce hydrocarbon vapor emissions into the atmosphere. During the cold months of winter, RVP is increased in northern areas to as high as 14 psi to make starting of cold engines possible and for the engines to warm up quickly. Normal butane not only increases RVP, but it also acts as an octane booster. During the summer months when less n–butane can be blended into gasolines, additional amounts of some other, more costly, octane booster must be added to compensate.

The cetane number determined in the laboratory by comparison with reference fuels is of value in that it is an indication of the ease of self–ignition. Diesel engines do not use a spark to ignite the fuel–air mixture in the cylinder but rely on the diesel fuel to self–ignite when injected with a high–pressure injection pump into the hot compressed air in the cylinder. The quicker the fuel self–ignites, the more efficient the engine. Many small refineries do not have cetane number determination equipment, so they substitute a cetane index equivalent to the cetane number and calculated from the fuel's gravity and average boiling point. The Environmental Protection Agency (EPA) requires that highway diesel fuels sold in the United States must have a cetane index (CI) of at least 40 and a sulfur content of less that 0.05% by weight (500 ppm). During 1998, almost 150 million gallons of diesel fuels and 60 million gallons of jet fuel were used in the United States per day.

REFINERY OPERATIONS

Refineries are unique businesses in that they operate on small profits per unit volume of product but produce very large volumes of product. This means that a regulatory adjustment (such as sulphur content) that creates a small change in operating costs per gallon of product (fractions of a cent) can mean the difference between profit and loss. The industry therefore strives to optimize refinery operations to lower costs and to improve efficiency.

One of the significant problems for refinery is the small amount of water emulsified in the crude oils coming into refineries. This water contains inorganic salts that decompose in the process units and form compounds that are corrosive to equipment and poison catalysts used in the processes. To prevent these costly problems, the first process in the refinery, desalting, reduces the water and salt content of the crude oils. Following desalting, the crude is sent to the distillation units (crude stills) to be separated into fractions by boiling range. The first of the crude stills, the atmospheric distillation unit, vaporizes and separates those components boiling below about 650°F (345°C) into four general categories; butane and lighter (methane, ethane, and propane) gases, light and heavy naphtha fractions, kerosene, and atmospheric gas oil. (Gas oil is a generic name for any hydrocarbon boiling above the end boiling point of gasoline that has been vaporized and condensed into a liquid.) The unvaporized material boiling above 650°F is removed from the bottom of the distillation column as vacuum resid or atmospheric reduced crude (ARC). The naphtha fractions are in the gasoline boiling range but, before blending into gasoline, their quality has to be improved by increasing their octanes and reducing the sulfur contents. The kerosene is hydrotreated (reacted with hydrogen) to reduce the sulfur content and blended into jet fuel. And the atmospheric gas oil is hydrotreated to reduce the sulfur content and is blended into diesel fuels and home heating oils (No. 1 and No. 2 heating oils).

The atmospheric reduced crude is the feedstock for the vacuum distillation unit. To prevent thermal decomposition (cracking) of the higher boiling point hydrocarbons in the crude oil, the pressure in the vacuum distillation fractionation column is reduced to about one-twentieth of an atmosphere absolute (one atmosphere pressure is 14.7 psia or 760 mm Hg). This effectively reduces the boiling points of the hydrocarbons several hundred degrees Fahrenheit. The components boiling below about 1050°F (565°C) are vaporized and removed as vacuum gas oils, the unvaporized material is removed from the bottom of the unit as vacuum reduced crude (vacuum resid or VRC). The vacuum gas oils are sent as feedstocks to either fluid catalytic cracking units or to hydrocrackers. The fluid catalytic cracking units crack the vacuum gas oils into gasoline blending stocks (up to 70 percent by volume on feed) and gas oil blending stocks for diesel fuels and home heating oils. The gasoline blending stocks have octane numbers in the high eighties and, if the sulfur contents are sufficiently low, can be blended directly into gasolines. The gas oil blending stocks have low cetane numbers and poor burning characteristics that limit the amounts that can be blended into diesel fuels and home heating oils.

If the crude oils from which the vacuum reduced crudes are made have the proper qualities (low wax and high asphaltenes), the vacuum-tower bottoms can be blended into asphalt products and sold to pave roads and make roofing materials and sealants. The vacuum reduced crudes also can be sold as heavy fuel oils or used as feedstocks to a severe thermal cracking unit (coker) to convert them into lower boiling hydrocarbons plus solid coke residues. The coker converts the vacuum tower bottoms into feedstocks for other refinery processing units to make gasolines, light heating oils, and, if used as hydrocracking unit feedstocks, produce gasoline, diesel fuel, and jet fuel blending stocks. Historically, the heavy fuel oils have sold, on a volume basis, for about 70 percent of the prices paid for the crude oils from which they were produced. When crude oil is selling for twenty dollars per barrel, heavy fuel oil will be selling for about fourteen dollars per barrel. This discounting is necessary because the amount of heavy fuel oil available exceeds demand. The alternate cost of processing the vacuum tower bottoms in the refinery is much higher than the cost of processing crude oil. The vacuum reduced crudes sell at a loss, or they are converted into salable products by processing the liquid products from a coker in other refinery processing units. The coke produced by the coker can be sold as a fuel, to make anodes for aluminum production, or to make electrodes for electric steel furnaces.

Fluid catalytic cracking units (FCC or FCCU) are the major processing units to reduce boiling ranges of those crude oil components that have boiling points higher than the final boiling points of the transportation fuels—typically above 650°F (343°C). These

Imperial Oil refinery in Edmonton, Alberta, Canada. (Jack Fields/Corbis)

units circulate finely divided (about the size of very fine sand particles) silica-alumina-based solid catalyst from the reactor to the regenerator and back. The reactor operates at temperatures between 950 and 1,100°F (510–595°C) to crack the large molecules in the gas oil feed selectively into smaller molecules boiling in the gasoline boiling range of 80–400°F (27–205°C). When the cracking takes place in the reactor, coke is also produced and is laid down on the surface of the catalyst. This coke layer covers the active cracking sites on the catalyst, and the cracking activity of the catalyst is reduced to a low level. Conveying the catalyst to the regenerator, where the coke is burned off with air, restores the catalyst activity. This cycle is very efficient: the heat generated by burning the coke off the spent catalyst in the regenerator supplies all of the heat needed to operate the unit. The regenerated catalyst leaving the regenerator has had its activity restored to its original value and at a temperature of about 1,300°F (705°C) is returned to the riser–reactor and mixed with fresh feed to continue the cracking reactions. By controlling the reactor temperature, the conversion of the feed to gasoline boiling range components can be maximized during the summer when gasoline demand is at its highest. In practice, up to 70 percent by volume of the product is in this boiling range. The product is rich in aromatic and olefinic compounds, which makes the high quality FCC gasoline very suitable for blending into the final refinery gasoline products. The remainder of the feed is converted into butane, lighter gases, and catalytic gas oils (typically called cycle gas oils) that are blended into diesel fuels or home heating oils, or used as feedstocks to hydrocrackers. During the winter, when gasoline demand is lower and home heating oil is needed, the severity of the operation is lowered by reducing the reactor temperature, and the yields of the cycle gas oils stocks used for blending into home heating oils and diesel fuels is more than doubled. The gasoline boiling-range products are reduced by one-quarter to one-third. The butane and lighter gases are separated into components in the FCC gas recovery plant. The propylenes, butylenes, and isobutane are used as feedstocks to alkylation units to make alkylate gasoline blending stocks. Alkylates have blending octanes in the nineties and have low sensitivities to driving conditions (they perform equally well for highway and city driving with RON and MON within two numbers of each other). The hydrocarbons in alkylates are also less environmentally damaging and show less health effects than olefins or aromatics. For those refineries without alkylation units, the propylenes can be used as feedstocks to polymerization units to produce polymer gasoline blending stocks. The polymer gasoline has a blending octane of approximately 90 but performs much more poorly under highway driving conditions than in the city (about sixteen numbers difference between the RON and MON).

Some refineries have hydrocracking units in addition to the FCC unit. Hydrocracking units operate at high hydrogen pressures and temperatures and are constructed of high-alloy steels. This makes them very expensive to build and operate. They are able to use feedstocks containing high concentrations of aromatics and olefins and produce jet fuel and diesel fuel products as well as naphthas for upgrading and blending into gasolines. The FCC units operate most efficiently with paraffinic feedstocks and produce high yields of high-octane gasoline blending stocks. Hydrocrackers give higher yields than FCC units, but the naphtha products have octanes in the seventies and the heavy naphtha fraction—180–380°F (82–195°C)—must be sent to the catalytic reformer to improve its octane. It is possible to make specification jet and diesel fuels as products from the hydrocracker.

In any cracking process, secondary reactions produce light hydrocarbons whose boiling points are too low to blend into gasolines. These include both olefins and paraffins. Olefins contain at least one double bond in each molecule and are very reactive whereas paraffin molecules contain only single bonds and are considered unlikely to undergo chemical reactions. Some of the light hydrocarbon olefins produced can be reacted with each other (polymerization) or with isobutane (alkylation) to produce high–octane hydrocarbons in the gasoline boiling. The olefins used are propylenes and butylenes; ethylene is also produced from cracking operations but is not used in refinery processing.

Crude oil components with boiling ranges in the gasoline blending range have low octanes, and the octanes must be increased in order to make a product suitable for modern automobiles. For light naphthas containing molecules with five and six carbons (pentanes and hexanes), octanes can be improved from thirteen to twenty numbers on the average by isomerization processes. These processes convert normal paraffins into isoparaffins that have more compact molecules with higher octane numbers. The heavy naphthas with hydrocarbons containing from seven to ten carbon atoms, have their octanes increased into the nineties by converting paraffin molecules into aromatic molecules by reforming the molecules into ring structures and stripping off hydrogen. This process is catalytic reforming. Catalytic reforming is the only refinery process in which the operator has any significant control over the octane of the product. The more severe the operation (the higher the temperature in the reactor), the greater the percentage of aromatics and the higher the octanes of the products—but the smaller the volume of products produced. Usually the units are operated to produce products (reformate) with research octane numbers (RON) in the range of 96 to 102 to be blended with other naphthas to give the desired quality (pool octanes) sold to consumers.

Other refinery processes used to improve product quality include hydrotreating (hydrodesulfurization, hydrodenitrogenation, and demetallization), sweetening, and visbreaking. Hydrotreating reduces impurity contents of all of the impurities except metals by reacting hydrogen with the impurity to produce low-boiling compounds that are gases and can be separated from the liquid material. These reactions are carried out at high temperatures and pressures in reactors filled with catalysts containing metals that promote the reaction of hydrogen with the impurity. Hydrogen reacts with sulfur to form hydrogen sulfide, with nitrogen to produce ammonia, and with the molecules containing metals to deposit the metals on the catalyst supports. The processes operate at temperatures between 600°F (315°C) and 750°F (400°C) and pressures from 400 psig (27 bars) to 1600 psig (110 bars). Because these processes operate with hydrogen at high temperatures, the equipment is made of very expensive high–alloy steels. Energy and hydrogen costs result in high operating costs, much higher per barrel of feed than the FCC unit.

Refiners use sweetening processes to remove mercaptans that give a very unpleasant odor to gasolines and middle distillates (the skunk uses mercaptans to protect itself). This is done by washing the hydrocarbon stream with a caustic solution followed by a wash with water to remove the caustic.

Visbreaking is a mild thermal cracking process that reduces the viscosity of heavy fuel oils and reduces the amount of low-viscosity blending stocks that must be added to the heavy residuals to meet viscosity specifications of the specific heavy fuel oil. The amount of heavy fuel oil production by a refinery is reduced by 20–30 percent if a visbreaker is used. The refinery profitability is improved with visbreaker operation, because heavy fuel oils are low value products.

Small refineries usually produce the products in greatest demand, such as gasolines, jet and diesel fuels, and heavy fuel oils. Larger refineries produce a much broader slate of products because, with much higher feed rates, they can make economical volumes of small volume products. In addition, environmental requirements have made it much more difficult for small refineries because of the large capital investments required for the equipment to reduce impurities. Published economic studies indicate that refineries of less than 80,000 barrels per day of capacity do not produce sufficient income to justify the expensive equipment necessary for reformulated fuel production. Having fewer large refineries increases the transportation costs of getting the products to the necessary markets, but the increased size gives economies of scale. There are always fixed costs of labor and management that are independent of equipment size, and the incremental costs of higher charge rates are much less per unit volume than the total of fixed and operating costs. Depreciation costs per unit volume of throughput also decrease with the size of the equipment. In 1998, U.S. refineries operated at above 95 percent of rated charge capacity. This is much higher than it has been in the past. Even so, they have not been able to keep up with domestic demand, and imports of finished products have been increasing.

Operating costs vary a great deal from one refinery to another. Factors include the types of processes and equipment, the amount of the crudes that have to be cracked and alkylated to achieve the products desired, the degree of hydrogenation needed to meet product and environmental specifications, and the complexities and locations of the refineries. Energy is the single largest component of operating cost, so the costs of crude oils are major factors. For a simple hydroskimming refinery, the energy needed to process the crude oil can be as low as 2 percent of that in the crude oil, while for a very complex refinery with a large number of hydrogenation processes, it can be 8 percent or more of that in the crude oil.

SAFETY AND ENVIRONMENTAL CONCERNS

The increasingly stringent environmental quality improvements for transportation fuels as specified by the EPA is putting constraints on present processing technologies. Process equipment needed for the newer technologies to meet future environmental restrictions is very costly and requires long lead times to design and build. Since the removal of low-cost lead compounds, used until the 1970s to increase the octanes of gasolines, refineries have been providing the replacement octane improvement needed by increasing the high-octane aromatic and olefin content of gasolines and also the amount of high-octane blending compounds containing oxygen (alcohols and ethers). The EPA is reducing the maximum aromatics and olefins content of reformulated gasolines for health and pollution reasons. There also is concern over the appearance of ethers in ground waters because of possible health effects, and it may be necessary to restrict their usage in gasolines. There is not enough ethanol available today to replace these components while providing the volumes of high-octane gasolines needed (over one million tons per day of gasolines used in the United States in 1998). Other ways must be used to provide the high octanes needed and to reduce degradation of the environment.

The United States is unique among the major countries in that supply and demand has determined price structures in the petroleum industry. Today, even though the products are much better than fifty years ago, the before-tax retail prices of gasolines, diesel fuels, and heating oils are much less on a constant-value dollar basis than they ever have been before. Even with the federal and state taxes included, the retail prices on a constant-value dollar basis are equivalent to those paid during the Depression years. In Europe, for example, taxes account for up to 80 percent of the costs of motor fuels that sell at retail for three to four times as much as they do in the United States.

U.S. production of crude oils has already peaked, and it is predicted that world production of crude oils will reach its maximum between 2010 and 2030. This does not mean that an adequate supply of crude oil will not be available, but supply and demand will create price structures that will make other sources of fuel more competitive. Global warming may also impose restrictions on the use of hydrocarbon fuels but, at the present time, there are no alternatives that do not require long lead times and very expensive and time consuming periods for building plants to produce the alternative fuels. If the alternatives cannot be marketed in existing petroleum facilities, it will also require that broad systems be built to market the fuels. It is difficult to conceive of the efforts that must be made to replace gasoline in the United States. Methanol has been mentioned as a replacement, but if the total annual production of methanol in the United States in 1998 was used in automobiles instead of gasoline, it would only be enough to operate them less than a week. During 1998, there were problems with electricity blackouts and restrictions of use in the United States without the added imposition of the power needed to charge batteries for replacement of gasolines in automobiles. conversion from petroleum fueled to electric powered vehicles will require the building of many more generating plants to supply the electricity necessary to meet transportation needs. When petroleum based fuels are replaced for transportation, several fuels probably will be used rather than one. Local deliveries, long distance deliveries, and travel will each use the most economical and environmentally friendly fuel.

James H. Gary

See also: Refining, History of.

BIBLIOGRAPHY

Chang, Thi. (1998). "Distillation Capacities Hold Steady, More Mergers Planned." *Oil & Gas Journal* 96(51):41–48.

Gary, J. H., and Handwerk, G. C. (1995). *Petroleum Refining, Technology and Economics*. New York: Marcel Dekker.

Radler, M. (1998). "1998 Worldwide Refining Survey." *Oil & Gas Journal* 96(51):49–92.

REFINING, HISTORY OF

The U.S. petroleum refining industry has played a central role in the expansion of this country's energy capacity over the last century. Currently, petroleum refiners generate products which account for approximately 40 percent of the total energy consumed in the United States (with respect to Btu units). The industry is characterized by a small number of large, integrated companies with multiple high-capacity refining facilities.

Between 1982 and 1997, the total number of U.S. refineries had declined from 300 to 164 operating companies. This contraction was due for the most part to the closing down of the smaller refining operations, i.e., refineries with less than 50,000 barrels of crude oil per day (BPD) capacity. While the smaller refineries still generally account for up to half of all U.S. facilities, in aggregate they control barely 14 percent of total U.S. crude refining capacity.

PROCESS DESCRIPTION: OVERVIEW

In general, refining consists of two major phases of production. The first phase of production acts on the crude oil once as soon as it enters the plant. It involves distilling or separating of the crude oil into various fractional components. Distillation involves the following procedures: heating, vaporization, fractionation, condensation, and cooling of feedstock.

Distillation essentially is a physical operation, i.e., the basic composition of the oil remains unchanged. The second phase of production, which follows distillation, is chemical in nature in that it fundamentally alters the molecular composition of the various oil fractions. These processes depend on heat and pressure and, virtually in all cases, the use of catalysts. These processes are designed to improve the octane number of fuel products. For example, in the process of isomerization, rearrangement of the molecules occurs, resulting in different chemical configurations (e.g., branched vis-à-vis linear structures) but with the atomic composition remaining constant. Polymerization involves the formation of larger molecules (polymers) from smaller ones. Reforming transforms hydrocarbons into other hydrocarbons, often to make aromatics from nonaromatic petroleum constituents (e.g., paraffins and naphthenes).

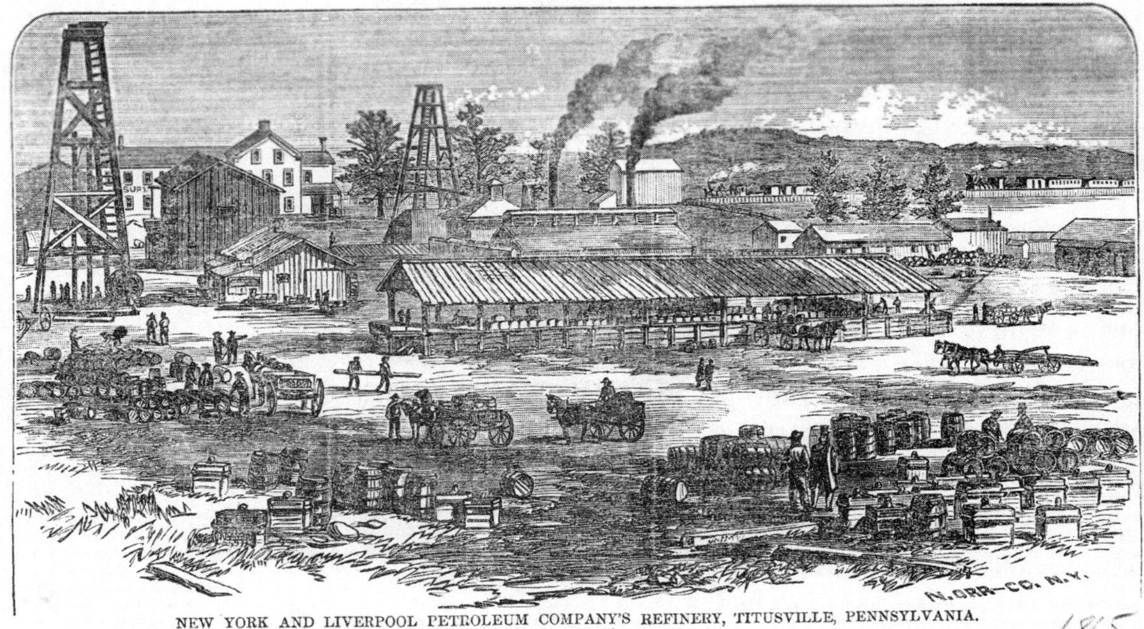

New York and Liverpool Petroleum Company refinery in Titusville, Pennsylvania, 1865. (Corbis Corporation)

The most important of the second phase refinery operations is cracking. Cracking breaks down large hydrocarbon molecules into smaller and lighter molecular units. At first performed using high temperatures and pressures only, by the 1930s, catalytic cracking was begun. Catalytic cracking went further than earlier cracking technology to reduce the production of less valuable products such as heavy fuel oil and cutter stock.

Cracking represents a fundamental advance in refining process technology. Innovation here was mostly responsible for increasing the throughput capability of the industry and heightening octane numbers for both motor and aviation fuel. Cracking technology was central to the success of the U.S. petroleum industry in delivering needed fuel for both the marketplace and the strategic requirements of World War II.

Cracking process technology has been a distinctly American achievement. Although Europeans helped transfer some important cracking innovations to the United States, it was the U.S.-based companies and their engineers who converted these yet unrealized prototypes into commercially viable, full-scale plants.

Cracking technology evolved out of the need to solve a series of increasingly complex technical problems. Ultimate success in handling such difficult problems as carbon buildup on the catalyst surface, non-uniform distribution of thermal energy, and catalyst breakage and equipment failure came with development of the fluid catalytic cracking process in the early 1940s.

EARLY REFINING TECHNOLOGY

In the late nineteenth and early part of the twentieth centuries, refining operations were in essence distillation procedures. The refinery did little more than separate the petroleum into various fractions for commercial use. Prior to World War I, gasoline was not a dominant product of the refinery. The most commercially useful products resided in the lower (e.g., denser) range of the petroleum fractions. In order of importance, these products consisted of kerosene (for heating and light), and a variety of oils, waxes, and lubricants for industrial and home use.

The machinery employed in the early refineries was rather small in scale and operated inefficiently. Generally, equipment consisted of a series of shell tubes or stills. These were placed in the horizontal position and were connected one to another from the top through the use of vapor pipes. These pipes directed the vapors from the stills into condensers which cooled the gases and so caused the products to separate out. These products were then collected in sequence, often at one point in the plant, as liquids of varying densities and properties.

Despite various mechanical attempts to increase throughput, operations were at first conducted in batches, which required the plant to be shut down for the still and ancillary equipment to be cleaned out. Until cracking technology entered the picture, refinery operations were inefficient as they captured for use barely 50 percent of the available petroleum.

View of cracking stills at a Sinclair oil refinery. (Corbis Corporation)

THERMAL CRACKING TECHNOLOGY

The initial impetus for development of a cracking process came in the years prior to World War I with the greater demand for gasoline products for the emerging automotive industry. Although early experimental work on the high-pressure cracking of petroleum was conducted in Europe in the late nineteenth century, the commercial breakthrough came in the years leading up to World War I by Standard Oil (Indiana). In 1913, William Burton, a petroleum chemist at Standard Indiana, conceived of commercially breaking down the molecular components comprising petroleum into smaller molecular units by the force of high pressure.

While his experiments yielded acceptable gasolines, serious problems arose as Burton attempted to scale up his process. The short cycles forced on producers by extensive coking worked against continuous operations. The carbon buildup interfered with heat transfer from the furnace to the petroleum and resulted in the formation of "hot spots" and in turn damage to the vessel. Pressure control was also a problem.

An assistant of Burton at Standard Indiana, E. M. Clark modified Burton's still in a fundamental way. Clark perceived that Burton's problems arose because heat and pressure were being applied to a static mass of petroleum within the autoclave. This situation was conducive to the formation of carbon at the bottom of the vessel and made it difficult to control pressure and obtain a uniform cracking throughout the charge. Clark retained Burton's use of high pressures but within the context of the tube still. In this design, oil flowed through, and was cracked within, banks of tubes. Each tube measured five feet in length. The hydrostatic pressures within the tubes produced uniform flow conditions. These banks of tubes were suspended in a furnace. Partially cracked charge was then directed upward to an overhead "soaking" tank. Here, cracking continued to completion under pressure but without the application of heat.

An important aspect of Clark's technology was that the oil being cracked, which flowed through the tubes and in a near vapor state, was maintained in a continually dynamic or turbulent condition. This meant that the coke particles which formed were prevented from adhering to the sides of the tubes. Further, as a result of the high surface/volume ratio of the still, heat transfer through the tubes was facilitated and higher pressures could be applied. Thus, whereas the Burton process operated at a maximum of 100 psi, Clark's tube still handled pressures of up to 1,000 psi.

The first commercial Tube-and-Tank cracking plant came on line in 1922. Overall, compared to the Burton Process, the Tube-and-Tank Process allowed larger volumes of petroleum to be processed under conditions of intense cracking and longer production cycles.

In the 1920s, a number of new thermal cracking technologies emerged that, in essence, were variations of the Burton-Clark designs. These included most notably the Cross and Holmes-Manley processes. These processes provided innovations in such parameters as operating conditions, methods of removing coke, and the number and configuration of the overhead soaking chambers.

These processes remained batch-type operations. As such they reached capacities in the range of 1,000–3,000 barrels per day (BPD). A truly novel approach to thermal cracking emerged from the facilities of the Universal Oil Products company in the late 1920s. To a greater extent than the other thermal cracking technologies, the so-called Dubbs Process approached continuous operations. Important elements of this technology were the continuous recycling of heavier byproducts back into the cracking section for further processing and the so-called bottom draw-off technique that continuously removed heavy liquid from the bottom portion of the soaking drum. This procedure reduced the rate of carbon buildup in the system and so increased the time over which the still could operate before being shut down for cleaning.

CATALYTIC CRACKING TECHNOLOGY

By the early 1930s, thermal cracking had achieved a fairly high level of operation. Both the Dubbs (UOP) and Tube-and-Tank (Jersey Standard) Processes represented the state of the art in the field. Between the end of World War I, when the Burton Process was still revolutionary, and the early 1930s, octane ratings of gasoline increased 36 percent. This improvement resulted from the existence of more advanced thermal plants and the increasing use of additives, especially tetraethyl lead. Moreover, the quantity of gasoline produced increased. Not only were the Dubbs and Tube-and-Tank technologies inherently more efficient than earlier thermal designs, but they were also readily scaled up in capacity. Through the 1930s, the size of thermal plants increased. The dimensions and amount of equipment used within a facility grew.

By the mid-1930s, thermal cracking reached its zenith in scale and sophistication. At this time the world's largest thermal cracking unit, built by Indiana Standard in Texas city, Texas, had a capacity of 24,000 BPD. This represents a capacity over 270 times larger than the first Burton stills and over forty-two times larger than the early Tube-and-Tank units.

Fixed Bed Processes: Houdry Catalytic Cracking

At this time, forward-looking companies understood that the ability of a refiner to control the highest octane fuels could corner critical and specialized niche automotive and aviation fuel markets. Refiners looked on catalytic cracking as a way to more finely tune their products for these market segments.

By the mid-1930s, catalytic technology entered into petroleum refining. To a greater extent than thermal cracking, catalysis permitted the close control of the rate and direction of reaction. It minimized the formation of unwanted side reactions, such as carbon formation, and overall improved the yield and quality of fuel output.

Within the United States, catalytic cracking was carried out on petroleum vapors. In contrast to thermal cracking, catalytic cracking did not require the application of ultra-high pressures.

The first attempt at a commercial catalytic process was the Macafee Process. In this design, cracking took place within a circular vessel packed with aluminum chloride catalyst. Cracked gasoline vapor products exited the reactor through a series of chimney-like structures. The Gulf Refining Company built a working plant at Port Arthur, Texas, just after World War I. However, the rapid accumulation of carbon deposits on the inside of the reactor and the high cost of the catalyst precluded further development in this direction.

Further catalytic cracking efforts did not take place in earnest in this country until the early 1930s. Sun Oil undertook the first significant commercial effort in catalytic cracking beginning in 1932. At this time, Sun Oil, based in Marcus Hook, Pennsylvania, was a relatively small company. It operated a number of oil tankers and had at its disposal a growing pipeline network in the Northeast and increasingly the South and West. In addition to a refinery at Marcus Hook, it operated a large refinery at Toledo, Ohio. Sun established its reputation as a producer of a variety of high-quality fuels and related oil products. Having taken thermal cracking to its limit through the use of high pressures, Sun embraced a promising new method developed by Eugene Houdry, a French mechanical and automotive engineer.

Through the 1920s Houdry had experimented in France on a number of possible catalytic routes to higher octane fuel. Finding little success in France, he came to the United States to further develop his process. After initial attempts at commercialization under the sponsorship of Sacony-Vacuum Company (currently Mobil) in Paulsboro, New Jersey, failed, Houdry and his development company, Houdry Process Corporation, moved to Sun.

Over the next four years, Houdry, working closely with Sun's engineering team headed by Clarence Thayer, worked to build a commercial plant. The limitations imposed by a static catalyst bed design imposed a major obstacle, particularly in the formation of carbon deposits that fouled the catalyst mass and impeded a continuous system of production.

The commercialization of the Houdry Process involved borrowing and integrating mechanical designs from the automotive, electrical, and metallurgical industries. The Houdry Process consisted of a series of catalyst cases containing the cracking catalyst. Each case went through a succession of operations: cracking (which over time resulted in carbon deposits on the catalyst), preparation of the catalyst for regeneration (via the purging of oil vapors), regeneration of the spent catalyst (burning off of carbon deposits from the catalyst surface), preparations for the next cracking operation (i.e., purging the newly regenerated catalyst of combustion or flue gases), cracking, and so on. To approximate continuous operation, the process placed the catalyst cases on different phases of the cycle (via a staggered arrangement) at any one time. In this way, cracking, purging, and regenerations were carried on simultaneously in different cases.

Internal tube components were an important part of the Houdry Process. Tubes were used to distribute oil vapors over the catalyst during cracking and to receive and direct cracked vapors and combustion gases (after regeneration) to various parts of the system. Tubes also played a central role in heat control. Fluid-filled, heat control tubes were placed deep within the catalyst bed so that they tapped and transferred heat which built up throughout the mass.

The hot fluid moved through the tubes to a boiler into which heat energy was transferred. Steam generated here was transported back through the tubes to heat the catalyst for the cracking operation. Initially, the heat transfer fluid used was water or superheated steam. Because these fluids tended to cause tubes to crack, a more advanced design was developed that utilized molten salts, chemically and physically more stable under the rigorous operating conditions.

By the early 1940s, a Houdry plant capacity ranged from 7,000 to 15,000 barrels per day. Houdry's contribution to the U.S. energy industries extends beyond fuels. The Houdry Process was the first industrial technology in the United States to employ on a large scale the gas turbine system, adapted from European designs, and the heart of the energy recycling process. Capturing and turning into useful mechanical energy the heat generated during combustion in the regeneration cycle was especially critical. The Houdry Process paved the way for energy recycling technology, including the process commonly called cogeneration, in such industries as power generation and iron and steel production.

But there were many problems in the operation of the Houdry Process that could not be resolved. Even with its sophisticated regeneration system, carbon continued to form on the catalyst over a period of time. The heat transfer system was not as efficient as it needed to be. The process never achieved fully continuous operations, limiting the quantity of oil that could be processed and the quality of the gasoline produced. At its height (late 1930s, early 1940s) the process did not produce gasoline with an octane rating above 87. It never captured more than 10 percent of the total U.S. cracking capacity, although in the early 1940s it controlled over 90 percent of the total catalytic cracking market.

Moving-Bed Processes: The Thermofor and Air-Lift Systems

Sun attempted to resolve these problems by improving upon the fixed-bed process. The Sacony-Vacuum Company (formerly the Standard Oil company of New York and the future Mobil Oil) was the first to develop a different type of catalytic cracking process. While involved with Sun Oil in development of the fixed-bed process, Sacony early on understood the limitations of that technology. Beginning in 1932, Sacony began conceiving of the moving-bed concept.

In its development, it adapted two existing technologies. In the agricultural sector, the mechanics of grain elevators provided a model for how to move solids vertical distances and in closed-loop flow arrangements. Sacony engineers modified the elevator bucket systems traditionally used by the grain industry to carry hot catalyst from the bottom to top of vessels and between vessels.

Then too, Sacony modeled its regenerator vessel after a certain type of metallurgical furnace (known as the Thermofor kiln). The vessel consisted of a series of semi-independent burning zones. Distribution channels delivered compressed oxygen to each zone to fuel combustion. A series of baffles and ducts within each combustion compartment produced a uniform distribution of air through the catalyst. Flue gases emitted by the regeneration process collected in common headers located between the burning zones.

The so-called Thermofor moving-bed process borrowed two important techniques from Houdry's technology: a molten salt cooling system for the kiln section and gas turbine technology that generated power to pressurize more air for the regenerator. The catalyst was stored in an overhead hopper, from which it was fed into the catalyst-filled reactor. The catalyst then traveled down the vessel under the influence of gravity. In the commercial plants, oil, injected as a liquid spray near the top of the reactor, moved down along with the catalyst. During this time, the latter transferred its heat, obtained and stored during the previous regeneration cycle, to the oil. The system conserved on energy by thus recycling heat units through the catalyst which simultaneously vaporized and cracked the oil particles that descended with it.

Following cracking, the spent catalyst and oil descended to a disengager that separated the gasoline from the catalyst. The catalyst, with oil residue entrained on its surface, then moved through a purging section where superheated steam thermally removed oil remnants. The oil-free catalyst, still laden with carbon deposits, was then lifted by elevator from the bottom of the reactor to the top of the regenerator.

Regeneration was carried out as the catalyst fell through various burning zones. The flue gases were recycled by being directed to the turbo-compressors that the pressurized air used as fuel in the combustion process.

The first Thermofor cracking unit came on line in late 1942. By March of 1943, twenty Thermofor units had been completed or were under construction. The larger Thermofor plants could circulate 100-150 tons of catalyst per hour and could process up to 20,000 barrels of petroleum per day.

But problems persisted. The catalyst, moving at rapid rates, tended to disintegrate as it impacted the inside surface of equipment . Dust particles formed, clogging pipes and transfer lines and disrupting the smooth flow of operation. This difficulty in turn prevented uniform heat distribution through the system and affected the rate and extent of both cracking and regeneration. Both the volume and quality of fuel produced suffered.

An improved design undertaken by Sacony used high-velocity gases to replace the mechanical elevator systems as catalyst carriers. These so-called "air-lift" units improved upon the Thermofor process both in terms of economies and octane numbers. It was, however, only with the fluid cracking process that catalytic technology realized fully continuous production.

Fluid Catalytic Cracking

Fluid catalytic technology addressed two major shortcomings of previous moving-bed systems: the slowness with which catalyst traveled through the vessels and the inability to sustain the cracking process continuously over an indefinite period of time. The central problem facing the industry was how to move the catalyst around the production circuit at rapid rates and at the same time fully control the intensity and duration of cracking (and regeneration). Jersey Standard (Exxon), up to this point not a major player in advanced catalytic cracking, addressed these issues as it continued to improve upon more traditional cracking processes.

By the time Jersey Standard moved into fluid research in the late 1930s, it possessed both cracking (thermal) and catalytic expertise. As noted above, Jersey developed its most important cracking technology up to that point with its Tube-and-Tank process, a thermal approach. And beginning in 1927, the company formed a patent-sharing agreement with the German giant, I. G. Farben, for development of a high-pressure catalytic hydrogenation process. Jersey's early catalytic work resulted in a number of reforming practices, such as hydroforming (1938) and steam reforming (1942) which supplemented cracking processes.

By the late 1930s, Jersey, associated with a group of companies looking for improved catalytic processes (the Catalytic Research Associates), combined its past cracking and catalytic experience in developing a fully continuous catalytic cracking system.

Prior to coming upon the notion of fluidization as a basic principle of continuous catalytic cracking, Jersey experimented with a series of fixed-and moving-bed systems. A unique approach emerging out of these efforts involved the design of "slurry" systems, by which a pumping mechanism propelled catalyst particles and oil fluid together in concurrent fashion along a horizontal reactor.

While problems with the circulatory (i.e., pumping) system limited the usefulness of this approach, it led to the next major leap for fluid cracking. Warren K. Lewis, head of the Department of Chemical Engineering at MIT and consultant to Jersey, was the leading creative force behind fluid cracking. He led the initial experiments (at MIT) identifying the commercial viability of, and establishing the preconditions for, fluidization phenomena, including the existing of the so-called turbulent bed. He also directed the design of the first semi-commercial units.

Fluid cracking retained the moving bed concept of the catalyst transported regularly between the reactor and regenerator. And as with the air-lift systems, the fluid plant rejected mechanical carrying devices (elevators) in favor of standpipes through which the catalyst fluid traveled.

Fluid technology exceeded gas-lift technique by incorporating two innovative concepts, that is, under certain conditions of velocity flow and solid/vapor concentration: (1) a catalyst-gas mixture traveling around the plant behaves in just the same way as a circulating liquid; and (2) within vessels, rather than the catalyst and oil falling under gravity, they closely intermingle indefinitely in a dense turbulent but stable, well-defined and carefully (and indefinitely) controlled cracking "bed."

The fact that a flowing catalyst-vapor mixture acted just as a traditional liquid was a crucial point. It meant that the cracking plant was in essence a hydrodynamic system readily controlled over a range of velocities and throughput by simple adjustment of pressure gradients and such variables as the height of the catalyst (i.e., the head of pressure) in the standpipe and the velocity of upward moving gas which carried the catalyst from the bottom of the standpipe to processing vessels. An important control element also was the "aeration" of the moving catalyst at strategic points along the standpipe (and transfer and carrier lines) by use of jets of air to adjust pressure gradients as required.

At the heart of the process was the turbulent bed. As catalyst rose with the carrier gas up the standpipe and into the reactor, it tended to "slip back" and concentrate itself into a boiling but well-delineated mass. The boiling action served to keep the catalyst particles tumbling about. Rather than falling via gravity, the boiling action in effect served to counteract the force of gravity and so maintain the catalyst in an indefinitely suspended state. Particles simply moved from one part of the bed—the turbulent mass—to another, all the while undergoing cracking. This design applied as well during combustion in the regenerator.

The plant was able to operate continuously. The continual and controlled state of turbulence in the bed assured close intermixing between solids and vapors and an even distribution of thermal energy throughout the bed, and the "liquid" catalyst flowed smoothly and rapidly from one vessel to the next.

An early fluid cracking unit removed spent catalyst from reactors (to be directed to the regenerator) using an overhead cyclone system. A more efficient technique, the so-called down-flow design, followed. It altered operating and flow conditions so that spent catalyst concentrated in the bottom part of vessels, where they could be removed, resulting in greater ease of catalyst recovery, simplification of plant layout, and improvement in operating flexibility.

Fluid catalytic cracking rapidly overtook its competitors as both a source of fuel and of critical organic intermediates. Prior to 1942, the Houdry Process controlled 90 percent of the catalytic fuel market. But only three years later, in 1945, fluid cracking led all other catalytic cracking processes in market share (40 percent). At this time Thermofor technology stood at 31 percent, and Houdry at less than 30 percent.

After 1942, fluid cracking technology increasingly dominated U.S. petrochemical production. It manufactured high-tonnage fuels for both motor vehicles

A fluid catalytic cracking unit in Joliet, Illinois, converts heavy components of crude oils into high octane gasoline and distillates. (Corbis Corporation)

and aircraft for the wartime effort. The quality of the gasoline was unprecedented. Octane ratings of fluid-produced fuels exceeded 95, an unheard-of figure only a few years before, and critical for aviation gasoline. Fluid cracking technology played a central role the U.S. synthetic rubber program: its byproduct gases supplied the butylenes which were essential in the making of butadiene, the essential rubber intermediate.

RECENT DEVELOPMENTS IN PETROLEUM REFINING

Since 1945, the fluid catalytic cracking process has rapidly overtaken fuel production and has become the central technology in the U.S. petrochemicals industry. With fluid cracking, the scale of petrochemical operations grew enormously. For the first time, refiners could process virtually any volume of oil rapidly and efficiently.

Accordingly, technological change in refining technology has centered on alterations and improvements made to the fluid cracking process. These modifications have included improvements made to catalysts, materials of construction, interior linings, and a variety of mechanical details. Since 1945, the process has operated on progressively higher temperatures for both cracking and regeneration. It has also incorporate newer generations of catalysts.

By the early 1960s, fluid cracking had become the workhorse of the refining industry. It was the central process in the production of over 70 percent of all high-octane fuel. From early 1940s to mid-1960s, capacities of fluid units have grown from less than 20,000 BPD to between 100,000–200,000 BPD.

The flexibility of the process was manifested in its ability to economically process smaller volumes of feedstock. Between the 1950s and 1980s, refiners and engineering firms developed smaller, lower investment units which could be readily scaled up as required. Recently, such fluid crackers have been simplified to the point that cracking and regeneration take place within a single, vertical standpipe or riser tube. The capacities of such units often do not exceed 1,000 BPD.

By the mid-1990s, the technology was virtually the only catalytic cracking process in operation in the major refineries. The technology accounted for over 95 percent of all high-octane fuel within the United States. By 1997, there were approximately 350 fluid cracking facilities in operation worldwide, most located within the United States. Between 8 and 9 percent of the fluid units existing worldwide (25-30 units) are owned and operated by Exxon, the original innovator of fluid cracking.

In the late 1990s, the growth of fluid cracking as a major petroleum refining process was about 2 percent per year. Total world fresh feed capacity for the fluid process now stands at more than 11 million barrels/day. Worldwide, FCC makes 80 billion gallons of high-grade gasoline annually. This represents nearly half of total world gasoline production.

A major recent development for the technology has been its closer integration with the large petrochemical complex. Essential petrochemical activity has been relying more on fluid technology and less on thermal units for their intermediate feedstock. The big development for the 1990s is increasing integration of FCC with the large petrochemical plant.

In the postwar period, fluidization influenced U.S. manufacture in other areas as well. Most critically, it has been applied in combustion processes for the making of metallurgical materials and by the utilities for generating heat and electricity for industrial and residential use. Moreover, the technology is a "cleaner" method for producing energy and so is an important means by power-based companies to comply with stricter environmental regulations.

Sanford L. Moskowitz

See also: Refineries.

BIBLIOGRAPHY

Enos, J. (1962). *Petroleum, Progress and Profits: A History of Process Innovation.* Cambridge, MA: MIT Press.

Frankenburg, W. G. et al., eds. (1954). *Advances in Catalysis and Related Subjects, Vol. 6.* New York: Academic Press.

Grace, J. R., and Matsen, J. M., eds. (1980). *Fluidization.* New York: Plenum Publishing.

Giebelhaus, A. W. (1980). *Business and Government in the Oil Industry: A Case Study of Sun Oil, 1876–1945.* Greenwich, CT: JAI Press.

Larson, H. M. et al. (1971). *History of Standard Oil Company (New Jersey): New Horizons, 1927–1950.* New York: Harper and Row.

Moskowitz, S. L. (1999). "Science, Engineering and the American Technological Climate: The Extent and Limits of Technological Momentum in the Development of the U.S. Vapor-Phase Catalytic Reactor, 1916–1950." Diss. Columbia University.

Murphree, E. V. et al. (1945). "Improved Fluid Process for Catalytic Cracking." *Transactions of the American Institute of Chemical Engineers.* 41:19–20.

Nelson, W. L. (1949). *Petroleum Refining Engineering.* New York: McGraw-Hill.

Othmer, D. F. (1956). *Fluidization.* New York: Reinhold Publishing Corp.

Popple, C. S. (1952). *Standard Oil Company (New Jersey) in World War II.* New York: Standard Oil Company (New Jersey).

Russell, R. P. (1944). "The Genesis of a Giant." *Petroleum Refiner* 23:92–93.

Spitz, P. (1988). *Petrochemicals: The Rise of an Industry.* New York: John Wiley and Sons.

Squires, A. M. (1986). "The Story of Fluid Catalytic Cracking: The First Circulating Fluid Bed." *Proceedings of the First International Conference on Circulating Fluid Beds, November 18-20, 1985.* New York: Pergamon Press.

Williamson, H. F., et al. (1963). *The American Petroleum Industry, Vol. II: The Age of Energy, 1899–1959.* Evanston, IL: Northwestern University Press.

REFRIGERATORS AND FREEZERS

With 120 million household refrigerators and freezers in operation in the United States, their consumption of electrical energy is of major concern not only to the consumer, but also to the power generating utilities that have to provide the power. The government, charged by Congress with guarding against air pollution, protecting the Earth's ozone layer, and fighting global warming, has a keen interest in refrigerator energy consumption.

Vapor compression, absorption refrigeration (instead of electric power, uses heat as the source of energy), and thermoelectric refrigeration (the direct conversion of electrical energy to cooling effect), are the principal means of refrigeration. Of the three methods, vapor compression, often referred to as mechanical refrigeration, is the most energy efficient, approximately two times more efficient than absorption refrigeration, and four times more efficient than thermoelectric refrigeration. Vapor compression is by far the most popular means for refrigerating household refrigerators and freezers, although the other two technologies have unique advantages in some specific applications.

VAPOR COMPRESSION BASICS

Figure 1 shows the four basic elements of the vapor compression system: (1) the evaporator, where the refrigerant vaporizes, and thus absorbs heat from the surroundings; (2) the compressor, where the refrigerant vapor is compressed (typically in the ratio of ten to one); (3) the condenser, where the refrigerant vapor of high pressure and high temperature is condensed by rejecting the heat absorbed by the evaporator, together with the heat of compression, to the atmosphere; and (4) the expansion device, be it an expansion valve or a capillary tube, that allows the liquid refrigerant arriving from the condenser at high pressure and at room temperature, to enter the evaporator, and repeat the refrigeration cycle.

Figure 2, the pressure-enthalpy plot of the standard vapor compression cycle, traces the state of the refrigerant through the refrigeration system. (Enthalpy represents the energy of the refrigerant as

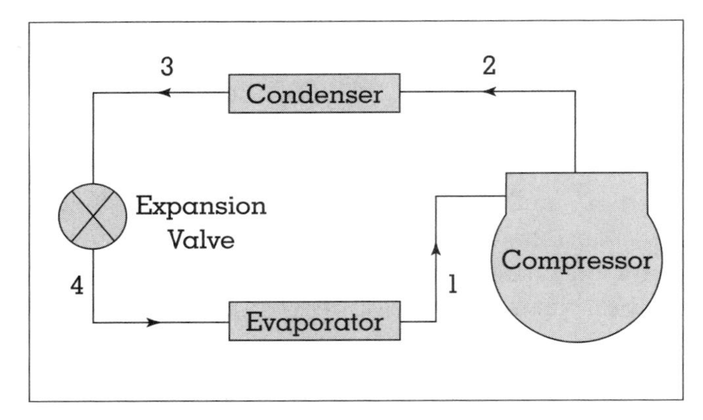

Figure 1.
Flow diagram of the standard vapor compression system.
SOURCE: W. F. Stoecker and J. W. Jones. (1982). *Refrigeration and Air Conditioning,* p. 201. New York: McGraw-Hill Book Company. Modified by author.

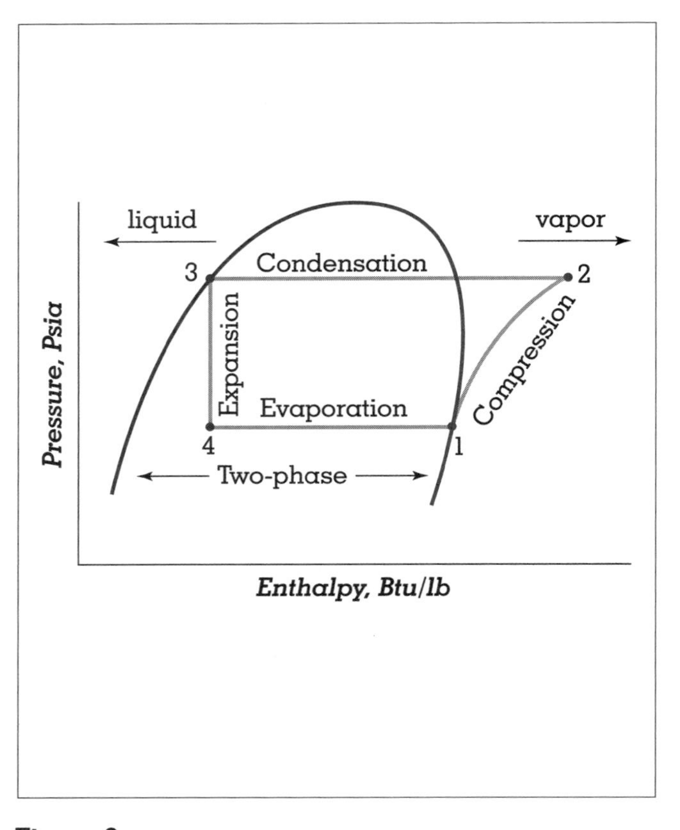

Figure 2.
The standard vapor compression cycle on the pressure-energy diagram.
SOURCE: W. F. Stoecker and J. W. Jones. (1982). *Refrigeration and Air Conditioning,* p. 201. New York: McGraw-Hill Book Company. Modified by author.

it circulates through the various components.) Note that the evaporator absorbs heat at low pressure, and the condenser rejects the heat absorbed by the evaporator and the work of compression, at high pressure. The curvature is the saturation curve of the refrigerant. It delineates the various *states* that the refrigerant passes through during the cycle, liquid to the left of the curve, two-phase mixture within the curve, *vapor* to the right of the curve.

The efficiency of any device is the ratio between what we are after to what we have to pay for it. In the case of a refrigerated appliance, we are after the refrigeration effect, (i.e., heat absorbed by the evaporator) and we pay for it with electric power that runs the compressor. In Figure 2, the evaporator absorbs more energy in the form of heat than the energy supplied to the *compressor* in the form of electricity. The ratio of the two is called the coefficient of performance (COP), a dimensionless number. Compressor efficiency is commonly expressed as the energy efficiency ratio (EER), the ratio of the refrigeration effect in Btu per lb of refrigerant, to the *energy input* to the compressor, in watt-hour per lb of refrigerant. The EER of a typical refrigerator compressor in the 1990s was about 5.5 Btu per watt-hour.

THE HISTORY OF VAPOR COMPRESSION

In 1748, William Cullen of the University of Glasgow in Scotland made the earliest demonstration of man-made cold when he evaporated ether in a partial vacuum. In 1824, Michael Faraday of the Royal Institute in London, found that ammonia vapor, when condensed by compression, would boil violently and become very cold when the pressure is removed. Only ten years later, in 1834, Jacob Perkins, an American living in London, patented the first closed vapor compression system that included a compressor, condenser, expansion device, and evaporator. This is the same basic system as in today's domestic refrigerator. He used ethyl ether as the refrigerant, and received British Patent 6662, dated 1834.

In 1844, John Gorrie, Florida, described in the Apalachicola Commercial Advertiser his new machine for making ice. The delays in ice delivery from the Boston lakes forced Gorrie to build his ice machine so that his hospital fever patients could be assured that ice would always be available.

In the early days ethyl ether and methyl ether were

The first refrigerator made by Frigidaire. (Corbis Corporation)

the refrigerants of choice. In 1869 the first ammonia system was introduced. Today ammonia is still a very common refrigerant not only in commercial vapor compression systems, but also in absorption refrigerators.

In 1915 Kelvinator marketed the first mechanical domestic refrigerator, followed shortly by Frigidaire. These models used belted compressors underneath wooden ice boxes that required frequent maintenance because of leaky shaft seals. In 1910 General Electric took out a license to build a sealed compressor that eliminated the need for shaft seals. The machine was originally invented in 1894 by Abby Audiffren, a French monk, who developed it for the purpose of cooling the monastery wine. After a major redesign it was introduced as the Monitor Top refrigerator in 1927, and was an instant success. Up to this time refrigerator cabinets were made principally of wood, with a great deal of carpentry, and not suitable for mass production. The Monitor Top had a steel cabinet, fabricated by punch presses and welding. The steel inner liner was attached to the outer shell by phenolic plastic breakerstrips. A more current production is the 27 cu ft refrigerator with through-the-door water and ice dispenser.

The *Monitor Top* used a toxic refrigerant, sulfur dioxide. In the late 1920s, Frigidaire Corporation, then a leading manufacturer of household refrigerators, asked the General Motors research laboratory to develop a refrigerant that is non-toxic and non-flammable. The result was a chlorofluorocarbon (CFC), namely dichlorodifluoromethane, commonly known as Refrigerant 12 of the Freon family. By the time of World War II, it completely replaced sulfur dioxide. In the 1980s, it was discovered that CFCs deplete the ozone layer surrounding the Earth, thus increasing the likelihood of skin cancer. In 1987 the United States joined other industrial nations in signing the

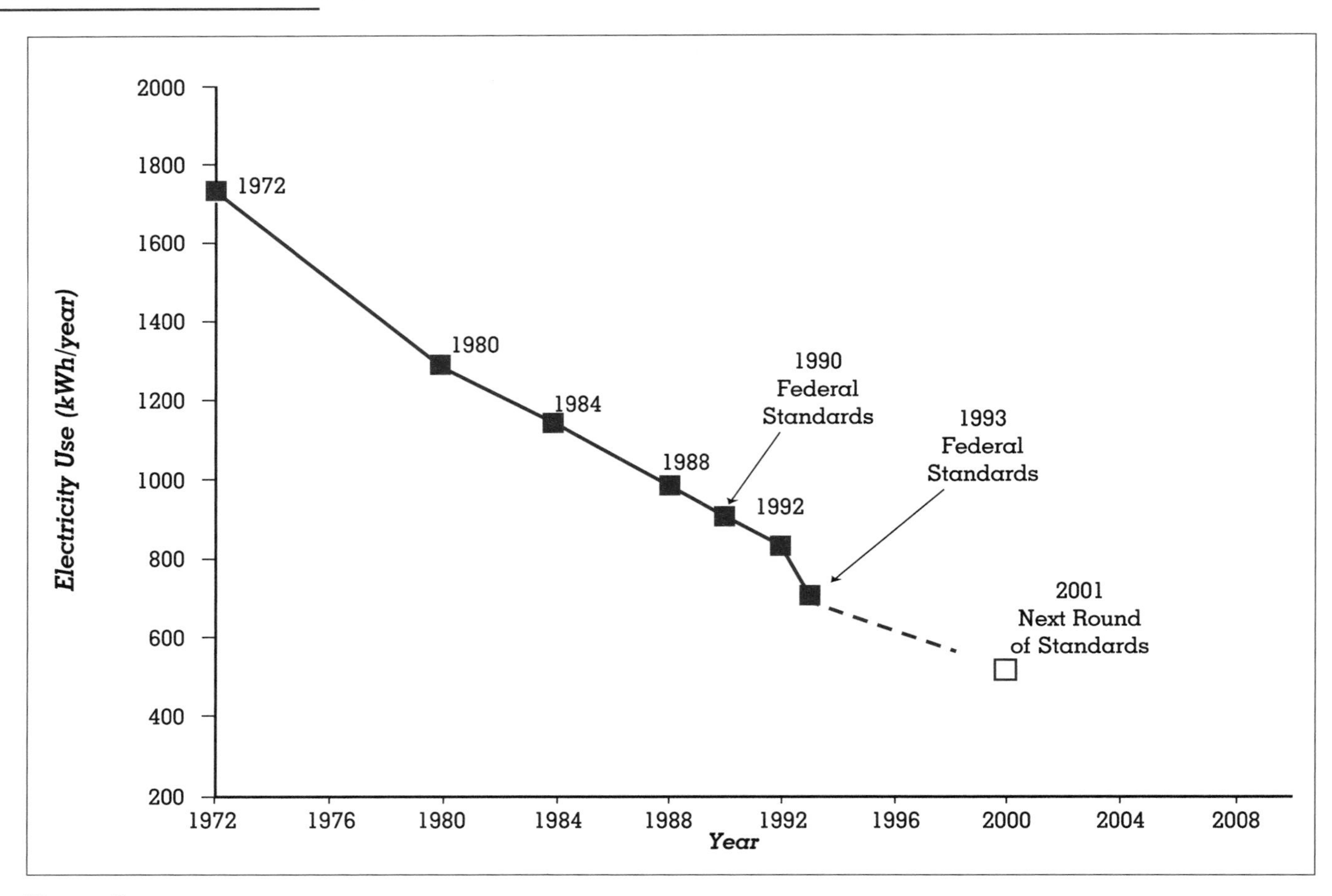

Figure 3.
Actual and projected refrigerator-freezer energy improvements from 1972 to 2001.
source: E. A. Vineyard, J. B. Horvay, and E. R. Schulak. "Experimental and Analytical Evaluation of a Ground-Coupled Refrigerator-Freezer." ASHRAE paper, p.2, presented in Toronto, Canada, on June 23, 1998.

"Montreal Protocol on Substances that Deplete the Ozone Layer." The phase-out of CFCs began on July 1, 1989, and by 1997, a hydrofluorcarbon, HFC134a, with zero ozone depletion potential, became the dominant refrigerant in the United States. The phase-out of CFCs in developing countries is on a slower schedule.

In the years after World War II many new customer-oriented features were introduced : plastic food liners, foam insulation, combination refrigerator-freezer, automatic defrosting, ice makers, and child-safe door closures. In the 1940s, the typical refrigerator was a single door cabinet, requiring periodic defrosting, with a storage volume of about 8 cu ft. By the 1990s, it developed into a 20 cu ft automatic defrosting two door combination refrigerator-freezer, with an ice maker.

ENERGY CONSIDERATIONS

The energy consumption of refrigerators and freezers is regulated by the Congress of the United States. The "National Appliance Energy Conservation Act of 1987" (NAECA) established an energy conservation standard "which prescribes a minimum level of energy efficiency or a maximum quantity of energy use". The Energy Standards were revised effective January 1, 1993, and again, effective July 2001. The first revision resulted in a cumulative 40 percent reduction in energy consumption when added to the initial standards. The result of the second revision in 2001 is an additional 30 percent reduction. A historical chart, Figure 3, shows actual and projected improvements in the use of electrical energy for refrigerator-freezers. In 1978 one manufacturer offered a 20 cu ft refrigerator-freezer that consumed 1,548 kWh per year. In 1997, the same manufacturer had a 22 cu ft refrigerator-freezer on the market that consumed 767 kWh per year. The 2001 target for the same product is 535 kWh per year. The latest mandatory energy reduction will add about $80 to the cost of the product, and result in an annual saving of $20 for the customer. Based on an annual production of 8.5 million refrigerators and freezers in the United States, this significant reduction of energy use will eliminate the need for building eight new power plants.

The energy consumption of a refrigerator is a function of two distinct elements: the heat that the evaporator needs to absorb to maintain the specified storage temperatures and the efficiency of the refrigeration system to reject that heat, and also the heat of compression, by the condenser. The heat load is determined by the storage volume of the appliance, the interior temperatures to be maintained, and the effectiveness of the insulation surrounding the storage space. The engineering challenge resides in reducing the heat load, and at the same time, improving the efficiency of the refrigeration system.

THE HEAT LOAD

Figure 4 shows the cabinet cross section of a typical automatic defrosting refrigerator. To defrost a refrigerator, forced convection heat transfer is needed between the evaporator and the load, thus the electric fan. The role of external heaters is to prevent the condensation of water vapor (sweating) on external cabinet surfaces. As the figure indicates, the heat flow through the walls of the cabinet represents 52 percent of the total heat load. Consequently, the thickness and thermal conductivity of the insulation has a major impact on the energy consumption of the appliance. Until the early 1960s the insulating material was glass fiber. Then the development of polyurethane insulation revolutionized the refrigeration industry. With an insulating value twice as good as glass fiber, wall thickness could be reduced by one-half, resulting in additional cubic feet of useful storage. In contrast to applying the glass fiber by hand, the assembly line for polyurethane foam is completely automated. The foam adds structural strength to the cabinet, allowing significant reduction in the thickness of the materials used on the inner and outer surfaces.

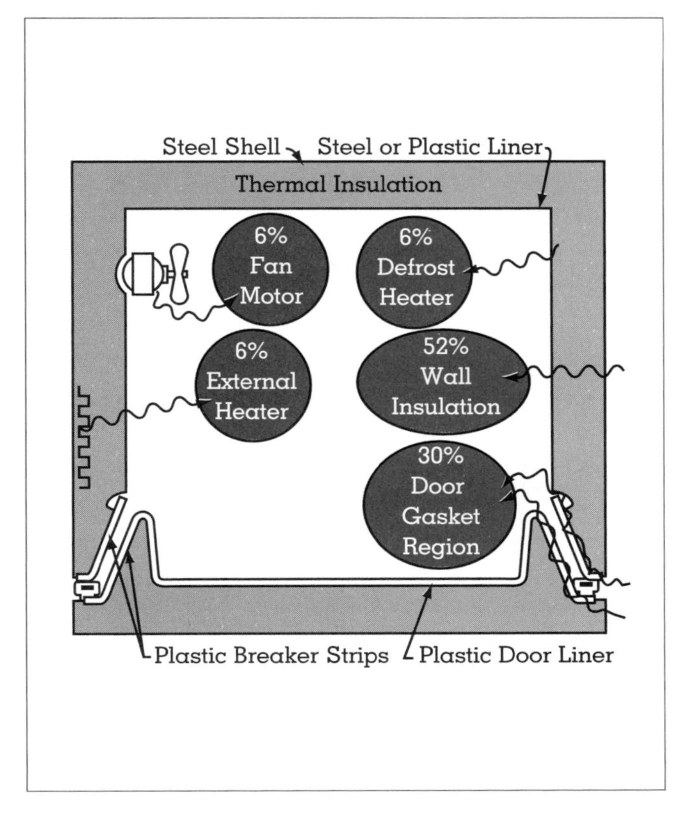

Figure 4.
Cabinet cross section showing typical contributions to the total basic heat load.
SOURCE: ASHRAE Handbook. (1998). *Refrigeration.* Atlanta, GA: ASHRAE.

In the 1980s, with the discovery of the ozone depletion potential of CFC refrigerants, the R11 blowing agent used in polyurethane foam was destined for phase-out in the United States by 1996, just like refrigerant R12 of the refrigeration system. A frantic search began for environmentally acceptable substitutes and Table 1 lists some of the more promising candidates. Note that in addition to ozone depletion potential (ODP), a relatively recent environmental concern, global warming potential (GWP) (sometimes referred to as the "greenhouse effect") is also listed. The numbers are all relative to the properties of CFC 11, the worst offender on the list. In their search for an alternative to R11, the United States and Western Europe went their separate ways. The American choice was hydrochlorofluorocarbon (HCFC) refrigerant R141b. Europe, with an annual production of some 17 million refrigerators and freezers, and under pressure from environmentalists,

Blowing Agent	*ODP*[1]	*GWP*[2]
CFC 11	1	1
HCFC 22	0,05	0,36
HCFC 123	0,02	0,01
HFC 134a	0	0,27
HCFC 141b	0,11	0,1
HCFC 142b	0,06	0,36
Pentane	**0**	**0,001**
Air	0	0
CO_2	0	0,00025

[1]Ozone Depletion Potential (ODP)
[2]Global Warming Potential (GWP)

Table 1.
Ozone depletion potential (ODP) and global warming potential (GWP) of various foam blowing agents.
SOURCE: Machinenfabrik Hennecke GmbH, Sankt Augustin, Germany, Brochure #302. Modified by author.

went to pentane.. While both substances have zero ODP, it is the difference in GWPs that tilted the Europeans toward pentane. As a matter of fact, R141b is scheduled to be phased out in the United States by January 2003, due to its residual ODP. Pentane is a hydrocarbon, like gasoline, and highly flammable. Due to the mandated safety measures, the conversion from R11 to pentane at the manufacturing sites is very extensive and expensive. Both foams, whether blown with R141b or with cyclopentane, have a higher thermal conductivity (poorer insulating value) than R11 blown foam. Consequently, these new formulations, require considerable redesign. For the energy consumption to remain the same, either the cabinet wall thickness needs to be increased or the efficiency of the refrigeration system needs to be improved.

To meet the 2001 U.S. energy standards and the 2003 phase-out of HCFCs, there is a great incentive to develop a significantly better thermal insulation. The most dramatic approach would use vacuum panels for insulating the cabinet. A number of U.S. and Japanese manufacturers have developed such panels and placed these kinds of refrigerators in homes. The panels consist of multilayer plastic envelopes filled with precipitated (fumed) silica. The claimed thermal conductivity is one-fourth that of polyurethane foam. The two major obstacles are cost and the maintenance of vacuum for twenty years.

THE REFRIGERATION SYSTEM

To meet the 1993 Energy Standards, the industry undertook , at considerable cost, the optimization of the various refrigeration system components. The most significant improvement was the increase in compressor efficiency, from an EER of about 4 to about 5.5. Other system improvements included more efficient fan motors, more effective heat transfer by the evaporator and the condenser, and less defrost energy. In the early 1980s, both the Whirlpool Corporation and White Consolidate Industries introduced electronic defrost controls. Heretofore, an electric timer initiated the defrost cycle, typically every twelve hours, whether the evaporator needed it or not. With the electronic control the defrost interval is more a function of frost accumulation than of time, and thus referred to as a "variable defrost control" or as "adaptive defrost."It saves energy by being activated only when needed.

Further improvements in system efficiency will be difficult to achieve with evolutionary changes. Following are some of the more promising areas of development:

Rotary Compressor. Inherently a rotary compressor is more efficient than the current reciprocating compressor. (room air-conditioners have been using rotary compressors for decades.) Several manufacturers in the U.S. and Japan have produced refrigerators with rotary compressors, but experienced long-term quality problems.

Dual Evaporators. A typical refrigerator-freezer, with automatic defrost, has a single evaporator for refrigerating both compartments, at a pressure that is determined by the temperature of the freezer. With separate evaporators in each compartment, the compressor would alternate between the two evaporators in providing refrigeration. The fresh food compartment has a significantly higher temperature than the freezer compartment, so it would requier less pressure to refrigerate, resulting in energy savings.

Variable Speed Compressor. Every time the compressor cycles on during the first few minutes of running, the compressor works on building up the pressure difference between the evaporator and the condenser, and is extremely inefficient. By letting the compressor run all the time, and modulating its speed according to the refrigeration needs, these periods of inefficiency can be eliminated.

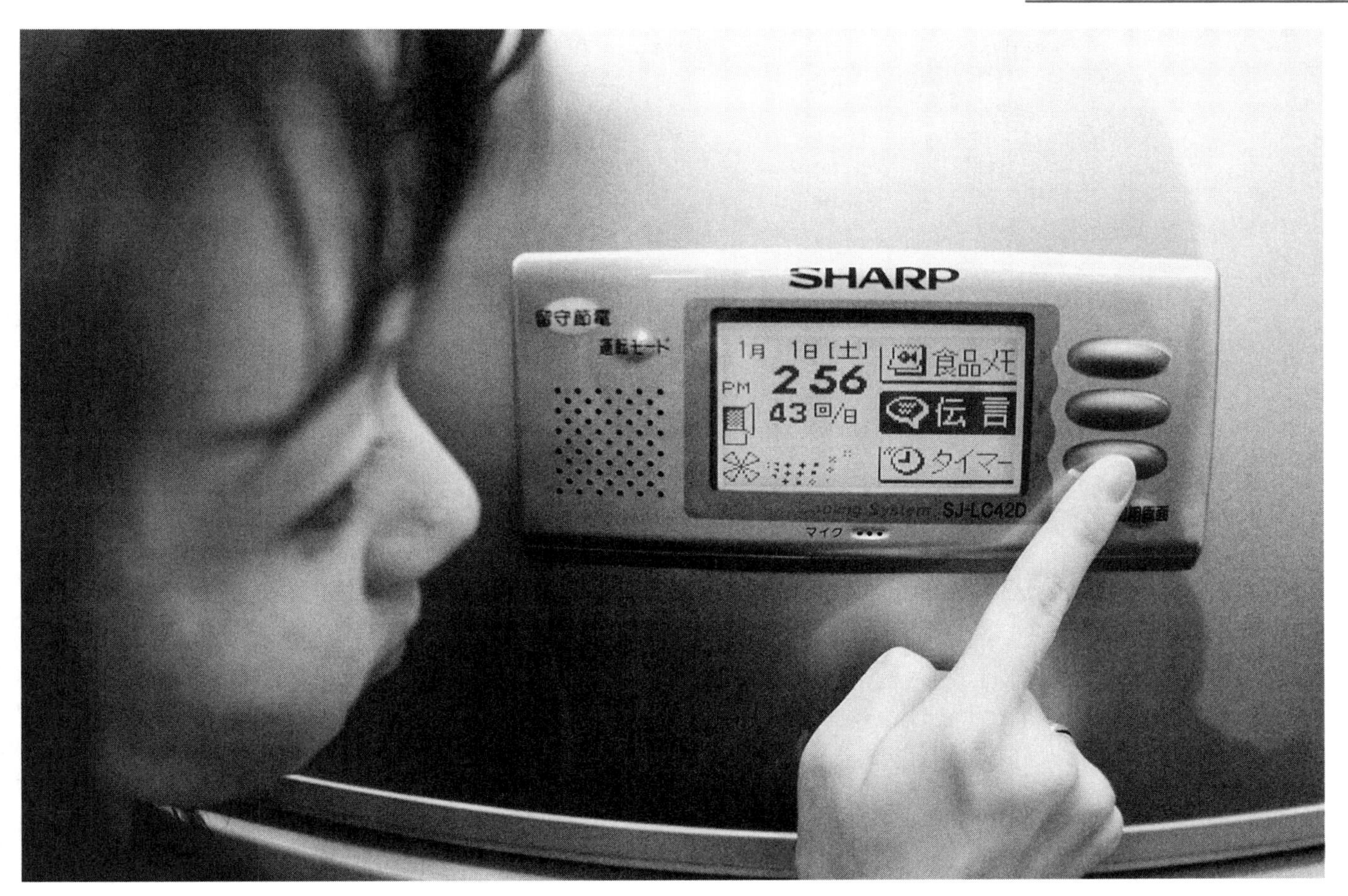

Sharp Corporation promotes its new refrigerator, which includes the first built-in LCD. (Corbis Corporation)

Sonic Compression. Sonic compression is appealing as a potential low-cost, high efficiency oil-free technology. It should work with a wide range of refrigerants, including hydrocarbons, fluorocarbons, and ammonia. Theoretical COP is comparable with current vapor compression refrigeration cycles. The initial prototypes have a target cooling capacity of 110 to 250 W.

ABSORPTION REFRIGERATION.

Absorption refrigeration is based on the great affinity between certain liquids and gases. For example, 1 cu ft of water is capable of absorbing 800 cu ft of ammonia gas. One might look at this process as compression of a sort. Starting out with a small volume of absorbent (water) and a large volume of refrigerant gas (ammonia), the process ends up with a small volume of liquid solution of the refrigerant in the absorbent. In comparison to the vapor compression system (Figure 1), the compressor is replaced by two components, the absorber and the generator (Figure 5). The condenser and the evaporator function just as in the vapor compression cycle. Upon entering the generator, the aqua-ammonia solution is heated, and the ammonia in the solution vaporizes first, due to it's lower boiling point. The ammonia vapor then enters the condenser as the water drains back to the absorber. The condensed liquid ammonia is allowed to expand into the evaporator. As it vaporizes it absorbs heat from the load, and then it is reabsorbed by the water in the absorber. The cycle is ready to repeat itself.

Vapor compression uses the highest form of energy, namely electrical energy. In absorption refrigeration, the energy input is any source of heat (e.g., electrical energy, bottled gas, kerosene, or solar energy).

The first practical absorption refrigerator was developed in 1850 by Edmund Carre of France, who

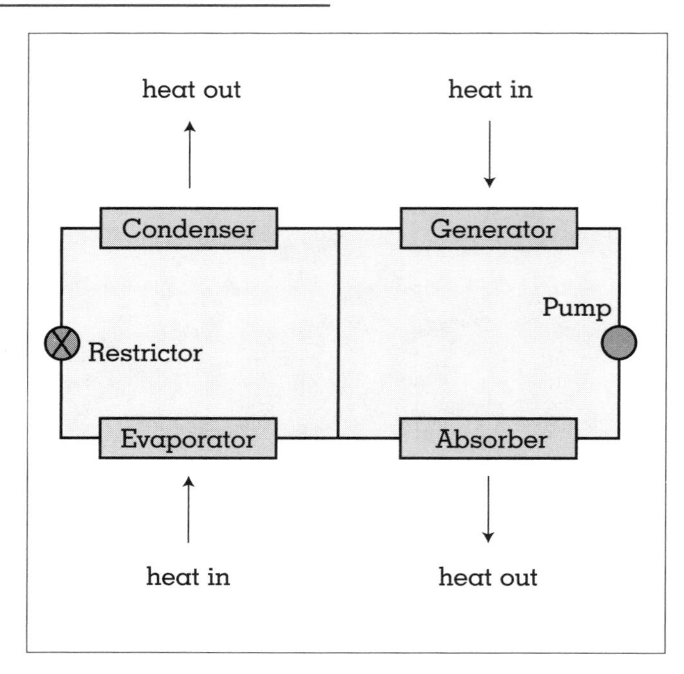

Figure 5.
Heat flow diagram of the absorption system.

was granted a U.S. patent in 1860. During the Civil War, the supply of ice from the North was cut off and Edmund's brother, Ferdinand Carre, shipped a 500 lb per day ice machine through the federal blockade to Augusta, Georgia, to be used in the convalescent hospital of the Confederate Army. Absorption refrigeration enjoyed it's heyday in the United States in the 1920s, with several manufacturers, like Servel and Norge, providing products. As larger refrigerators were demanded by the public, absorption refrigerators were unable to compete due to their inefficiency. The only remaining role for absorption refrigeration in the United States is in remote areas that are without electricity, and in boats and recreational vehicles that have bottled gas. There are many under-developed countries where absorption refrigeration is still the principal means of food preservation, due to lack of electricity.

THERMOELECTRIC REFRIGERATION

In 1821, Thomas Seebeck, an Estonian physician, discovered the existence of an electric current in a closed circuit consisting of unlike conductors, when the junctions between the conductors were at different temperatures. This discovery is the basis for thermocouples, used for all kinds of temperature measurements. In 1834, Jean Peltier, a French watchmaker, discovered the reverse of the "Seebeck Effect"; namely, if a direct current passes through a junction of two dissimilar metals in the appropriate direction, the junction will get cold. When thermocouples are put in series, one face of the assembly will get cold, the other hot.

The "Peltier Effect" was a laboratory curiosity until 1954, when Goldsmid and Douglass of England achieved a temperature difference of 47°F between the two junctions using a semiconductor, bismuth telluride, types negative and positive. It was predicted at that time that by the 1970s thermoelectric refrigeration would replace the vapor compression system in refrigerators and freezers. The 1966 Sears catalog offered a "Coldspot Thermo-electric Buffet Bar" that "chills foods and drinks to 40 degrees...." The buffet bar took advantage of an exciting feature of thermoelectric devices; namely, by reversing the electric current flow, the cold junctions will get hot, and the hot junctions cold. By the flick of a switch, the 2 cu ft compartment would change from a refrigerator to an oven. Due to its inefficiency, the thermoelectric refrigerator did not replace vapor compression.

The key to future advances in thermoelectric refrigeration is the discovery of a semiconductor more efficient than bismuth telluride. Until then, the only marketable applications are small portable refrigerators for cars and boats. They take advantage of the readily available direct current provided by the battery. Just plug it into the cigarette lighter and it begins to refrigerate.

J. Benjamin Horvay

See also: Air Conditioning; Appliances.

BIBLIOGRAPHY

American Home Appliance Manufacturers. (1988). *AHAM Standard for Household Refrigerators and Household Freezers.* Chicago: Author.

Appliance. (June 1992). "Tomorrow's Insulation Solution?" Oak Brook, IL: Dana Chase Publication.

ASHRAE Handbook (1981). *Fundamentals.* Atlanta, GA: American Society of Heating, Refrigerating and Air-Conditioning Engineers.

ASHRAE Handbook (1998). *Refrigeration.* Atlanta, GA: American Society of Heating, Refrigerating and Air-Conditioning Engineers.

ASHRAE Journal. (January 1999). "Washington Report—Montreal and Kyoto Protocols' Relevance to HFC's."

Bansal, P. K., and Kruger, R. (1995), "Test Standards for Household Refrigerators and Freezers I: Preliminary Comparison." *International Journal of Refrigeration* 18(1):4–20
Environmental Protection Agency. (1993). *Multiple Pathways to Super-Efficient Refrigerators.* EPA-430-R-93-008. Washington, DC: Author.
Holladay, W. L. (1994), "The General Electric Monitor Top Refrigerator." *ASHRAE Journal* September. 49–55.
National Appliance Energy Conservation Act. (1987). *Public Law 100-12.* March 17 (as amended by National Conservation Policy Act, Public Law 95-619), Part B—Energy conservation for consumer products other than automobiles. Washington, DC: U.S. Department of Energy.
United Nations Environmental Program. (1995). *Montreal Protocol on Substances that Deplete the Ozone Layer—Report on the Refrigeration, Air Conditioning and Heat Pumps Technical Options Committee.* New York: Author.
Stoecker, W. F., and Jones, J. W. (1982). *Refrigeration and Air Conditioning.* New York: McGraw-Hill Book Company.
Woolrich, W. R. (1969). "The History of Refrigeration: 220 Years of Mechanical and Chemical Cold: 1748–1968." *ASHRAE Journal* July:31–39.

REGULATION AND RATES FOR ELECTRICITY

Electric utilities have historically been franchise monopolies, vertically integrated from power production through transmission, distribution, and customer service with no competition from other electric utilities. However, in many parts of the country, electric and gas utilities do compete. Rates charged by these utilities were determined in a regulatory proceeding: Electric utilities proposed rates that compensated them for their expenses and allowed them to earn a reasonable return; state regulatory commissions reviewed and approved the proposals.

This historical relationship between electric utilities and their regulators is undergoing a dramatic change. Policymakers are restructuring and deregulating portions of the electric industry. Restructuring of the electric industry is consistent with the deregulation of other U.S. industries since the 1980s. The objective in "restructuring " is to increase efficiency, lower costs, increase customer choices, and lower the prices paid by consumers in restructured industries.

As in other industries, restructuring of the electric industry is a response to underlying conditions in the industry. Policymakers are responding to two phenomena. First, there is a disparity in retail electric prices among states and regions. For example, the average price of electricity in New England states such as New Hampshire (nearly 12 c/kWh) is almost three times as large as the price in low-cost states such as Idaho and Washington (approximately 4c/kWh). Second, advances in information technology will make it possible to perform complex real-time functions.

In restructured electric markets, the vertical electric monopoly will no longer be the sole provider of electricity. The generation, transmission, distribution, and customer service functions will be separated. The upstream generation function will be competitive, allowing new, any power producer to produce and sell electricityin any service territory. The transmission and distribution functions will continue to be regulated, but will be required to allow access to power suppliers and marketers. This separation or "unbundling" of the industry is necessary to provide nondiscriminatory access for all suppliers of electricity. Customers will have their choice of electric suppliers.

In these restructured electric markets, prices will be determined more by market forces and less by regulatory proceedings. To some, introducing competition will promote more efficient markets, providing the proper financial incentives for firms to enter or leave the industry. In this way, consumers will benefit from lower production costs and, hence, reduced electricity prices. To others, restructuring will increase electricity prices for some customers, sacrifice the current environmental and social benefits, and jeopardize system reliability of the status quo.

ADVANTAGES OF RESTRUCTURING

Consider a typical rate-making proceeding for a regulated utility. Electric utilities can recover all prudently incurred operating and maintenance costs plus an opportunity to earn a fair return on their investment. This process involves three steps: (1) determining the total amount of revenues (i.e., required revenues) that an electric utility needs; (2) allocating the total to individual customer groups (e.g., residential, commercial, and industrial customers); and (3) designing a rate structure for each customer group that allows the utility to recoup costs.

Required revenues are the sum of operation and maintenance expenses, depreciation, taxes, and a return on rate base. The rate base is the total amount of fixed capital used by the utility in producing, transmitting, and delivering electricity. The return on rate base is the weighted average cost of capital, including debt and equity sources.

Allocating required revenues to customer groups involves four steps: (1) categorizing customers into groups with similar characteristics (e.g., low-voltage customers); (2) functionalizing costs into those pertaining to production, transmission, distribution, and administration; (3) classifying functionally assigned costs into those attributable to the customer (e.g., metering costs), energy (e.g., amount of consumption), and capacity (e.g., instantaneous demand); and (4) allocating costs to customer groups. The result of this process is the allocation of all of a utility's revenue needs into customer groups. The number of groups depends on the characteristics of a utility's service territory.

Rates are then designed to recoup the revenues for each of the customer groups. Rates can be based on the amount of consumption or the type of service. Consumption-based rates are either flat, or increasing or decreasing in steps. Service-based rates depend on the type of service a utility offers its customer classes: firm rates, lifeline rates, interruptible rates, stanby rates, various incentive rates, or time-of-use rates.

The problems with this cost-of-service approach are the incentives and opportunities given to electric utilities. They are not the type of incentives that characterize an efficient market and that balance the additional risk of operating in an efficient market.

Industry restructuring advocates believe that a competitive market will create incentives to operate more efficiently and make better economic investments in utility plant and equipment. In contrast to a competitive market, cost-based ratemaking does not generally reward utilities with a higher return for making an especially good investment in plant and equipment, or penalize them for making an especially bad investment. The return that a utility earns on a highly successful investment is generally the same as its return on a less successful one. The opportunity to prosper—or, alternatively, to go bankrupt—is generally not part of this regulatory process.

In many cases, the regulators themselves distort the ratemaking process. Historically they have allowed or even encouraged utility investments that would not otherwise have been made in competitive markets.

Makers of public policy in the U.S. Department of Energy (DOE) and the Federal Energy Regulatory Commission (FERC) believe that industry restructuring will change these incentives and introduce more efficient practices and technologies in the industry. Proper pricing of electricity is one such change. Time-of-use (TOU) rates is one example. TOU rates vary over the course of a year: hour by hour, day by day, or season by season. They are theoretically appealing: Consumers who cause daily peaks bear the burden of paying higher costs during those periods. Likewise, consumers contributing to seasonal peaks—such as using air conditioners—bear the cost of building the capacity needed to meet the peaks. In practice, TOU pricing has proved effective in shaping load for electric utilities, reducing peak demand, and lowering production costs.

When initially adopted, TOU rates were based on projections of future costs by season, month, day, or hour. However, advances in metering and communications technology now afford utilities the ability to transmit prices to customers based on actual operating costs and to read meters in real time. This "real-time TOU pricing" is one of the most important aspects of many of the restructuring efforts to date. They can provide customers with direct access to the prices arising in competitive electric markets.

Under restructuring, customer exposure to market-based electric rates will broaden in other ways. Because deregulation allows customers to choose their own electric suppliers, potential suppliers are becoming more innovative in attracting customers. Although the Internet is in its infancy now, entrepreneurs are in the process of harnessing it for electricity sales. Some Internet companies are purchasing wholesale power from generating plants and reselling it to customers on the Net. Other companies are aggregators. They enlist customers on the Net and create buying pools from which they can extract lower prices from suppliers.

Restructuring will promote the introduction of other advanced technologies and practices as well. For example, the use of combined-cycle, gas turbine power plants are expected to proliferate under restructuring. These plants are generally more efficient and more environmentally benign than many fossil fuel plants currently in use.

DRAWBACKS OF RESTRUCTURING

Competition can promote efficiency and lower average prices for electricity. However, there is no guarantee that all customers will benefit equally from lower prices. Larger commercial and industrial customers generally have the wherewithal to obtain better rates in competitive markets than do smaller residential customers. Because of fewer options, low-income households are especially vulnerable to competitive markets for electricity.

Environmental and social benefits could also be jeopardized under restructuring. The elimination of integrated resource planning (IRP) is particularly of concern. In an IRP process, energy-efficiency programs and renewable-energy technologies "compete" with conventional generating plants for the resource investment expenditures of electric and gas utilities. The competition takes into account the environmental consequences of producing electricity from fossil fuels. At IRP's peak in the early 1990s, more than thirty-three states mandated the use of IRP processes. These mandates came from state legislation or from state regulatory commission orders.

As the electric industry undergoes reorganization, the retail price of electricity will be determined more and more in markets with the participation of multiple parties who do not themselves generate electricity. The "wires" portion of utilities that historically ran energy-efficiency programs argue that they are unfairly burdened by running these programs if their competitors are not also obligated to do so. Therefore, IRP processes are jeopardized in restructured electric markets. Without IRP, the environmental consequences of relying more on fossil fuels—and less on energy efficiency and renewables—are obvious.

Finally, a number of industry engineering experts and industry engineering organizations voiced concern that the electric grid may become less reliable after restructuring. The operation and maintenance of the North American electric grid depends on the coordinated interaction of more than a hundred control areas. The incentives to buy and sell power, retain adequate surplus capacity, and maintain the grid will change in a restructured electric industry. The reliability of the electric grid may suffer as a consequence.

Lawrence J. Hill

See also: Capital Investment Decisions; Eco-nomically Efficient Energy Choices; Economic Growth and Energy Consumption; Electric Power, System Protection, Control, and Monitoring of; Energy Management Control Systems; Government Intervention in Energy Markets; Subsidies and Energy Costs; Supply and Demand and Energy Prices; Utility Planning.

BIBLIOGRAPHY

Fox-Penner, P. (1997). *Electric Utility Restructuring: A Guide to the Competitive Era.* Vienna, VA: Public Utilities Reports.

Phillips, C. F., Jr. (1993). *The Regulation of Public Utilities.* Arlington, VA: Public Utilities Reports.

RENEWABLE ENERGY

Renewable energy is energy that can be replenished on a time scale appropriate to human use. Solar energy for growing plants for food is renewable because light flows continuously from the sun, and plants can be reproduced on a time scale suitable for human needs. Coal is produced continually in some geologic formations, but the time scale is on the order of hundreds of thousands of years. Accordingly, coal is considered a nonrenewable energy.

The Department of Energy categorizes sources of nonrenewable energy production as fossil fuels, nuclear electric, and pumped hydroelectric. Sources of renewable energy are delineated as conventional hydroelectric, geothermal, biofuel, solar, and wind. Conventional hydroelectricity and energy from biofuels are considered mature sources of renewable energy; the remaining types are thought of as emerging. The chart in Figure 1 depicts energy production for 1997.

More electric energy is invested in pumping water for pumped storage hydroelectricity than is produced by the electric generators. Therefore, the net production of electricity from pumped hydro sources is slightly negative and does not appear on the chart in

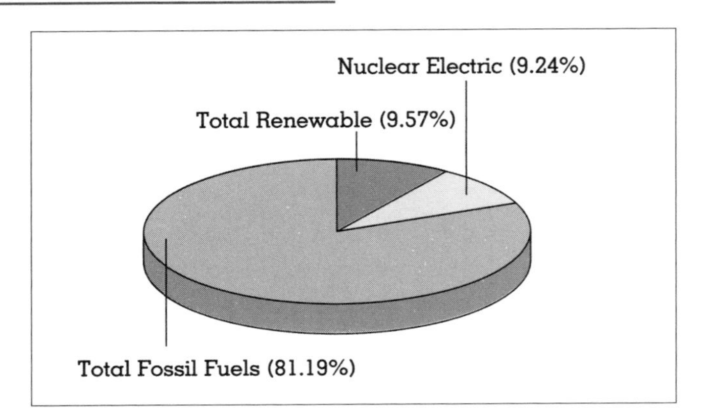

Figure 1.
Percentage of total energy production for 1997.
SOURCE: U.S. Energy Information Agency

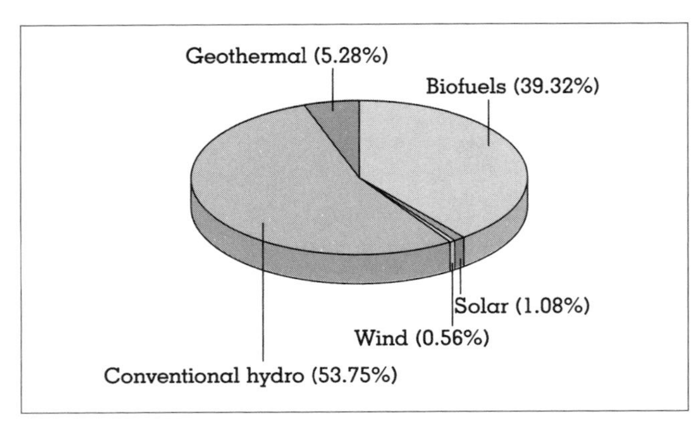

Figure 2.
Breakdown of energy production from renewable sources.
SOURCE: U.S. Energy Information Agency

Figure 1. A breakdown of the 9.57 percent contribution from renewable sources is shown in Figure 2. The great bulk of the energy produced from renewable sources is from biofuels and conventional hydroelectric systems, both of which are considered mature.

Nonrenewable energy such as coal, petroleum, and natural gas are vast but not inexhaustible. A nonrenewable resource can be preserved if it can be replaced with a renewable one. This is the prime attraction of renewable energy. Sometimes a renewable resource offers an environmental advantage. If a house heated by burning fuel oil could instead use a solar heating system, then products from burning that might contribute to environmental problems would not be released into the atmosphere. On the other hand, damming a river to make a reservoir for a hydroelectric power plant might inflict damage to the environs by destroying plant and animal habitat.

Most use of renewable energy involves the sun. Solar energy is converted to thermal energy for space heating, heating hot water for domestic use and, to some extent, for generating electricity. A photovoltaic cell, which functions like a battery, converts solar energy directly to electric energy. Biomass energy has its origin in plants grown with the help of solar radiation. Wind energy is due to unequal heating of Earth's surface by the sun. A hydroelectric plant converts the gravitational energy of water into electricity, but the mechanism that replenishes the water is powered by the sun. Roughly, 95 percent of all renewable energy is solar in origin. The remainder is from geothermal energy. Tidal energy, having its origin mainly in the gravitational force between Earth and Earth's moon, is used in some parts of the world (France, for example), but is yet to be exploited in the United States.

Figure 2 shows that biomass and hydroelectricity account for about 93 percent of renewable energy. Biomass includes wood, ethanol, and municipal solid waste. Wood pellets, manufactured from finely ground wood fiber, represent a growing market for biomass fuels for specially designed stoves and furnaces. Paper products, a biomass fuel, are the major component of municipal solid wastes. Disposal of these wastes is a major societal problem. Burning the wastes for their energy content was a popular option during the 1980s because of federal, state, and local policies promoting the construction of waste-to-energy facilities. Since then, environmental policies encouraging recycling and requiring costly pollution control have made the economics of operating a waste-to-energy facility far less favorable. Accordingly, the use of waste-to-energy facilities has declined. The technology for converting corn to ethanol is very well developed, but the degree to which ethanol becomes a substitute for gasoline will depend strongly on economics.

Solar and wind energy account for less than 2 percent of the use of renewable energy (see Figure 2). Nevertheless, there are some important developments. The greatest increase in wind energy is outside the United States; however, in 1997 the United States led the world for new wind generating systems by adding 1,620 megawatts of capacity. A power output of 1,620 megawatts is the equivalent of the power output of two large electric power plants. However, wind systems do not operate as continuously as large electric power plants, so in the long run they would not deliver nearly as much energy as two large electric power plants. Most wind energy projects are in California, but significant projects are in Texas and Minnesota.

When assessing ways to take advantage of solar energy, one should not necessarily think of sophisticated technological schemes such as roof-top collectors, photovoltaic cells, and wind-powered electric generators. Designing a house or a building to maximize the input of solar radiation during the heating season can produce significant savings in energy. If in the southern United States all the worn-out, dark-colored roofs were replaced with white ones, ten times more energy would be saved in air conditioning than is produced by all wind generators in the United States (Rosenfeld 1997).

The major facility for generating electricity from geothermal sources is at The Geysers in Northern California. Generation at The Geysers is declining both for economic reasons and because of reduced steam pressure. However, other facilities continue to produce steady quantities of electricity.

Joseph Priest

See also: Solar Energy; Hydroelectric Energy; Turbines, Wind.

BIBLIOGRAPHY

Energy Efficiency and Renewable Energy Network (EREN) Home Page. Energy Efficiency and Renewable Energy Network. August 1, 2000 <http://www.eren.doe.gov/>.

National Renewable Energy Laboratory (NREL) Home Page. National Renewable Energy Laboratory. August 1, 2000 <http://www.nrel.gov/>.

Oak Ridge National Laboratory Energy Efficiency and Renewable Energy Program Web Site. Oak Ridge National Laboratory. August 1, 2000 <http://www.ornl.gov/ORNL/Energy_Eff/Energy_Eff.html>.

Renewable Energy Technologies. Sandia National Laboratories, Renewable Energy Technologies Division. August 1, 2000 <http://www.sandia.gov/Renewable_Energy/renewable.html>.

Rosenfeld, A. H. et al. (1997). "Painting the Town White—And Green." *Technology Review* 100:52–59.

RESERVES AND RESOURCES

We need to know about the quantity and quality of oil and gas reserves because the prosperity of the world is dependent upon petroleum-based fuels. The debate about forthcoming oil shortages—as soon as 2004 or perhaps later in the twenty-first century—hinges on our understanding of petroleum reserves and future resources. Global catastrophic changes are predicted by some if a shortage occurs early in the twenty-first century; however, others are less concerned, because of new estimates of reserves and potential petroleum resources.

Our concepts of petroleum reserves and resources and their measurements are changing to reflect the uncertainty associated with these terms. Petroleum reserves have been largely calculated deterministically (i.e. single point estimates with the assumption of certainty). In the past decade, reserve and resource calculations have incorporated uncertainty into their estimates using probabilistic methodologies. One of the questions now being addressed are such as "how certain are you that the reserves you estimate are the actual reserves and what is the range of uncertainty associated with that estimate?" New techniques are required to address the critical question of how much petroleum we have and under what conditions it can be developed.

The goal of most industry and financial groups is to forecast future production rates, cash flow, net present values (NPV) and other measures designed to prudently manage financial and production aspects of the petroleum reserve. The determination of the appropriate category to which a reserve is assigned is a critically important decision. These groups emphasize the least risky category of reserves—proved (measured) reserves.—because industry and financial groups can determine the present value of a reserve from existing infrastructure, whereas other resources require additional investment and increased uncertainty regarding the ultimate asset value.

Revisions of original reserve estimates, that generally increase the earlier estimates, have caused several groups to undertake studies of these changes. Many such studies focus on the original underestimation of reserves and the supporting analysis of reserve growth. However, there are instances of original over-estimation of reserves. U.S. Geological Survey (USGS) studies on reserve growth of the world's oil fields demonstrate the significant potential of reserve growth, a potential that may be greater than that of undiscovered resources. Close estimation of the true petroleum reserves is critically important to both producers and consumers of hydrocarbons, and opinions about estimates are numerous, conflicting, and often contentious. To understand the current flux of reserves and resources, definitions are provided followed by a discussion. Definitions given are those used in western literature. Countries of the former Soviet Union (FSU) use different reserve and resource definitions. Our analysis of databases in the Volga-Ural and West Siberian Basins shows that when FSU data are converted to our western definitions, similar uncertainties exist.

DEFINITIONS

There are many definitions of reserves and resources, and we need some historical perspective to understand the current debates.

U.S. Geological Survey/U.S. Bureau of Mines

In 1976 the USGS/U.S. Bureau of Mines (USBM) defined resources and reserves (Figure 1) as follows.

Resources: A concentration of naturally occurring solid, liquid, or gaseous materials in or on the Earth's crust in such form that economic extraction of a commodity is currently or potentially feasible.

Reserves: That portion of the identified resource from which a usable mineral and energy commodity can be economically and legally extracted at the time of determination. According to the American Geological Institute (AGI), by using this resource/reserve scheme, we can update the definition: an estimate within specified accuracy limits of the valuable metal or mineral content of known deposits that may be produced under current economic conditions and with present technology. The USGS and USBM do not distinguish between extractable and recoverable reserves and include only recoverable materials.

Securities and Exchange Commission

The critical linkage between financial and accounting considerations and reserve definitions requires an understanding of the definitions given in Securities and Exchange Commission Regulation (SEC) Regulation 4-10. The SEC was empowered to develop energy accounting standards for the United States by the Energy Policy and Conservation Act of 1975 (EPCA), which was passed largely in response to the 1973 oil embargo. This act required the establishment of a national energy database including financial information. The SEC's definitions may be inherently conservative or restrictive thereby causing initial conservative estimates of reserves. The SEC definitions largely focus on proved oil and gas reserves, but are divided into two subcategories: (1) proved developed oil and gas reserves and (2) proved undeveloped reserves.

From SEC Regulation 4-10, Section 210.4-10 and as used by the industry, proved oil and gas reserves are the estimated quantities of crude oil, natural gas, and natural gas liquids that geological and engineering data demonstrate with reasonable certainty to be recoverable in future years from known reservoirs under existing economic and operating conditions (i.e., prices and costs as of the date the estimate is made). It goes on to define proved reserves if economic producibility is supported by either actual production or conclusive formation test and include that portion delineated by drilling and defined by gas-oil and/or oil-water contacts, and immediately adjoining portions not yet drilled, but which can be reasonably judged as economically productive on the basis of available geological and engineering data. In the absence of information on fluid contacts, the lowest known structural occurrence of hydrocarbons controls the lower proved limit of the reservoir. Proved reserves do not include: oil that may become available from known reservoirs but is classified separately as "indicated additional reserves"; crude oil, natural gas, and natural gas liquids, the recovery of which is subject to reasonable doubt because of uncertainty as to geology, reservoir characteristics or economic factors; crude oil, natural gas, and natural gas liquids that may occur in undrilled prospects; and crude oil, natural gas, and natural gas liquids, that may be recovered from oil shales, coal, gilsonite, and other such sources.

Proved developed oil and gas reserves are reserves that can be expected to be recovered through existing wells with existing equipment and operating methods. Additional oil and gas expected to be obtained through the application of fluid injection or other improved recovery techniques for supplementing the natural forces and mechanisms of primary recovery

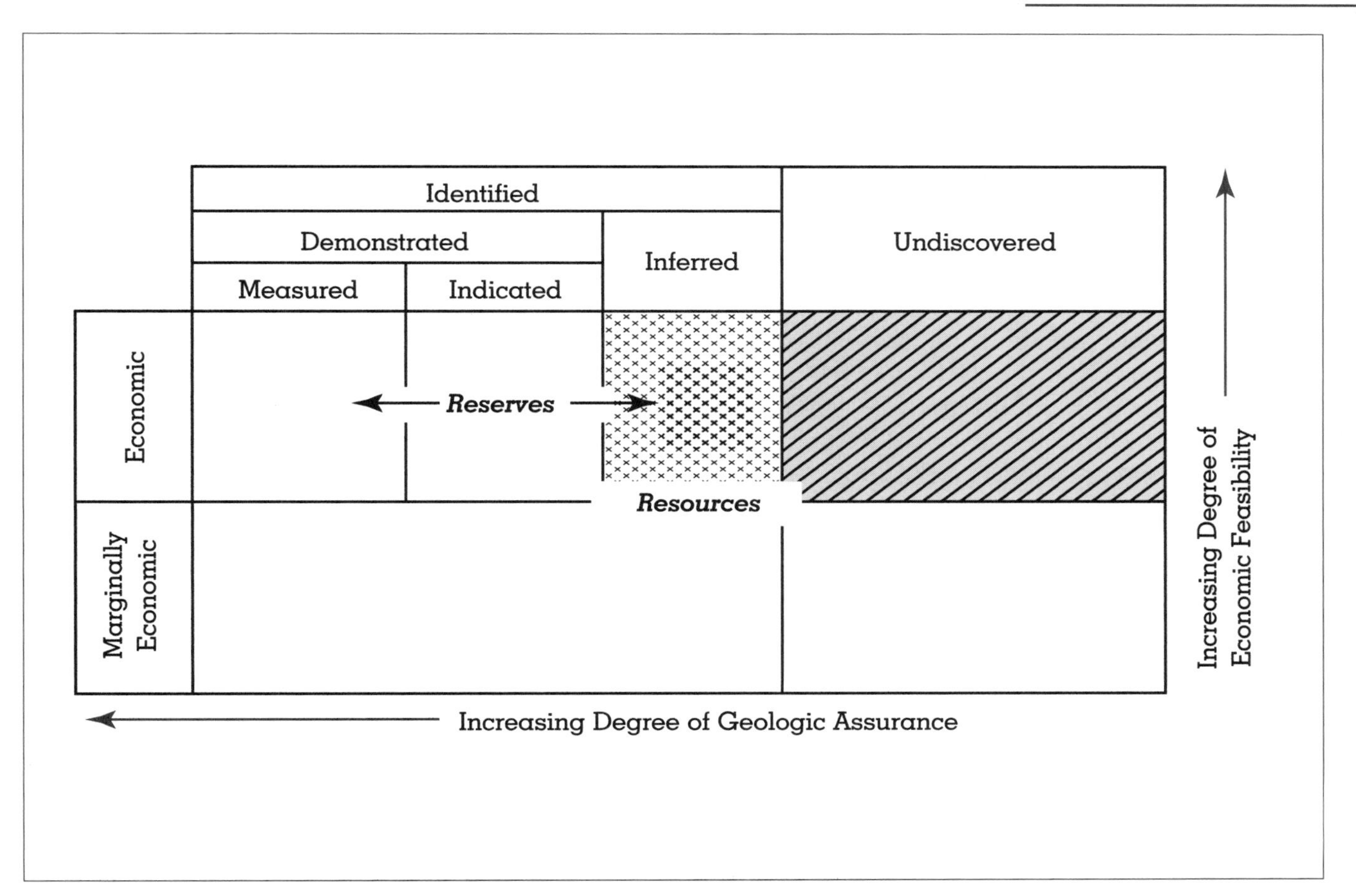

Figure 1.
Reserves vs. Resources Chart.
SOURCE: USGS/U.S. Bureau of Mines, 1976

should be included as "proved developed reserves" only after testing by a pilot project or after the operation of an installed program has confirmed through production response that increased recovery will be achieved.

Proved undeveloped oil and gas reserves are reserves that are expected to be recovered from new wells on undrilled acreage, or from existing wells where a relatively major expenditure is required for recompletion. Reserves on undrilled acreage shall be limited to those drilling units offsetting productive units that are reasonably certain of production when drilled. Proved reserves for other undrilled units can be claimed only where it can be demonstrated with certainty that there is continuity of production from the existing productive formation. Under no circumstances should estimates for proved undeveloped reserves be attributable to any acreage for which an application of fluid injection or other improved recovery technique is contemplated, unless such techniques have been proved effective by actual tests in the area and in the same reservoir.

The World Petroleum Congress/Society of Petroleum Engineers

In 1997, the World Petroleum Congress (WPC) and the Society of Petroleum Engineers (SPE) jointly published petroleum reserve definitions that added the element of probability to the deterministic definitions in common use. The WPC/SPE definitions build on the SEC definitions by including probabilistic estimates.

Reserves are those quantities of petroleum which are anticipated to be commercially recovered from known accumulations from a given date forward.

Proved reserves are those quantities of petroleum which, by analysis of geological and engineering data, can be estimated with reasonable certainty to be commercially recoverable, from a given date forward, from known reservoirs and under current economic conditions, operating methods, and government regulations. Proved reserves can be categorized as developed or undeveloped. If deterministic methods are used, the term reasonable certainty is intended to express a high degree of confidence that the quantities will be recovered. If probabilistic methods are used, there should be at least a 90 percent probability that the quantities actually recovered will equal or exceed the estimate.

Unproved reserves are based on geologic and/or engineering data similar to that used in estimates of proved reserves; but technical, contractual, economic, or regulatory uncertainties preclude such reserves being classified as proved. Unproved reserves may be further divided into two subcategories: probable reserves and possible reserves.

Probable reserves are those unproved reserves which analysis of geological and engineering data suggest are more likely than not to be recoverable. In this context, when probabilistic methods are used, there should be at least a 50 percent probability that the quantities actually recovered will equal or exceed the sum of estimated proved plus probable reserves.

Possible reserves are those unproved reserves which analysis of geological and engineering data suggests are less likely to be recoverable than probable reserves. In this context, when probabilistic methods are used, there should be at least a 10 percent probability that the quantities actually recovered will equal or exceed the sum of estimated proved plus probable plus possible reserves.

The reserves identified in the upper left part of Figure 1, measured reserves, have greater economic feasibility and greater geologic certainty than indicated or inferred reserves. Undiscovered resources have even greater uncertainty than the previous categories. Many industry groups use the terms proved, probable, possible and speculative that roughly correlate to USGS categories of measured, indicated, inferred, and undiscovered. Proved, probable and possible reserves are commonly called P1, P2, and P3 respectively (Figure 2) and considerable analysis has been applied to relationships among these reserve categories to estimate ultimate recoverable resources.

Technological advances, particularly 3-D seismic techniques and deepwater drilling technology, have revolutionized our ability to discover and develop petroleum reserves. Technology has accelerated the rate of discovery of new global petroleum resources throughout the 1990s. Deepwater offshore drilling technology (up to 3,000 m water depths) permit exploration of many areas that were previously inaccessible, such as offshore west Africa and offshore South America. This geographic expansion of petroleum resource development into offshore areas provides opportunities for many non-OPEC nations. The discovery of significant oil resources in the Western Hemisphere, particularly in deepwater areas of the south Atlantic Ocean and Gulf of Mexico, are important to the United States, which seeks oil from more widely distributed potential supply sources. Any regional optimism springing from this discovery is mitigated by the realization that future contributions from field growth, which according to the USGS are nearly as large as undiscovered petroleum resource potential, will come from the areas where fields are already discovered.

Natural gas is an underutilized resource worldwide relative to oil but will play an increased role in the twenty-first century. New technologies such as gas-to-liquids and large-scale liquefied natural gas projects will also dramatically alter the fossil fuel landscape.

DISCUSSION

The definitions above are an abbreviated version of those used in a very complex and financially significant exercise with the ultimate goal of estimating reserves and generating production forecasts in the petroleum industry. Deterministic estimates are derived largely from pore volume calculations to determine volumes of either oil or gas in-place (OIP, GIP). This volume when multiplied by a recovery factor gives a recoverable quantity of oil or natural gas liquids—commonly oil in standard barrels or natural gas in standard cubic feet at surface conditions. Many prefer to use barrels of oil equivalency (BOE) or total hydrocarbons for the sum of natural gas, natural gas liquids (NGL), and oil. For comparison purposes 6,000 cubic feet of gas is considered to be equivalent to one standard barrel on a British thermal unit (Btu) basis (42 U.S. gallons).

Accommodation of risk (or uncertainty) in reserve estimations is incorporated in economic decisions in several ways. One common method

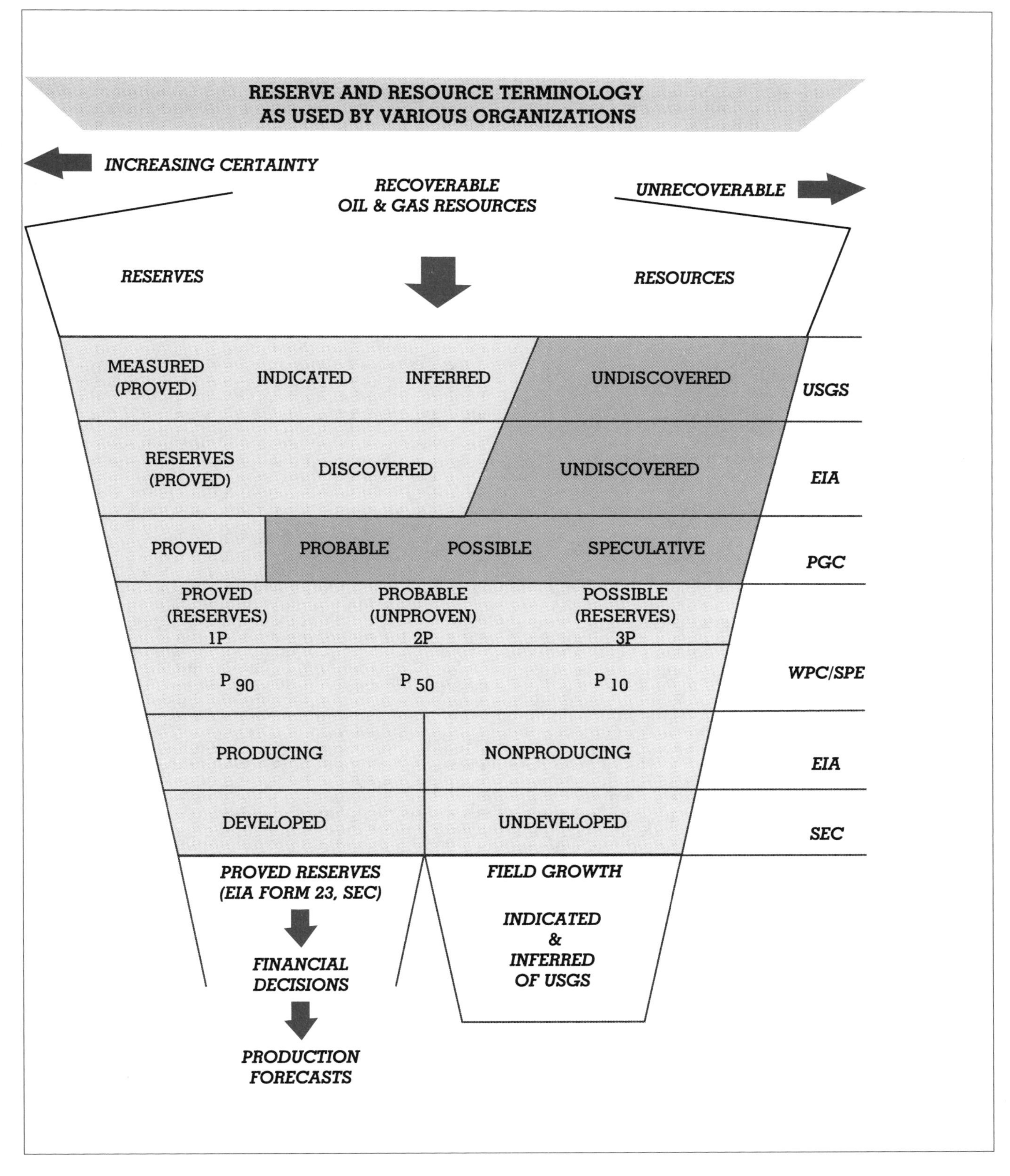

Figure 2.
Reserve and Resource Terminology.

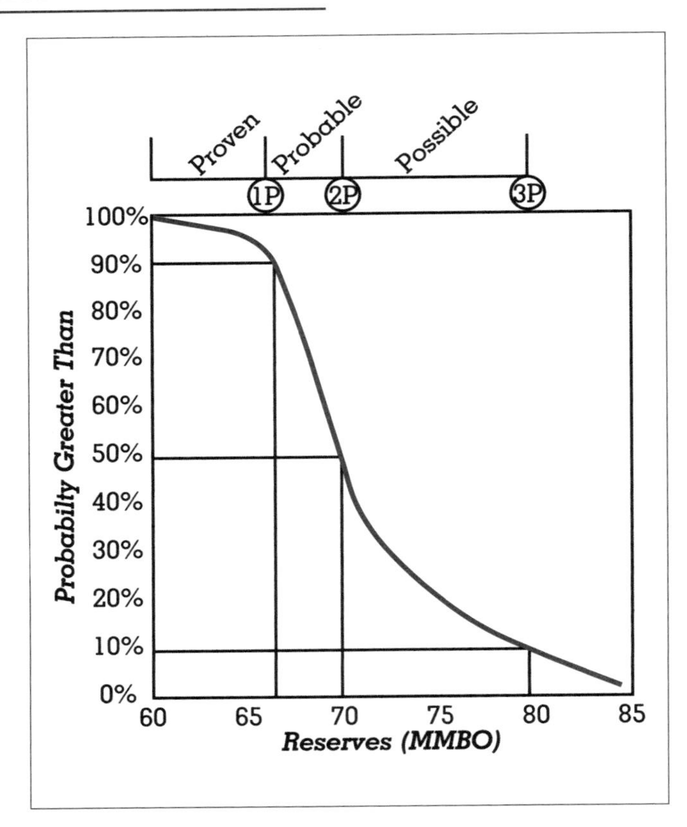

Figure 3.
Proven, Probable, and Possible Reserves and Associated Probabilities
As defined by the World Petroleum Congress/Society of Petroleum Engineers (modified from Grace, J. D. et al. (1993). "Comparative Reserves Definitions: U.S.A., Europe, and the Former Soviet Union." *Journal of Petroleum Technology* 45(9):866-872.

incorporates the expected value (EV) which is the probability-weighted average of all possible outcome values. This process is called "decision analysis"; however, the estimation or production forecast process has become so important that even more sophisticated analysis is commonly undertaken. In regions such as the North Sea, operating costs are high and critical infrastructure decisions are needed early in order to maximize the value of the reserve. Oil price, transportation costs, capital expense, operating expenditures, production rates, and closely estimated reserve volumes are the principal parameters used in developing an evaluation strategy. Reserve estimates can change quite dramatically. The United States is by far the most intensely developed petroleum region in the world with more than 3.5 million wells drilled and a long history of reserve estimation that shows substantial increases in ultimate recoverable reserve estimates. However, negative revisions also occur; either overestimation or underestimation of reserves can be costly. Petroleum reserve estimates are one of the parameters used in determining the EV of a field's reserves. In order to accommodate uncertainty in the various reserve categories, the reserve estimates are probability-weighted and used to determine the EV as shown in the following example and in Figure 3. Although a company may believe they have 3,525,000 barrels of recoverable oil, their probability-weighted reserve is only 31 percent of that figure or 1,093,750 barrels of recoverable oil. This procedure, repeated over and over in many fields throughout the world, has led to underestimation of ultimate recoverable reserve estimates. The subsequent upward revision of these initial conservative estimates is known as reserve growth or field growth. In the United States field growth accounts for twice the volume of future potential reserves of oil (i.e., 60 billion barrels) as undiscovered oil resources (i.e., 30 billion barrels) according to USGS 1995 estimates. In the world as a whole, field growth, based on analysis of the PetroConsultants datafile for a sixteen year period (1981–1996), accounts for nearly 500 billion barrels of oil for just those fields discovered since 1981.

Although, initially, deterministic or point value estimates of reserves were largely believed to be adequate, it is now clear that many factors complicate reserve estimation for recoverable oil or gas. Some complicating geologic and reservoir factors include: variations in reservoir architecture and quality (low complexity or high complexity), microscopic displacement efficiency, volumetric sweep efficiency (the effectiveness of hydrocarbon recovery from secondary and tertiary recovery processes), data quality and analysis. In the North Sea study, these variables are quantified and scored to improve field reserve estimates. In another study of North Sea fields, 65 percent of the responding countries used deterministic ultimate recovery estimates, whereas 53 percent used probablistic estimation methods. Inaccuracies of ultimate recovery and production forecasts were very expensive to those North Sea operators who either underestimated or overestimated facilities requirements. The conclusion that more unified and refined reserve definitions are needed is almost uni-

Reserve Category	Total Company (boe)	Probability	Probability Weighted Reserve (boe)
Proved producing	600,000	1.00	600,000
Proved developed non-producing (waiting on pipeline connection)	50,000	0.90	45,000
Proved behind pipe	125,000	0.75	93,750
Proved undeveloped (within 1 kilometer of producing well)			
Basin A	200,000	0.50	100,000
Basin B	50,000	0.40	20,000
Probable	500,000	0.35	175,000
Possible (2,000 hectares of exploratory leases)	2,000,000	0.03	60,000
	3,525,000		**1,093,750**

Table 1.
Probability Weighted Reserve Estimates
Modified from Seba, R. D. (1998). *Economics of Worldwide Petroleum Production.* Tulsa, OK: Oil and Gas Consultants International Inc.

versally supported. Appraisals at reservoir level are on a smaller scale than at field level. There are three basic ways to describe uncertainty in relation to reservoir and estimation models: fuzziness, incompleteness, and randomness. Fluctuations in the price of oil or natural gas affect the recoverability of the resource base and as prices increase, reserves increase as they become more economically viable. Advances in exploration and production technology also make recoverability more viable and enhance the amount of reserves that can be extracted from any given reservoir.

IMPLICATIONS

It is well known that reserve estimates made early in the development of a field are often wrong and that there's a 50 percent change in estimated ultimate recovery in many fields during the first ten years. In addition, the average field lifetime has been a decade longer than initially expected in the North Sea. If you believe in reserve growth, you would conclude that there will not be a petroleum crisis anytime soon. If, on the other hand, you believe that reserves have been overstated and you make negative revisions to reserve estimates, there will be a crisis soon. Current known petroleum reserves (proved + P50 probable) in the world are 1.44 trillion barrels of oil, 5,845 trillion cubic feet of gas, and 80 billion barrels of natural gas liquids based on 1996 PetroConsultants data. Depending upon the volume of undiscovered and field growth reserves added to these known reserves, you can develop either pessimisitic or optimistic scenarios.

In 1997, net U.S. petroleum imports of 8.9 million barrels of oil per day were worth $67 billion and exceeded U.S. petroleum production of 8.3 million barrels of oil per day. Concerns about oil shortages caused the United States to build a strategic petroleum reserve in 1977 that currently holds 563 million barrels of oil or about two months of net imported petroleum. Although domestic oil and gas production declined from 1970 to 1993, natural gas production has increased since the mid-1980s and energy equivalent production from natural gas exceeded domestic oil production in the late 1980s. Estimates in 1999 of per well oil production for the United States have fallen to 11.4 barrels a day, the lowest value in the last forty-five years. Considering this decline, the world's reserves and resources are critical to our future because the United States is the largest consumer of petroleum resources (~22% of world consumption). Increasingly, the United States relies on foreign oil supplies. Determining how much oil the world has and where it is located will continue to focus our attention on reserve estimation.

Thomas S. Ahlbrandt

See also: Energy Economics; Oil and Gas, Drilling for; Oil and Gas, Exploration for; Oil and Gas, Production of; Supply and Demand and Energy Prices.

BIBLIOGRAPHY

Ahlbrandt, T. S. (1999). "USGS New Millennium World Oil and Gas Assessment." *Energy Mix of the Future Session, Abstracts with Program*, vol. 31, no.7 (October). Denver, CO: Geological Society of America Annual Meeting.

American Geological Institute. (1997). *Glossary of Geology*, 4th ed., edited by J. A. Jackson. Alexandria, VA: Author.

Attanasi, E. D. and Root, D. H. (1994). "The Enigma of Oil and Gas Field Growth." *American Association of Petroleum Geologists* 78:321–332.

Brock, H. R.; Jennings, D. R.; and Feiten, J. B. (1996). *Petroleum Accounting: Principles, Procedures and Issues*, 4th ed. Denton, TX: Price Waterhouse Coopers, Professional Development Institute.

Campbell, C. J. (1997). *The Coming Oil Crisis*. Brentwood, United Kingdom: Multi-Science Publishing Company and Petroconsultants

Curtis, J. B. (1999). "Comparison of Estimates of Recoverable Natural Gas Resources in the United States: A Report of the Potential Gas Committee." *Gas Resource Studies Number 7*. Golden, CO: Colorado School of Mines.

Dromgoole, P., and Speers, R. (1997). "Geoscore: A Method for Quantifying Uncertainty in Field Reserve Estimates." *Petroleum Geoscience* 3:1–12.

Edwards, J. D. (1997). "Crude Oil and Alternate Energy Production Forecasts for the Twenty-first Century: The End of the Hydrocarbon Era." *American Association of Petroleum Geologists Bulletin* 81:1292–1305.

Energy Information Administration. (1998). *International Energy Outlook 1998*. Washington, DC: Department of Energy/Energy Information Administration.

Energy Information Administration. (1999). *Annual Energy Review 1998*. Washington, DC: Department of Energy/Energy Information Administration.

Foley, L.; Ball, L.; Hurst, A.; Davis, J.; and Blockley, D. (1997). "Fussiness, Incompleteness and Randomness: Classification of Uncertainty in Reservoir Appraisal." *Petroleum Geoscience* 3:203–209.

Grace, J. D.; Caldwell, R. H.; and Heather, D. I. (1993). "Comparative Reserves Definitions: U.S.A., Europe, and the Former Soviet Union." *Journal of Petroleum Technology* 45(9):866–872.

International Energy Agency. (1998). *World Energy Outlook*. Paris, France: International Energy Agency/OECD.

Klett, T. R.; Ahlbrandt, T. S.; Schmoker, J. W.; and Dolton, G. L. (1997). *Ranking of the World's Oil and Gas Provinces by Known Petroleum Volumes*. Denver, CO: U.S. Geological Survey, Open File Report 97–463, CD-ROM.

Masters, C. D.; Attanasi, E. D; and Root, D. H. (1994). *World Petroleum Assessment and Analysis: Proceedings of the 14th World Petroleum Congress*. London, England: John Wiley and Sons.

Petroconsultants, Inc. (1996). *Petroleum Exploration and Production Database*. Houston, TX: Author.

Potential Gas Committee. (1999). "Potential Supply of Natural Gas in the United States. A Report of the Potential Gas Committee." *Potential Supply of Natural Gas, 1998*. Golden, CO: Colorado School of Mines.

Schmoker, J. W., and Klett, T. R. (1999). "U.S. Geological Survey Assessment Model for Undiscovered Conventional Oil, Gas, and NGL Resources—The Seventh Approximation." *U.S. Geological Survey Electronic Bulletin 2165*. Golden, CO: U.S. Geological Survey.

Schuyler, J. (1999). "Probabilistic Reserves Definitions, Practices Need Further Refinement." *Oil and Gas Journal*, May 31.

Seba, R. D. (1998). *Economics of Worldwide Petroleum Production*. Tulsa, OK: Oil and Gas Consultants International Inc.

U.S. Geological Survey. (2000). *U.S. Geological Survey World Petroleum Assessment 2000—Description and Results*. Washington, DC: U.S. Government Printing Office.

U.S. Geological Survey. (1995). "National Assessment of United States Oil and Gas Resources." *U.S. Geological Survey Circular 1118*. Washington, DC: U.S. Government Printing Office.

U.S. Geological Survey and U.S. Bureau of Mines. (1976). "Principles of the Mineral Resource Classification System of the U.S. Bureau of Mines and U.S. Geological Survey." *U.S. Geological Survey Bulletin 1450-A*. Washington, DC: U.S. Government Printing Office.

RESIDUAL FUELS

A residual fuel oil is any petroleum-based fuel which contains the undistilled residue from atmospheric or vacuum distillation of crude oil and may be called Bunker Fuel Oil, No. 6 Fuel Oil, or heavy Fuel Oil. When diluted with distillate to reduce viscosity, it may be called Marine Diesel Intermediate Fuel, No. 5 Fuel Oil, No. 4 Fuel Oil, or Light Fuel Oil) (Table 1). It is higher in viscosity than a distillate fuel, so it normally requires preheating for pumping and atomization. Asphaltenes, which are high molecular weight condensed aromatics in residual fuel, are difficult to burn completely, so combustion tends to give particulate emissions. Residual fuel usually contains organometallic compounds of vanadium, which leave an ash residue that may cause deposits and corrosion in boilers and engines because residual fuels are higher in sulfur and nitrogen than distillates. The emission of sulfur and nitrogen oxides during combustion is much greater. Carbon content is also higher than for distillate fuels or gas, so it emits more CO_2 than they do, but less than coal. Basically residual fuels are more difficult to handle and tend to pollute more than distillate or gaseous fuels, but less than solid fuels such as coal.The major advantage of residual fuel is that it is the cheapest liquid fuel available, in large part because it is a byproduct of the refining of gasoline and diesel fuel. As such, it sells for less per gallon than crude oil, the raw material from which it is made. Since heavy fuel prices are usually 50 percent to 75 percent of the price of crude, refiners try to minimize production of heavy fuel oil and maximize the production of higher-value gasoline and distillate fuels.

In order to use this low-priced fuel, consumers must make a significant investment in handling equipment. Heavier grades need preheat equipment to maintain 50°C (122°F) for pumping in the distribution system, and 135°C (275°F) for atomization in boilers or diesel engines. A relatively long residence time is needed in the boiler or engine to give complete burnout of the fuel asphaltenes, which tend to produce particulates. Thus, residual fuels are normally used only in relatively large installations.

No. 6 Fuel Oil, Bunker Fuel Oil	
Viscosity, Cst @ 50C	500
Density, Kg/Cu.Meter	985
Flash Point, Deg C	60
Energy Content, KJoule/Kg.	43,000
Chemical Composition	
Carbon, Wt.%	86
Hydrogen, Wt.%	11.5
Sulfur, Wt.%	2.5
Ash Content, Wt.%	0.08
Vanadium, PPM	200
Lower Viscosity Blends	
Marine Diesel Fuel	
Intermediate Fuel, IF 180 Cst @ 50 C	180
Intermediate Fuel, IF 380 Cst @ 50 C	380
No. 5 Fuel Oil, Cst @ 50 C	40
No. 4 Fuel Oil, Cst @ 40 C	20

Table 1.
Typical Residual Fuel Properties
SOURCE: ASTM T.P.T. Course Notes "Marine Fuels: Specifications, Testing, Purchase, and Use."

In the earliest days of refining, kerosene was the principal refinery product. All the crude heavier than kerosene was sold as residual fuel and used initially for industrial and commercial steam generation. Residual fuel soon began to displace coal in steamships and railroad locomotives because of its portability and relatively low ash content. In the 1920s, residual fuel underwent a drastic change in composition because of the installation of thermal cracking (a refinery process in which a heavy distillate or residual fraction of petroleum is subjected to high temperature, which causes large, high boiling molecules to crack into smaller ones which boil in the gasoline or heating oil range) to meet gasoline demand. Most heavy fuels contained large amounts of thermal tars and visbreaker products (products of a thermal racking process). Great care had to be taken when blending these fuels to avoid sediment formation due to incompatibility.

With the development of catalytic cracking in the 1930s and 1940s, residual fuel composition changed again. Vacuum distillation came into use to provide additional clean feed for catalytic cracking. Vacuum distillates, which had previously been part of residual fuel, were now converted to gasoline and heating oil in the catalytic cracker.

The residuum from vacuum distillation became, and still is, the basic component of residual fuel oil. It contains the heaviest fraction of the crude, including all the ash and asphaltenes. It is extremely high in viscosity and must be diluted with light distillate flux (a low viscosity distillate or residual fraction which is blended with a high viscosity residual fraction to yield a fuel in the desired viscosity range) to reach residual fuel viscosity. The lowest value distillates, usually cracked stocks, are used as flux. In some cases the vacuum residuum is visbroken to reduce its viscosity so that it requires less distillate flux.

By the end of World War II the use of residual fuel oil in the United States had reached about 1.2 million barrels per day. The bulk of this use was in industrial/commercial boilers, railroad locomotives, and steamships. Shortly thereafter, railroad use declined rapidly as diesel engines, which used distillate fuel, replaced steam locomotives. In the 1950s and 1960s residual fuel oil use for marine and industrial applications, as well as for electric power generation, con-

	United States		Worldwide	
	Electric Power Generation	Industrial & Commercial	Diesel Ships	Steam Ships
1973	1406	1421	2591	1085
1978	1612	1411	2101	700
1983	652	769	1418	245
1988	646	732	1803	240
1993	424	656	2171	227
1997	301	496	2451	210

Table 2
Residual Fuel Oil Consumption
SOURCE: Source: U.S. Energy Information Administration. (1998). *Monthly. Energy Review* (October).

tinued to grow. During this period, the use of fuel oil for marine propulsion gradually shifted from steamships to low and medium speed diesel engines, which were gradually displacing steam turbine propulsion because they were more efficient.

By 1973 about 1.4 million barrels per day of residual fuel oil were used for electric power generation in the United States. This accounted for 16.8 percent of U.S. electricity generation, mostly in areas where cheap, foreign heavy fuel could be delivered by tanker. That same year, another 1.4 million barrels per day of heavy fuel oil were used in the United States for industrial and commercial applications. Worldwide during 1973 about 2.6 million barrels per day of residual fuel oil were used in marine diesel engines, and another 1.1 million barrels per day were used for steamship propulsion.

After the energy crisis of the 1970s, the price of heavy fuel oil rose dramatically and demand dropped drastically. By 1983, heavy fuel oil for electric power generation in the United States fell to 650 thousand barrels per day, providing just 6.2 percent of the total generated. Industrial/commercial demand fell almost as far, to 770 thousand barrels per day. Worldwide demand for marine diesel fuel had dropped to 1.4 million barrels per day and steamship demand to 245 thousand barrels per day. Since 1983, the demand in each of these sectors except marine diesel fuel has continued to decrease. Diesel engines, which are more efficient than steam turbine systems, have become the dominant means of ship propulsion. By 1997, the demand for marine diesel fuel was back to about 2.4 million barrels per day. It is forecast to stay at about that level or slightly higher.

Refiners have coped with this decrease in demand for residual fuel oil by making as little as possible and by shifting their production to heavier grades. They no longer make lighter grades, such as No. 4 or No. 5 fuels, except by special arrangement. Consumers who previously used these grades now use No. 6 fuel oil or gas. Similarly, most marine diesel fuel oil is now supplied as IF 180 (180 Cst @ 50°C) or IF 380 (380 Cst @ 50°C). Prior to 1973, many marine diesel engines were operated on lighter grades. Switching from IF 60, which is roughly 29 percent distillate and 71 percent bunker fuel (the residual fuel burned in boilers on steamships), to IF 380, which is 2 percent distillate and 98 percent bunker fuel, would increase bunker fuel consumption by 38 percent and reduce distillate consumption by 93 percent. Such switches have been beneficial to both refiners and consumers since the energy crisis. Refiners supply more residual and less distillate to a given consumer, and the consumer pays a lower price for the heavier grade of fuel. In most cases, the diesel engine operates properly on the heavier fuel.

While some refiners have reduced residual fuel production by supplying heavier grades, others have eliminated residual fuel completely by installing cokers or hydrocrackers. These are process units that convert residua to gasoline or distillate. They are very expensive to install and operate, but can be justified when there is an oversupply of residual fuel.

Charles W. Siegmund

BIBLIOGRAPHY

Guthrie, V. B. (1960). *Petroleum Products Handbook.* New York: McGraw-Hill.

Energy Information Administration. (1998). *Monthly Energy Review* (October). Washington, DC: U.S. Depart-ment of Energy.

Siegmund, C. W. (1997). *Marine Fuels: Specifications, Testing, Purchase, and Use.* ASTM Technical & Professional Training Course Notes. West Conshocken, PA: ASTM.

REST ENERGY

See: Einstein, Albert; Matter and Energy

RIDESHARING

See: Traffic Flow Management

RISK ASSESSMENT AND MANAGEMENT

By their very nature, operations in the energy industry are characterized by high risk. Operating an oil well, a power plant, or petrochemical plant is considerably more complex and costly than running most other operations. Losses may be infrequent, yet when they occur they could be substantial. The risks involved in these kinds of operations can be classified under two categories—technical (engineering) risk and financial (price) risk. This article focuses on Financial Risk.

PRICE RISK

The price risk can be defined and understood in alternative ways. One can view the risk as the probable fluctuation of the price around its expected level (i.e., the mean). The larger the deviation around the mean the larger is the perceived price risk. The volatility around the mean can be measured by standard deviation and be used as a quantitative measure for price risk. At the same time, in the industry it is common to define "risk" referring only to a price movement that would have an adverse effect on the profitability. Thus, one would talk about an "upward potential and downside risk."

ENERGY PRICE DYNAMICS

Unpredictable movements in the price level are uncommon in energy markets. The magnitudes of these price shocks can be substantial: 1973 and 1979 oil price shocks, electricity price swings in June 1998, and 2000 oil price increases exemplify the potential magnitudes of these price fluctuations. Energy price dynamics usually consist of three components: deterministic part, seasonal and cyclical influences, and noise. In a market situation with no demand or supply shocks, the only uncertainty observed in the market will be due to noise. A number of variables, such as import policies, policy changes of industry organizations (e.g., OPEC); unexpected weather and atmospheric conditions, tax and regulatory policy changes, legal, environmental, and economic problems, and political and currency crises can affect the demand/supply for energy products and thus lead to large price swings.

DIVERSIFIABLE AND NONDIVERSIFIABLE RISK

Market-wide (governmental, political, legal, environmental, or economic) events may affect individual firms in the energy industry differently. The magnitude of the individual impact will be determined by the co-movement of the market and individual firms (systematic risk). At the micro level, companies are not only exposed to price and systematic risk, but also to firm-level, nonsystematic risk factors. At the market level, these individual firm-level risks can be reduced or diversified, making it possible to focus on price management only. But at the individual firm level, risk managers also have to address a number of engineering risk factors, such as windstorms that may damage offshore platforms, explosions or fire in the well, and vapor cloud explosions.

MANAGING RISK

Beause energy firms face a variety of potential risk factors that may lead to substantial price fluctuations, risk managers in energy industries combine traditional insurance strategies with financial instruments and hedging policies to manage the risk. The standard financial hedging instruments that are employed are futures and options contracts traded on several energy products, such as crude oil, heating oil, gasoline, natural gas, and electricity.

Energy companies' property/casualty insurance programs typically address a wide spectrum of exposures, such as political risk, earthquake, and workers' compensation. Traditional insurance contracts have been used in energy markets for years against property damage and casualty, for reducing coastal hurricane liability and similar cost reduction purposes. Due to the many great risks that energy companies face, they commonly have to reduce some of the total

The seismic braces seen here in the Diablo Canyon Nuclear Power Plant reduce the risks associated with an area prone to earthquakes. (Corbis Corporation)

risk so that they can afford the insurance to undertake a project.

A recent financial concept that is becoming popular is reinsurance, a transaction whereby one insurance company agrees to indemnify another insurance company against all or part of the loss that the latter sustains under a policy or policies that it has issued. For this service, the ceding company pays the reinsurer a premium. The purpose of reinsurance is the same as that of insurance (i.e., to spread risk). With financial reinsurance in place, energy companies can use traditional excess insurance to address their higher levels of coverage. The combined risk management strategy uses a financial instrument in combination with an insurance policy. This improves the ability of energy companies to fund their own losses during periods of high prices and profits, yet in unfavorable market conditions, the companies can tap an arranged insurance capacity to decrease their financial vulnerability. This strategy is attractive for energy companies because it limits the probability of having to pay for losses to cases where the insurance is actually needed.

There are several other financial risk management tools used in energy markets. For example, option-trading strategies are employed to avoid the adverse effects of price movements. Companies can also protect themselves from volatile (or stable) prices by trading various option strategies. They can trade on the cost of processing products by buying and selling contracts on the raw and processed goods. Recent contributions to risk management in energy industries comprise instruments and strategies such as bidding hedges and weather swaps. A bidding hedge is used to protect against the financial exposures in a competitive bid to construct an energy facility (e.g., currency risks, interest rate, and price risks). Weather swaps allow companies to exchange a series of variable cash flows for a series of fixed cash flows dependent on an index based on weather statistics. One common application of this class of instruments is the temperature swap that attracts many energy companies' interest whose products are sensitive to weather conditions and temperature (e.g., heating oil).

Recent trends in energy finance can be outlined as follows: The focus of main energy sources is turning away from coal and other high CO_x emitting sources to natural gas. Investment in crude oil is slowing. Nuclear power plants remain unpopular due to high operational risk and high waste clean-up costs. On the other hand, investments in the electricity sector are growing at a rapid rate; they have increased from $500 million in 1988 to over $10 billion in 1997.

Ahmet E. Kocagil

See also: Efficiency of Energy Use; Energy Economics; Futures; Supply and Demand and Energy Prices.

BIBLIOGRAPHY

Buljevich, E. C., and Park, Y. S. (1999). *Project Financing and the International Financial Markets.* Boston: Kluwer Academic Press.

Fusaro, P. (1998). *Energy Risk Management.* New York: McGraw-Hill.

Helbling, C. P.; Fallagger, G.; Hill, D. (1996). *Rethinking Risk Financing.* Zurich. Switzerland: Swiss Reinsurance Company.

McKechnie, G. (1983). *Energy Finance.* London: Euromoney Publications.

The Petroleum Economist. (1991). *Energy Finance Handbook,* London: Author.

Risk Books. (1999). *Energy Modelling and the Management of Uncertainty.* New York: Author.

ROCKET PROPELLANTS

Rocket propellant is a mixture of combustible substances that is burned inside the combustion chamber of a rocket engine. Burning is the chemical process of decomposition and oxidation of the propellant. The resulting highly heated and compressed gas (propulsive mass) is ejected from a combustion chamber and facilitates propulsion—movement of the aggregate attached to the rocket engine. In physical terms, combustion converts chemical energy into kinetic energy.

Rocket propellants possess unique properties, such as a capability to self-sustain the burning process, generate thermal energy, and simultaneously produce propulsive mass. Some types of propellants are even able to self-ignite (initiate burning without outside power input). Unlike most other combustible chemicals, rocket propellants can burn in vacuum. This is because a propellant consists of two integral components: a fuel that burns and produces propulsive mass and an oxidizer that facilitates and sustains oxidation. In this respect, rocket propellants are more like explosives than like automobile and aviation fuels used that require atmospheric air for oxidation. The major difference between explosives and rocket propellants is the gas expansion rate which is much slower in propellants and makes it possible to contain and control the process of burning inside the rocket engine. Most modern rockets use solid and liquid propellants.

SOLID PROPELLANTS

A solid propellant is a mechanical (heterogeneous) or a chemical (homogeneous, or colloidal) mixture of solid-state fuel and oxidizer-rich chemicals. Specially-formed charges of solid propellant (grains) are placed in the combustion chamber of the solid rocket motor (SRM) at a production facility. Once assembled, the engine does not require additional maintenance, making it simple, reliable and easy to use.

The earliest known rocket engines used gunpowder as a solid propellant. Composed of 75 percent potassium nitrate, 10 percent sulfur, and 15 percent charcoal, gunpowder was probably invented in China in the tenth century. One of the first references to rockets can be found in Chinese chronicles from 1232. Rockets began to appear in Europe in thirteenth century. A century later, they were used for various purposes, mostly for fireworks, in France and Italy. The term "rocket," which became common in fifteenth or sixteenth century, originated from the Italian word *rocchetta* (spindle), a reference to the shape of early rockets. In seventeenth and eighteenth centuries, rockets spread all around Europe. In the nineteenth century they were widely used in Great Britain, Russia, France, Austria,

The space shuttle *Challenger* lifting off. (NASA)

Germany and other European countries as a weapon: an auxiliary light-weight artillery. During the same period, British rockets were brought to the American continent. Solid rocket motors evolved with the introduction of smokeless gunpowder, a far more energy-efficient propellant. Unlike the black gunpowder that was a mechanical mixture, smokeless gunpowder was a complex compound developed chemically from cellulose. Rockets of this type first appeared in the early twentieth century but were not widespread until the 1940s. Unguided short-range barrage missiles with smokeless gunpowder propellant were used extensively in World War II by the Soviet Union, United States and Germany. Germans built the first long-range solid propellant missile, called *Rheinbote*. A composite solid propellant developed in the United States in 1941 facilitated a real revolution in solid rocket technology in the 1950s and1960s. It allowed the production of very large and powerful solid propellant engines that became suitable for long-range ballistic missiles and space launch systems.

Modern composite solid propellant is a mechanical mixture of the powder-like chemicals and a binding resin. The propellant used for the Space Shuttle solid rocket boosters (SRBs) is a typical example of such mixture:

- Ammonium perchlorate (oxidizer), 69.93%
- Aluminum powder (fuel), 16.00%
- Rubber-based polymer, called PBAN (binder), 12.04%
- Epoxy-based curing agent, 1.96%
- Iron oxide powder (burning rate catalyst), 0.07%

After these components are mixed, the propellant looks like a thick syrup. It is poured into the engine's casing, where it cures and solidifies. During solidification process, the propellant glues itself to the walls of the casing and special tools are used to produce a cylindrical or star-shaped channel in the center of the grain. Due to changing area of the burning surface as propellant is consumed, the channel's diameter and shape determine the initial thrust and how it changes during the burn time. Since the burning starts inside the channel, the propellant itself shields the engine's casing from the hot gases until the very last seconds of the burn time. This is the most common configuration, but other shapes of the propellant grains can be used, if necessary.

In its final state, the composite solid propellant looks like a gray rubbery material and must be flexible enough to resist stress and vibration. Even small imperfections of the propellant surface, such as air bubbles, caverns and cracks, are very dangerous. Imperfections may cause a sudden radical increase of internal pressure (due to increase of the burning surface area) and rupture the engine's casing, leading to a catastrophic explosion. Integrity of the engine's body is especially important for very large solid rocket motors, which are assembled from separate sections. In 1986, loss of the engine's casing integrity caused a national tragedy: hot gases leaked through a faulty joint between two sections of the SRB and destroyed the Space Shuttle *Challenger*, killing seven astronauts. Production of the large solid propellant engines require a substantial amount of time, high precision equipment, and a tightly controlled environment. As a result, the modern powerful solid rocket motors are as expensive as the much more complex liquid propellant engines.

Robert H. Goddard standing to the left of the first flight of a liquid-propellant rocket. (Public Domain)

Ammonium perchlorate (oxidizer)	69.93%
Aluminum powder (fuel)	16%
Rubber-based polymer, called PBAN (binder)	12.04%
Epoxy-based curing agent	1.96%
Iron oxide powder (burning rate catalyst)	0.07%

Table 1.
Energy Efficiency of Modern Rocket Propellants

LIQUID PROPELLANTS

A liquid propellant consists of two liquid chemicals, fuel and oxidizer, which are delivered from separate tanks into the combustion chamber of a liquid propellant rocket engine (LPRE).

The use of a liquid propellant, namely liquid oxygen and liquid hydrogen, was proposed by the Russian space pioneer Konstantin Tsiolkovsky in 1903. He hypothesized that in addition to supplying more energy than gunpowder, such propellant would also supply water and oxygen to the crew of the future spacecraft. Space proponents from other countries independently came to the same conclusion. One of them was the American physics professor Robert Goddard, who built and launched the world's first liquid propellant rocket in 1926. LPRE technology received a major boost in the 1940s, when German engineers under Wernher von Braun developed the first operational long-range ballistic missile, *A-4* (better known as the *V-2*).

Despite the variety of liquid propellants, all of them can be divided into two major categories: storable and cryogenic. Physical properties of the storable propellants are very attractive from a technological standpoint. They have high density, do not boil under normal temperature, and degrade slowly. That makes it possible to keep the propellant inside a rocket for long periods, which is especially important for military applications and lengthy space missions. Among numerous possible combinations, the derivatives of nitric acid and hydrazine are the most widely used storable propellants today. Such propellants are self-igniting (hypergolic): the rocket engine fires as soon as the propellant components are mixed in a combustion chamber. An added advantage is that no electrical ignition mechanism is required. A typical example of a hypergolic propellant is a propellant for the Russian space launch vehicle Proton. It contains 73 percent of nitrogen tetroxide (NT) as an oxidizer, and 27 percent of unsymmetrical dimethyl hydrazine (UDMH) as a fuel.

Convenient physical properties cannot overshadow the extremely dangerous nature of storable propellants. Practically all of them are highly toxic, corrosive toward many materials, and cause severe burns on human skin. The history of the world rocketry reveals numerous cases of injures and deaths among rocket service personnel due to mishandling of these deadly chemicals. The most tragic of such accidents happened in the Soviet Union on October 24, 1960, when a ballistic missile filled with storable propellant exploded while being serviced by a ground crew. About a hundred people were fatally burned, or suffocated in toxic fumes. The founder of the Soviet space program, Dr. Sergei Korolev, called storable propellants "the devil's venom."

PROPELLANT	*SPECIFIC IMPULSE AT SEA LEVEL*	*COMMENTS*
Composite solid propellant	270 seconds	Used for modern solid propellant engines in military missiles, space launch vehicles and spacecraft. Pollutes environment by toxic exhaust.
Nitrogen Tetroxide + UDMH (liquid storable)	296 seconds	The best modern storable propellant. Widely used in military missiles, space launch vehicles and spacecraft. Both components are toxic before burning and pollute environment by toxic exhaust.
LOX + Kerosene (cryogenic, non-storable)	311 seconds	Widely used in space launch vehicles. Non-toxic, environmentally clean.
LOX + Liquid Hydrogen (cryogenic, non-storable)	388 seconds	The best modern cryogenic propellant. Used in the most advanced space launch vehicles. Non-toxic, environmentally clean.
Liquid Fluorite + Liquid Hydrogen (cryogenic, non-storable)	411 seconds	Not Used. Oxidizer is very toxic before burning and pollutes environment by toxic exhaust.

Table 2
Specific impulses of Rocket Propellants.

At first glance, the physical properties of the cryogenic propellants seem far less suitable for rocket technology since they consist of volatile liquified gases, such as liquid oxygen (LOX). Due to extremely low temperature (-183° Celsius for LOX), cryogenic propellant components require special care and complex equipment. These components are constantly boiling and evaporating under normal conditions and cannot be stored for a long time. After the cryogenic rocket is fueled and prepared for launch, it must constantly be fed by new portions of propellant to compensate for evaporation losses. This feeding can only be stopped just seconds before the takeoff. In spite of these disadvantages, cryogenic propellants provide superior energy characteristics. Liquified gases represent pure chemical elements that fully participate in a process of oxidation. For example, nitrogen tetroxide contains only 70 percent of oxygen compared to 100 percent for LOX. Thus, the energy output of complete oxidation is greater than in the case of the complex chemical compounds. Cryogenic components cost much less to produce than solid composite and liquid storable propellants. Natural resources of oxygen is also unlimited: the Earth's atmosphere. Yet the complexity of the cryogenic rocket hardware makes this technology very expensive.

Liquid oxygen is the most common cryogenic material today and the oldest liquid oxidizer in the rocket technology: it was used for the first liquid propellant rocket of Robert Goddard. It remains unsurpassed in its combination of effectiveness, low cost, and environmental "friendliness." Although liquid oxygen can be used with a variety of storable fuels (alcohol, kerosene, hydrazine, etc.), the best result is achieved with another cryogenic component, liquid hydrogen (LH_2). This combination is the most energy-efficient propellant of today's rocket technology. Yet it is probably one of the most difficult propellants to use. The major problem with liquid hydrogen is its low density (only 7% of water density). Thus, LH_2 occupies large volume and requires fuel tanks to be so spacious and heavy that it sometimes makes the use of LH_2 prohibitive. Rocket engineers are trying to resolve this problem by various methods. For example, the introduction of the new lightweight alloys decreased the mass of the Space Shuttle External

Tank (containing LOX/LH_2 propellant) by 13 percent, and NASA is willing to make it 26 percent lighter. Another solution is the "hydrogen slush," which is liquid hydrogen hyper cooled almost to the point of solidification (–259° Celsius compared to –252° Celsius for "normal" LH_2). Due to 15 percent greater density, hydrogen slush occupies lesser volume and has lower evaporation losses.

Although modern chemistry allows development of even more effective rocket propellants, energy efficiency is not the only consideration factor. For example, fluorine and its derivatives are better oxidizers than oxygen, but their extreme toxicity make them environmentally dangerous. The same concerns prevent the use of beryllium hydride—an excellent fuel that combines high density with the energy efficiency comparable to liquid hydrogen.

An increasing number of space launches worldwide makes environmental and economic problems more acute even with the existing technology. Most modern space launch vehicles use solid and storable propellants that pollute air, water and soil by toxic exhausts and by the remains of unburned propellant inside the ejected stages. One possible solution is replacement of the toxic propellants with the less expensive non-toxic components on the existing launch vehicles. Russian engineers, for example, proposed the use of liquid methane (or another natural gas) in combination with liquid oxygen. This new propellant would completely eliminate pollution and allow a lower cost per launch. At the same time its energy-efficiency would be second only to the liquid hydrogen/liquid oxygen propellant.

Apart from its chemical content and physical properties, the performance of a rocket propellant depends on engine design. Assuming that such design is optimal for a particular propellant, it is possible to compare various propellants using the energy-efficiency criteria of rocket engines, the specific impulse. Greater specific impulse indicates better propellant. Table 2 shows specific impulses of several typical propellants achievable by the best modern rocket engines.

Peter A. Gorin

See also: Propulsion; Spacecraft Energy Systems.

BIBLIOGRAPHY

Gatland, K., ed. (1984). *The Illustrated Encyclopedia of Space Technology*. 4th ed. London: Salamander Books.

Rycroft, M., ed. (1990). *The Cambridge Encyclopedia of Space*. Cambridge, Great Britain: Cambridge University Press.

ROCKETS

See: Spacecraft Energy Systems

ROTARY ENGINES

See: Engines

S

SAKHAROV, ANDREI DMITRIEVICH (1921–1989)

Andrei Sakharov was a Soviet physicist who became, in the words of the Nobel Peace Prize Committee, "a spokesman for the conscience of mankind." He made many important contributions to our understanding of plasma physics, particle physics, and cosmology. He also designed nuclear weapons for two decades, becoming "the father of the Soviet hydrogen bomb" in the 1950s. After recognizing the dangers of nuclear weapons tests, he championed the 1963 U.S.-Soviet test ban treaty and other antinuclear initiatives.

From the 1960s onward, at great personal risk, Sakharov severely criticized the Soviet regime and ardently defended human rights against it. He won the Nobel Peace Prize in 1975.

Andrei Sakharov was born in Moscow, Russia, to a family of the intelligentsia on May 21, 1921. His father, Dmitri, taught college physics and wrote textbooks and popular science books. Sakharov studied at home until the seventh grade. Dmitri, a man of warmth and culture, was his first physics teacher.

In 1938 Sakharov entered the Physics Department of Moscow State University. When World War II began, his academic prowess exempted him from military service. He and the remaining students and teachers were evacuated to Ashkhabad in Soviet Central Asia. Sakharov graduated with honors in 1942, with the war at its height, but joined a factory rather than continue school. While doing routine laboratory work at a munitions factory in Ulyanovsk on the Volga River, his engineering talent showed through a number of inventions. He also met Klavdia Vikhireva, a laboratory technician. They married in 1943.

Sakharov returned to Moscow in early 1945, as a graduate student at FIAN, the Physical Institute of the USSR Academy of Sciences. Igor Tamm, head of FIAN's Theoretical Physics Department, influenced him greatly. In 1947, Sakharov received his Ph.D. for work on particle physics.

In June 1948 Tamm was commissioned to help Yakov Zeldovich and his research team, which for two years had studied the feasibility of a thermonuclear, or hydrogen, bomb. Tamm and several of his students, including Sakharov, formed a special auxiliary group at FIAN to work on an H-bomb proposal. Sakharov suggested a radically new scheme for the bomb, and Vitaly Ginzburg added a key idea concerning thermonuclear explosive material. The U.S.-Soviet arms race took off. In the spring of 1950, Tamm and Sakharov moved to the Installation, a secret city in the central Volga region of the USSR. They worked on Sakharov's scheme and successfully tested the first prototype Soviet H-bomb on August 12, 1953.

Also in 1950 Sakharov and Tamm proposed an idea for a controlled thermonuclear fusion reactor, the TOKAMAK (acronym for the Russian phrase for "toroidal chamber with magnetic coil"), which achieved the highest ratio of output power to input power of any fusion device of the twentieth century. This reactor grew out of interest in a controlled nuclear fusion reaction, since 1950. Sakharov first considered electrostatic confinement, but soon came to the idea of magnetic confinement. Tamm joined the effort with his work on particle motion in a magnetic field, including cyclotron motion, drifts, and magnetic surfaces. Sakharov and Tamm realized that

Andrei Sakharov with his granddaughter Anya, in the 1970s. (Library of Congress)

destructive drifts could be avoided either with current-carrying rings in the plasma or with an induction current directly in the plasma. The latter is essentially the TOKAMAK concept.

At the Installation, Sakharov worked with many colleagues, in particular Yakov Zeldovich and David Frank-Kamenetskii. Sakharov made key contributions to the Soviets' first full-fledged H-bomb, tested in 1955. He also made many contributions to basic physics, perhaps the most important being his thesis that the universe is composed of matter (rather than all matter having been annihilated against antimatter) is likely to be related to charge-parity (CP) noninvariance.

Sakharov received many honors. He was elected as a full member of the Soviet Academy of Sciences in 1953 (at only age thirty-two); he was awarded three Hero of Socialist Labor Medals; he received a Stalin Prize; and he was given a country cottage. Sakharov's anti-Soviet activism cost him these rewards.

In May 1968 he completed the essay "Reflections on Progress, Peaceful Coexistence, and Intellectual Freedom." It proposed Soviet cooperation with the West, which the Soviets flatly rejected. The manuscript circulated in several typewritten copies known as *samizdat* ("self-print" in Russian) and was widely read outside the USSR. Sakharov was summarily banned from military-related research.

Also in May 1968, Sakharov accepted an offer to return to FIAN to work on academic topics. He combined work on fundamental theoretical physics with increased political activism, developing contacts to the emerging human rights movement. His wife, Klavdia, died of cancer in March 1969. In 1970 Sakharov and Soviet dissidents Valery Chalidze and Andrei Tverdokhlebov founded the Moscow Human Rights Committee. In the movement he met Elena Bonner, who became his companion-in-arms. They married in 1972.

Although Sakharov won the 1975 Nobel Peace Prize, and was the only Soviet ever to win it, he was barred from leaving Russia to receive it. The Nobel Committee's official citation praised Sakharov for his "fearless personal commitment in upholding the fundamental principles for peace.... Uncompromisingly and with unflagging strength Sakharov has fought against the abuse of power and all forms of violation of human dignity, and he has fought no less courageously for the idea of government based on the rule of law."

The Soviet regime persecuted Sakharov for his activism on behalf of dissidents and those seeking to emigrate. After he spoke out against the Soviet invasion of Afghanistan in 1979, he was picked up by the KGB and exiled to Gorky, under house arrest. There was no trial. In 1986 he was released by Premier Mikhail Gorbachev and returned to Moscow.

Sakharov made his first trip outside the Soviet Union in late 1988. In 1989 he was elected to the Congress of People's Deputies, the supreme legislative body of the Soviet Union. He died on December 14, 1989.

Andrew M. Sessler

See also: Ethical and Moral Aspects of Energy Use; Military Energy Use, Historical Aspects of; Nuclear Energy, Historical Evolution of the Use of; Nuclear Fusion.

BIBLIOGRAPHY

Anderson, R. H. (1989). "Andrei D. Sakharov, 68, Nobel Laureate and Wellspring of the Soviet Conscience." *New York Times*, December 16.

Andrei Sakharov: Soviet Physics, Nuclear Weapons, and Human Rights. American Institute of Physics. Niels Bohr Library. August 1, 2000 <http://www.aip.org/history/sakharov/>.

Kapitza, S. P., and Drell, S., eds. (1991). *Sakharov Remembered,* New York: American Institute of Physics.

Sakharov, A. (1990). *Memoirs.* New York: Alfred A. Knopf.

SANDIA NATIONAL LABORATORY

See: National Energy Laboratories

SAVERY, THOMAS (1650–1715)

In the study of seventeenth-century life, mystery often pervades the lives and deaths of even the most famous people and Thomas Savery is no exception. There is no record of his birth in local registers, but it is believed that he was born near Plymouth in Shilston, England, around 1650. There is no known portrait of him available today, although there was purportedly a drawing of him in a reprint of *A Miner's Friend,* published in 1827. After 1700 he became known as Captain Savery, with little evidence as to why, although he had done service as a military engineer, working as a Trench Master in 1696. According to folklore, many who were placed in charge during that era were likely to be called "Captain."

What *is* known is that Savery is sometimes left out of listings of the great mechanical inventors, even though he played a decidedly major role in the invention of the first steam engine. In fact, he was called "the most prolific inventor of his day" by the very book that set out to give proper due to the oft-neglected Thomas Newcomen as the steam engine's true inventor—*The Steam Engine of Thomas Newcomen* (Rolt and Allen, 1997).

Savery's career was hardly preordained, although he came from a family of prosperous merchants in Totnes, who acquired the manors of Shilston and Spriddlescombe in the parish of Modbury in the early seventeenth century. Savery's family was well known in the West Country and he may have been a merchant in Exeter for a period of time. At the age of twenty-three, he was protected by a writ from Charles II forbidding the local politicians to molest him. The writ in 1673 cited Savery's "many losses, particularly in the last Dutch wars," while serving as a freeman for the Merchant Adventurers Company.

Continuing a career that began in military engineering, his first patent was granted on July 25, 1698, for his historic work entitled "Raising water by the impellent force of fire." He called it "The Miner's Friend." Captain Savery's most important contribution was to evaluate the advances and frustrations of more than a century of experiments with steam power and create important innovation by combining steam with the effects of atmospheric pressure. His inventiveness was stimulated by his keen knowledge of copper and tin mining operations.

The principle of his pump was that steam was passed from a boiler into a closed receiver filled with water where its pressure forced the water through a nonreturn valve and up an ascending delivery pipe. When all the water was expelled, the steam supply was shut off and a new supply of cold water was poured over the outside walls of the receiver. This cooled the receiver and condensed the steam within. A vacuum was therefore created in the receiver, forcing water up a suction pipe, though a second nonreturn valve by atmospheric pressure. When the receiver was refilled the cooling water was shut off, steam turned on, and the cycle was repeated.

In subsequent years, Savery made important improvements that benefited future steam inventions. In June 1699 he demonstrated to the Royal Society a pump with two receivers, each with a separate, hand-controlled steam supply. This ensured improved continuity of operation, allowing one receiver to operate in its vacuum stage and the other under steam pressure. In 1701, he added two more critical steps: a second boiler, avoiding the need to shut down the fire and pump, between stages; and he replaced the two interconnected steam cocks with a single valve, run with a manually operated long lever. This may have been the inspiration for the modern slide valve and his inventiveness created, in effect, the world's first feed-water heater.

Savery appeared to be the first to take the huge step out of the lab and into the practical workshop. His equipment was made of brass and beaten copper, using firebrick furnaces. It was said that Salisbury Court (extending from Fleet Street to the river Thames) was the site of the world's first steam pump factory, although there is evidence that Savery abandoned his project in 1705. The limitations of his progress became known, literally under fire.

When tested, his engines generated too much heat and steam pressure for the technology of the times. Soldering would melt or machine joints would split. In order to solve the need for pumping water from mines, the necessary system of pumps would be far too costly and dangerous. In light of the dangers, both historians and scientists might find it intriguing that Savery never saw fit to add a safety valve, invented around that time.

In one of the more amusing assessments of this danger, Steven Switzer commented in *Hydrostatics and Hydraulics* (1729) about the steam power, "How useful it is, in gardens and fountain works ... in the garden of that noble peer, the Duke of Chandos, where the engine was placed under a delightful banqueting house, and the water was being forced up into a cistern on top, used to play a fountain in a delightful manner." Rival steam inventors derisively chided that the guests might forego the delight of the water fountain, had they known exactly what pressures were building under their feet.

Savery's work at the turn of the century preceded Newcomen's radically different and more successful engine in 1712. It also followed on the heels of noteworthy early experimenters, such as Giambattista della Porta in Italy in the early 1600s, Salomon de Caus in England in the 1620s, David Ramsay of Scotland in 1631, Edward Somerset of England in 1663, Sir Samuel Morland in 1667, Otto von Guericke of Magdeburg in 1672, and Denis Papin in the 1690s.

Savery's name and work remained in the public eye, followed by many scientists and inventors during the eighteenth century. The most important was Englishman Thomas Newcomen, who is believed to have known Savery and also to have seen his engines at work. Although Newcomen created a very different engine to solve the needs of the mines, the similarities were sufficient and Savery had earned his patent first. As a result, Newcomen and Savery entered into a joint patent agreement, whereby each shared the benefits of the other's work, while saving themselves the expense of a costly battle over separate patents.

The practical, if limited, use of Savery's incrementally successful efforts were permanently inscribed in history, as highly important advancements of his time.

Dennis R. Diehl

See also: Steam Engines; Turbines, Steam.

BIBLIOGRAPHY

Cannon, J. (1997). *The Oxford Companion to British History*. Frome, Somerset: Butler & Tanner, Ltd.

Rolt, L. T. C. and Allen, J. S. (1997). *The Steam Engine of Thomas Newcomen*. Ashbourne, UK: Landmark Publishing.

SCIENTIFIC AND TECHNICAL UNDERSTANDING OF ENERGY

The word "energy" entered English and other European languages in the sixteenth century from Aristotle's writings, and was restricted to meanings close to his until the nineteenth century. The total entry for "Energy" in the first edition (1771) of the *Encyclopaedia Britannica* is as follows: "A term of Greek origin signifying the powers, virtues or efficacy of a thing. It is also used figuratively, to denote emphasis of speech." These meanings survive; Swift's energetic pronouncement in 1720 that "Many words deserve to be thrown out of our language, and not a few antiquated to be restored, on account of their *energy* and sound" still rings a bell. But they are no longer dominant. Stronger by far is the scientific meaning, fixed between 1849 and 1855 by two men, William Thomson (the future Lord Kelvin) and W. J. Macquorn Rankine, professors, respectively, of natural philosophy and engineering at the University of Glasgow in Scotland.

Though Rankine and Thomson gave the term "energy" its modern scientific meaning, they did not originate the concept. The idea that through every change in nature some entity stays fixed arose in a complex engagement of many people, culminating between 1842 and 1847 in the writings of four men: Robert Mayer, James Prescott Joule, Hermann Helmholtz, and Ludvig Colding. The later choice of a word, easy as it may seem, is no mere detail. In the 1840s the lack of a word for the concept struggling to be born was a heavy impediment. The two nearest "Kraft" in German, and "force" in English—were riddled with ambiguity. With "energy" everything fell into place.

Three issues are involved in the coming-into-being of both concept and term: (1) the eighteenth-century debate on Leibniz's notion of *vis viva* ("living force"), (2) the unfolding eighteenth- and nineteenth-century understanding of steam engines, and (3) the search after 1830 for correlations among "physical forces." Concept must come first, with (3) treated before (2). On force as a term, compare (1) and (3) with Newton's definition. For the Newtonian, a force is a push or a pull, which, unless balanced by another, makes objects to which it is applied accelerate. "Living force" was not that at all: It corresponded to the quantity that we, under Thomson's tutelage, know as kinetic energy. As for forces being correlated, force in that context bore a sense not unlike the modern physicist's four "forces of nature," gravity, electromagnetism, and the two nuclear ones. It referred to the active principle in some class of phenomena, vital force, chemical force, the forces of electricity, magnetism, heat, and so on. To correlate these—to unify them under some overarching scheme—was as much the longed-for grail in 1830 as Grand Unification would be for physicists after 1975.

CORRELATION OF FORCES AND CONSERVATION OF ENERGY

Sometime around 1800 there came over the European mind a shadow of unease about Newtonian science. Beauty, life, and mystery were all expiring in a desert of atoms and forces and soulless mechanisms—so the poets held, but not only they. Science also, many people felt, must seek higher themes, broader principles, deeper foundations—symmetries, connections, structures, polarities, beyond, across, and above particular findings. Elusive, obscure, sometimes merely obscurantist, remarkably this inchoate longing actualized itself in a few decades in a succession of luminous discoveries, and flowed forward into the idea of energy.

In Germany it took shape in a much-praised, much-derided movement, interminable in discourse about Urprinciples and life forces, spearheaded by Lorenz Oken and Friedrich Schelling under the name *Naturphilosophie*, turgid indeed, but source of a powerful new credo. Nature is one, and so must all the sciences be; the true philosopher is one who connects. Thus in 1820 Hans Christian Oersted joined magnetism to electricity with his amazing discovery that an electric current deflects a magnet. Readers of Oersted baffled by strange talk of "polarities" and the "electric conflict" should know that it stems from Schelling: Oersted was a Naturphilosoph. But there were other voices. Similar hopes expressed in a more down-to-earth English tone mark John Herschel's *Preliminary Discourse on the Study of Natural Philosophy* (1830), Mary Somerville's *Connexion of the Physical Sciences* (1839), and W. R. Grove's *On the Correlation of Physical Forces* (1843). Fluent in German and Hanoverian in descent, Herschel has special interest because his "natural philosophy" had so resolutely English a cast, remote from Naturphilosophie. His master was Bacon, yet Herschel sought connection. So above all did the man whose first published paper (1816) was on the composition of caustic lime from Tuscany, and his last (1859) on effects of pressure on the melting point of ice, the profoundly English Michael Faraday.

From worldviews about the connectedness of everything to the correlation of specific "forces" to enunciating the law of energy was a long journey requiring two almost antithetical principles: great boldness in speculation and great exactness in measurement. Two eras may be identified: an era of correlation from the discovery of electrochemistry in 1800 through much of Faraday's career, and the era of energy beginning in the late 1840s. The two are connected: Without correlation there might have been no energy, and with energy new light was thrown on correlations, but they are not the same. In the great correlations of Oersted and Faraday, the discovery, when finally made, often embodied a surprise. An electric current exerts a magnetic force, as

Oersted had expected, but it is a transverse force. A magnet generates an electric current, as Faraday had hoped, but it must be a moving magnet, and the current is transient. A magnetic field affects light traversing glass, as Faraday had surmised, but its action is a twisting one far from his first guess. Not number but vision was the key, an openness to the unexpected. Energy is different. To say that energy of one kind has been transformed into energy of another kind makes sense only if the two are numerically equal.

Between Faraday and Joule, the man after whom very properly the unit of energy, the joule, is named, lay an instructive contrast. Both master experimenters with a feeling for theory, they stood at opposite conceptual poles: geometry versus number. For Faraday, discovery was relational. His ideas about fields of force reveal him, though untrained, as a geometer of high order. His experiments disclose the same spatial sense; number he left mainly to others. "Permeability" and "susceptibility," the terms quantifying his work on magnetism are from Thomson; the quantity characterizing his magnetooptical effect is (after the French physicist Marcel Verdet) Verdet's constant. For Joule, number was supreme. His whole thinking on energy originated in the hope that electric motors run from batteries would outperform steam engines. His best experiments yielded numbers, the electrical and mechanical equivalents of heat. His neatest theoretical idea was to compute one number, the extraordinarily high velocity gas molecules must have if pressure is to be attributed to their impacts on the walls of the container.

The doctrine of energy came in the 1840s to the most diverse minds: to Mayer, physician on a Dutch ship, off Java in 1842; to Joule, brewer turned physicist, at Manchester in 1843; to Colding, engineer and student of Oersted's, at Copenhagen in 1843; to Helmholtz, surgeon in the Prussian Army, at Potsdam in 1847. In a famous text, "Energy Conservation as an Example of Simultaneous Discovery" (1959), Thomas Kuhn listed eight others, including Faraday and Grove, directly concerned with interconvertibility of forces, and several more peripherally so, arguing that all in some sense had a common aim. Against this it is necessary to reemphasize the distinction between correlation and quantification (Faraday, the geometrist, found more "correlations" than any energist) and the desperate confusion about force. In exchanges in 1856 with the young James Clerk Maxwell, Faraday discussed "conservation of force"—meaning what? Not conservation of energy but his geometric intuition of a quite different law, conservation of *flux*, the mathematical result discovered in 1828 by M. V. Ostrogradsky, afterward Teutonized as Gauss's theorem. In 1856, energy—so natural to the twenty-five-year-old Maxwell—was utterly foreign to the sixty-four-year-old master from another age.

Physicists regard energy, and the laws governing it, as among the most basic principles of their science. It may seem odd that of the four founders two were doctors, one an engineer, and one only, Joule (and he only inexactly), could have been called then a physicist. The surprise lessens when we recognize that much of the impetus of discovery came from the steam engine, to which one key lay in that potent term coined and quantified in 1782 by James Watt, "horsepower." Seen in one light, horses and human beings are engines.

WORK, DUTY, POWER, AND THE STEAM ENGINE

The steam engine was invented early in the eighteenth century for pumping water out of mines. Later Watt applied it to machinery, and others to trains and ships, but the first use was in mines, especially the deep tin and copper mines of Cornwall, in southwestern England. Crucial always, to use the eighteenth-century term, was the *duty* of the engine: the amount of water raised through consuming a given quantity of fuel: so many foot-pounds of water per bushel of coal. To the mineowner it was simple. Formerly, pumping had been done by horses harnessed to a rotary mill. Now (making allowance for amortization of investment), which was cheaper: oats for the horses or coal for the engine? The answer was often nicely balanced, depending on how close the metals mine was to a coal mine.

Improving the duty meant using fuel more efficiently, but to rationalize that easy truth was the work of more than a century. Theoretically, it depended, among other things, on the recognition in 1758 by Adam Black—Watt's mentor at Glasgow—of a distinction between heat and temperature, and the recognition by nineteenth-century chemists of an absolute zero of temperature. On the practical side lay a series of brilliant inventions, beginning in 1710 with Thomas Newcomen's first "atmospheric" engine, preceded (and impeded) by a patent of 1698 by Thomas Savery. Eventually two kinds of efficiency were distinguished, thermal and thermodynamic. Thermal efficiency means avoiding unnecessary loss, heat going up the chimney, or heat dissipated in the walls of the cylinder. Thermodynamic efficiency, more subtly, engages a strange fundamental limit. Even in an ideal engine with no losses, not all the heat can be converted into work.

Both Savery's and Newcomen's engines worked by repeatedly filling a space with steam and condensing the steam to create a vacuum. Savery's sucked water straight up into the evacuated vessel; Newcomen's was a beam engine with a pump on one side and a driving cylinder on the other, running typically at ten to fifteen strokes a minute. The horsepower of the earliest of which anything is known was about 5.5. Comprehensively ingenious as this engine was, the vital difference between it and Savery's lay in one point: how the vacuum was created. Savery poured cold water over the outside, which meant that next time hot steam entered, much of it was spent reheating the cylinder. Newcomen sprayed water directly into the steam. The higher thermal efficiency and increased speed of operation made the difference between a useless and a profoundly useful machine.

Newcomen engines continued to operate, especially in collieries, almost to the end of the eighteenth century. Yet they, too, were wasteful. The way was open for a great advance, James Watt's invention in 1768 of the separate "condenser." Next to the driving cylinder stood another cylinder, exhausted and permanently cold. When the moment for the downstroke came, a valve opened, the steam rushed over and condensed; the valve then closed and the cycle was repeated. The driving cylinder stayed permanently hot.

Watt and Matthew Boulton, his business partner, were obsessed with numbers. Testing Clydesdale dray horses, the strongest in Britain, Watt fixed their lifting power (times 1.33) at 33,000 ft-lbs per minute. Thus was horsepower defined. Comparing his first engines with the best Newcomen ones, he found his to have four times higher duty. His next leap was to make the engine double acting and subject to "expansive working." Double action meant admitting steam alternately above and below the piston in a cylinder closed at both ends; for an engine of given size it more than doubled the power. Expansive working was one of those casual-seeming ideas that totally reorder a problem. Steam expands. This being so, it dawned on Watt that the duty might be raised by cutting off the flow into the cylinder early—say, a third of the way through the stroke. Like something for nothing, the steam continued to push without using fuel. Where was the effort coming from? The answer, not obvious at the time, is that as steam expands, it cools. Not pressure alone, but pressure and temperature together enter the lore of engines.

Behind the final clarifying insight provided in 1824 by Sadi Carnot, son of the French engineer-mathematician and revolutionary politician Lazare Carnot, lay three decades of invention by working engineers. One, too little remembered, was a young employee of Boulton and Watt, James Southern. Sometime in the 1790s he devised the indicator diagram to optimize engine performance under expansive working. In this wonderfully clever instrument a pressure gauge (the "indicator") was mounted on the cylinder, and a pen attached by one arm to it and another to the piston rod drew a continuous closed curve of pressure versus volume for engines in actual operation. The area of the curve gave the work (force times distance) done by the engine over each cycle; optimization lay in adjusting the steam cutoff and other factors to maximize this against fuel consumption. Known to most purchasers only by the presence of a mysterious sealed-off port on top of the cylinder—and exorbitant patent royalties for improved engine performance—the indicator was a jealously guarded industrial secret. The engineer John Farey, historian of the steam engine, had heard whispers of it for years before first seeing one in operation in Russia in 1819. To it as background to Carnot must be added three further points: (1) the invention in 1781 by one of Watt's Cornish rivals, J. C. Hornblower, of compounding (two stages of expansion successively in small and large cylinders); (2) the use of high-pressure steam; and (3) the recognition, especially by Robert Stirling, inventor in 1816 of the "Stirling cycle" hot air engine, that the working substance in engines need not be steam. Of these, high pressure had the greatest immediate impact, especially in steam traction.

Because Watt, who had an innate distaste for explosions, insisted on low pressures, his engines had huge cylinders and were quite unsuitable to vehicles. No such craven fears afflicted Richard Trevithick in Cornwall or Oliver Evans in Pennsylvania, each of whom in around 1800 began building high-pressure engines, locomotive and stationary, as a little later did another Cornish engineer, Arthur Woolf. Soon a curious fact emerged. Not only were high-pressure engines more compact, they were often more efficient. Woolf's stationary engines in particular, combining high pressure with compounding, performed outstandingly and were widely exported to France. Their success set Carnot thinking. The higher duty comes not from pressure itself but because high-pressure steam is hotter. There is an analogy with hydraulics. The output of a waterwheel depends on how far the water falls in distance: the output of a heat engine depends on how far the working substance falls in temperature. The ideal would be to use the total drop from the temperature of burning coal—about 1,500°C (2,700°F)—to room temperature. Judged by this standard, even high-pressure steam is woefully short; water at 6 atmospheres (a high pressure then) boils at 160°C (321°F).

Carnot had in effect distinguished thermal and thermodynamic inefficiencies, the latter being a failure to make use of the full available drop in temperature. He clarified this by a thought experiment. In real engines, worked expansively, the entering steam was hotter than the cylinder, and the expanded steam cooler, with a constant wasteful shuttling of heat between the two. Ideally, transfer should be isothermal—that is, with only infinitesimal temperature differences. The thought experiment—put ten years later (1834) into the language of indicator diagrams by the French mining engineer Émil Clapeyron—comprised four successive operations on a fixed substance. In two it received or gave up heat isothermally; in the other two it did or received work with no heat transfer. Neither Carnot nor Clapeyron could quantify thermodynamic efficiency; Thomson in 1848 could. He saw that the Carnot cycle would define a temperature scale (now known as the Kelvin scale) identical with the one chemists had deduced from gas laws. For an ideal engine working between temperatures T_1 (cold) and T_2 (hot) measured from absolute zero, the thermodynamic efficiency is $(T_2 - T_1)/T_2$. No engine can surpass that.

Nothing in nature is lost; that was the message of the 1840s. Chemical, mechanical, and electrical processes are quantifiable, exact, obey a strict law of energy. Equally exact are the amounts of heat these processes generate. But the reverse does not hold. The history of steam was one long struggle against loss; not even Carnot's engine can extract all the energy the fuel has to give. This was the conceptual paradox three men, Thomson and Rankine in Scotland, and Rudolf Clausius in Germany, faced in 1849–1850 in creating the new science of thermodynamics. Out of it has arisen a verbal paradox. For physicists conservation of energy is a fact, the first law of thermodynamics. For ordinary people (including off-duty physicists), to conserve energy is a moral imperative.

Heat as energy is somehow an energy unlike all the rest. Defining how takes a second law, a law so special that Clausius and Thomson found two opposite ways of framing it. To Clausius the issue was the age-old one of perpetual motion. With two Carnot engines coupled in opposition, the rule forbidding perpetual motion translated into a rule about heat. His second law of thermodynamics was that heat cannot flow without work uphill from a cold body to a hot body. Refrigerators need power to stay cold. Thomson's guide was Joseph Fourier's great treatise on heat *Théorie analytique de la chaleur* (1827), and the fact that unequally heated bodies, left to themselves, come always to equilibrium. His version was that work has to be expended to maintain bodies above the temperature of their surroundings. Animals need food to stay warm.

In a paper with the doom-laden title "On a Universal Tendency in Nature to a Dissipation of Mechanical Energy," Thomson in 1852 laid out the new law's dire consequences. How oddly did the 1850s bring two clashing scientific faiths: Darwin's, that evolution makes everything better and better; and Thomson's, that thermodynamics makes everything worse and worse. Thirteen years later Clausius rephrased it in the aphorism "The energy of the world is constant, the entropy of the world strives toward a maximum." After Maxwell in 1870 had invented his delightful (but imaginary) demon to defeat entropy, and Ludwig Boltzmann in 1877 had related it formally to molecular disorder through the equation which we now, following Max Planck, write $S = k \log W$, the gloom only increased. With it came a deep physics problem, first studied by Maxwell in his earliest account of the demon but first published by Thomson in 1874, the "reversibility paradox." The second law is a consequence of molecular collisions, which obey the laws of ordinary dynamics—laws reversible in time. Why then is entropy not also reversible? How can reversible laws have irreversible effects? The argument is sometimes inverted in hope of explaining another, even deeper mystery, the arrow of time. Why can one move forward and backward in space but only forward in time? The case is not proven. Most of the supposed explanations conceal within their assumptions the very arrow they seek to explain.

In the history of heat engines from Newcomen's on, engineering intuition is gradually advancing scientific knowledge. The first engines strictly designed from thermodynamic principles were Rudolf Diesel's, patented in 1892. With a compression ratio of fifty to one, and a temperature drop on reexpansion of 1,200°C (~2,160°F), his were among the most efficient ever built, converting 34 percent of the chemical energy of the fuel into mechanical work.

SOLIDIFICATION OF THE VOCABULARY OF ENERGY

Two concepts entangling mass and velocity made chaos in eighteenth-century dynamics, Newton's *vis motrix* mv and Leibniz's *vis viva* mv^2 (now written ½ mv^2, as first proposed by Gaspard Coriolis in 1829). It was a battle of concepts but also a battle of terms. Then casually in 1807 Thomas Young in his *Course of Lectures on Natural Philosophy and the Mechanical Arts*, dropped this remark: "The term energy may be applied with great propriety to the product of the mass or weight of a body times the number expressing the square of its velocity." Proper or not, *energy* might have fallen still born in science had not John Farey also, with due praise of Young, twice used it in 1827 in his *History of the Steam Engine* for *vis viva*. There, dormant, the word lay, four uses in two books, until 1849, when Thomson in last-minute doubt added this footnote to his second paper on Carnot's theorem, "An Account of Carnot's Theory on the Motive Power of Heat": "When thermal agency is spent in conducting heat through a solid what becomes of the mechanical effect it might produce? Nothing can be lost in the operation of nature—no energy can be destroyed. What then is produced in place of the mechanical effect that is lost?"

In that superb instance of Thomson's genius for producing "science-forming" ideas (the phrase is Maxwell's), thermodynamics is crystallized and energy is given almost its modern meaning. His next paper, "On the Quantities of Mechanical Energy Contained in a Fluid in Different States as to Temperature and Density" (1851), elevated energy to the title and contained the first attempt to define it within the new framework of thought. The next after was the sad one on dissipation. The torch then passed to Rankine, who in two exceptionally able papers, "On the General Law of the Transformation of Energy" (1853) and "Outlines of the Science of Energetics" (1855), completed the main work. The first introduced the term (not the concept) "conservation of energy," together with "actual or sensible energy" for *vis viva*, and "potential or latent energy" for energy of configuration. "Potential energy" has stuck; "actual energy" was renamed "kinetic energy" by Thomson and P. G. Tait in their *Treatise on Natural Philosophy* (1867); their definition (vol. 1, sec. 213) nicely explains the Coriolis factor: "The *Vis Viva or Kinetic Energy* of a moving body is proportional to the mass and the square of the velocity, conjointly. If we adopt the same units for mass and velocity as before there is particular advantage in defining kinetic energy as half the product of the mass and the square of the velocity."

A paper in French by Thomson, "Mémoire sur l'energie mécanique du système solaire" (1854), was the first use in a language other than English; but it was Clausius who by adapting the German *Energie* to the new scientific sense assured its future. He met much opposition from Mayer and some from Helmholtz, both of whom favored Kraft. Mayer indeed held fiercely that Kraft or force in the sense "forces of nature" had conceptual priority, so that Newton's use of "force" should be disowned. More was at stake here than words. The dispute goes back to the three meanings of "force." For Mayer, conservation and correlation were two sides of the same coin, and the single word "Kraft" preserved the connection. For advocates of "energy," "Kraft/force" in this general sense was a vague term with no quantitative significance. "Energy" was so precise, explicit, and quantitative a concept that it required a term to itself.

Beyond the expressions "conservation of energy" and "potential energy," and the notion and name of a science of energetics (revived in the 1890s by Wilhelm

Ostwald), Rankine gave far more in concept and term than he receives credit for. He supplied the first theory of the steam engine, the important term "adiabatic" to denote processes in which work is done with no transfer of heat, and—by a curious accident—the name "thermodynamics" for the science. In 1849 Thomson, seeking a generic name for machines that convert heat into work, had used "thermo-dynamic engine." Rankine followed suit, introducing in 1853 the vital concept of a thermo-dynamic function. Then in 1854 he submitted to the Philosophical Transactions of the Royal Society a long paper, "On the Geometrical Representation of the Expansive Action of Heat and the Theory of Thermo-Dynamic Engines." Papers there carried a running head, author's name to the left, title to the right. Faced with Rankine's lengthy title, the compositor set "Mr. Macquorn Rankine" on the left and "On Thermo-Dynamics" on the right. So apt did this seem that "thermodynamics" (without the hyphen) quickly became the secure name of the science. Later critics, unaware of its roots, have accused the inventors of thermodynamics of muddled thinking about dynamics; in fact, this and the history of Rankine's function provide one more instance of the power and subtlety of words. The "thermo-dynamic function" is what we, after Clausius who rediscovered it and gave it in 1865 its modern name, call entropy. In its original setting as the mathematical function proper to the efficient functioning of engines, Rankine's name was a good one. In the long view, however, Clausius's name for it, formed by analogy with energy, was better, even though the etymological reading he based it on (energy = work content, entropy = transformation content) was false. It was shorter and catchier but, more important, stressed both the parallel and the difference between the two laws of thermo-dynamics.

In 1861, at the Manchester meeting of the British Association for the Advancement of Science, two engineers, Charles Bright and Latimer Clark, read a paper "On the Formation of Standards of Electrical Quantity and Resistance," which had an influence on science and everyday life out of all proportion to the thirteen lines it takes in the Association Report. Its aim was simple. As practical telegraphers, Bright and Clark wished to attain internationally agreed definitions, values, and names for working quantities in their field. Their proposal led to the setting up under Thomson of a famous British Association committee to determine—from energy principles—an absolute unit of electrical resistance. Another, and fascinating, topic was Clark's suggestion made verbally at Manchester that units should be named after eminent scientists. He proposed "volt," "ohm," and "ampere" for units of potential, resistance, and current, though on ampere there was dispute, for Thomson's committee preferred to name the unit of current for the German electrician Weber.

International blessing and the setting of many names came at the Paris Congress of Electricians in 1881, but with no choice yet for units of energy and power. A year later C. William Siemens, the Anglicized younger brother of Ernst Werner von Siemens, was president of the British Association. In his address reported in the British Association report in 1882 he proposed "watt" and "joule"—the one "in honour of that master mind in mechanical science, James Watt ... who first had a clear physical conception of power and gave a rational method of measuring it," the other "after the man who has done so much to develop the dynamical theory of heat." According to the *Athenaeum* in 1882, Siemens's suggestions were "unanimously approved" by the Physics Section of the British Association a few days later. Joule, who was sixty-four at the time and had another seven years to live, seems to have been the only man so commemorated during his lifetime.

QUANTIZATION, $E = MC^2$ AND EINSTEIN

Accompanying the new nineteenth-century vista on heat were insights into its relation to matter and radiation that opened a fresh crisis. Gases, according to the "kinetic" theory, comprise large numbers of rapidly moving colliding molecules. As remade by Clausius and Maxwell in the 1850s it was an attractive—indeed, a correct—picture; then something went horribly wrong. From the Newtonian mechanics of colliding bodies, Maxwell deduced a statistical theorem—later called the equipartition theorem—flatly contrary to experiment. Two numbers, 1.408 for the measured value of a certain ratio, 1.333 for its calculated value, failed to agree. So upset was he that in a lecture at Oxford in 1860 he said that this one discrepancy "overturned the whole theory." That was too extreme, for the kinetic theory, and the wider science of statistical mechanics, continued to flourish and bring new discoveries, but there was truth in it. Equipartition of energy ran on, an ever-worsening riddle, until finally in 1900 Max Planck, by focusing not on molecules but on radiation, found the answer. Energy is discontinuous. It comes in quantized packets $E = h\nu$, where h is a constant ("Planck's constant") and n the frequency of radiation. Expressed in joule-seconds, h has the exceedingly small value 6.67×10^{-34}. Quantum effects are chiefly manifested at or below the atomic level.

The forty years from equipartition to Planck deserve more thought than they ordinarily get. The theorem itself is one hard to match for sheer mathematical surprise. It concerns statistical averages within dynamical systems. As first put by Maxwell, it had two parts: (1) if two or more sets of distinct molecules constantly collide, the mean translational energies of the individual ones in each set will be equal, (2) if the collisions produce rotation, the mean translational *and rotational* energies will also be equal. This is equipartition. Mathematically, the result seemed impregnable; the snag was this. Gases have two specific heats, one at constant pressure, the other at constant volume. Calculating their ratio C_p/C_v from the new formula was easy; by 1859 the value was already well known experimentally, having been central to both acoustics and Carnot's theory. Data and hypothesis were excellent; only the outcome was wrong.

Worse was to come. Boltzmann in 1872 made the same weird statistical equality hold for *every* mode in a dynamical system. It must, for example, apply to any internal motions that molecules might have. Assuming, as most physicists did by then, that the sharp lines seen in the spectra of chemical elements originate in just such internal motions, any calculation now of C_p/C_v would yield a figure even lower than 1.333. Worse yet, as Maxwell shatteringly remarked to one student, equipartition must apply to solids and liquids as well as gases: "Boltzmann has proved too much."

Why, given the scope of the calamity, was insight so long coming? The answer is that the evidence as it stood was all negative. Progress requires, data sufficiently rich and cohesive to give shape to theoretical musings. Hints came in 1887 when the Russian physicist V. A. Michelson began applying equipartition to a new and different branch of physical inquiry, the radiation emitted from a heated "black body"; but even there, the data were too sparse. Only through an elaborate interplay of measurement and speculation could a crisis so profound be settled.

Gradually it was, as Wilhelm Wien and others determined the form of the radiation, and Planck guessed and calculated his way to the new law. Found, it was like a magic key. Einstein with dazzling originality applied it to the photoelectric effect. Bohr, after Rutherford's discovery of the nucleus, used it to unlock the secrets of the atom. The worry about gases evaporated, for quantization limits which modes of energy affect specific heats. As for that later worry of Maxwell's about solids, trouble became triumph, a meeting ground between theory and experiment. Among many advances in the art of experimentation after 1870, one especially was cryogenics. One by one all the gases were liquefied, culminating with hydrogen in 1898 and helium in 1908, respectively at 19.2 K and 4.2 K above absolute zero. Hence came accurate measurements near zero of the specific heats of solids, started by James Dewar. They fell off rapidly with temperature. It was Einstein in 1907 who first saw why—because the modes of vibrational energy in solids are quantized. More exact calculations by Peter Debye, and by Max Born and Theodore von Kármán, both in 1912, made C at low temperatures vary as T^3. It was a result at once beautifully confirmed by F. A. Lindemann (Lord Cherwell) in experiments with Walther Nernst in Berlin.

From Planck's h to Bohr's atom to the austere reorganization of thought required in the 1920s to create quantum mechanics is a drama in itself. Crucial to its denouement was Heisenberg's "uncertainty principle," a natural limit first on knowing simultaneously the position x and momentum p of a body ($\Delta x \, \Delta p \sim h$) and second, no less profound, on energy. The energy E of a dynamical system can never be known exactly: measured over a time interval Δt it has an uncertainty $\Delta E \sim h/\Delta t$. But whether quantum uncertainty is an epistemological truth (a limit on knowledge) or an ontological one (a limit in nature) is an endless, uncertain debate.

If the $E = h\nu$ of quantum theory is one beguilingly simple later advance in the concept of energy, the supremely famous mass-energy law $E = mc^2$ is another. This and its import for nuclear energy are usually credited solely to Einstein. The truth is more interesting.

It begins with J. J. Thomson in 1881 calculating the motion of an electric charge on Maxwell's electromagnetic theory, a theme Maxwell had barely touched. To the charge's ordinary mass Thomson found it necessary to add an "electromagnetic mass" connected with electrical energy stored in the surrounding field. It was like a ship moving through water—some of the water is dragged along, adding to the ship's apparent mass. Had the calculation been perfect, Thomson would have found $E = mc^2$, as Einstein and others did years later. As it was, Thomson's formula yields after a trivial conversion the near-miss $E = \frac{3}{4} mc^2$.

After Thomson nothing particular transpired, until a number of people, of whom Joseph Larmor in 1895 was perhaps first, entertained what historians now call the "electromagnetic worldview." Instead of basing electromagnetism on dynamics, as many had tried, could one base dynamics on electromagnetism? How far this thought would have gone is unclear had it not been for another even greater leap of Thomson's, this time experimental: his discovery of the electron. In 1897, by combining measurements of two kinds, he proved that cathode rays must consist of rapidly moving charged objects with a mass roughly one one-thousandth of the hydrogen atom's. Here was revolution—the first subatomic particle. Hardly less revolutionary was his conjecture that its mass was *all* electromagnetic. A shadow ship plowing through an electromagnetic sea: That was Thomson's vision of the electron.

The argument then took two forms: one electromagnetic, the other based on radioactivity. The first was tied to a prediction of Maxwell's in 1865, proved experimentally in 1900 by Petr Lebedev, that light and other electromagnetic radiations exert pressure. This was the line Einstein would follow, but it was Henri Poincaré who, in 1900 at a widely publicized meeting in honor of H. A. Lorentz, first got the point. The Maxwell pressure corresponds to a momentum, given by the density of electromagnetic energy in space divided by the velocity c of light. Momentum is mass times velocity. Unite the two and it is natural to think that a volume of space containing radiant energy E has associated with it a mass $m = E/c^2$. Meanwhile, as if from nowhere, had come radioactivity—in Rutherford's phrase, "a Tom Tiddler's ground" where anything might turn up. In 1903 Pierre Curie and his student Albert Laborde discovered that a speck of radium placed in a calorimeter emitted a continuous flux of heat sufficient to maintain it at 1.5°C above the surroundings. Their data translated into a power output of 100W/kg. Where was the power coming from? Was this finally an illimitable source of energy, nature's own perpetual-motion machine? It was here that Einstein, in a concise phrase, carried the argument to its limit. In his own derivation of $E = mc^2$ in 1905, five years after Poincaré's observation, he remarked that any body emitting radiation should lose weight.

The events of World War II, and that side of Einstein that made credit for scientific ideas in his universe a thing more blessed to receive than to give, have created a belief that atomic energy somehow began with him. In truth the student of early twentieth-century literature notices how soon after radioactivity there arose among people who knew nothing of $E = mc^2$ or Einstein a conviction that here was a mighty new source of energy. Geologists hailed it with relief, for as the Irish physicist John Joly demonstrated in 1903, it provided an escape from Kelvin's devastating thermodynamic limit on the age of Earth. But not only they, and not always with relief. In Jack London's nightmare futurist vision *The Iron Heel*, published in 1908, atomic gloom permeates the plot. No less interesting is this, from a 1907 work London may have read that remains among the most illuminating popularizations of the events of that era, R. K. Duncan's *The New Knowledge*: "March, 1903, ... was a date to which, in all probability, the men of the future will after refer as the veritable beginning of the larger powers and energies that they will control. It was in March, 1903, that Curie and Laborde announced the heat-emitting power of radium. The fact was simple of demonstration and unquestionable. They discovered that a radium compound continuously emits heat without combustion, or change in its molecular structure.... It is all just as surprising as though Curie had discovered a red-hot stove which required no fuel to maintain it on heat." But there *was* fuel: the fuel was mass.

The highest (and most puzzling) of Einstein's offerings is not $E = mc^2$, where his claim is short, but the spectacular reinterpretation of gravitational energy in the new theory he advanced in 1915 under the name general relativity. Owing much to a brilliant paper of 1908 by his former teacher Hermann Minkowski, this, of course, is not the popular fable that everything is relative. It is a theory of relations in which two paired quantities—mass-energy and space-time—are each joined through the velocity of light mass-energy via $E = mc^2$, space-time via a theorem akin to Pythagoras's where time multiplied by c enters as a fourth dimension, not additively but by subtraction $-(ct)^2$. When in this context Einstein came to develop a theory of gravity to replace Newton's, he met two surprises, both due ultimately to the curious—indeed deeply mysterious—fact that in gravitation, unlike any other force of nature, mass enters in two ways. Newton's law of acceleration $F = ma$ makes it the receptacle of inertia; his law of gravity $F = GMm/r^2$ makes it the origin of the force. If, therefore, as with electromagnetism, the energy is disseminated through space, it will have the extraordinary effect of making gravitation a source for itself. Consider Earth. Its mass is 6×10^{21} tons; the mass of the gravitational field around it, computed from $E = mc^2$, though tiny by comparison, has the by-human-standards enormous value of 4×10^{10} tons. It, too, exerts an attraction. For denser bodies this "mass of the field" is far greater. Because of it any relativistic theory of gravitation has to be nonlinear.

Even deeper, drawn from his much-discussed but too little understood "falling elevator" argument, was Einstein's substituting for Newton's equivalence between two kinds of mass an equivalence between two kinds of acceleration. How innocent it seems and how stunning was the transformation it led to. The theory could be rewritten in terms not of mass-energy variables but of space-time variables. Gravitation could be represented as warping space-time—in just the right nonlinear way. For this one form, energy had been transmuted into geometry. But was the geometrization limited to gravity, or might it extend to energy of every kind, beginning with electromagnetic? That great failed vision of the geometrization of physics would charm and frustrate Einstein and many others for the next forty years, and it retains its appeal.

Looking back over more than a century since energy began as concept and term, it is impossible not to be struck both by the magnitude of the advance and the strenuousness of the puzzles that remain. In the design of engines, the practice of quantum mechanics, the understanding of elementary particles, the application of relativity, detailed prescriptions and laws exist that work perfectly over the range of need. Beyond lies doubt. To physicists with their conviction that beauty and simplicity are the keys to theory, it is distressing that two of the most beautifully simple conceptions proposed in physics, J. J. Thomson's that all mass is electromagnetic, Einstein's that all energy is geometric, both manifestly verging on truth, have both failed, with nothing either as simple or as beautiful arising since to take their place. Then there are issues such as quantum uncertainty and the problems of dissipation, reversibility, and the arrow of time, where to a certain depth everything is clear, but deeper (and the depth seems inescapable), everything is murky and obscure—questions upon which physicists and philosophers, such as Milton's theologians, sit apart in high debate and find no end in wandering mazes lost.

C. W. F. Everitt

BIBLIOGRAPHY

Born, M. (1951). *Atomic Physics*, 5th ed. New York: Hafner.

Brush, S. G. (1976). *The Kind of Motion We Call Heat.* Amsterdam: North Holland.

Brush, S. G. (1983). *Statistical Physics and the Atomic Theory of Matter.* Princeton, NJ: Princeton University Press.

Cantor, G. (1991). *Michael Faraday: Sandemanian and Scientist: A Study of Science and Religion in the Nineteenth Century.* London: Macmillan.

Cardwell, D. S. L. (1971). *From Watt to Clasuius.* Ithaca, NY: Cornell University Press.

Cardwell, D. S. L. (1989). *James P. Joule.* Manchester, Eng.: Manchester University Press.

Carnot, S. (1960). *Reflections on the Motive Power of Fire by Sadi Carnot; and Other Papers on the Second Law of Thermodynamics, by E. Clapeyron and R. Clausius,* ed. E. Mendoza. New York: Dover.

Duncan, R. K. (1907). *The New Knowledge: A Popular Account of the New Physics and the New Chemistry in Their Relation to the New Theory of Matter.* London: Hodder and Stoughton.

Elkana, Y. (1974). *The Discovery of the Conservation of Energy.* London: Hutchinson Educational.

Everitt, C. W. F. (1975). *James Clerk Maxwell: Physicist and Natural Philosopher.* New York: Charles Scribner's Sons.

Ewing, J. A. (1894). *The Steam-Engine and Other Heat-Engines.* Cambridge, Eng.: Cambridge University Press.

Farey, J. (1827). *A Treatise on the Steam Engine.* London: Longman, Rees, Orme, Brown, and Green.

Harman, P. (1998). *The Natural Philosophy of James Clerk Maxwell.* New York: Cambridge University Press.

Jammer, M. (1957). *Concepts of Force.* Cambridge: Harvard University Press.

Joly, J. (1909). *Radioactivity and Geology: An Account of the Influence of Radioactive Energy on Terrestrial History.* London: A. Constable.

Kuhn, T. S. (1977). "Energy Conservation as an Example of Simultaneous Discovery." In *The Essential Tension.* Chicago: University of Chicago Press.

Kuhn, T. S. (1978). *Black-Body Theory and the Quantum Discontinuity, 1894–1912.* New York: Oxford University Press.

Lorentz, H. A.; Einstein A.; Minkowski, H.; Weyl, H. (1923). *The Principle of Relativity: A Collection of Original Memoirs on the Special and General Theory of Relativity, with Notes by A. Sommerfeld.* London: Dover.

Miller, A. I. (1987). "A précis of Edmund Whittaker's 'The Relativity Theory of Poincaré and Lorentz.'" *Archives Internationales d'Histoire de Science* 37:93–103.

Pais, A. (1982). *Subtle Is the Lord: The Science and the Life of Albert Einstein.* New York: Oxford University Press.

Rankine, W. J. M. (1881). "On the General Law of the Transformation of Energy." In *Miscellaneous Scientific Papers: From the Transactions and Proceedings of the Royal and Other Scientific and Philosophical Societies, and the Scientific Journals; with a Memoir of the Author* by P. G. Tait, ed. W. J. Millar. London: C. Griffin.

Rankine, W. J. M. (1881). "Outlines of the Science of Energetics." In *Miscellaneous Scientific Papers: From the Transactions and Proceedings of the Royal and Other Scientific and Philosophical Societies, and the Scientific Journals; with a Memoir of the Author by P. G. Tait,* ed. W. J. Millar. London: C. Griffin.

Smith, C., and Wise, N. (1889). *Energy and Empire : A Biographical Study of Lord Kelvin.* New York: Cambridge University Press.

Thomson, W. (1882–1911). "An Account of Carnot's Theory on the Motive Power of Heat." In *Mathematical and Physical Papers: Collected from Different Scientific Periodicals from May 1841 to the Present Time, in 6 Vols.*, vol. 1, pp. 156–164. Cambridge, Eng.: Cambridge University Press.

Thomson, W. (1882–1911). "On a Universal Tendency in Nature to the Dissipation of Mechanical Energy." In *Mathematical and Physical Papers: Collected from Different Scientific Periodicals from May 1841 to the Present Time, in 6 Vols.*, vol. 1, pp. 511–514. Cambridge, Eng.: Cambridge University Press.

Thomson, W. and Tait, P. G. (1888–1890). *Treatise on Natural Philosophy.* Cambridge, Eng.: Cambridge University Press.

Truesdell, C. (1980). *The Tragicomical History of Thermodynamics, 1822–1854.* New York : Springer-Verlag.

Whittaker, E. T. (1951–1953). *A History of the Theories of Aether and Electricity,* rev. ed. New York: T. Nelson.

Williams, L. P. (1965). *Michael Faraday.* New York: Basic Books.

SEEBECK, THOMAS JOHANN (1770–1831)

Thomas Johann Seebeck, German physician, physicist, and chemist, was born April 9, 1770, in the Hanseatic town of Revel (now Tallinn Estonia). Seebeck studied medicine in Berlin and Goettingen, where in 1802 he took a physician's doctor degree. He lived in Jena (Thuringia) between 1802 and 1810 as an independent scholar not practicing medicine. In the Jena period Seebeck met poet Johann Wolfgang Goethe, who carried on his own studies on chromatics. From 1806 untill 1818 Seebeck consulted and informed Goethe about his optical observations. He discovered entoptical colors in 1813. These are color figures which originate inside glass volume after cooling. The phenomenon of entoptical colors was broadly documented by Goethe in his treatises on theory of colors. Seebeck became a member of the Academy of Sciences in Berlin in 1814. After stays in Bayreuth and Nuernberg (Bavaria) he moved to Berlin in 1818. Goethe hoped that Seebeck, in the prestigious position of an academician, would help to lend respectability to his theory of colors. Instead, the contact between the two men loosened in the 1820s, as Seebeck seems to have distanced himself cautiously from Goethe's theory. In Berlin Seebeck devoted most of his time to experiments measuring the influence of heat on magnetism and electricity. Seebeck died December 10, 1831, in Berlin.

Seebeck's outstanding scientific achievement was the discovery of one of the three classical thermoelectric effects, which are the Seebeck, the Peltier, and the Thomson effects. Seebeck's discovery was the first, dating from 1822–1823, followed by that of Jean-Charles-Athanase Peltier in 1832 and that of William Thomson in 1854. Seebeck observed that an electric current in a closed circuit comprised different metallic components if he heated the junctions of the components to different temperatures. He noted that the effect increases linearly with the applied temperature difference and that it crucially depends on the choice of materials. Seebeck tested most of the available metallic materials for thermoelectricity. His studies were further systematized by the French physicist

Antoine-Cesar Becquerel (1788–1878). It has become common after Becquerel to classify materials according to their so-called thermoelectric power. The difference in thermoelectric power of the applied materials determines the strength of thermoelectricity. Since the Seebeck effect can only be observed if two dissimilar conductors are employed, one obtains therewith only access to differences of values of thermoelectric power. Absolute values can be ascertained through the Thomson effect, which states that in a single homogeneous material heat develops (or is absorbed) in case electric current flows parallel (or antiparallel) to a temperature gradient. The strength of this effect is expressed by the Thomson coefficient, which is a temperature-dependent quantity varying from material to material. The absolute value of the thermoelectric power of a material is given by the temperature integral of the Thomson coefficient multiplied by the inverse temperature.

Seebeck understood that his effect might be used for precision measurements of temperature differences, and indeed it is exploited for this purpose in modern thermoelements. The basic device in these thermoelements is an electric circuit comprising two metallic components of different thermoelectric power, a thermocouple. The temperature difference to which the thermocouple is exposed is measured through the voltage built up in the circuit, according to Seebeck. Thermoelements with various combinations of materials for their thermocouples are used in the temperature range from –200°C to 3,000°C. their simplicity and robustness makethem an almost universal element in the fields of temperature recording and temperature regulation. In addition to their higher precision as compared to conventional thermometers, these thermoelements have the additional advantage that they can be introduced into apertures fitting to the thinnest technologically realizable wires. Thermoelements also have a much smaller heat capacity than conventional thermometers (e.g. mercury thermometers) and therefore interfere much less with the body to be measured.

Another application of the Seebeck effect is to be found in detectors of small quantities of heat radiation. These sensitive detectors comprise a thermopile, a pile of thermocouples (small pieces of two different metals connected in V form and put into series). Half of the junctions of the thermopile are shielded within the detector, whereas the other half are exposed to external heat radiation which is recorded through a voltmeter in the thermopile circuit.

Barabara Flume-Gorczyca

BIBLIOGRAPHY

Goethe, J. W. (1949). *Naturwissenschaftliche Schriften* Vol. 16. Zurich: Artemis.

Hecker, M. (1924). *Jahrbuch der Goethe-Gesellschaft* 178. Weimar.

Seebeck, T. J. (1822). *Denkschriften der Berliner Akademie.*

Seebeck, T. J. (1826). *Poggendorff's Annalen* VI:133.

Streit, H. (1901–1902). *Programme des Progymnasiums zu Schlawe.*

SEISMIC ENERGY

Seismic energy is energy in the form of elastic waves that are propagated through the earth's interior or along its surface, or have the form of standing waves (free oscillations) in the planet as a whole. The source of seismic waves may be either natural—an earthquake—or artificial, for example, a chemical or nuclear explosion, or a weight dropped on the earth's surface. The energy source, whatever it is, imposes a strain—a change in shape and volume—on the surrounding earth material, which resists the strain because of its elastic properties. As the strained material tries to recover its original shape and volume, oscillations are produced, which are transferred to material slightly farther away from the source. In this way, seismic waves are produced that spread out from the source in all directions. In detail, the mechanics of seismic sources can be quite complex.

Several types of seismic waves are recognized. Longitudinal waves, also called compressional waves or P waves, consist of oscillations back and forth in the direction of wave propagation. Transverse waves, also called shear waves or S waves, consist of oscillations at right angles to the direction of wave propagation. Because S waves depend upon the rigidity of materials for their existence, they cannot propagate through a fluid medium. L waves, so-called from their long wavelengths, propagate along the free surface of the earth, or within layers near the surface. They are called surface waves, in contrast to the body waves (P and S waves) that travel through the interior body of the earth. Free oscillations of the earth, which exist in several modes, are standing seismic waves produced by very large earthquakes.

EARTHQUAKES

Although earthquakes may be caused by natural events such as rockfalls or the movement of magma in volcanic regions, most are caused by a sudden release of energy at a fault. Before being released in the forms of heat and seismic waves, this energy was stored as elastic strain—that is, recoverable distortions of the size or shape of a volume of rock. Strain develops in response to tectonic forces acting in the lithosphere, the relatively rigid uppermost 100 kilometers (60 miles) or so of the earth. This may be likened to the energy stored in a stretched rubber band, which then is released when the band snaps. The ultimate source of the tectonic forces, and therefore of the energy released by earthquakes, is to be sought in the heat of the deeper portions of the earth. This heat is thought to be a combination of heat produced by natural radioactivity, of residual heat from the earth's accretion, and of heat generated by the separation of iron into the earth's core.

At some point during the accumulation of strain, a certain volume of rock suddenly lurches back toward its original, unstrained condition. The level of stress, which has accumulated along with the strain, decreases rapidly, the rock slips a certain distance along a fracture plane (fault), and seismic waves propagate away from the source region. At sufficiently large distances, the source appears to be a single point within the earth, called the hypocenter (or focus) of the earthquake. The epicenter is a point on the earth's surface located directly above the hypocenter.

The total energy of the earthquake is partitioned into heat and seismic-wave energy. The latter has been estimated to account for only one-sixth to one-quarter of the total energy for tectonic earthquakes, and much less—perhaps only 1 percent—for rockfalls and explosions. The energy of the seismic waves is responsible for the destruction wrought by large earthquakes, either directly through ground shaking, or indirectly by inducing ground failure. The size of a tectonic earthquake is a function of the area of the fault that has slipped, the average amount of slip, and the rigidity of the rock in the vicinity of the fault. The product of these three factors is called the moment of the earthquake, and is related to the more familiar concept of earthquake magnitude.

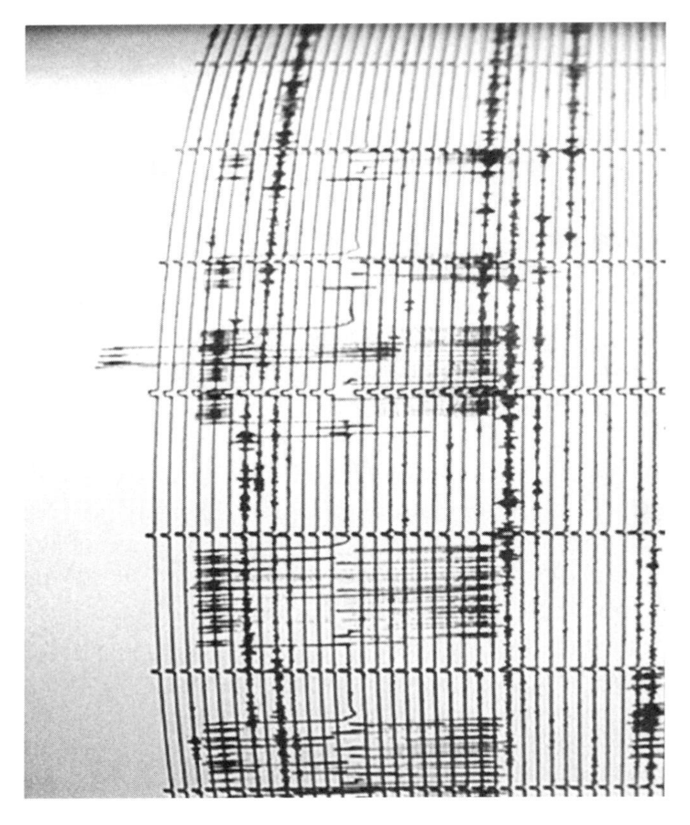

Seismograph recording seismic activity of Volcano Island, Philippines. (Corbis Corporation)

Earthquake magnitude is a term coined in the 1930s by Charles Richter, a seismologist at the California Institute of Technology. Richter based his original magnitude scale on the amplitudes of seismic waves as recorded on a common type of seismograph then in use in southern California, after correction for the distance to the epicenter. Subsequently, many other methods of measuring earthquake magnitude were developed, not all of which gave results that closely coincided with those of Richter's original method. Today the method that gives the most consistent results across the entire range of magnitudes is based on the earthquake moment, which can be found by analyzing the wavelengths and amplitudes of the seismic waves. Moment is then converted into a magnitude number, known as moment magnitude. On this scale, the very largest earthquakes have magnitudes about 9.5, while the smallest have values of magnitude that are negative. In principle, seismic-wave energy and earthquake moment are directly proportional: $E = AM_0$. In this equation E is energy, M_0 is moment, and A is a constant with an approximate value of

Workers conduct seismic explosions in search of oil and gas deposits on private islands in Bob Marshall Wilderness Area, Montana. (Corbis Corporation)

5×10^{-5} if E and M_0 are expressed in the same units. For example, the moment magnitude of the great Alaska earthquake of 1964 has been estimated at 9.2. This corresponds to a seismic moment of nearly 10^{23} joules, or seismic-wave energy of 5×10^{18} joules. The energy generated by this one earthquake was ten times greater than the amount of seismic-wave energy generated in the entire earth in an average year, which is estimated to be 5×10^{17} joules. That is why high-magnitude earthquakes are the most destructive of natural phenomena. Each whole step on the moment magnitude scale corresponds to an increase of about thirty-two times in seismic-wave energy. Seismologists cannot yet predict when an earthquake of a given size will occur on a given segment of a given fault, but many years of observation have revealed that the frequency of occurrence of earthquakes decreases by about a factor of ten for every whole step increase in magnitude. Thus, in a typical year southern California might experience about a hundred earthquakes in the magnitude range 3 to 4, about ten in the range 4 to 5, and perhaps one of magnitude 5 or greater.

USE OF SEISMIC ENERGY

Since about the beginning of the twentieth century, seismologists have made use of natural seismic waves to investigate the internal structure of the earth. Because of changes in the density and elasticity of earth materials with depth and, to a lesser extent, horizontally, seismic waves are refracted, reflected, and dispersed in fairly complex ways as they propagate through and around the earth. Analysis of free oscillations also provides valuable information about the state of the earth's interior. It is primarily through these techniques that we know that the earth is subdivided into a crust, mantle, outer core, and inner core, as well as details about each of these internal regions.

Artificially generated seismic waves also may be used to explore below the earth's surface, for either scientific or commercial purposes. Seismic energy produced by explosions, weight drops, vibrators, or, at sea, by electrical sparkers and air guns is used extensively to explore for geologic structures that may contain petroleum. The most common method makes use of P waves that are recorded as they arrive back at the earth's surface after having been reflected from boundaries separating rocks of differing properties. Intensive computer-assisted processing converts the raw data so collected into a detailed picture of the subsurface region. Similar techniques are used on a smaller scale in civil engineering projects to find the depth to bedrock beneath soil cover, to investigate the quality of the bedrock, or to find the depth to a groundwater aquifer.

Charles K. Scharnberger

See also: Energy Intensity Trends; Geography and Energy Use; Geothermal Energy; Thermal Energy.

BIBLIOGRAPHY

Bolt, B. A. (1982). *Inside the Earth*. San Francisco: W. H. Freeman.

Bolt, B. A. (1993). *Earthquakes*. New York: W. H. Freeman.

Howell, B. F., Jr. (1990). *An Introduction to Seismological Research*. Cambridge, Eng.: Cambridge University Press.

Kearey, P., and Brooks, M. (1991). *An Introduction to Geophysical Exploration*. Oxford, Eng.: Blackwell Scientific Publications.

Richter, C. F. (1958). *Elementary Seismology*. San Francisco: W. H. Freeman.

SEMICONDUCTORS

See: Electricity

SHIPS

From the earliest times, ships and boats were propelled by human power and wind power. In the nineteenth century very large and efficient sailing ships (many 70 to 100 m long) were transporting passengers and cargo to ports all over the world. In the Western world most of the cargoes were carried by sailing ships until the 1890s. By the middle of the nineteenth century, with the development of reliable steam engines, the designers of ships began to use coal-burning steam power plants to propel ships, using paddle wheels and then propellers. Other power plants were developed in the late nineteenth century and during the twentieth century, including steam turbines, spark-ignition engines, diesel engines, gas turbines, and nuclear power plants. These ship power plants were developed from their land-based counterparts to operate in the severe ocean environment. A relatively new power source, the fuel cell, is being developed to provide electrical power for underwater vessels and also is being evaluated for future applications in surface ships.

High-speed ships and large-displacement slower ships can have power requirements of 30,000 to 75,000 kW (approximately 40,000 to 100,000 hp) for propulsion. Some nuclear-powered aircraft carriers have power plants that develop 200,000 kW. In addition to the power required to propel the vessel, some energy and power have to be provided for various ship services, such as heating, lighting, and cooling; this is often described as the "hotel load." Power also is required to refrigerate the cargo in some vessels and to pump oil in tankers.

SHIP RESISTANCE

Experiments by William Froude in the nineteenth century indicated that ship resistance has two main components: frictional, and wave-making resistance or drag. Frictional resistance is caused by the movement of the underwater hull (wetted surface area) through the viscous fluid, water. Wave-making drag is the component of resistance resulting from the energy expended on the waves generated by the moving ship. Short, bluff ships have less wetted surface area than long, slender ships for the same displacement and therefore have lower frictional drag. The long, slender ships, however, tend to generate smaller waves and thus have lower wave-making drag at high ship speeds, where wave-making drag predominates. Placing bulbous bows on large commercial ships such as tankers has been shown to reduce wave-making drag. In addition to frictional and wave-making drag, there is form drag induced by the complex fluid flow around the hull near the stern of many ships, and there is also the wind resistance of the structure of the ship above the sea surface.

Information on ship resistance has been determined from large numbers of tests on scale models of ships and from full-size ships, and compilations of these experimental results have been published. For a new and innovative hull form the usual procedure is to construct a scale model of the ship and then to conduct resistance tests in a special test facility (towing tank). Alternatively, analytical methods can provide estimates of ship resistance for a range of different hull shapes. Computer programs have been written based on these theoretical analyses and have been used with success for many ship designs, including racing sailboats.

Methods for reducing wave-making drag are often introduced for ships that operate at high speed. Lifting the hull partly or completely out of the water by using hydrodynamic lift or air pressure has been investigated for high-speed ship designs. Fast but short-range ferries have been built on these principles using planing hulls, air-cushion designs, and hydrofoil configurations. In air cushion and surface-effect ships, 15 to 30 percent of the total power has to be supplied to the fans that provide the air pressure to lift the hulls.

Another concern is the ability of a ship to operate in storms. The pounding of waves on the hull increases the resistance; this can cause structural damage, and the violent ship motions can harm the passengers, cargo, and crew of the ship. Some designers have proposed special configurations that can maintain the ship speed in storms.

PROPULSORS

Propellers are the predominant propulsive devices driving ships, although water jets are now used in some high-speed ships. An experimental installation in a small ship of a magnetohydrodynamic propulsor has been tested, but it achieved rather low propulsive efficiency. Fish-like propulsion also has been examined for possible application to ships and underwater vehicles.

In the middle of the nineteenth century some ship designers began to replace paddle wheels with propellers (or screws). A propeller has a series of identical blades placed around a hub that is driven by the engine. W. J. M. Rankine presented a theoretical model of the ideal propeller in his classic paper "On the Mechanical Principles of the Action of Propellers" (1865). He demonstrated that the efficiency of an ideal frictionless propeller to produce a specified thrust was improved as the propeller diameter was increased. There are, of course, practical limitations to the propeller diameter, depending on the ship geometry and to some extent on the fabrication capabilities of the propeller manufacturers. Designers have developed the shapes of the propeller blades to achieve good efficiency.

Methods for defining the details of the geometry of the propeller and its blading to achieve high efficiency are well established and are continuously being improved. Digital computer programs have been developed that are widely used to provide the design and analysis of propulsors. These programs were originally used to design conventional propellers, and the methods have been extended to apply to propellers in ducts and also to the pumps for waterjets.

To produce thrust it is necessary to apply torque to the propeller; this results in a swirling flow downstream of the propeller from the reaction of this torque acting on the water. The swirling flow is not used to produce thrust and is therefore a source of energy loss. Several methods have been developed to raise the propulsive efficiency by canceling or removing this swirling flow. Stationary blades (stators) upstream or downstream of the propeller have been installed in a few propellers for this purpose, and counterrotating screws have been used. An additional device downstream of the propeller that rotates freely, called a Grim wheel or a vane wheel, has been used in some applications. The inner portion of the vane wheel absorbs the swirling energy from the propeller wake while the outer portion acts as an additional propeller and produces thrust.

In the 1930s ducts or nozzles surrounding highly loaded propellers were introduced. Experiments had shown that propeller efficiency was improved, and many of these devices (often referred to as Kort nozzles) have been used in tugs and fishing boats. Theoretical analysis of ducted propellers has shown that the efficiency can be improved compared to conventional propellers when some of the thrust is produced by the fluid flow around the nozzle.

The designer is expected to develop a propeller design to minimize the power required by the power

Ship Type	*Large Tanker*	*Container Ship*	*Fast Ferry*	*Navy Destroyer*	*Aircraft Carrier*	*Submarine*
Length, m	334	290	45	142	330	110
Displacement, t	344,000	82,000	200	9,000	1000,000	7,000
Operating Speed, kt	15.3	24	44	10-32	10-30+	32
Range, n.miles	27,000	21,000	125	6,000	Unlimited	Unlimited
Power Plant Type	Large Diesel	Large Diesel	2 Gas Turbines	4 Gas Turbines	2 Nuclear Reactors	Nuclear Reactor
Propulsion Power, kW	25,000	38,000	8,400	78,000	194,000	26,000
Ship Service Power, kW	3,000	3,000	100	3,000	8,000	2,000
Fuel	Heavy Oil	Heavy Oil	Light Diesel	Light Diesel	Nuclear	Nuclear
Fuel Load, t	8,400	6,600	7.5	1,00		

Table 1.
Representative Ship and Power Data for Various Ship Types

plant to propel the ship at the design speed. At this speed the effective power, P_e, is the product of ship resistance and ship speed and would be determined by model tests, correlations of experimental results, or computed using theoretical predictions, discussed earlier. The shaft power developed by the engine, P_s, is the power required to drive the propeller through the transmission and shafting. The propulsive efficiency or propulsive coefficient (PC) is the ratio of effective power to shaft power: $PC = P_e/P_s$. The propulsive coefficient is usually between 60 and 75 percent for most ship designs.

The propulsor, in a steady situation, has to provide the thrust equivalent to the resistance of the ship. The prediction of this balance between ship resistance and propeller thrust is a complex process because the water flow in the stern region of the ship interacts with the flow through the propeller. The ship resistance may be modified by the action of the propeller, while the propeller efficiency usually is influenced by the disturbed flow regime near the stern of the ship.

The phenomenon of propeller cavitation often occurs at high ship speed and on highly loaded propellers. It provides a very serious limitation to propeller performance because it can damage the propeller and ship components near it. In addition, the efficiency may be reduced and the phenomenon can cause noise and vibration. Cavitation occurs when the local pressure of the water on the blade surface is reduced below the vapor pressure of the water passing over the blade surface. The high velocity of the water passing over regions of the propeller blades creates this reduction in pressure. The water boils when the local pressure becomes very low, and pockets of water vapor are formed at the blade surface. These vapor pockets and their subsequent collapse cause the problems associated with cavitation. Modern methods of propeller design have allowed designers to predict the onset of cavitation and to define the performance limitations.

WATERJETS

Waterjets have been developed for application to high-speed ships. The waterjet has an inlet usually on the side or bottom of the ship in the region of the stern, which allows water to flow into a water pump. The pressure of the water is raised in the pump, and the water is expelled as a jet to produce the desired thrust. The direction of the jet flow can be controlled to provide maneuvering forces, eliminating the need for rudders. The propulsive coefficient of modern waterjets usually is lower than propellers for the same application, but they have been used where severe cavitation would occur in conventional propellers.

SHIP POWER PLANTS

Most ships are powered by thermal power plants, including diesel engines, gas turbines, and nuclear systems. Merchant ships usually have diesel engines, while gas turbines or a combination of diesel engines and gas turbines often power naval vessels. Some of the larger ships in the U.S. Navy have nuclear power plants.

Large, low-speed diesel engines are remarkably efficient and convert about 45 percent of the energy of the fuel-air reaction into power. In addition, the cooling water and engine exhaust can be used to generate steam to provide heat and power for the ship. The steam has been used in some designs to generate power through steam turbines.

Improvements also have been made to the gas turbine for naval applications. An intercooled recuperative (ICR) gas turbine has been designed to improve the fuel consumption of naval power plants. The engine has a recuperator to take the heat that would otherwise be wasted in the exhaust and transfers it to the air entering the combustor. The new engine is expected to save about 30 percent of the fuel consumed, compared to the simple gas turbine. The ICR engine is, however, larger and more expensive than the simple gas turbine.

In the early steam ships the paddle wheels and propellers were directly connected to reciprocating steam engines. The large low-speed diesel engines of today are able to operate efficiently at the low propeller speeds and usually are directly connected by shafts to the propellers. Other ship power plants, such as steam turbines, gas turbines, and medium-speed diesel engines, operate most efficiently at higher rotational speeds than propellers. Speed converters have to be provided to reduce the rotational speed from the high engine speed to the low propeller speed. Mechanical reduction gearboxes are generally used to provide this conversion. Electric drives, consisting of high-speed electric generators driven by steam turbines coupled to low-speed electric motors, were used during World War II, when there was a shortage of gearboxes. A modern version of electric drive sometimes is used in cruise ships and also is being evaluated for the U.S. Navy.

Submarines, submersibles, and other underwater vehicles have a difficult operational problem, namely the absence of atmospheric oxygen, which is necessary for all thermal engines except nuclear power plants. At about the beginning of the twentieth century rechargeable batteries were developed, and these batteries, coupled to electric motors and propellers, were used to propel submarines under the sea. Spark ignition engines and later diesel engines provided the power for operation at the surface and to recharge the batteries. Since 1985, stored oxygen has been used in some thermal power plants and in fuel cells for experimental underwater vehicles. Nuclear power plants, which do not require oxygen for power generation, were introduced in 1954 to power U.S. military submarines. Other nations also have developed nuclear power for naval submarines.

ENVIRONMENTAL PROBLEMS

Two of the main environmental concerns are the atmospheric pollutants generated by ship's engines, and the possibility of damaging oil spills from accidents and ship operations.

Commercial ships have traditionally used the least expensive and therefore the most polluting oil fuels. Large diesel engines can now operate successfully with the lowest grade of fuel oils, and these fuels have a relatively high sulfur content. As a result the harmful SO_x by-products are released to the atmosphere. In addition, the combustion process in diesel engines produces harmful NOX pollutants. International, national, and local restrictions have encouraged engine manufacturers to develop remedial methods to reduce pollution.

The reduction of SO_x is best achieved by reducing the sulfur content of the fuel oil. Ship engineers are required to replace high-sulfur fuel with expensive higher-grade oil fuel as their ships approach coastlines where strict pollution restrictions apply.

Higher-grade fuels also reduce NO_x production in diesel engines, but more stringent methods are required to satisfy pollution regulations in some areas (e.g., California). Experiments conducted on diesel engines indicate that the concentration of NO_x in the exhaust gases increases as the engine power is reduced as ships approach coasts and harbors. Two approaches have been evaluated to improve the situation. First, changes to the combustion process have been tried by such measures as recirculating some of the exhaust and by the addition of water to the fuel to form an emulsion. Tests on engines have shown that about a 25 percent reduction in NO_x can be attained with such methods. The second approach has involved attempts to clean up the exhaust by catalytic reactions. Catalytic converters, as used in automobiles, have not been successful up to now, because diesel engines operate with too much oxygen in the exhaust products. An alternative approach using ammonia vapor as the catalyst has been shown to reduce the NO_x in the exhaust by more than 80 percent. This has been termed the selective catalytic reduction (SCR) method; it relies on the excess oxygen in the exhaust products to react with the ammonia and the NO_x to produce harmless water vapor and nitrogen. Careful measurement of the pollutants in the exhaust gases ahead of the ammonia injector has to be provided so the optimum quantity of ammonia can be injected.

In the last 30 years of the twentieth century, there were very damaging oil spills from the grounding of tankers filled with crude oil. In addition, there have been many smaller accidents, as well as routine operations, that have resulted in significant amounts of oil being released to the environment.

The International Maritime Organization (IMO) has issued a series of rules to tanker designers in an attempt to minimize the outflow of oil after accidental side or bottom damage. New tankers are required to have double hulls or other structural innovations to minimize tanker spills.

When cargo ships and fishing boats are involved in accidents, there are various measures that have to be carried out to limit the flow of fuel oil into the water. Accidents in harbors and close to shore are treated with great care. Oil booms would be placed around the vessel to prevent the spread of oil, and skimmers would be brought to the area to collect the oil released to the environment. In extreme cases the fuel oil cargo may be burned when the oil could not be pumped out or when a stranded vessel could not be refloated.

Routine operations of many ships have resulted in oil pollution. Cleaning up minor spills on deck and in the engine room is now treated very carefully. Fuel-oil and crude-oil tanks are cleaned from time to time. In the past the polluted oil-water mixture was dumped overboard, but now it must be pumped ashore for treatment.

A. Douglas Carmichael
Clifford A. Whitecomb

See also: Aerodynamics; Diesel Cycle Engines; Diesel Fuel; Engines; Environmental Problems and Energy Use; Nuclear Fission Fuel; Thermodynamics; Transportation, Evolution of Energy Use and; Turbines, Gas; Turbines, Steam; Waves.

BIBLIOGRAPHY

Giles, D. L. (1997). "Faster Ships for the Future." *Scientific American* 277(4):126–131.

Rankine, W. J. M. (1865). "On the Mechanical Principles of the Action of Propellers." *Trans Institute of Naval Architects* 6:13–39.

van Manen, J. D., and van Oossanen, P. (1988). "Propulsion." In *Principles of Naval Architecture, Vol. 2*, ed. E. V. Lewis. Jersey City, NJ: The Society of Naval Architects and Marine Engineers.

van Manen, J. D., and van Oossanen, P. (1988). "Resistance." In *Principles of Naval Architecture, Vol.2*, ed. E. V. Lewis. Jersey City, NJ: The Society of Naval Architects and Marine Engineers.

SIEMENS, ERNST WERNER VON (1816–1892)

Werner Siemens, German inventor, engineer, and entrepreneur was born on December 13, 1816, in Lenthe near Hanover, Lower Saxony, as the fourth child of fourteen children. His father was a tenant farmer. Siemens's education was first undertaken by his grandmother Deichmann, and it continued under a series of home tutors. In 1823 the family moved to Mecklenburg where the father took over the running of the estate of Menzendorf. From Easter 1832 until Easter 1834, Siemens attended the Katharineum, a gymnasium (secondary school) in the Hanseatic town of Lübeck, which he left without final examination. Because his interests were mathematics and natural sciences, rather than classical languages, he took some private lessons in mathematics to improve his education.

The family's precarious financial situation prompted Siemens to choose a military career as a way to advance his education in mathematics, physics and chemistry. In 1835 after basic training at the Artillery Corps in Magdeburg, he moved to Berlin where he stayed for three years as an officer candidate and a student of the Prussian Artillery School. There he was able to study under the guidance of the physicist Gustav Magnus. The death of both parents in 1840 put Siemens, as the oldest son, in charge of his nine surviving siblings. He took his brother Wilhelm to Magdeburg, where he continued his service in an artillery regiment. Passing a term of confinement in fortress, because of assisting a duel, Siemens smuggled in chemicals and experimented with galvanoplasty, a procedure discovered by Moriz Hermann Jacobi. He successfully developed a galvanic method of gold and silver plating. His brother Wilhelm went to England, patented and sold the discovery. From Siemens's electrochemical interests originated also a method of production of ozone and of electrolytic production of clean copper.

TELEGRAPHY

Siemens met the precision engineer Johann Georg Halske at Magnus's Institute of Physics in 1846. Together the two men founded "Siemens & Halske Telegraph Construction Co" in 1847 to exploit the new technology of electric telegraphy, which had been established three years earlier with the first public telegraph line of Samuel Morse. Siemens improved the dial-telegraph, invented the method for a seamless insulation of copper wire with gutta-percha, and discovered the multiple telegraphic use of copper conductors. The first long-line telegraph line, which was 660 km long and laid by Siemens & Halske between Berlin and Frankfurt was inaugurated on March 28, 1849, the day at which the first German national parliament in Frankfurt passed a constitution. His industrial career ensured, Siemens retired from the army. In 1851 Siemens & Halske commenced installation of a telegraph line between Moscow and St. Petersburg. For the Russian project Siemens put his brother Carl in charge. In 1855 the telegraph network for European Russia was installed, and through a Polish line a crucial link with Western Europe was established.

To compete worldwide, Siemens founded several industry subsidiary companies: in St. Petersburg, in London, in France, and in Vienna. In these endeavors, Siemens placced great trust in family members. After cousin Johann Georg helped to establish the company financially, Siemens incorporated other relatives: brother Friedrich and nephew Georg, later the first director of Deutsche Bank. In 1868 Siemens Brothers Co. of London started a cooperation with the Indo-European Telegraph Co. to lay mainly overhead cables through Southern Russia and Persia, effectively joining London and Calcutta. An improved printer telegraph with a punched tape synchronized with current pulses of an inductor with a double T armature operated from 1870 on this line.

Between 1874 and 1880 Siemens Brothers laid transatlantic cables using their own cable-steamer *Faraday*.

ELECTRIC GENERATORS AND MACHINES

The only source of electric energy available until the 1830s was in the form of galvanic elements. The situation changed in 1831 with the discovery of the law of electromagnetic induction by Michael Faraday. The law states that an electric current is generated in a conducting circuit if the latter is exposed to a fluctuating magnetic field. As a consequence of Faraday's discovery, numerous engineers—among them Siemens—designed and improved electric generators. The basic idea behind the first generators was to move wire coils into and out of the magnetic field of a permanent magnet, thereby generating an electric current in the coils according to the induction law. A hand-turned generator with magnets revolving around coils was constructed as early as 1832 in Paris. A year later, the reversed principle to rotate the coils in a fixed field of a permanent magnet was realized in a generator constructed in London. Around 1850, industrial production of electric generators was set up in several European countries. In 1856, Siemens developed a double T armature, cylinder magnet constructed on the basis of Faraday's law of induction to provide a source of electricity for telegraphs cheaper than Zinc galvanic batteries. Though more powerful than galvanic elements, the first generators did not yet provide sufficient electric energy for heavy mechanical work. The limitation was imposed through the rather modest size of the available permanent magnets. To overcome this shortcoming Siemens proposed in 1866, almost simultaneously and independently of some others (Soren Hjorth, Anyos Jedlik, Alfred Vaarley, and Charles Wheatstone) to substitute an electromagnetic effective magnet for the permanent magnet. Siemens is credited with having worked out the principle underlying the new type of generator: a rotating coil inside a stationary coil, both forming a single circuit connected to the external consumer circuit. By starting the rotation of the movable coil, a weak current is induced through a residual magnetic field delivered from the iron armature of the apparatus. The current in the stationary coil causes an increase of the magnetic field, which in turn gives rise to an increased current output from the rotating coil. A self-enhancing process of current generation is so initiated. Siemens called this scheme of a self-excited generator the "dynamo electrical principle." Its formulation, presented to the Academy of Science on January 17, 1867, and subsequent demonstration of the electric light from the dynamo-machine moved through a steam-machine at the factory side in Markgrafenstrasse, Berlin, laid the groundwork for the modern technology of electric generators. This discovery is considered Siemens's most important contribution to science and engineering.

Siemens constructed the first electric railway shown at the Berlin Trade Fair of 1879 and the first electrically-operated lift in 1880, and the first electric trams began operating in Berlin in 1881. Siemens received many honors for his work: an 1860 honorary doctorate from the University of Berlin, an 1873 membership in the Royal Academy of Science, and an 1888 knighthood from Emperor Friedrich III. Siemens died in Berlin, Charlottenburg, on December 6, 1892.

Barbara Flume-Gorczyca

See also: Magnetism and Magnets.

BIBLIOGRAPHY

Feldenkirchen, W. (1992). *Werner von Siemens*. Zurich: Piper München.

Kundt, A. (1979). *Physiker über Physiker*, Vol. 2. Berlin: Akademie-Verlag.

Siemens, W. (1881). *Gesamelte Abhandlungen und Vorträge*. Berlin: Springer.

Weiher, S. von. (1975). *Werner von Siemens*. Goettingen: Musterschmidt.

SMEATON, JOHN (1724–1792)

John Smeaton was one of the first great British civil engineers, the first to use the title "Civil Engineer" (seeing himself as a professional ranking alongside doctors and lawyers), and the first to achieve distinction as an engineering scientist.

Smeaton was born on June, 8, 1724 at Austhorpe Lodge, near Leeds. He was the son of a successful Leeds lawyer, and from an early age Smeaton developed an interest in mechanics (which his father indulged by providing a workshop), building a working model of a steam engine being erected at a nearby colliery (coal mine). It was the desire of Smeaton's father that his son should succeed him, and at the age of sixteen Smeaton was employed in his father's office. In an attempt to wean him from his mechanical pursuits and partly to give him a good legal education, Smeaton was sent to London. In London, Smeaton's desire to pursue a mechanical career was not suppressed, and his father finally agreed to finance his training as a philosophical instrument maker, to make instruments or apparatus for study of natural philosophy or the physical sciences. The young Smeaton did not however, live the life of a workman, but attended meetings of the Royal Society and was often in the company of educated people. He was soon able to set up as a mathematical instrument maker (making instruments used by draftsmen, navigators, or land surveyors) in London, and at the age of twenty-six presented a paper before the Royal Society on improvements in the marine compass. More papers quickly followed, and in 1753 he was elected a fellow.

In 1759 Smeaton presented to the Royal Society a paper entitled "An Experimental Enquiry concerning the natural Power of Water and Wind to turn mills, and other machines, depending on a circular motion." This paper was the result of a series of experiments carried out in 1752 and 1753, Smeaton having delayed its publication until he had put his deductions into practice. At the time there was a lively debate as to the merits of undershot and overshot waterwheels, but little published data to substantiate various claims. Smeaton's solution to this problem depended solely on experiments made with working model waterwheels. He wanted to compare the "power" of the water delivered to the wheel with the useful "effect" produced by the wheel.

Smeaton's elegant experimental technique enabled him to deal with both hydraulic and mechanical friction losses, allowing him to calculate water velocity at the wheel and thereby determine an "effective" or "virtual" head. Smeaton's experimental apparatus was a brilliant device that enabled him to measure the efficiency of the waterwheel, alone rather than the overall efficiency of the experiment. Smeaton was able to conclusively show that a water-wheel when driven by the weight of water alone, is about twice as efficient as when driven by the impulse of water. This demonstration ensured that British mills, wherever possible, from then on would be fitted with overshot or breastshot waterwheels, rather than undershot.

Smeaton also turned his attention to windmills, but with inconclusive results. This may have been because there was no clear-cut issue to be resolved as was the case with the waterwheel regarding which form was the best type. However, he did produce useful guidelines on the construction of windmills. Smeaton was careful to distinguish the circumstances in which a model test rig differs from the full-size machine, cautioning, "otherwise a model is more apt to lead us from the truth than towards it." Smeaton's experiments demonstrated that work done by a prime mover was a good measure of its performance. In 1765, seeing a need for a practical measure, he fixed the power of a horse working an eight-hour day as 22,000/ft. lb/min.

Britain in the mid-eighteenth century experienced rapid expansion in the number and scale of public works. This enabled Smeaton, in 1753, to give up instrument-making and become a consulting civil engineer, building both wind and water mills, canals, bridges and, from 1756 to 1759, the Eddystone lighthouse. In preparing his designs, Smeaton adopted a scientific approach. It was his practice to prepare a report and have the work carried out by a resident engineer.

In 1766 Smeaton designed an atmospheric steam-pumping engine for the New River Company; however, when the engine was set to work, its performance was less than Smeaton had expected. Realizing his knowledge of steam engines was deficient, he approached the problem in the same manner as he did his study of water and wind power—by

John Smeaton. (Library of Congress)

deciding to carry out a systematic, practical study. His first action was to have drawn up a list of all steam engines working in the district around Newcastle. Of these, fifteen were chosen and a set of engine tests made. Smeaton was the first to gather such comprehensive data, from which he calculated for each engine the "great product per minute" or power, and "effect per minute of one bushel of coal per hour" or performance.

Smeaton constructed on the grounds of his house, a small experimental atmospheric engine with a ten-inch cylinder. Beginning in 1770 he made some 130 tests over two years. All relevant measurements were made, including temperature, pressure—both internal and barometric, evaporation of water per bushel of coal, etc. Smeaton's method of testing was to adjust the engine to good performance and take measurements, then alter one of the parameters and take further measurements. By this means he was able to optimize valve timing, piston loading, and size of injector nozzle. He also carried out experiments to test the evaporative powers of various types of coal.

From the knowledge gained, Smeaton drew up a table for the proportions of parts for Newcomen-type engines. Although Smeaton's experiments added nothing to the invention of the Newcomen engine, establishing proper proportions for engines enabled him to built more efficient and powerful engines. The great leap forward in the operation of atmospheric steam engines came with Watt's separate condenser.

Smeaton was a born mechanic and incessant experimentor, but a man of simple tastes and wants. He limited his professional engagements in order to devote a certain portion of his time to scientific investigations. One of Smeaton's rules was not to trust deductions drawn from theory when there was an opportunity for actual experiment. In 1771 Smeaton founded a club for engineers, which later came to be call the "Smeatonian Society."

In 1756 Smeaton married Ann Jenkinson, the daughter of a merchant tailor and freeman of the city of York. They had two daughters. Ann Smeaton fell ill and died in 1784. Smeaton continued working until 1791, when he retired to Austhorpe Lodge to prepare for publication descriptions of his various engineering works. While walking in his gardens he suffered a stroke and died six weeks later, on October 28, 1792, in his sixty-ninth year.

Robert Sier

See also: Newcomen, Thomas; Watt, James.

BIBLIOGRAPHY

Farey, J. (1827). *A Treatise on the Steam Engine.* London: Longmans.

Ress, A. (1813). *The Cyclopedia; or Universal Dictionary of Arts, Science and Literature.* London: Longman.

Smeaton, J. (1759). "An Experimental Enquiry Concerning the Natural Power of Water and Wind to Turn Mills and Other Machinery, Depending on a Circular Motion." *Philosophical Transactions of the Royal Society* 51:100–174.

Smeaton, J. (1791). *A Narrative of the Building and a Description of the Construction of the Eddystone Lighthouse.* London.

Smeaton, J. (1797, 1812). *Reports of the Late John Smeaton, FRS Made on Various Occasions in the Course of His Employment as a Civil Engineer,* Vol. 1 (1797), Vol. 2–3 (1812). London: Longman, Hurst, Rees, Orme, and Brown.

Smiles, S. (1904). *Lives of the Engineers. John Smeaton.* London: John Murrey.

Wilson, P. N. (1955). "The Waterwheels of John Smeaton." *Transactions of the Newcomen Society* 30:25–43.

SMOG

See: Air Pollution

SOLAR ENERGY

Energy from the sun is abundant and renewable. It is also the principal factor that has enabled and shaped life on our planet. The sun is directly or indirectly responsible for nearly all the energy on earth, except for radioactive decay heat from the earth's core, ocean tides associated with the gravitational attraction of Earth's moon, and the energy available from nuclear fission and fusion. Located approximately ninety-three million miles from Earth, the sun, an average-size star, belongs to the class of dwarf yellow stars whose members are more numerous than those of any other class. The energy radiated into outer space by the sun (solar radiation) is fueled by a fusion reaction in the sun's central core where the temperature is estimated to be about 10 million degrees. At this temperature the corresponding motion of matter is so violent that all atoms and molecules are reduced to fast-moving atomic nuclei and stripped electrons collectively known as a plasma. The nuclei collide frequently and energetically, producing fusion reactions of the type that occur in thermonuclear explosions.

While about two-thirds of the elements found on earth have been shown to be present in the sun, the most abundant element is hydrogen, constituting about 80 percent of the sun's mass (approximately two trillion trillion million kilograms). When hydrogen nuclei (i.e., protons) collide in the sun's core, they may fuse and create helium nuclei with four nucleons (two protons and two neutrons). Roughly 20 percent of the sun's mass is in the form of helium. Also created in each fusion reaction are two neutrinos, high-energy particles having no net electrical charge, which escape into outer space, and high-energy gamma radiation that interacts strongly with the sun's matter surrounding its core. As this radiation streams outward from the core, it collides with and transfers energy to nuclei and electrons, and heats the mass of the sun so that it achieves a surface temperature of several thousand degrees

Celsius. The energy distribution of the radiation emitted by this surface is fairly close to that of a classical "black body" (i.e., a perfect emitter of radiation) at a temperature of 5,500°C, with much of the energy radiated in the visible portion of the electromagnetic spectrum. Energy is also emitted in the infrared, ultraviolet and x-ray portions of the spectrum (Figure 1).

The sun radiates energy uniformly in all directions, and at a distance of ninety-three million miles, Earth's disk intercepts only four parts in ten billion of the total energy radiated by the sun. Nevertheless, this very small fraction is what sustains life on Earth and, on an annual basis, is more than ten thousand times larger than all the energy currently used by Earth's human inhabitants. Total human energy use is less than 0.01 percent of the 1.5 billion billion kilowatt-hours (kWh) of energy per year the sun delivers to Earth. A kilowatt hour is one thousand watt hours and is the energy unit shown on your electric bill. A one-hundred-watt light bulb left on for ten hours will use one kWh of electric energy.

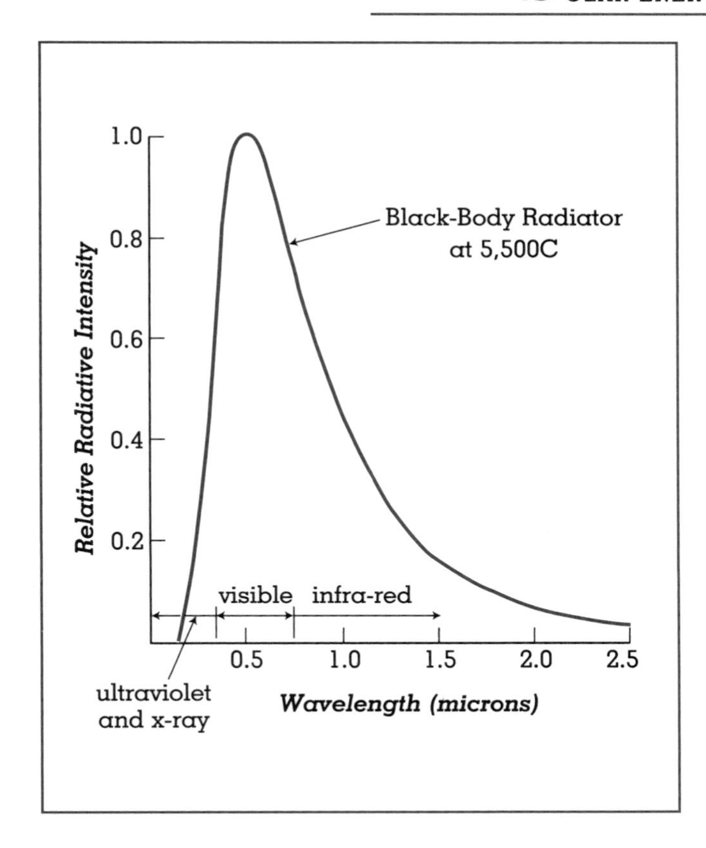

Figure 1.
Relative radiative intensity as a function of wavelength.

AVAILABILITY OF SOLAR ENERGY

While the amount of energy radiated by the sun does vary slightly due to sunspot activity, this variation is negligible compared to the energy released by the sun's basic radiative process. As a result, the amount of energy received at the outer boundary of Earth's atmosphere is called the Solar Constant because it varies so little. This number, averaged over Earth's orbit around the sun, is 1,367 watts per square meter (W/m^2) on a surface perpendicular to the sun's rays. If, on average, Earth were closer to the sun than ninety-three million miles, this number would be larger; if it were farther from the sun, it would be smaller. In fact, Earth's orbit about the sun is not circular, but elliptical. As a result, the Solar Constant increases and decreases by about 3 percent from its average value at various times during the year. In the northern hemisphere the highest value is in the winter and the lowest is in the summer.

A non-negligible fraction of the solar radiation incident on the earth is lost by reflection from the top of the atmosphere and tops of clouds back into outer space. For the radiation penetrating the earth's atmosphere, some of the incident energy is lost due to scattering or absorption by air molecules, clouds, dust and aerosols. The radiation that reaches the earth's surface directly with negligible direction change and scattering in the atmosphere is called "direct" or "beam" radiation. The scattered radiation that reaches the earth's surface is called "diffuse" radiation. Some of the radiation may reach a receiver after reflection from the ground, and is called "albedo." The total radiation received by a surface can have all three components, which is called "global radiation."

The amount of solar radiation that reaches any point on the ground is extremely variable. As it passes through the atmosphere, 25 to 50 percent of the incident energy is lost due to reflection, scattering or absorption. Even on a cloud-free day about 30 percent is lost, and only 70 percent of 1,367 W/m^2, or 960 W/m^2, is available at the earth's surface. One must also take into account the earth's rotation and the resultant day-night (diurnal) cycle. If the sun shines 50 percent of the time (twelve hours per day, every day) on a one square meter surface, that surface receives no more than (960 W/m^2) × (12 hours/day) × (365 days/year) =

4,200 kWh of solar energy per year. Since, on average, the sun actually shines less than twelve hours per day at any location, the maximum solar radiation a site can receive is closer to 2,600 kWh per square meter per year. To put this number into perspective, the average person on earth uses about 18,000 kWh of all forms of energy each year.

HISTORY

For centuries the idea of using the sun's heat and light has stimulated human imagination and inventiveness. The ancient Greeks, Romans, and Chinese used passive solar architectural techniques to heat, cool, and provide light to some of their buildings. For example, in 100 C.E., Pliny the Younger built a summer home in northern Italy that incorporated thin sheets of transparent mica as windows in one room. The room got warmer than the others and saved on short supplies of wood. To conserve firewood, the Romans heated their public baths by running the water over black tiles exposed to the sun. By the sixth century, sunrooms on private houses and public building were so common that the Justinian Code introduced "sun rights" to ensure access to the sun.

In more recent times, Joseph Priestly used concentrated sunlight to heat mercuric oxide and collected the resulting gas, thus discovering oxygen. In 1872, a solar distillation plant was built in Chile that provided six thousand gallons of fresh water from salt water daily for a mining operation. In 1878, at an exhibition in Paris, concentrated sunlight was used to create steam for a small steam engine that ran a printing press.

Many further solar energy developments and demonstrations took place in the first half of the twentieth century. However, only one solar technology survived the commercial competition with cheap fossil fuels. That exception was solar water heaters that were widely used in Japan, Israel, Australia and Florida, where electricity was expensive and before low-cost natural gas became available.

Also, passive solar, or climate responsive, buildings were in such demand by 1947 as a result of energy shortages during World War II, that the Libbey-Owens-Ford Glass Company published a book entitled *Your Solar House*, which profiled forty-nine solar architects. In the mid-1950s, architect Frank Bridgers designed the world's first commercial office building using passive design and solar water heating. This solar system has since been operating continuously, and is now in the U.S. National Historic Register as the world's first solar-heated office building.

It was the oil embargo of 1973 that refocused attention on use of solar energy, when the cost of fossil fuels (coal, oil and natural gas) increased dramatically and people began to appreciate that fossil fuels were a depletable resource. More recently, concerns about energy security (the dependence on other countries for one's energy supply), and the local and global environmental impacts of fossil fuel use, have led to the growing realization that we cannot project today's energy system, largely dependent on fossil fuels, into the long-term future. In the years following the oil embargo, much thought and effort has gone into rethinking and developing new energy options. As a result, a growing number of people in all parts of the world believe that renewable (i.e., non- depletable) energy in its various forms (all derived directly or indirectly from solar energy except for geothermal and tidal energy) will be the basis for a new sustainable energy system that will evolve in the twenty-first century.

USING SOLAR ENERGY

Solar energy can be used in many direct and indirect ways. It can be used directly to provide heat and electricity. Indirectly, non-uniform heating of Earth's surface by the sun results in the movement of large masses of air (i.e., wind). The energy associated with the wind's motion, its kinetic energy, can be tapped by special turbines designed for that purpose. Winds also create ocean waves, and techniques for tapping the kinetic energy in waves are under active development. Hydropower is another indirect form of solar energy, in that the sun's radiant energy heats water at Earth's surface, water evaporates, and it eventually returns to Earth as rainfall. This rainfall creates the river flows on which hydropower depends. In addition, sunlight is essential to photosynthesis, the process by which plants synthesize complex organic materials from carbon dioxide, water and inorganic salts. The solar energy used to drive this process is captured in the resulting biomass material and can be released by burning the biomass, gasifying it to produce hydrogen and other combustible gases, or converting it to liquid fuels for use in transportation and other applications.

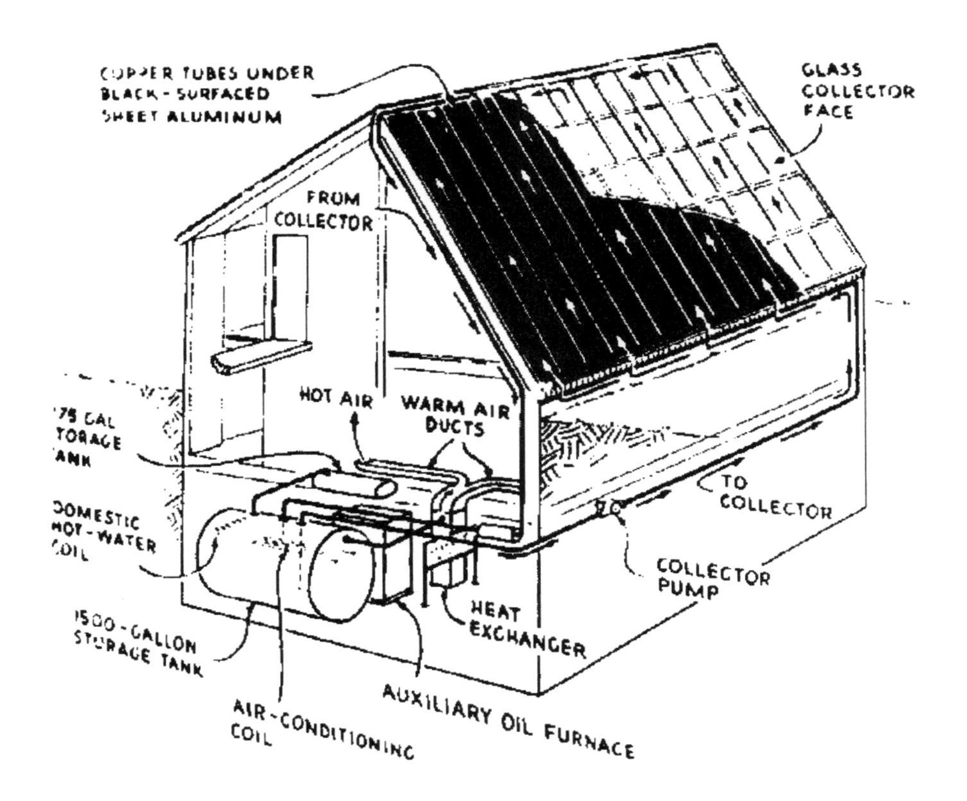

A 1958 diagram of sun-heated home designed by the Massachusetts Institute of Technology showing air ducts and tanks. (Library of Congress)

In principle, energy can also be extracted from the kinetic energy of large currents of water circulating beneath the ocean's surface. These currents are also driven by the sun's energy, much as are wind currents, but in this case it is a slower-moving, much more dense fluid. Research into practical use of this energy resource is in its very earliest stages, but is looking increasingly promising.

Finally, there are renewable sources of energy that are not direct or indirect forms of solar energy. These include geothermal energy in the form of hot water or steam derived from reservoirs below the surface of Earth (hydrothermal energy), hot dry rock, and extremely hot liquid rock (magma). These resources derive their energy content from the large amount of heat generated by radioactive decay of elements in Earth's core. Energy can also be tapped from tidal flows resulting from Earth's gravitational interaction with the moon.

SOLAR HEATING AND COOLING

Solar heating and cooling systems are classified either as passive or active. Passive systems make coordinated use of traditional building elements such as insulation, south-facing glass, and massive floors and walls to naturally provide for the heating, cooling and lighting needs of the occupants. They do not require pumps to circulate liquids through pipes, or fans to circulate air through ducts. Careful design is the key. Active solar heating and cooling systems circulate liquids or air through pipes or other channels to move the necessary heat to where it can be used, and utilize other components to collect, store and control the energy from the solar energy source. Systems that combine both passive and active features are called hybrid systems. Solar heating systems are used primarily for hot water heating, interior space heating, and industrial and agricultur-

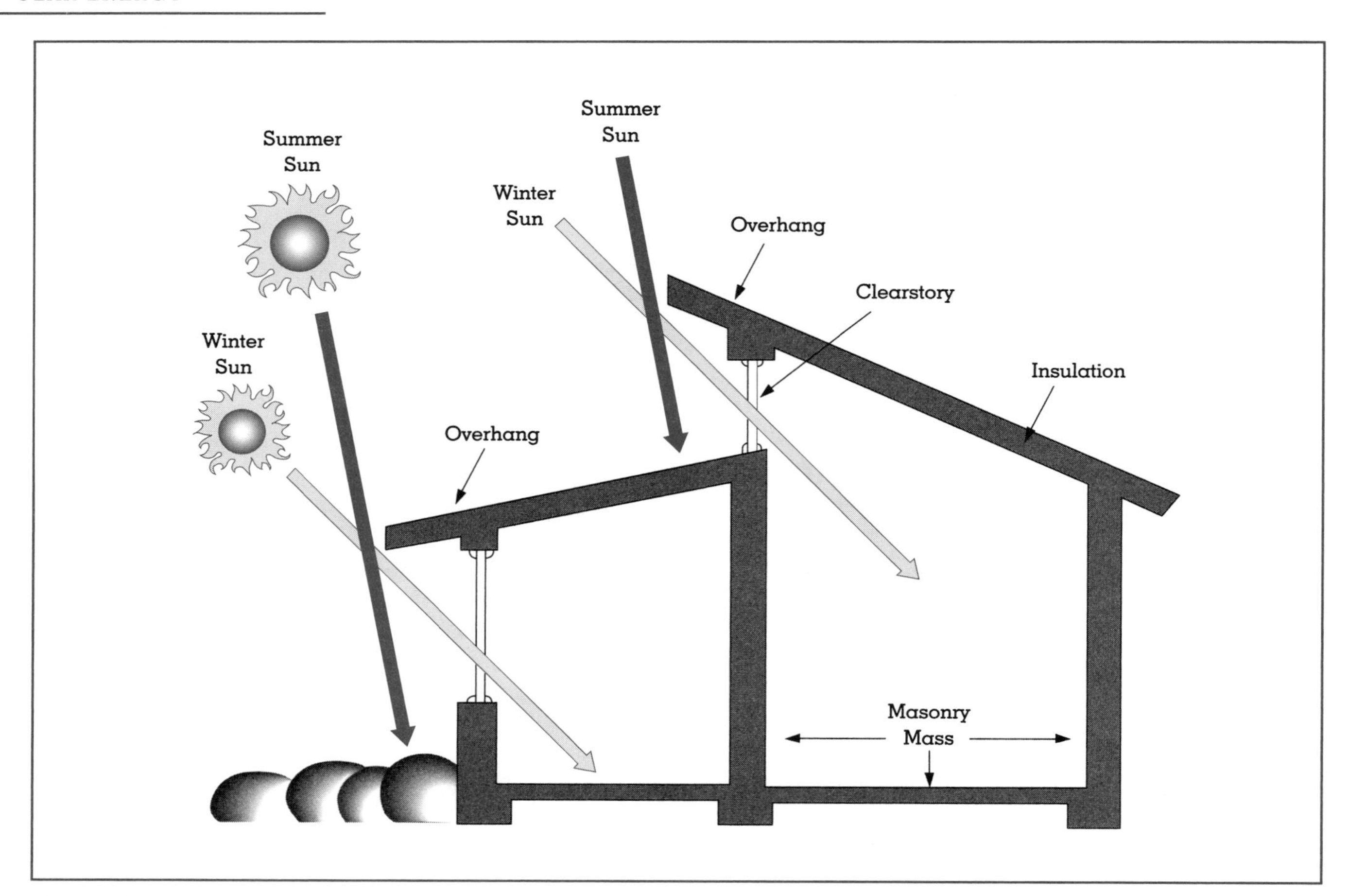

Figure 2.
Direct Gain.

al process heat applications. Solar cooling systems use solar heat to energize heat-driven refrigeration systems or to recycle dessicant-cooling systems. They are most often combined with solar hot water and space heating systems to provide year-round use of the solar energy system.

PASSIVE SYSTEMS

There are several types of passive system designs: direct gain, thermal storage, solar greenhouse, roof pond, and convective loop. The simplest is the direct gain design (Figure 2), which in the northern hemisphere is an expanse of south-facing glass (usually double glazed). The energy in the sunlight entering directly through the windows is absorbed, converted to heat, and stored in the thermal mass of the floors and walls. A thermal storage wall system consists of a massive sunlight-absorbing wall behind south-facing double glazing. The wall may contain water or masonry to store the energy during the day and release it during the night. Another type of passive system is the solar greenhouse (or sunspace), which combines direct gain in the greenhouse and a thermal storage wall between the greenhouse and the rest of the house. Solar energy provides the heat for the greenhouse and a good share of the energy for heating the living space in the house.

A fourth type of passive solar heating and cooling system is the roof pond, in which containers of water are used to collect and store the sun's energy. Movable insulating panels are used to control the gain and loss of heat. In winter, the insulation is opened during the day to allow collection of the sun's heat; at night the insulation is closed to minimize loss of heat that is to be released to the house. In summer, the insulation is closed during the day to block the sun, and opened to the sky at night to provide radiative cooling.

Solar collector panels. (Digital Stock)

The fifth type of passive system is the natural convective loop, in which the collector is placed below the living space and the hot air that is created rises to provide heat where it is needed. This same principle is used in passive solar hot water heating systems known as thermosiphons. The storage tank is placed above the collector. Water is heated in the collector, becomes less dense, and rises (convects) into the storage tank. Colder water in the storage tank is displaced and moves down to the collector where it is heated to continue the cycle.

More than one million residences and twenty thousand commercial buildings across the United States now employ passive solar design.

ACTIVE SYSTEMS

An active solar heating and cooling system consists of a solar energy collector (flat plate or concentrating), a storage component to supply heat when the sun is not shining, a heat distribution system, controls, and a back-up energy source to supply heat when the sun is not shining and the storage system is depleted.

The simplest type of active solar collector, the flat-plate collector, is a plate with a black surface under glass that absorbs visible solar radiation, heats up, and reradiates in the infrared portion of the electromagnetic spectrum. Because the glass is transparent to visible but not to infrared radiation, the plate gets increasingly hotter until thermal losses equal the solar heat gain. This is the same process that occurs in a car on a sunny day, and in the atmosphere, where gases such as carbon dioxide and methane play the role of the glass. It is called the greenhouse effect. Careful design can limit these losses and produce moderately high temperatures at the plate—as much as 160°F (71°C). The plate transfers its heat to circulating air or water, which can then be used for space heating or for producing hot water for use in homes, businesses or swimming pools.

Higher temperatures—up to 350°F (177°C)—can be achieved in evacuated tube collectors that encase both the cylindrical absorber surface and the tubes carrying the circulating fluid in a larger tube containing a vacuum which serves as highly efficient insulation. Still another high-temperature system is the parabolic-trough collector that uses curved reflecting surfaces to focus sunlight on a receiver tube placed along the focal line of the curved surface. Such concentrating systems use only direct sunlight, and can achieve temperatures as high as 750°F (400°C), but do require tracking of the sun.

Water heating accounts for approximately 25 percent of the total energy used in a typical single-family home. An estimated 1.5 million residential and commercial solar water-heating systems have been installed in the U.S. In Tokyo—there are nearly that many buildings with solar water heating. Large numbers of solar water heaters have also been installed in Greece, Israel, and other countries. More than 300,000 swimming pools are heated by solar energy in the United States.

SOLAR THERMAL POWER SYSTEMS

A solar thermal power system, increasingly referred to as concentrating solar power, to differentiate it from solar heating of residential/commercial air and water, tracks the sun and concentrates direct sunlight to create heat that is transferred to water to create steam, which is then used to generate electricity. There are three types of solar thermal power systems: troughs that concentrate sunlight along the focal axes of parabolic collectors (the most mature form of solar thermal power technology), power towers with a central receiver surrounded by a field of concentrating mirrors, and dish-engine systems that use radar-type reflecting dishes to focus sunlight on a heat-driven engine placed at the dish's focal point. Solar ponds are also a form of solar thermal power technology that does not require tracking.

Trough Systems

Trough systems currently account for more than 90 percent of the world's solar electric capacity. They use parabolic reflectors in long trough configurations to focus and concentrate sunlight (up to one hundred times) on oil-filled glass tubes placed along the trough's focal line. The dark oil absorbs the solar radiation, gets hot (up to 750°F (400°C)), and transfers its heat to water, creating steam that is fed into a conventional turbine-generator. Troughs are modular and can be grouped together to create large amounts of heat or power. Troughs operating in the Mojave Desert in the United States have been feeding up to 354 MW of electrical energy reliably into the Southern California power grid since the early 1990s, usually with minimal maintenance. While not currently cost competitive with fossil fuel-powered sources of electricity (trough electricity costs are about 12 cents per kWh), construction of the trough systems was encouraged by generous power purchase agreements for renewable electricity issued by the State of California, and Federal tax incentives.

Power Towers

Power towers, also known as central receivers, have a large field of mirrors, called heliostats, surrounding a fixed receiver mounted on a tall tower. Each of the heliostats independently tracks the sun and focuses sunlight on the receiver where it heats a fluid (water or air) to a very high temperature (1,200°F (650°C)). The fluid is then allowed to expand through a turbine connected to an electric generator. Another version heats a nitrate salt to 1,050°F (565°C), which provides a means for storage of thermal energy for use even when the sun is not shining. The heat stored in this salt, which becomes molten at these elevated temperatures, is then transferred to water to produce steam that drives a turbine-generator. Ten megawatt electrical power towers have been built and tested in the United States, one using water as the material to be heated in the receiver, and one using nitrate salt. A smaller unit that heats air has also been operated at the Platforma Solar test site in Almeria, Spain. While there are no commercial power tower units operating today, technical feasibility has been established and, in commercial production, electricity production costs are projected to be well below ten cents per kWh. At such costs, power towers will begin to compete with more traditional fossil fuel-powered generating systems. It is believed that power towers will be practical in sizes up to 200 MW (electrical).

Dish-Engine Systems

A dish-engine system uses a dish-shaped parabolic reflector, or a collection of smaller mirrors in the shape of a dish, to focus the sun's rays onto a receiv-

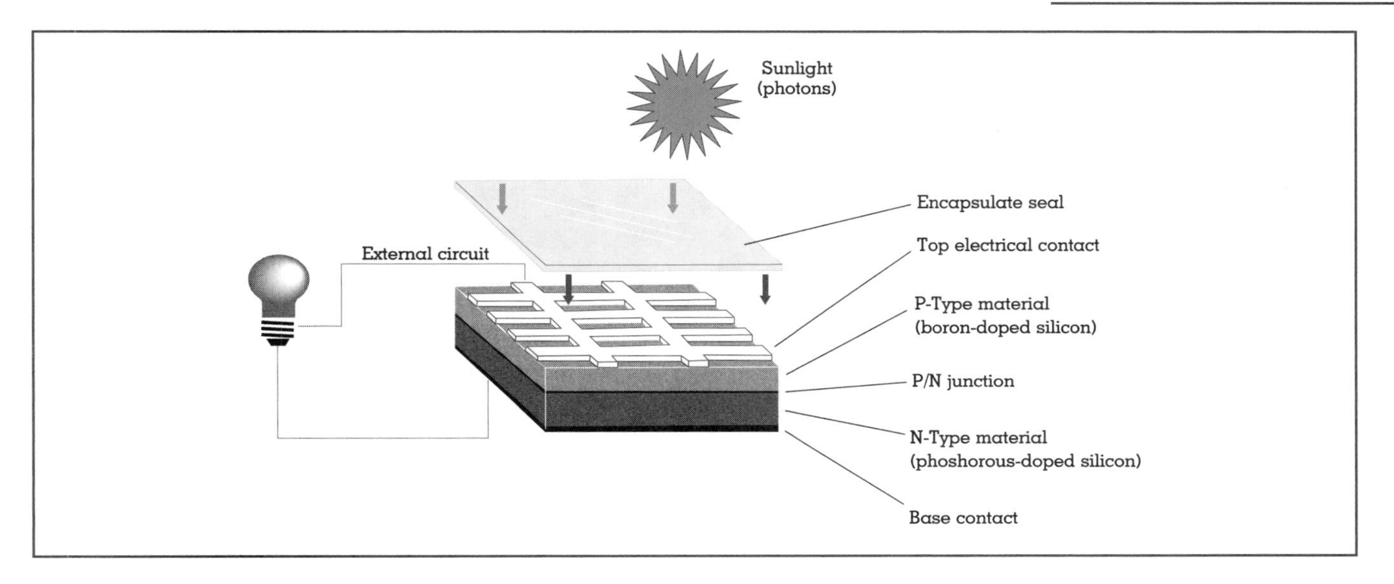

Figure 3.
Basic PV cell construction.

er mounted above the dish at its focal point. Two-axis tracking is used to maximize solar energy capture, and most often a Stirling engine (a sealed engine that can be driven by any source of external heat) is mounted on the receiver. Such systems are not yet commercial, but have been tested in sizes ranging from 5 to 50 kW (electrical). Conversion efficiencies of up to 29 percent have been achieved. Dish-engine systems are also modular, and can be used individually or in large assemblies.

Power-tower and dish-engine systems have higher concentration ratios than troughs, and therefore the potential to achieve higher efficiencies and lower costs for converting heat to electricity. All three systems can be hybridized (i.e., backed up by a fossil fuel system for supplying the heat), an important feature that allows the system to produce power whenever it is needed and not just when the sun is shining (dispatchability).

Solar Ponds

A solar pond does not concentrate solar radiation, but collects solar energy in the pond's water by absorbing both the direct and diffuse components of sunlight. Solar ponds contain salt in high concentrations near the bottom, with decreasing concentrations closer to the surface. This variation in concentration, known as a salt-density gradient, suppresses the natural tendency of hot water to rise, thus allowing the heated water to remain in the bottom layers of the pond while the surface layers stay relatively cool. Temperature differences between the bottom and top layers are sufficient to drive an organic Rankine-cycle engine that uses a volatile organic substance as the working fluid instead of steam. Temperatures of 90°C are routinely achieved in the pond bottom, and solar ponds are sufficiently large to provide some degree of energy storage.

The largest solar pond in the United States is a tenth of an acre experimental facility in El Paso, Texas, which has been operating reliably since 1986. The pond runs a 70 kW (electrical) turbine-generator and a 5,000 gallon per day desalination unit, while also providing process heat to an adjacent food processing plant. The potential of solar ponds to provide fresh water, process heat and electricity, especially for island communities and coastal desert regions, appears promising, but has not been fully investigated.

PHOTOVOLTAICS

Photovoltaics (photo for light, voltaic for electricity), often abbreviated as PV, is the direct conversion of sunlight to electricity. Considered by many to be the most promising of the renewable electric technologies in the long term, it is an attractive alternative to conventional sources of electricity for many reasons: it is silent and non-polluting, requires no special

Two huge solar energy panels track a photovoltaic system at the National Renewable Energy Lab in Golden, Colorado. (U.S. Department of Energy)

training to operate, is modular and versatile, has no moving parts, and is highly reliable and largely maintenance-free. It can be installed almost anywhere, uses both direct and diffuse radiation, and can be incorporated into a variety of building products (roofing tiles and shingles, facades, overhangs, awnings, windows, atriums).

The photoelectric effect (the creation of an electrical current when light shines on a photosensitive material connected in an electrical circuit) was first observed in 1839 by the French scientist Edward Becquerel. More than one hundred years went by before researchers in the United States Bell Laboratories developed the first modern PV cell in 1954. Four years later, PV was used to power a satellite in space and has provided reliable electric power for space exploration ever since.

The 1960s brought the first terrestrial applications for PV. At that time the technology was very expensive, with PV collectors, called modules, costing upwards of $1,000 per peak watt. The term "peak watts" is used to characterize the power output of a PV module when incident solar radiation is at its peak. Nevertheless, PV was a preferred choice in remote locations where no other form of power production was feasible. Over the past three decades, steady advances in technology and manufacturing have brought the price of modules down more than 200-fold, to $4 per peak watt. Further reductions in module cost to $1 to $2 per peak watt are expected within the next decade.

A photovoltaic cell (often called a solar cell) consists of layers of semiconductor materials with different electronic properties. In most of today's solar cells the semiconductor is silicon, an abundant element in the earth's crust. By doping (i.e., chemically introducing impurity elements) most of the silicon with boron to give it a positive or p-type electrical character, and doping a thin layer on the front of the cell with phosphorus to give it a negative or n-type character, a transition region between the two types

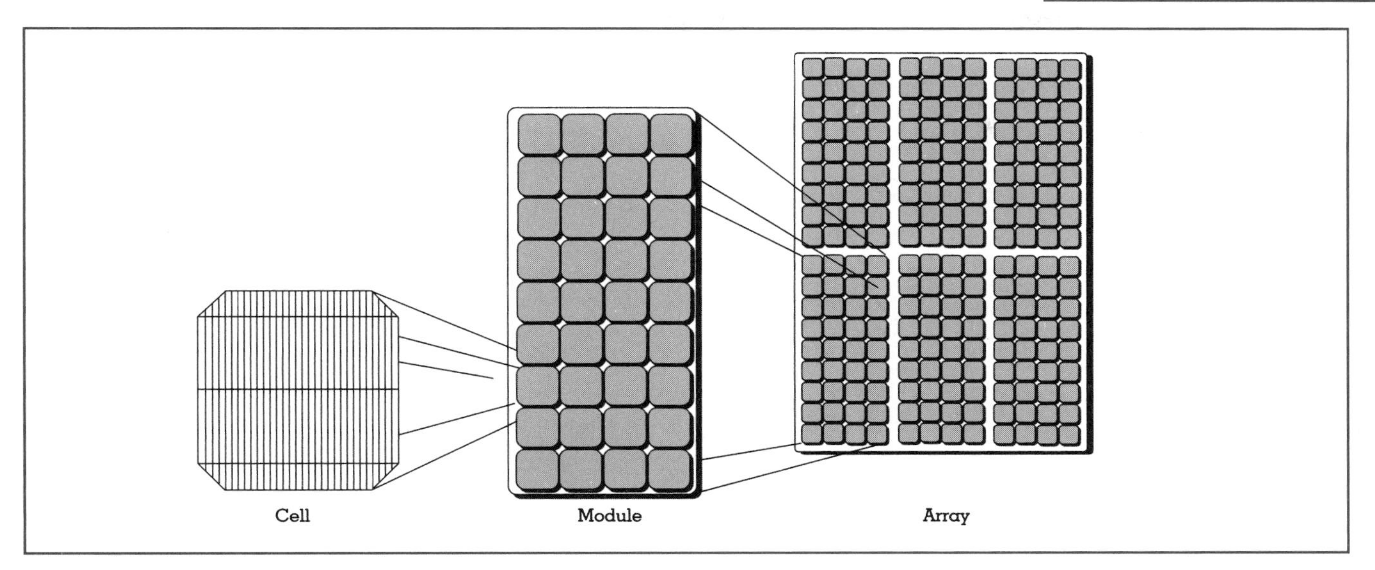

Figure 4.
PV cell, module, and array.

of doped silicon is formed that contains an electric field. This transition region is called a junction.Light consists of energy packets called photons. When light is incident on the cell, some of the photons are absorbed in the region of the junction and energy is transferred to the semiconductor, freeing electrons in the silicon. If the photons have enough energy, the electrons will be able to overcome the opposing electric field at the junction and move freely through the silicon and into an external circuit. As these electrons stimulate current flow in the external circuit, their energy can be converted into useful work (Figure 3).

Photovoltaic systems for specific applications are produced by connecting individual modules in series and parallel to provide the desired voltage and current (Figure 4). Each module is constructed of individual solar cells also connected in series and parallel. Modules are typically available in ratings from a few peak watts to 250 peak watts.

Commercially available PV systems most often include modules made from single-crystal or polycrystalline silicon or from thin layers of amorphous (non-crystalline) silicon. The thin-film modules use considerably less semiconductor material but have lower efficiencies for converting sunlight to direct-current electricity. Cells and modules made from other thin-film PV materials such as copper-indium-diselenide and cadmium telluride are under active development and are beginning to enter the market. Most experts consider thin-film technology to be the future of the PV industry because of the reduced material requirements, the reduced energy required to manufacture thin film devices, and the ability to manufacture thin films on a mass-production basis.

Amorphous silicon modules experience a conversion efficiency loss of about 10 percent when initially exposed to sunlight, but then stabilize at the reduced figure. The mechanism for this reduction is being actively investigated, but is still not well understood. Individual modules made with other PV materials do not exhibit such loss of conversion efficiency, but combinations of modules in arrays do exhibit systematic reductions in power output over their lifetimes. Estimated at about 1 percent per year on average, based on data to date, these reductions are most likely associated with deteriorating electrical connections and non-module electrical components.

Complete PV systems require other components in addition to the PV modules. This “balance-of-system” generally includes a support structure for the modules to orient them properly to the sun, an inverter to convert direct-current electricity to alternating-current electricity, a storage system for electrical energy (usually batteries), an electronic charge regulator to prevent damage to the batteries by protecting against overcharging by the PV array and excessive discharge by the electrical load, and related wiring and safety features (Figure 5).

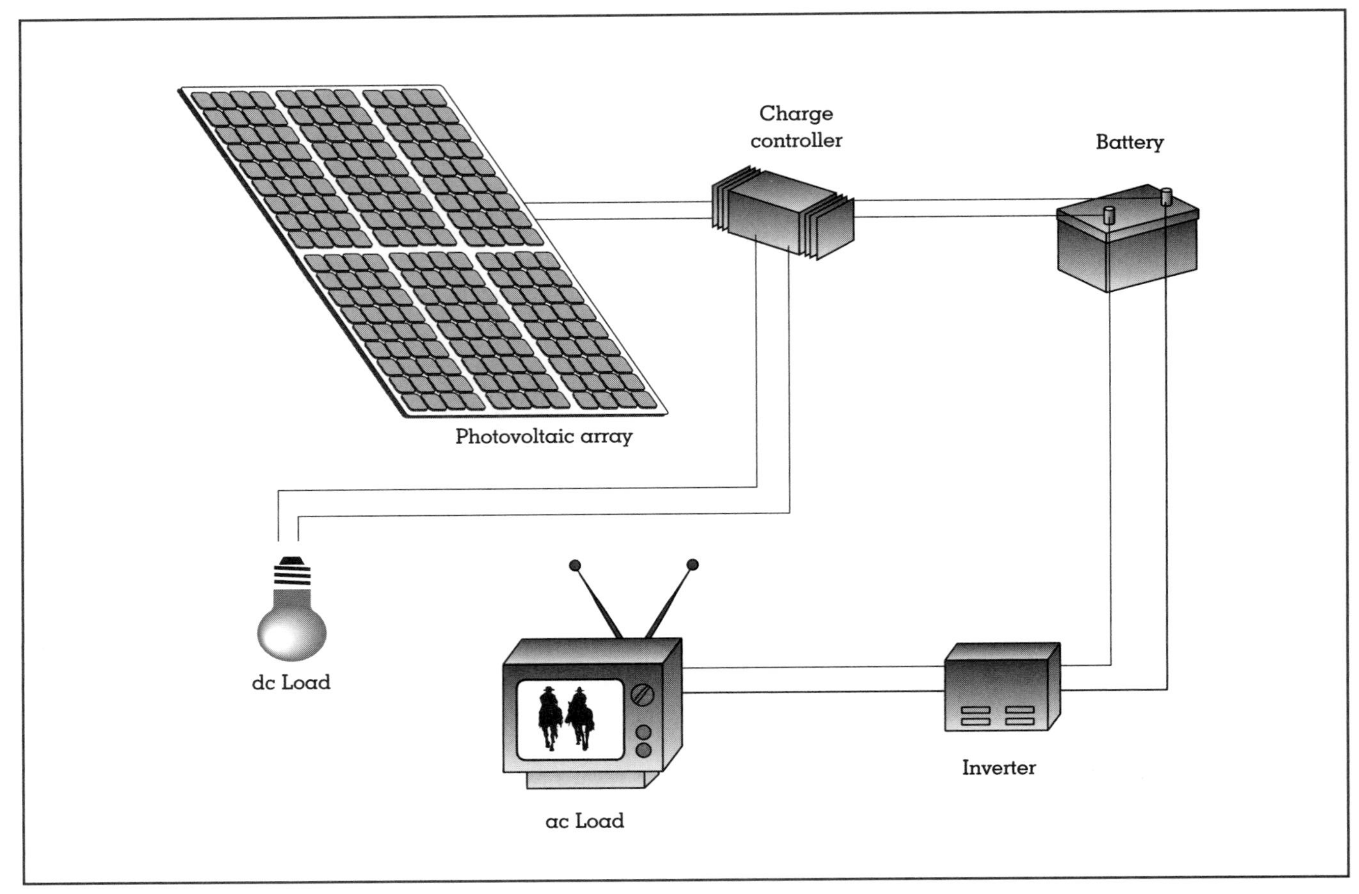

Figure 5.
Components of a typical off-grid PV system.

The output of a solar cell can be increased significantly by concentrating the incident sunlight onto a small area. Commercially available concentrator systems use low-cost lenses, mirrors or troughs in conjunction with tracking systems to focus sunlight on a small, high efficiency but expensive solar cell. One-axis trackers follow the sun during the day; two-axis trackers track daily and yearly variations in the sun's position. The higher cost of the cell and the associated tracking equipment, and the inability to use diffuse radiation, are offset by the higher conversion efficiencies achieved.

The cost of PV electricity is largely determined by four factors: cost of the PV modules, the efficiency of converting sunlight to electricity, the balance-of-system costs, and the lifetime of the systems. There are many combinations of these factors that provide about the same cost of electricity. Modules are now available that are designed to last at least thirty years.

Another scheme that has been proposed for use of photovoltaics is the solar satellite power system. Such a system would place large individual PV arrays (up to ten megawatts peak power) in synchronous orbit around the earth, and beam the collected and converted solar energy through the atmosphere in the form of microwaves to large receiving antennas on earth. The microwave generators would be powered by the PV electricity. The microwave energy collected at the surface would then be converted back to electricity for broad distribution. While of some interest, there are major issues associated with cost, environmental impact, and energy security.

PHOTOVOLTAIC APPLICATIONS

By its very nature PV can be used wherever electricity is needed and sunlight is available. For many years PV systems have been the preferred power sources for

Researchers at Sandia National Labs use a trough of parabolic mirrors to focus sunlight on a long glass tube in order to detoxify water. (U.S. Department of Energy)

buildings or other facilities in remote areas not served by the utility grid. As the cost of PV systems continues to decline, there is growing consensus that distributed PV systems on buildings that are grid-connected may be the first application to reach widespread commercialization. Typical off-grid applications for PV today include communications (microwave repeaters, emergency call boxes), cathodic protection (pipelines, bridges), lighting (billboards, highway signs, security lighting, parking lots), monitoring equipment (meteorological information, water quality, pipeline systems), warning signals (navigational beacons, railroad signs, power plant stacks), and remote loads (village power, parks and campgrounds, water pumping and irrigation, vacation cabins). On-grid applications include support for utility transmission and distribution (T&D) systems, where the benefits derive from avoiding or deferring T&D system investments, and from improving the quality of electrical service.

More than 200,000 homes in the U.S. currently use some type of PV technology, and more than 10,000 U.S. homes are powered entirely by PV. A large number of homes in Japan, where consumer electricity rates are three times higher than in the U.S., have roof-integrated PV systems. Half a million PV systems have been installed in developing countries by the World Bank. More than 10,000 have been installed in Sri Lanka, 60,000 in Indonesia, 150,000 in Kenya, 85,000 in Zimbabwe, 40,000 in Mexico, and 100,000 in China. Global use of PV is expected to grow rapidly in the coming decades.

OTHER DIRECT APPLICATIONS OF SOLAR ENERGY

In addition to the applications already discussed, solar energy can also be used for solar cooking (use of a reflecting surface to focus sunlight on pots used for food preparation), distillation (use of solar heat to evaporate and desalinate seawater), purification (use of concentrated ultraviolet radiation to kill organisms in contaminated water), and agriculture (use of solar heat to dry crops and grains, space and water heating, heating of commercial greenhouses, and use of PV electricity to power water pumping and other remote facilities). Many other applications will also become feasible as PV costs continue to decline.

Production of PV modules is still relatively small, but has been growing at a steady and significant rate. Global production in 1998 reached 150 megawatts peak. As costs come down and new manufacturing facilities are placed in operation, this number will grow rapidly.

SOLAR ENERGY'S POTENTIAL AND PROBLEMS

The amount of solar energy available on earth is many times greater than all the energy collectively used by people around the world. When one takes into account all the various forms in which solar energy is directly or indirectly available, along with energy available from geothermal and nuclear sources, it becomes clear that there is no global shortage of energy. The only shortage is that of low-cost energy. In principle there is no reason that solar energy in its various forms could not supply all the world's energy. To give but one example, all the electricity used by the United States (a little more than 3 trillion kWh) could be generated by commercially available PV modules covering half of a 100 mile by 100 mile area in the Nevada desert.

Greater use of solar and other forms of renewable energy is increasingly seen as the long-term response to growing demand for electricity and transportation fuels in developed and developing countries. For example, electricity derived from renewable resources can be used to electrolyze water and create hydrogen, which can then be used in fuel cells to provide electricity or power electric vehicles. As a result of the issues raised by the oil embargo of 1973, and increasing awareness of the local and global environmental damage associated with use of fossil fuels, many people now believe that the world must undergo a transition to a clean and sustainable energy system during the twenty-first century. There is great hope that renewable energy will be the basis for that new, sustainable system.

This is not to suggest that widespread use of solar energy is not facing major barriers. For many applications, today's high cost of solar energy systems is an important limitation. Nevertheless, significant progress has been and will continue to be made in improving system performance and reducing associated energy costs. A particular challenge to solar energy research and development efforts is that incident solar energy is not highly concentrated and is intermittent. The former necessitates use of concentrator systems, or large collection areas that can have impacts on ecosystems due to land use and disturbances during the construction stage of large-scale power plants. Centralized, multi-megawatt power facilities can also result in significant visual impacts.

The fact that solar energy is an intermittent energy resource means that energy storage systems (e.g., batteries, ultracapacitors, flywheels, and even hydrogen) will be required if solar energy is to be utilized widely. In addition, a variety of toxic chemicals are used in the manufacture of PV cells; however, studies of the risks associated with their manufacture and disposal indicate little threat to surroundings and the environment.

CONCLUSIONS

Solar energy offers a clean, sustainable alternative to continued use of fossil fuels. In its various forms it is already providing useful amounts of energy on a global basis, and will provide steadily increasing amounts in the twenty-first century, especially as developing countries require more energy to improve their economies.

In February 1979 the Carter Administration released a 30-federal agency study entitled *Domestic Policy Review of Solar Energy*. This study found that, with increasing oil prices and "comprehensive and aggressive initiatives at the Federal, State and local levels," renewable energy sources "could provide about 20 percent of the nation's energy by the year 2000." This did not occur, as oil prices actually dropped in the 1980s, Federal funding for renewables was reduced, and, for a period of time, energy issues largely disappeared from public view. Today, renewable energy sources supply about 12 percent of U.S. electricity and 8 percent of total U.S. energy consumption.

Continuing research and development efforts by the public and private sectors have significantly reduced the costs of solar energy, and further significant reductions are expected. Together with growing public concern about global climate change, it is anticipated that the 21st century will see widespread deployment of solar energy and other renewable energy systems, in both developing and developed countries.

Allan R. Hoffman

See also: Solar Energy, Historical Evolution of the Use of.

BIBLIOGRAPHY

Brinkworth, B. J. (1972). *Solar Energy for Man.* New York: John Wiley & Sons.

Center for Renewable Energy and Sustainable Technology (CREST). <http://www.solstice.crest.org/>

Daniels, F. (1972). *Direct Use of the Sun's Energy.* New Haven, CT: Yale University Press.

Energy Efficiency and Renewable Energy Network (EREN) home page U.S. Department of energy, Energy Efficiency and Renewable Energy Network. August 9, 2000 <http://www.eren.doe.gov/>.

Flavin, C., and Lenssen, N. (1994). *Power Surge: Guide to the Coming Energy Revolution.* New York: W.W. Norton & Company.

International Energy Agency. (1997). *Energy Technologies for the 21st Century.* Paris: OECD Publications.

International Energy Agency. (1998). *Benign Energy?: The Environmental Implications of Renewables.* Paris: OECD Publications.

International Energy Agency. (1999). *The Evolving Renewable Energy Market.* Paris: OECD Publications.

International Solar Energy Society (ISES) web site International Solar Energy Society. August 9, 2000 <http://www.ises.org/>.

Johansson, T. B.; Kelly, H.; Reddy, A. K. N.; and Williams, R. H., eds. (1993). *Renewable Energy: Sources for Fuels and Electricity.* Washington, DC: Island Press.

Marion, W., and Wilcox, S. (1994). *Solar Radiation Data Manual for Flat-Plate and Concentrating Collectors.* Golden, CO: National Renewable Energy Laboratory.

National Renewable energy Laboratory (NREL) Home Page. National Renewable Energy Laboratory. August 9, 2000 <http://www.nrel.gov/>.

Passive Solar Industries Council. <http://www.psic.org/>

Pegg, D. *CADDET Renewable Energy—International demonstration projects* (July 24, 2000). International Energy Agency, Center for the Analysis and Dissemination of Demonstrated Energy Technologies: Renewable. August 9, 2000 <http://www.caddet-re.org/>.

President's Committee of Advisors on Science and Technology, Panel on Energy Research and Development. (1997). *Federal Energy Research and Development for the Challenges of the Twenty-First Century.* Washington, DC: U.S. Government Printing Office.

U.S. Congress, Office of Technology Assessment. (1995). *Renewing Our Energy Future.* Washington, DC: U.S. Government Printing Office.

U.S. Department of Energy, Energy Information Administration. (1999). *Renewable Energy Annual 1998.* Washington, DC: U.S. Government Printing Office.

U.S. Solar Energy Industries Association. <http://www.seia.org/>.

Welcome to the American Solar Energy Society American solar Energy Society. August 9, 2000 <http://www.ases.org/>.

SOLAR ENERGY, HISTORICAL EVOLUTION OF THE USE OF

Every ninety-five minutes an object the size of a bus circles Earth. It represents an unparalleled achievement in space observation technology and is called the Hubble Space Telescope (HST). Since its deployment in 1990, the HST has brought images to the astronomy community that traditionally had been reserved for expensive long-range space missions by unmanned craft. One of the essential technological developments that make the HST such an attractive investment is its use of one "natural resource" found in space, the sun. The HST utilizes photovoltaic technology to capture the sun's rays and convert them directly into electricity. This procedure is commonly referred to active solar generation.

Active solar generation is nothing new to the space industry. Photovoltaics have been used to power satellites since the late 1950s and are an essential element to the new international space station. Closer to home, active solar technologies are used on houses and buildings to generate heat, hot water, and electricity. Spacecraft use of the sun is indeed a technological marvel. However, strategic use of the sun for heating and cooling is not a new discovery. In fact, 2,400 years ago the Greek philosopher Socrates observed:

> Now in houses with a south aspect, the sun's rays penetrate into the porticos in winter, but in the sum-

The operation of a solar-powered printing press. Experiment conducting in the Garden of Tuileries, Paris, on Aug 6, 1882, for the festival of "L'Union Francaises of Jeuenesse." (Corbis Corporation)

> mer, the path of the sun is right over our heads and above the roof, so that there is shade. If then this is the best arrangement, we should build the south side loftier to get the winter sun and the north side lower to keep out the winter winds. (Xenophon, 1994, chap. 2, sec. 8)

Socrates has described the strategic placement of a home to best utilize the rotation of Earth, to receive maximum sunlight during the winter and minimal sunlight in summer. The strategic utilization of the seasonal course of the sun in home design is called passive solar architecture. Passive solar use is best described as the greenhouse effect. This simple phenomenon is illustrated by the experience of returning to your car on a sunny, cool day and finding it heated. Hubble and Socrates represent opposite ends of the solar historical spectrum. However, solar technologies have historically been met with mixed results. While the technology has consistently proved itself to be useful and environmentally friendly, it has often been pushed into the shadows by cheaper, less environmentally friendly energy technologies.

Historically the most common problems associated with solar technologies are efficiency and economics compared to competing energy technologies. During the Industrial Revolution coal was relatively cheap and, for the most part, readily available in most of the world. These facts, which still hold true today, make coal a powerful competitor against alternative power technologies. However, solar technologies have been able to survive in niche markets long enough to increase efficiency and lower costs. Although sunlight is free and also readily available, solar technologies have experienced little success in large-scale applications. Solar technologies also have been able to take advantage of advances in silicon technology. An initial niche market for solar technology was the colonies of Europe.

EARLY APPLICATIONS: SOLAR THERMAL

One of the earliest forays in the use of solar technologies for power generation was in 1861. French engineer Auguste Mouchout utilized solar thermal technology by focusing the sun's rays to create steam, which in turn was used to power engines. Mouchout's design incorporated a water-filled cauldron surrounded by a polished metal dish that focused the sun's rays on the cauldron, creating steam. Mouchout was granted the first solar patent for an elementary one-half horsepower steam engine powered by a solar reflective dish. Several of Mouchout's solar motors were deployed in Algeria, then a French protectorate, where coal was expensive due to transportation costs. One-half horsepower may not seem like much, but it was a considerable advance in alternative power generating. At the time, the major source of energy was man and draft animals, the steam engine was in its infancy, and water power was not very portable. Mouchout was able to expand his design to a larger scale, but the project was abandoned due to new innovations in coal transportation that made coal a more efficient investment.

Solar thermal technology survived Mouchout and was further developed by Swiss-born engineer John Ericsson. In 1870 Ericsson created a variation of Mouchout's design that used a trough instead of a dish to reflect the Sun's rays. Ericsson's basic design would later become the standard for modern-day parabolic troughs. He allegedly refined his design to the point where it would be able to compete actively with conventional generation techniques of the time but he died without leaving any plans or blueprints.

Englishman William Adams designed another interesting variation of Mouchout's work in 1878. Adams's design used mirrors that could be moved to track sunlight and focus the rays on a stationary cauldron. In theory, a larger number of mirrors would generate increased heat levels and greater steam output for larger engines. Adams was able to power a two-and-a-half-horsepower engine during daylight hours and viewed his invention as an alternative fuel source in tropical countries. Although Adams was more interested in proving that it could be done than in developing the technology, his basic design would become the prototype for contemporary large-scale centralized solar plants called "power towers" that are currently in operation in the United States and other countries.

CONTEMPORARY USES: SOLAR THERMAL

Solar thermal technology varied little in design from 1870 to 1970. Advances were made in the materials, such as lower-boiling-point fluids and higher-efficiency motors, but the basic design of using mirrors to focus sunlight remained the same. The aforementioned introductory ventures in the use of solar technology provided the intellectual foundation for growth and extensive technological development of solar use in the twentieth century. These pioneers helped to establish solar as a viable alternative to conventional power generation. However, many of the same barriers resurfaced, preventing widespread implementation. Many of the previously mentioned designs survived by gaining support within environmental communities as environmentally friendly alternatives to fossil fuels, as well as a safer alternative to nuclear power. In addition, solar technologies were able to establish themselves within new niche markets and to gain further interest during the oil crisis of the 1970s.

As previously mentioned, both Ericsson's and Adams's original designs were upgraded and implemented as modern solar power generating facilities. Contemporary versions of the solar trough and power tower were constructed during the 1980s in the Mojave Desert of California by LUZ International Ltd. Currently the corporate successor of LUZ, which went bankrupt in 1991, partly due to dependence on a federal subsidy (tax credits) not renewed in 1991, has nine solar thermal plants in California and has generated more energy from solar thermal than any other company in the world.

EARLY APPLICATIONS: PHOTOVOLTAICS

Use of solar panels or photovoltaics (PVs) is another popular way to generate solar electricity. The space program is perhaps the most recognized user of PVs and is responsible for most of the advancements in PVs. Many people are familiar with PVs through small applications such as calculators and perhaps solar water heaters, but early forays in PV experimentation were little more than noted side observations in non-PV experiments.

In 1839 Alexandre-Edmund Becquerel, a French experimental physicist, did the earliest recorded experiments with the photovoltaic effect. Becquerel discovered the photovoltaic effect while experimenting with an electrolytic cell made up of two metal

Buckminster Fuller stands in front of a solar-powered geodesic dome house. (AP/Wide World Photos)

electrodes placed in an electricity-conducting solution. When exposed to sunlight, the generation increased. Between 1873 and 1883 several scientists observed the photoconductivity of selenium. American inventor Charles Fritts described the first experimental selenium solar cells in 1883. Notable scientists such Albert Einstein and Wilhelm Hallwachs conducted many experiments with various elements to explore it further. In 1921 Einstein won the Nobel Prize in Physics for his theories explaining the photoelectric effect.

Many elements were found to experience the photoelectric effect. Germanium, copper, selenium, and cuprous oxide comprised many of the early experimental cells. In 1953 Bell Laboratories scientists Calvin Fuller and Gerald Pearson were conducting

Parabolic solar energy concentrating dishes sit in a row at the power station at White Cliffs, Australia. The plant, operated by Solar Research Corp. of Melbourne, is Australia's first experimental solar station. (Corbis Corporation)

experiments that brought the silicon transistor from theory to working device. When they exposed their transistors to lamplight, they produced a significant amount of electricity. Their colleague Darryl Chaplin was looking for better ways to provide power for electrical equipment in remote locations. After a year of refinements Bell Laboratories presented the first working silicon solar cell, which would later become the core of contemporary PV applications.

CONTEMPORARY USE: PHOTOVOLTAICS

The utility industry has experimented with large-scale central station PV plants but concluded that in most cases they are not cost-effective due to their perceived low efficiency and high capital costs. However, PV has proven useful in niche markets within the utility system as remote power stations, peak generation applications, and in a distributed generation scheme. Distributed generation is the inverse of central generation. Instead of constructing large generating facilities energy has to be sent to consumers, energy is generated where it is used. Highway signs are a good example of distributed generation. They use solar panels and batteries on the signs to collect and store the power used to illuminate the signs at night. Because very energy-efficient LED lighting replaced incandescent bulbs, these signs could be powered much more economically by PVs than by traditional gasoline-powered generators. Perhaps the most prolific use of PVs is in the space program. The sun, the only natural energy source available in space, is the logical choice for powering space vehicles. Photovolttaics are safer and longer-lasting energy sources than nuclear power is.

An instance where large-scale grid-connected PV generation has occurred is the Sacramento Municipal Utility District (SMUD). SMUD built several dozen "solar buildings" and placed PV systems directly on their clients' homes. SMUD purchased one megawatt of distributed PV systems each year until the year 2000.

SMUD's approach introduced another distributed application concept for PV, called building integrated photovoltaics or BIPV. Currently Japan has one of the largest BIPV programs, started in 1996. The simplest application of BIPV is the installation of solar panels directly on a building. While this method is perhaps the easiest, it does not reflect the best use of PV technology. The best potential for the growth of PV is for PV to double as building materials. Examples include exterior insulation cover, roofing shingles, windows, and skylights that incorporate solar cells. Incorporating PV into construction materials is very complex and requires joint cooperation among traditionally separate entities of the building sector. Although no actual products exist, many of the major manufacturers began experimenting with prototypes in the late 1990s and hope to introduce cost-effective products on the market by 2005. Additional uses for PV include desalinization of seawater for fresh water, solar water heaters, and solar ovens.

IMPLEMENTATION AND BARRIERS

Solar energy is used worldwide in many applications, from niche markets in developed countries to primary village power in rural and developing communities. Its attractiveness can be attributed to several factors:

- Solar power works well in small facilities, and it may be comparatively cost-effective when all factors (transmission, fuel transportation costs, service life, etc.) are taken into consideration. In many instances solar energy is the *only* viable energy option in the region.
- Solar distributed energy systems provide security and flexibility during peak use and outages associated with environmental disasters. In addition, programs that allow consumers to sell surplus energy they generate back to the utility, called "net metering," will make residential solar technologies more attractive.
- Solar power is environmentally friendly and will help mitigate global environmental concerns associated with climate change.

These claims have been the primary call to arms for many solar advocates and industries since the 1970s, but solar technologies have not enjoyed widespread use except within niche markets, especially among developed countries. According to a U.S. Department of Energy report by the Energy Information Administration, solar technologies have only experienced modest sector growth domestically because domestic solar thermal technologies face stiff competition from fossil-fueled energy generation. PV growth faces similar challenges, since deregulation is likely to lower the cost of electric power from natural gas and coal.

Because of the wide range of ideological leanings of various administrations, direct subsidies and federally funded research and development have fluctuated wildly since the late 1970s. In 1980 the United States spent $260 million on solar research and development, but then the amount fell to $38 million by 1990. Between 1990 and 1994 support doubled, reaching $78 million.

Because of the problems associated with nuclear power and burning fossil fuels, solar technologies may again be viewed as a viable alternative. However, solar technology faces several challenges

First. the energy expended in the production of PV panels is considerable. Second, solar cells become less efficient over time (degradation is slow, but eventually they lose most of their conductivity). Third, contemporary solar cells use mercury in their construction and present toxic waste disposal problems. Finally, solar technologies are part-time power sources. Solar thermal technology requires full, direct sunlight, and PVs do not work at night (although batteries and supplementary generating applications such as wind power can be combined to provide full-time generation).

The need for long-term clean power will be a controversial but necessary issue in the twenty-first century. The debate will focus on the deregulation of the power industry and the needs of consumers, coupled with what is best for the environment. For certain applications, solar technology has proven to be efficient and reliable. The question remains whether solar will be wholly embraced as energy options for a sustainable future in a climate of cheap fossil fuels and less than enthusiastic public support.

J. Bernard Moore

See also: Batteries; Becquerel, Alexandre-Edmund; Einstein, Albert; Electric Power, Generation of; Fuel Cells; Heat and Heating; Heat Transfer; Power; Renewable Energy; Spacecraft Energy Systems; Solar Energy; Thermal Energy; Water Heating.

BIBLIOGRAPHY

Berger, J. J. (1997). *Charging Ahead: The Business of Renewable Energy and What It Means for America.* New York: Henry Holt.

Flavin, C., and Lenssen, N. (1994). *Power Surge: Guide to the Coming Energy Revolution.* New York: W. W. Norton.

Fitzgerald, M. (July 14, 2000). *The History of PV.* Science Communications, Inc. August 9, 2000 <http://www.pvpower.com/pvhistory.html>

Hoffman, J. S.; Well, J. B.; and Guiney, W. T. (1998). "Transforming the Market for Solar Water Heaters: A New Model to Build a Permanent Sales Force." *Renewable Energy Policy Project* No. 4. <http://www.repp.org>.

Maycock, P. (1999). "1998 World PV Cell Shipments Up 20.5 Percent to 151.7 MW." *PV News* 18:2.

National Renewable Energy Laboratory. (1996). *Photovoltaics: Advancing Toward the Millenium.* DOE/GO-100095-241, DE96000486.

Perlin, J. (1999). "Saved by the Space Race." *Solar Today* (July-August).

Smith, C. (1995). "Revisiting Solar Power's Past." *Tech-nology Review* 98(5):38–47.

U.S. Department of Energy, Energy Information Administration. (1996). *Annual Energy Outlook 1997 with Projections to 2015.* Washington, DC: U.S. Government Printing Office.

Wan, Yih-huei. (1996). *Net Metering Programs.* NREL/SP-460-21651. Golden, CO: National Renewable Energy Laboratory.

Xenophon. (1994). *Memorabilia,* tr. A. L. Bonnette. Ithaca, NY: Cornell University Press.

SPACECRAFT ENERGY SYSTEMS

Energy systems in space technology are devices that convert one kind of energy into another to ensure the functioning of automated and piloted satellites, interplanetary probes, and other kinds of spacecraft. Multiple functions of any spacecraft require two distinctly different energy sources: propulsion for launch and maneuvers, and electricity supply to power the onboard equipment.

PROPULSION SYSTEMS

Propulsion generates kinetic energy to facilitate motion of spacecraft. Although propulsion is provided by various types of devices, the most common propulsion system of modern space technology is a rocket engine, a device that propels a rocket by a force of reaction. In its simplest form, the rocket engine could be described as a container (combustion chamber) with an opening at one end (exhaust). Burning of combustible chemicals inside the container generates a large amount of a compressed gas—a propulsive mass—that is ejected through the exhaust. According to the laws of mechanics, the force of pressure that ejects the gas creates the reaction force, called thrust, that pushes the container in the direction opposite to the exhaust. The International System of Measurements designates thrust in a unit of force called a newton (N= 0.098 kg[f] = 0.216 lb[f]), while more often it is measured in tons, kilograms, or pounds. The thrust is greater if a bell-shaped nozzle is installed at the engine's exhaust. The nozzle's specially designed profile (the so-called Laval nozzle) creates a strong pressure differential at the exhaust. This differential facilitates expansion of the ejected gas with a substantially higher ejection velocity, thus releasing greater kinetic energy.

The energy efficiency of a rocket engine is determined by specific impulse, commonly measured in seconds (although scientifically correct designations are N s/kg, or m/s). Specific impulse is the time in which a rocket engine consumes the amount of propellant equal to the engine's thrust. For example, compare two engines with an equal thrust of 100 kg each. If the first engine consumed 100 kg of propellant in 20 seconds, and the second one consumed the same amount in 30 seconds, the second engine generated greater kinetic energy per kilogram of propellant—it worked longer. Thus the higher specific impulse indicates the more efficient engine.

The most common chemical rocket engines use gaseous propulsive mass produced through the burning of special chemicals called propellants. Thermal energy released in this process accelerates and ejects the propulsive mass, converting itself into kinetic energy of motion. Solid and liquid propellant motors are the two major types of chemical rocket engines. The solid rocket motor (SRM) burns propellant that consists of fuel and oxidizer mixed together in a solid-state compound. Solid propellant is stored inside a combustion chamber of the SRM, making the engine relatively simple and reliable due to a lack of moving parts (see Figure 1). However, because of lower specific impulse and difficulty controlling the final speed, since the SRM usually cannot be shut down and restarted until all propellant is consumed,

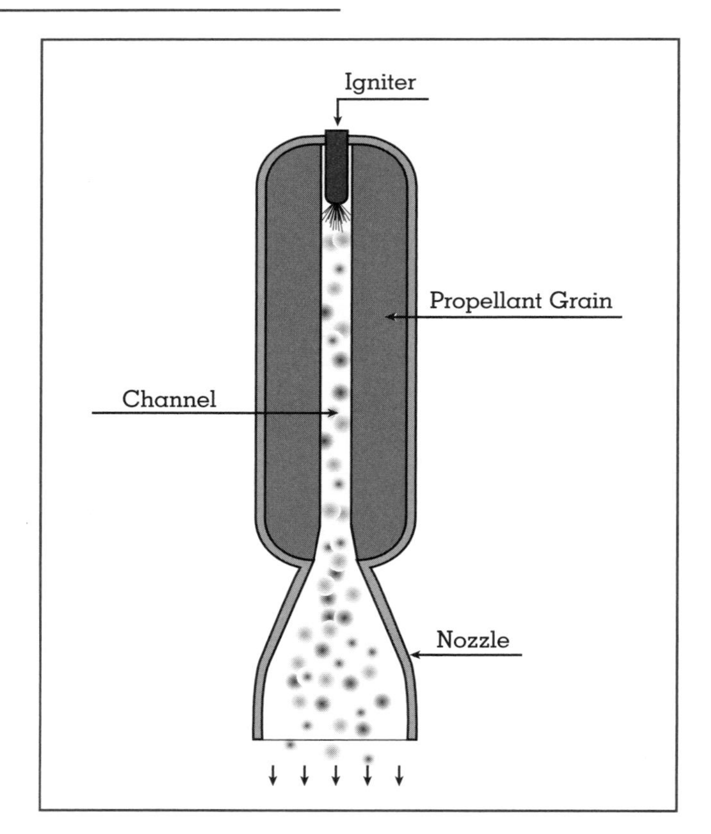

Figure 1.
Solid rocket motor.

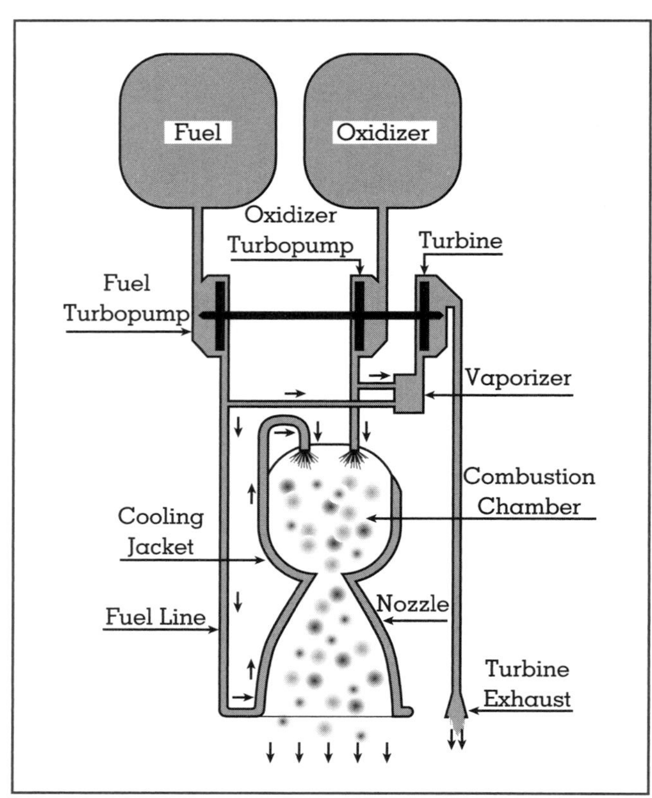

Figure 2.
Liquid propellant rocket engine.

the solid rocket is primarily for short duration but powerful impulse.

A liquid propellant rocket engine (LPRE) generates propulsive mass through burning of two liquid chemicals—fuel and oxidizer—which are pumped into a combustion chamber from separate tanks. LRPE has a very complex design: It contains numerous components with a large number of moving parts. The combustion chamber and nozzle have intricate structures that allow them to withstand high internal pressure while being lightweight. One of the propellant components is driven between the thin double walls of the chamber and nozzle to cool them and prevent hot exhaust from burning through (see Figure 2). LPRE complexity is compensated by advantages of better energy efficiency (higher specific impulse); precision control over burn time, thrust and final speed; and the possibility to restart the engine.

There are two major types of liquid propellants: a storable propellant (that does not boil under a normal temperature) and a cryogenic propellant (a liquefied gas under very low temperature). Storable propellant engines are simpler and more reliable. They are useful for spacecraft that require multiple propulsion impulses and a long stay in space. Although more complex and demanding, cryogenic engines are regarded as best for launch vehicles due to their superior energy efficiency.

In each space mission, the maximum amount of energy is used for spacecraft launching. This is because the spacecraft had to be accelerated to a velocity of 28,500 km/h (17,700 miles/h) in fewer than 10 minutes, while overcoming forces of gravity and atmospheric drag. Due to the enormous energy requirements for this task, propellant accounts for up to 92 percent of the total mass of modern launch vehicles, leaving no more than 4 percent for the useful payload (the rest is the "dry mass" of a rocket). That dictates the importance of energy efficiency—better engines allow for the launching of heavier payloads with less propellant. Although the overall efficiency of a launch vehicle depends on many fac-

Designation & Country	*Number of Stages & Propellant*	*Launch Mass (LM)*	*Payload Mass in low orbit (PM)*	*PM/LM Ratio*	*Comments*
Delta II 7925 USA	I - liquid cryogenic + 9 solid boosters II - liquid storable	230 tons (506,000 lb)	5,089 kg (11,220 lb)	2.21%	One of the newer models in a family of SLVs converted from a ballistic missile. Operational since 1990.
Soyuz-U Russia	I - 4 liquid cryogenic boosters II - liquid cryogenic III - liquid cryogenic	310 tons (682,000 lb)	7,000 kg (15,400 lb)	2.25%	One of the models in a family of SLVs converted from a ballistic missile. Operational since 1973.
Atlas II AS USA	I - liquid cryogenic + 4 solid boosters II - liquid cryogenic	234 tons (516,000 lb)	8,640 kg (19,000 lb)	3.69%	One of the newer models in a family of SLVs converted from a ballistic missile. Operational since 1993.
CZ-2E China	I - 4 liquid storable boosters II - liquid storable III - liquid storable	464 tons (1,021,000 lb)	8,800 kg (19,360 lb)	1.90%	One of the newer models in a family of SLVs converted from a ballistic missile. Operational since 1990.
Arian-44L Arianspace European Consortium	I - liquid storable + 4 liquid storable boosters II - liquid storable III - liquid cryogenic	470 tons (1,034,000 lb)	9,600 kg (21,120 lb)	2.04%	One of the newer models in a family of the purpose-built SLVs. Operational since 1989.
Zenit-2 Russia Ukraine	I - liquid cryogenic II - liquid cryogenic	459 tons (1,010,000 lb)	13,740 kg (30,300 lb)	2.99%	Purpose-built SLV. A modified version is intended for Sea Launch international project. Operational since 1985.
Proton-K Russia	I - liquid storable II - liquid storable III - liquid storable	689 tons (1,516,000 lb)	20,000 kg (44,000 lb)	2.90%	One of the models in a family of SLVs converted from a ballistic missile. Operational since 1971.
Titan IV USA	I - liquid storable + 2 solid boosters II - liquid storable III - liquid cryogenic	860 tons (1,900,000 lb)	21,640 kg (47,600 lb)	2.51%	One of the newer models in a family of SLVs converted from a ballistic missile. Operational since 1989.
Space Shuttle USA	I - 2 solid boosters II - liquid cryogenic	2,040 tons (4,500,000 lb)	29,500 kg+ orbiter 94,000 kg (65,000 + 207,000 lb)	1.45%	The world's only operational reusable launch vehicle (RLV). Purpose-built. Operational since 1981.
Venture Star (project) USA	Single Stage-to-Orbit liquid cryogenic	993.6 tons (2,186,000 lb)	20,500 kg + RLV 117,000 kg (45,100 + 257,000 lb)	2.06%	Future US reusable launch vehicle (RLV). Under development. Expected to start operations in 2005.

Table 1.
Space Launch Vehicles (SLVs)

tors, it could be roughly estimated by a simple percentage ratio between the rocket's maximum payload mass (PM) and of its total mass at launch (LM). The greater percentage would indicate a more efficient launch vehicle (see Table 1).

Energy efficiency is one of the major factors that determines the launch vehicle layout and choice of engines. Currently, about 98 percent of all spacecraft worldwide are launched by liquid-propellant rockets. While the purpose-built launchers mostly use cryogenic engines to ensure the best performance, many current space rockets utilize storable propellants. This is because the majority were developed from military ballistic missiles. Although not ideal from the energy efficiency standpoint, such conversion is less expensive than developing a brand-new launch vehicle. Solid-propellant engines are primarily used as auxiliary boosters for some liquid propellant rockets. Only a few purely solid-propellant space launch vehicles exist today in the world because they are capable of placing fewer than 1,000 kg (2,200 lb) into low orbit, while liquid-propellant rockets can launch much more—up to 29,500 kg (64,900 lb).

Some unique launchers were even more powerful. Such was *Saturn V*—a gigantic rocket capable of launching a payload of 140,000 kg (308,000 lb) that allowed American astronauts to land on the moon. Russian *N-1* and *Energia* rockets were capable of launching 95,000 kg (209,000 lb) and 105,000 kg (231,000 lb), respectively. All these launch vehicles are not in use anymore due to their enormous cost. These giants were created for specific national prestige goals, such as manned lunar expedition, and for launching of the expected massive civilian and military payloads. The shifts in political and economic priorities in the space programs of the United States and Russia have left these rockets without any feasible payload assignments. At the same time, the cost of the launch vehicles' preservation along with their production and maintenance facilities for possible future use was proven to be prohibitive and unjustifiable economically.

Space launchers usually consist of several stages, which are the individual rockets stacked together. This is another method to increase efficiency—the stages are ejected as soon as their propellant is spent, and a rocket does not have to carry empty tanks. The number of stages also depends on engines' effectiveness. Liquid-propellant rockets usually consist of two to three stages, while solid ones have to have three to five stages.

Rocket engines are also used for maneuvers in space. Some operations, such as a onetime transfer of a satellite from lower to higher orbit, could be performed by a solid-propellant engine. Yet many complex maneuvers, such as rendezvous and docking with another spacecraft, require multiple engine firings and variable power impulses. Hence modern spacecraft are equipped with an assortment of attitude control engines that usually use liquid storable propellant.

A good illustration of how all these engines work together in a single system is the American Space Shuttle. Two very large solid rocket boosters (SRBs) represent the first stage of the launch vehicle. The SRBs are the most powerful rocket engines in existence: Each produces more than 1,500 tons (15.3 MN, or 3,310,000 lb) of thrust for 122 seconds. The second stage consists of a large external tank with cryogenic propellant (liquid hydrogen and liquid oxygen) and three Space Shuttle main engines (SSMEs) mounted on the orbiter. Each SSME produces more than 170 tons (1.73 MN, or 375,000 lb) of thrust at sea level. At launch, SRBs and SSMEs ignite just a few seconds apart and are used to break through the most dense layers of atmosphere. After a separation of the SRBs at the altitude of about 40 km (25 mi), the Space Shuttle continues its acceleration with the SSMEs only. Just before the Space Shuttle reaches orbital velocity (520 seconds after launch), the SSMEs are shut down and the external tank is ejected. Now two smaller engines of the orbital maneuvering system (OMS) aboard the orbiter provide the final acceleration. The OMS engines are also used for orbital maneuvers and for deceleration before the spacecraft reentry. Unlike the SSMEs, they utilize storable liquid propellant, and each produces thrust of 3 tons (30.6 kN, or 6,700 lb). Very fine Space Shuttle orbital maneuvering is performed by the reaction control system (RCS), which consists of 44 small liquid-propellant engines assembled in three arrays.

The immense advantages to humanity of modern space technology come at a hefty price. High risks and environmental hazards of rocketry, uniqueness of spacecraft equipment, and the necessity for specialized launching facilities make space technology very expensive. Today the average cost to launch just 1 kg (2.2 lb) of a payload in orbit could be as high as $10,000. Because of that, only six countries—the United States, Russia, China, Japan, India, and Israel—and the France-dominated European Space Agency (ESA) maintain independent space launching capabilities.

One of the seemingly obvious ways to cut down the cost is a reusable launch vehicle, such as the Space Shuttle. NASA hoped that the enormous cost of the Space Shuttle development would have been compensated by the expected tenfold reduction in the price of putting a payload in orbit. Unfortunately, that goal was not achieved. Despite the Shuttle's unique capabilities, the cost of its payload delivery

The Space Shuttle *Discovery*'s robotic arm deploys the SPARTAN-201 satellite. (Corbis-Bettmann)

appeared to be at least two times higher than that of expendable rockets. The Shuttle's maintenance after each mission turned out to be much more expensive than had been expected. As a reusable spacecraft, the orbiter has to spend energy carrying into space additional massive components that are not essential for payload deployment, such as the main engines, wings, and heat protection. Thus the Shuttle's efficiency factor is two times lower than of the best expendable launchers. The very large size of the orbiter and the necessity to accommodate a crew even for routine satellite launches substantially increase the overall cost of Space Shuttle operations. Projected high expenses recently forced Russia and

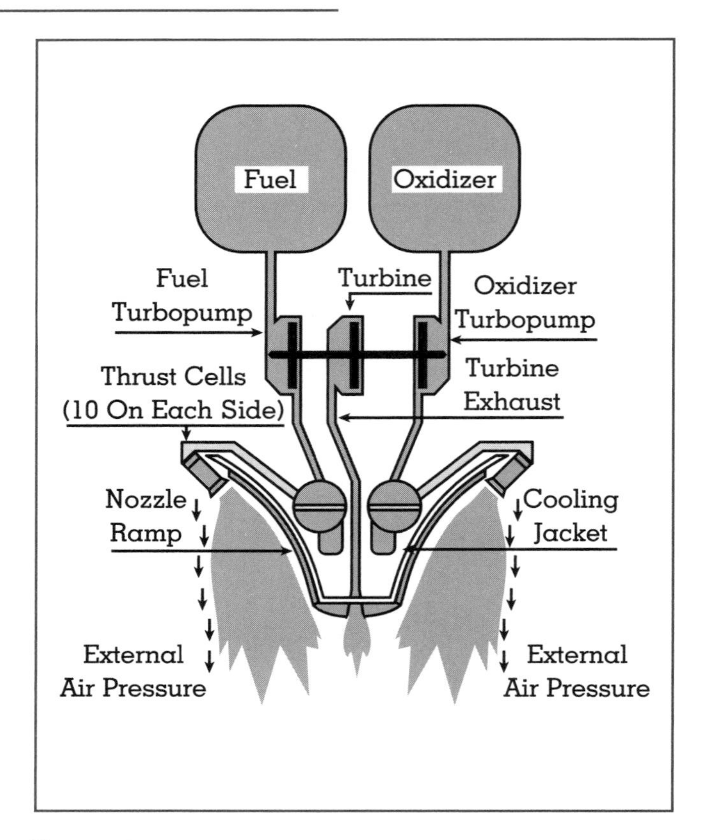

Figure 3.
Aerospike rocket engine.

the ESA to abandon their own nearly completed space shuttle programs.

The numerous projects of the reusable launch vehicles that have appeared from the mid-1980s through the 1990s are intended to overcome the Space Shuttle drawbacks. Some vehicles are designed as conventional rockets that would use parachutes, helicopter rotors, or rocket engines for landing. Others retain the airplane configuration but are planned to be more compact. Regardless of configuration, almost all these projects rely on new types of rocket engines that have yet to be developed.

The most advanced among today's projects is the future American reusable launch system known as *Venture Star*. It utilizes two new concepts that are expected to boost efficiency and cut down the cost of a payload. The first is the single stage-to-orbit concept—the launch vehicle carries the entire propellant load and does not have any expendable parts. The latter is planned to be achieved by a revolutionary rocket engine design called *Aerospike*. The bell-shaped nozzle of a conventional rocket engine controls the expansion of exhaust gases and influences the engine's thrust and efficiency. Since the optimal gas expansion is different at various altitudes, the lower and upper stages of conventional rockets have engines with nozzles of different shapes. In *Aerospike*, an array of combustion chambers fire along the curved central body, called a ramp or a spike. The spike plays the role of an inner wall of a nozzle, while the outer wall is formed by the pressure of the incoming air flow. Changes in atmospheric pressure upon the vehicle's ascent immediately change the exhaust expansion, allowing the engine to perform optimally at any altitude (see Figure 3). The vehicle will take off vertically but land like an airplane, using its lifting body configuration. The *Venture Star* is planned to be about 30 percent smaller than the Space Shuttle. The *Venture Star*'s payload will be 20,500 kg (45,100 lb), compared of the Shuttle's 29,500 kg (64,900 lb). If successful, the *Venture Star* is to make its first flight in the first decade of the twenty-first century. Meanwhile, NASA prepares to fly a scaled-down technology demonstrator, *X-33*, which will test the *Venture Star* design.

Another method to increase efficiency is to launch a rocket eastward close to Earth's equatorial plane. It would be additionally accelerated by Earth's rotation, achieving greater payload capability without any extra energy consumption. Small rockets could be launched from an airplane, which makes building expensive launching facilities unnecessary. This concept is currently utilized by the American Pegasus launch system. Larger rockets could take off from floating launchers, such as the San Marco platform, which had been used by Italy for the American Scout rockets in the 1960s. The end of the Cold War allowed the use of existing military technology for this purpose. For example, the Russian Navy successfully launched a small satellite by a ballistic missile from a submerged strategic submarine in 1998. The largest floating launch platform was built in 1998 in Norway for the Sea Launch international consortium. It operates from the United States, making commercial launches of heavy rockets jointly built by Russia and Ukraine.

In their quest for energy efficiency, some designers are seeking radically new methods of launching spacecraft. For example, leaving the energy source on the ground instead of being carried by a spacecraft would make the launching more efficient. One such method is firing a satellite from a large cannon—something that revives a space travel idea from Jules Verne's science fiction classic *From the Earth to the Moon* (1865). In his joint experiments with the U.S. Navy in the late 1960s, the Canadian ballistics expert Gerald V. Bull proved that small satellites could indeed survive tremendous shock and acceleration in a cannon barrel and achieve orbital velocity. Instead of conventional cannons, very long chemical accelerators with numerous combustion chambers along the barrel are considered. In this case, the pressure of expanding gas would not decrease, but continuously be built behind a projectile, providing it with greater velocity. Among variations of that concept are electromagnetic accelerators, where the force of chemical explosion is replaced by electromagnetic energy. A typical example is a coil gun, which accelerates a projectile by magnetic fields created in sequence by electric coils wrapped around a barrel. Unlike cannons, electromagnetic accelerators promise a smoother rate of acceleration and better control over the projectile's velocity. They also are much "friendlier" to the environment in terms of lack of excessive noise and chemical exhaust pollution. Although attractive, such launchers would require consumption of enormous amounts of electricity, as well as construction of large and expensive facilities. Experiments with chemical and electromagnetic accelerators have been under way in the United States, Russia, and some other countries for many years.

The choice of exotic propulsion systems is even broader in open space, where the lack of gravity and atmospheric drag allows a wide selection of low-thrust but economical engines. For example, thermal energy produced by a controlled chain reaction in a nuclear reactor can heat a propulsive mass to a higher extent than in chemical rocket engines. Higher temperature means greater ejection velocity and higher specific impulse. This concept of a nuclear rocket engine was thoroughly researched in the United States and the USSR from the 1960s to the 1980s, and working prototypes were ground-tested in both countries. For instance, the American NERVA (nuclear engine for rocket vehicle application) engine used liquid hydrogen as a propulsive mass and had a sea-level thrust of 29.4 tons (300 kN, or 64,680 lb). Figure 4 shows a principal design of the nuclear rocket engine with a solid-state reactor.

The heating elements in modern nuclear reactors are made of solid radioactive materials, but theoretically they could exist in a colloidal (semiliquid) or gaseous state. That would allow a radical increase in propulsive mass temperature and ejection velocity. Research has shown that the colloidal and gaseous-state nuclear reactors would respectively produce specific impulses from 20 percent to 50 percent higher than the solid-state reactor. Some very advanced concepts of gaseous-state reactors promise even greater efficiency.

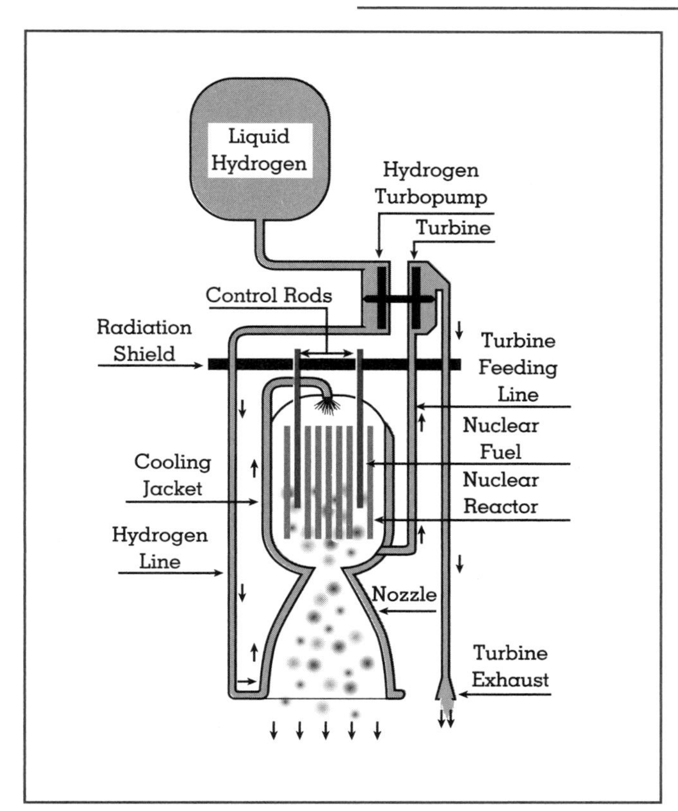

Figure 4.
Nuclear rocket engine.

Another attractive idea is to use electric power for propulsion. Having many design variations, the electric rocket engines (ERE) could roughly be divided into three groups:

- thermal electric engines, which eject propulsive mass heated by the electricity;
- magneto-hydrodynamic (plasmic) engines, which convert propulsive mass into plasma and accelerate it by a magnetic field;
- ion engines, where an ionized propulsive mass is ejected by an electrostatic field.

Type	Maximum Thrust	Continuous Firing Time	Specific Impulse	Comments
Chemical Solid Propellant Rocket Engines	up to 1,500,000 kg	dozens of seconds	270 seconds	Widely used in launch vehicles and spacecraft as auxiliary boosters and, sometimes, for main propulsion.
Chemical Liquid Propellant Rocket Engines	up to 800,000 kg	dozens of minutes	388 seconds	Main type of propulsion system of modern space technology.
Nuclear Solid State Rocket Engines	up to 30,000 kg	dozens of minutes	700 - 900 seconds	Under development. Prototypes were ground tested in the 1960s - 1980s.
Electric Rocket Engines	up to 0.15 kg	thousands of hours	1,200 - 25,000 seconds	Under development. Used to stabilize some satellites and as a main propulsion of a planetary probe.

Table 2.
Space Propulsion Systems

Experiments showed that the EREs can utilize a variety of chemical elements for propellant: from metals (cesium, rubidium, mercury) to gases (hydrogen, argon, xenon). The main advantage of these engines is in unsurpassed efficiency. They produce very low thrust but can continuously work for thousands of hours, consuming very little propellant due to their unparalleled specific impulse. In open space, slow but steady acceleration eventually produces the same result as a powerful but short-duration impulse—the spacecraft achieves a desired velocity. It is just a matter of time. Yet the low power and high electricity consumption so far remain the main difficulties in the ERE development.

The first experiments with the thermal electric engine were conducted in Russia in 1929 by its inventor, Valentin P. Glushko, who later became a world-famous authority in rocket propulsion. For more than forty years, the United States and Russia have devoted many resources to research and development of various kinds of EREs. First tested in space by the Russians in 1964, these engines have found some limited applications in modern space technology. For more than two decades Russian weather and communication satellites have regularly used electric rocket engines for orbital stabilization. The first spacecraft to employ ERE for main propulsion was the American asteroid exploration probe *Deep Space 1*, launched in 1998. The performance of its ion engine is illustrated by the fact that after 2.5 months of continuous work it increased the spacecraft velocity by 2,415 km/h while consuming only 11.5 kg (25.3 lb) of propellant (xenon).

Table 2 compares some of the modern propulsion systems.

POWER SUPPLY SYSTEMS

Power supply systems provide electricity to feed the onboard spacecraft equipment and can be classified by the primary energy sources: chemical, solar, and nuclear.

Dry cells (batteries) and fuel cells are the main chemical electricity sources. Dry cells consist of two electrodes, made of different metals, placed into a solid electrolyte. The latter facilitates an oxidation process and a flow of electrons between electrodes, directly converting chemical energy into electricity. Various metal combinations in electrodes determine different characteristics of the dry cells. For example, nickel-cadmium cells have low output but can work for several years. On the other hand, silver-zinc cells are more powerful but with a much shorter life span. Therefore, the use of a particular type of dry cell is determined by the spacecraft mission profile. Usually these are the short missions with low electricity consumption. Dry cells are simple and reliable, since they lack moving parts. Their major drawbacks are the bulk and deterioration of current stability during the life span.

In fuel cells, electricity is generated by the same chemical process of oxidation between liquid fuel and oxidizer in the presence of a catalytic agent and electrolyte. While various chemical solutions could be utilized, the most common fuel cells use the oxygen/hydrogen combination. The oxygen/hydrogen fuel cell consists of a chamber with three compartments separated by electrodes (nickel plates with tiny holes and a catalytic agent). The middle chamber is filled with electrolyte. Oxygen and hydrogen (fuel) are pumped into the side chambers. As a result of a chemical reaction, the cell produces electricity, heat, and water. Fuel cells provide 75 percent higher output than the best dry cells, and their electric current does not deteriorate with time. At the same time they are much bulkier and more complex in design, with numerous moving parts. Their life time is limited by the supply of fuel and oxidizer onboard a spacecraft. Fuel cells usually are used for short-duration missions with high electricity consumption. They were installed aboard the American *Gemini* and *Apollo* manned spacecraft and later—on the Space Shuttle.

Solar energy systems utilize an unlimited natural source of energy—the sun. Solar arrays directly convert visual light into electricity. Such devices are the most widely used sources of electric power in modern space technology. They consist of a large number of specially developed semiconducting platters (cells) that develop electrical current when exposed to light. Silicon semiconducting cells are most common, while the newest and more expensive gallium and cadmium cells are more efficient. The cells are connected and assembled in arrays on the body of the spacecraft, or on specially deployed external panels. The panels are usually made of a number of hinged plates that are stacked during the launch and open after the spacecraft reaches orbit. However, in the late 1980s a new type of panel—based on a thin, flexible film—was introduced, making solar arrays lighter in weight and easier to store and deploy.

Solar array efficiency depends on its orientation to the sun. The best result is achieved if the array faces the sun at an angle of 90° to 85°. Precise orientation can be achieved either through the rotation of the whole spacecraft, or the rotation of the solar arrays only. To ensure constant electricity flow regardless of changes in spacecraft orientation, solar arrays charge a secondary power source—the rechargeable booster batteries—which in turn power the spacecraft equipment.

Large solar arrays can produce high power, since the total electrical output depends on the overall area of panels. The capability to work for several years with stable current makes solar panels more effective than chemical sources for long-duration space missions. Spacecraft power supplies could be upgraded if solar arrays are enlarged, or replaced in flight. Such operations were performed by cosmonauts and astronauts aboard the Russian *Salyut* and *Mir* space stations and the American Hubble Space Telescope. At the same time, the solar arrays and related equipment are bulky and massive. Large solar arrays are susceptible to erosion by space radiation, micrometeorites, and damage from space debris. Large solar arrays also have limited applications for low-flying satellites (due to increase in atmospheric drag) and for deep space probes, since their efficiency diminishes at greater distances from the sun. To solve the latter problem, NASA recently pioneered a new concept—the refractive solar concentrator array. The solar panel in this case contains a layer of tiny lenses that concentrate solar light on the cells and compensate for energy losses in distant space. The first spacecraft to test this new design was *Deep Space 1*.

Modern nuclear electric systems are represented by two different types of devices: radioisotope generators and nuclear reactors. Radioisotope generators utilize the natural decay of radioactive materials, which could last for several years. The simplest among radioisotope generators—the so-called nuclear battery—develops electricity in electrodes directly exposed to isotope radiation. This is a low output but very compact device that could be used as an auxiliary power supply. Unlike nuclear batteries, thermal radioisotope generators produce electricity from the heat emitted by the isotope via static thermal converters of two major types. The most common are the radioisotope thermoelectric generators (RTGs), where electricity appears in arrays of semiconductors heated at one end and cooled at the other. RTGs have been intensively studied in the United States and Russia for almost four decades and have reached a high level of sophistication and reliability. The early models were flight-tested by both countries in the mid-1960s. Since then, the United States has used RTGs on more than two dozen military and scientific space missions, mostly for deep-

Type	Prime Energy Source	Advantages	Disadvantages	Comments
Dry Cells	Chemical.	-Simplicity.	-Low output; -Short life-span; -Bulk; -Current deterioration.	In use for short duration space missions. Various designs provide different life-span and output.
Fuel Cells	Chemical.	-High output; -Stable current; -Water supply as a sub-product.	-Complex design; -Bulk; -Life-span is limited by onboard fuel supply.	Used in some piloted spacecraft for short duration missions.
Solar Arrays	Visual solar radiation.	-Long life-span; -High output; -Stable current.	-Require precision Sun orientation; -Bulk; -Require booster batteries for electricity storage; -Ineffective in low-flying and deep space spacecraft.	The most widely used power supply system in modern spacecraft.
Atomic Batteries	Radioisotope decay.	-Long life-span; -Stable current; -Simplicity; -Compactness.	-Very low output; -Emit harmful radiation; -High cost.	Used as auxiliary power sources in some spacecraft.
Radioisotope Generators	Thermal from radioisotope decay.	-Long life-span; -Stable current; -Compactness.	-Low output; -Require additional devices to produce electricity; -Emit harmful radiation; -High cost.	Used for some long-duration missions, mostly deep-space probes.
Nuclear Reactors	Thermal from nuclear chain reaction.	-High output; -Long life-span; -Stable current; -Compactness.	-Complex design; -Require additional devices to produce electricity; -Emit harmful radiation; -High cost.	Used for some space missions with high power consumption. May produce radioactive pollution of the environment in case of disintegration at launch, or de-orbiting.
Solar Concentrators	Thermal from visual solar radiation.	-High output; -Long life-span; -Stable current.	-Complex design; -Require additional devices for electricity production and storage; -Bulk; -Require precision Sun orientation.	Not used. Considered as prime power supply systems for large space stations.

Table 3.
Power Supply Systems

space and planetary probes. Automated scientific stations installed by astronauts on the moon, *Viking* landers on Mars, *Pioneer* missions to Jupiter and Saturn, and *Voyager* explorations of the outer planets of the solar system all owe their many years of uninterrupted power supplies to radioisotope generators. The *Cassini* probe, launched toward Saturn in 1997, was the latest spacecraft equipped with RTGs. However, since the 1980s closer attention has been paid to the development of the more complex thermionic generators, since they are considered more efficient than RTGs. In thermionic converters electricity is developed in a chain of metal electrodes enclosed in a sealed container and exposed to a high temperature. The electrodes are placed very close to each other (up to 0.01 mm), with a special dielectric solution between them.

Radioisotope generators cannot produce high output, but their main advantages are compactness, self-containment, and unsurpassed durability. The lack of moving parts makes them highly reliable. At the same time, radioisotope generators are expensive and require protection of the onboard electronic equipment and ground personnel from harmful radiation. The use of radioisotope generators is justifiable in small, long-duration, and low-power-consumption spacecraft.

Unlike radioisotope generators, nuclear reactors utilize the much more intense process of nuclear chain reaction. Since this process is controlled in the reactor, the energy output could be regulated depending on the system's requirements. It actually could produce twice its nominal power, if necessary. Nuclear reactors can provide greater electrical output than radioisotope generators using the same types of thermal converters. This output is comparable to that of fuel cells and solar arrays, while nuclear reactors are more durable and compact.

Reactor-based electrical systems were studied in the United States and the USSR for many years. After testing in space of its first nuclear reactor, SNAP-10A, in 1965, the United States concentrated mostly on developing RTGs. On the contrary, the Russian engineers put more effort into researching nuclear reactors. The Soviet Union was the first nation to start regular use of spaceborne nuclear reactors, called Buk, with thermoelectric converters. That system powered the military ocean radar surveillance and targeting satellite US-A ("US" stands for universal satellite), better known in the West as RORSAT. The satellite's size limitations, low orbit, and radar's high power consumption made the nuclear reactor the only possible energy source. A total of thirty-one such satellites were launched from 1970 to 1987. At the end of their life span, the satellites eject the reactors to a higher "storage" orbit, were they will exist for several hundred years. By the time of their natural reentry, the nuclear material should decay beyond the danger of atmospheric contamination. Unfortunately, that did not always work: Three times the reactors accidentally reentered along with satellites and caused air and water pollution in Canada and the Pacific. Later the USSR developed a series of more advanced reactors with thermionic converters and successfully flight-tested them in 1987. In 1992 one of those reactors, Topaz 2, was purchased by the United States to study its possible military application.

Future spacecraft probably will require much more power than today. This problem could be solved by dynamic thermal converters, in which a high-temperature gas spins the electrical generator's turbine. Obviously, such a generator would depend on a supply of the gas-producing material onboard a spacecraft. In this respect most effective would be a closed-cycle generator, where the used gas is not discharged into space, but continuously recycled through a special radiator. Theoretically, dynamic converters can produce 45 percent higher output than static ones. As a source of thermal energy, dynamic generators may use nuclear reactors, or solar concentrators—large parabolic mirrors that produce a high temperature zone in a focal point. Unlike solar arrays, the mirrors are less susceptible to erosion and theoretically could work for a longer time. Solar concentrators have been studied in the United States and Russia but have never been used in space flight.

Table 3 compares various spacecraft power supply systems.

Peter A. Gorin

See also: Aviation Fuel; Batteries; Engines; Fuel Cells; Fuel Cell Vehicles; Military Energy Use, Historical Aspects of; Rocket Propellants; Storage Technology.

BIBLIOGRAPHY

Gatland, K., ed. (1984). *The Illustrated Encyclopedia of Space Technology*, 4th ed. London: Salamander Books.

Rycroft, M., ed. (1990). *The Cambridge Encyclopedia of Space*. Cambridge, Eng.: Cambridge University Press.

SPACE HEATING

See: Heat and Heating

SPEED AND ENERGY CONSUMPTION

See: Automobile Performance

SPERRY, ELMER AMBROSE (1860–1930)

Elmer Ambrose Sperry was born near Cortland, New York, on October 12, 1860. Sperry, whose mostly absent father was a farmer and itinerant worker and whose mother died hours after his birth, was raised by his aunt Helen until she married, as well as by his paternal grandparents. The young Sperry often studied the many water-driven mills, a steam-driven mill, and the railway station in Blodgett Mills. Later, he also became a familiar presence at the mills, factories, blacksmithy, machine shop, printing press, foundry, railway yard, and pottery of the larger Cortland.

Elmer Ambrose Sperry. (Library of Congress)

A turning point for Sperry was his attendance at the Philadelphia Centennial exposition in 1876. Sperry was particularly fascinated by the Machinery Hall, observing for hours, for example, a Jacquard loom incorporating an early version of card punch automatic control.

The YMCA greatly influenced Sperry, who avidly read the latest scientific magazines, including the *Official Gazette of the U.S. Patent Office*, in its periodical room. The organization also sponsored lectures given by professors at the Cortland Normal School, which Sperry attended, as well as lectures by leaders in many fields, including Alexander Graham Bell. While growing up, he was active in the Baptist church in Cortland. After moving to Chicago in 1882, Sperry became heavily involved at the First Baptist Church where Zula Goodman, whom he later married, played the organ. That marriage, beginning June 28, 1887, was long and happy and produced three sons and a daughter.

In 1883, Sperry began his first business, the Sperry Electric Light, Motor, and Car Brake Company. This company, which pursued dynamo-driven arc lighting during a time of the rapid development of incandescent lighting and alternating current, folded in 1887. Out of the ashes of the former company, Sperry formed the Sperry Electric Company, whose principal assets were Sperry's arc-light patents. His attention, however, was divided between this company and the Elmer A. Sperry Company, a research and development enterprise formed in 1888 and almost entirely owned by Sperry, in which he was freed from engineering and business matters. It was the Elmer A. Sperry Company that allowed him to develop and patent his ideas, with the aid of about thirty workmen in the shop. Sperry developed a series of labor-saving devices, including a highly successful electric locomotive for mine haulage (introduced in 1891), to ease miners' work and increase mine productivity. Sperry also created the Sperry Electric Mining Machine Company (1889; sold in 1892) and the Sperry Electric Railway Company (1892). In addition, a separate company was developed (reorganized in 1900 as Goodman

Manufacturing Company and in which Sperry acquired a large block of stock) that bought mining industry patents, as well as manufacturing several products Sperry's company had originally manufactured. Throughout his life, Sperry stayed involved in policy decisions and technical consulting as a vice president of the Goodman Company, which became a good source of income for Sperry to apply to his future invention projects.

Sperry's method of invention was to examine a field, determine market need, and focus on creating an innovative device to meet that need. He was amazingly visually oriented rather than mathematically inclined. When an invention was patented and ready to market, its patents were often used as the principal capital assets of a specially formed company. Sperry then relegated management of its manufacture and sales to others, freeing him to concentrate on the next invention. Sperry invented his streetcar during the early 1890s, assigning the resulting patents to General Electric in 1895. With its invention, Sperry combined his in-depth grasp of both mechanical and electrical engineering. By 1898, he had developed a superior electric car. Running on a battery he created, it could drive a reputed 87 miles per charge, compared to around 30 miles for other electric cars. The jigs and tools to turn out "Sperry Electric Carriages" in lots of 100 were being readied; instead, in 1900, the American Bicycle Company acquired some of the key patents related to Sperry's electric car, using these to protect its own lines of steam, petroleum, and electric automobiles. Heartened by his success with the battery, Sperry then focused on electrochemistry, spurred in part by the harnessing of electrical power at Niagara Falls in 1895, which made electric power cheaply available. By 1904, Sperry and Clinton P. Townsend had refined an electrolysis process to the point where Elon Huntington Hooker decided to purchase the patents for the process—the beginning of the Hooker Electrochemical Company of Niagara Falls, which became a major U.S. chemical manufacturer.

Sperry turned his attention to the American Can Company's huge amount of scrap metal remaining after round can tops were pressed from square sheets. He and his colleagues refined American Can's electrolytic detinning process to deal with this scrap so that absolutely pure tin powder resulted. In 1907 and 1908, Sperry was involved in patent interference cases concerning this detinning process. The outcome was that the Goldschmidt Detinning Company, formed by American Can in 1908 using the detinning process pioneered by the German Goldschmidt company, bought Sperry's patents, and in 1909 sold the rights to these patents to same Goldschmidt company in Germany (attesting to their value). Meanwhile Sperry's interest since 1907 had been moving into the nascent and promising area of gyroscopic technology.

After attempting unsuccessfully to convince the automobile industry of the safety value of his car gyrostabilizer, Sperry continued his concentration, begun in late 1907, on a gyrostabilizer for ships, seeking to improve the one developed by Ernst Otto Schlick in Germany. By May 1908, Sperry had completed his invention and filed a patent application, marking the beginning of Sperry's most famous period—as inventor of an *active* (as opposed to Schlick's *passive*) ship gyrostabilizer. It was active in that an actuator (later called a sensor), consisting of a pendulum that, upon sensing small rolls of the ship, activated the switches or valves or other means that controlled the motor; the motor then stimulated the much less sensitive gyro to overcome its greater inertia (compared to the pendulum) and begin precession *before* the gyro actually began to feel the torque of the ship's roll. The pendulum system incorporated a feedback control mechanism, one of many Sperry had developed, and would continue to develop, as essential components of his inventions—making Sperry a pioneer in the area that came to be known as cybernetics.

By 1911, Sperry had in addition invented and developed a gyrocompass that was installed for testing on one of the U.S. Navy's state-of-the-art Dreadnoughts. After refining its design, Sperry fulfilled his company's Navy contract by installing the gyrocompass on three more of the Navy's Dreadnoughts and by June 1912 on two submarines. One refinement was a mechanical analog computer that automatically corrected compass errors.

Between 1909 and 1912, aircraft invention was burgeoning, but lack of adequate stabilization resulted in many lost lives. Sperry, his brilliant son Lawrence (who would die in 1924 in a crash while flying across the English Channel), and Sperry's hand-picked engineers developed an airplane automatic stabilizer, redesigned it into the main component of an automatically controlled missile during World War I, and after the war transformed it again into an automatic pilot system. During the war, when the Sperry Gyroscope Company (formed in 1910) became almost exclusively devoted to wartime development, several other inventions, such as a superior high-intensity searchlight, were introduced. Also introduced and installed on several Navy battleships was a highly effective improved gunfire control system, whose basic component was the gyrocompass.

Following the war, the gyrocompass was adapted for merchant-marine vessels, providing unmatched ease of use and maintenance, far outstripping the magnetic compass in navigational economies and safety. By 1932, the Sperry gyrocompass had been installed on more than 1,000 merchant-marine ships. The gyrostabilizer had also been installed on a few huge ocean liners. Another major achievement during this time was the automatic pilot developed for ships, incorporating the most sophisticated guidance and control mechanisms Sperry had yet invented. Nicknamed "Metal Mike," the gyropilot was installed in around 400 merchant-marine ships worldwide by 1932. In addition, Sperry developed and marketed a helm indicator and course recorder that proved valuable adjuncts to the gyropilot and gyrocompass.

Sperry's output continued undiminished until his wife died on March 11, 1930, after which his health declined until he died on June 16, 1930.

Sperry applied for his first patent—a dynamo-electric machine—when he was twenty. His final patent application—for a variable pitch propeller—was submitted in 1930, shortly after his death. In total, 355 of his over 400 patent applications matured as actual granted patents.

Robin B. Goodale

See also: Batteries; Electric Vehicles.

BIBLIOGRAPHY

Hughes, T. P. (1993). *Elmer Sperry: Inventor and Engineer.* Baltimore: Johns Hopkins University Press.

SPRING

See: Elastic Energy

STANFORD LINEAR ACCELERATOR CENTER

See: National Energy Laboratories

STEAM ENGINES

A steam engine is a device that converts a portion of the heat energy absorbed by a liquid, after it has vaporized, into mechanical work.

For a classic reciprocating steam engine, pressurized steam from a boiler is admitted into cylinder, driving a piston attached to a connecting rod and crankshaft. Steam admission is typically cut off in midstroke. This cutoff steam continues to expand against the piston until the stroke ends, converting more heat energy into work.

The most efficient performance is achieved when the engine expands steam to achieve the greatest absolute difference between steam admission and exhaust temperatures. *Superheaters* are devices in the boiler that raise the temperature of steam above that of the water from which it was generated. Efficiency is improved because this added temperature, or superheat, permits greater expansion (and thus temperature drop) before the steam begins condensing and is exhausted. Further efficiency gains can be realized by exhausting the steam not to the atmosphere but rather to a separate apparatus called a *condenser.* This is a sealed chamber cooled by air or water. As entering steam cools, it condenses, and its volume is greatly reduced, producing a partial vacuum. Exhausting into this partial vacuum permits yet further expansion and temperature drop.

Steam *turbines*, which generate more than 80 percent of the world's electric power, differ from steam engines in that steam drives blades and not pistons. Steam turbines expand pressurized steam through nozzles that accelerate the steam at the expense of heat energy and pressure. Work is created by transferring a portion of steam velocity to blades, buckets, or nozzles affixed to a rotor to move at high speeds. Steam turbines are relatively compact in relation to steam engines. Turbines are generally operated at higher pressure, resulting in greater output for the size. Their rotary nature also permits continuous admission, expansion, and exhaust so that power production is uninterrupted. Steam engines, by contrast, only develop maximum output during the period before cutoff and none at all during the exhaust period.

Steam power was a cornerstone of the Industrial Revolution. Steam freed humans from limits imposed by natural energy sources as muscle, water, wind, and sun. Human and beast-of-burden muscle power was limited. Most humans turning a crank affixed to an electric generator cannot sustain enough effort to illuminate a 150-watt lightbulb for more than a few minutes. Water was a fairly reliable energy source, but only obtainable along rivers capable of sustaining suitable dams. Windmills were also an early energy source. Although effective, they lacked a means of storing energy, and were only useful for jobs that could be profitably performed intermittently, such as grinding grain.

Starting in eighteenth-century England, steam engines suddenly made concentrated power available on demand anywhere. Earliest uses included pumping water from iron and coal mines, driving iron mill furnaces and hammers, and, a bit later, propelling locomotives to and from the mines. In a synergistic relationship, steam engines made iron and coal affordable, which in turn lowered the cost of building and operating steam engines. Previously costly iron products soon became affordable to all, and engine-driven industry and transportation spread rapidly across the globe.

PREDECESSORS OF THE MODERN STEAM ENGINE

In 1606 Gioavanni Battista della Porta of Naples performed two experiments that formed the basis for steam engines. He proved that steam could develop pressure by expelling water from a sealed container. In the next experiment a steam-filled flask was immersed neck down in water. As the cooling steam condensed, water flowed upward into the flask, proving that a vacuum had been created.

By 1698 Captain Thomas Savery exploited these principles with a patented steam pump. Alternately admitting pressurized steam into a vessel forced water upward through a one-way (check) valve, while the vacuum created by cooling the vessel drew water from below through a second valve. French inventor Denis Papin demonstrated the first steam-driven piston in 1690. His experiment consisted of a cylinder with a closed bottom and a piston. A weight attached to a cord passing over two pulleys applied upward force on the piston. Flame beneath the cylinder heated a small quantity of water, creating steam pressure to raise the piston. Condensation followed removal of heat, and the difference in air pressure on the piston top and the vacuum below caused the piston to descend, raising the weight. By 1705 Thomas Newcomen introduced his pumping engine to England. It built upon the work of Papin but was made practical by use of a separate boiler. The Newcomen engine consisted of a vertical cylinder with a piston attached to a counterweighted crossbeam. A piston pump to drain mines was attached to the crossbeam. The combined efforts of low-pressure steam and counterweight forced the piston upward, to the cylinder top. Spraying cold water into the cylinder condensed the steam, which created a partial vacuum, returning the piston to the bottom.

Both the Newcomen and the Papin pumps derived most of their power from this atmospheric force rather than from steam pressure. Newcomen engines consumed perhaps 32 pounds of coal each hour per horsepower developed, about 1 percent of the energy that burning coal emits.

MODERN STEAM ENGINES

Scottish doctor Joseph Black discovered in 1765 that applying heat to water would steadily raise temperature until reaching the boiling point, where no further temperature increase occurred. However, boiling did not commence immediately but only after a relatively large additional hidden; or latent heat was added. Learning of this discovery, Scottish instrumentmaker James Watt realized that the cold water spray in the Newcomen cylinder wastefully removed latent heat. Maintaining the cylinder at steam temperature would conserve heat and reduce energy consumption. Watt conceived of the condenser, a separate vessel cooled by water immersion and connected to the cylinder through an automatic valve. Steam entering the vessel rapidly condensed, and the vacuum created was applied to the piston, as in the Newcomen engine. Early Watt cylinders were open at the top, like the Newcomen engine. Later, Watt created the double-acting engine by enclosing both cylinder ends and connecting them to the boiler and condenser with automatic valves. Air no longer wastefully cooled the cylinder, and steam rather than air pressure pushed the piston in both directions.

While Watt engines admitted steam throughout the full stroke, he did patent the idea of cutoff to allow the steam to work expansively and more efficiently. Higher pressures were needed to exploit expansion, and it fell to other inventors, notably Richard Trevithick of Wales and Oliver Evans of the United States, to make this leap. Their design eliminated the pull by a vacuum and introduced a push by high pressure. Further efficiency gains occur by superheating the steam, heating it to a greater temperature than by which it was generated. This added temperature permits greater expansion and work to occur before the steam condenses. Watts's use of separate boiler, engine, and condenser, double-acting engines, and insulation all resulted in the first modern steam engine. He tripled efficiency over the Newcomen engine, reaching about 3 percent efficiency overall.

In 1781 Jonathan Hornblower invented the compound engine, but conflicting patent rights prevented development. Arthur Woolf revived the idea in 1804 and invented a valve mechanism often associated with compounding. The compound engine partially expands steam in one cylinder and then further expands it in one or more succeeding cylinders, each expansion being termed a stage. Compound expansion reduces temperature swings in each cylinder, improving efficiency since less energy is lost reheating metal on each piston stroke. While simple engines with condensers reached 10 to 15 percent efficiency on superheated steam, compounds with up to four stages readily reach 20 percent and in a few cases as much as 27 percent.

Reciprocating steam engine development reached its pinnacle with the development of the uniflow engine. Invented by T. J. Todd of England in 1885 and refined some twenty years later by German engineer Johann Stumpf, the uniflow features a single-inlet valve, and exhaust ports at the end of the stroke. Steam is cut off relatively quickly and allowed to expand; at the end of the stroke, the piston exposes the exhaust port. As the piston returns and covers the port, the piston it recompresses the steam remaining in the cylinder back to the admission pressure. This recompression heats the steam in much the same fashion as it cooled off while expanding; consequently little energy is lost reheating the metal. Uniflow engines are more simple and economical to build than compounds and generally approach compounds in efficiency; at least a few examples may have surpassed compounds, with one engine reportedly testing at 29 percent efficiency. Uniflow engines never became prevalent due to rapid concurrent development of the steam turbine.

Steam engine development preceded understanding of the principles governing their operation. Steam engines were widely used in locomotives, automobiles, ships, mills and factories. It was not until 1859, when the Rankine cycle (named after Scottish engineer and physicist William John Macquorn Rankine) was postulated, that steam engine design finally received a sound theoretical footing.

STEAM TURBINES

Curiously, although practical steam turbines are a more recent development than piston steam engines, two of the earliest steam devices were turbines. The earliest generally recognized steam-powered mechanism was the aeolipile, built by Hero of Alexandria in the first century C.E. It was essentially a pan-shaped boiler connected to a hollow rotating shaft on which was mounted a hollow sphere. Jets of steam exiting the sphere through two tangentially placed nozzles spun the device. The action was similar to a modern rotating lawn sprinkler. The Branca engine of 1629 consisted of a boiler fancifully shaped into a human head and torso. Steam exiting a nozzle placed at the mouth impinged upon a paddle wheel. Although efficiency and performance were very low, cog wheels provided sufficient torque for a small mortar and pestle to powder drugs.

Modern steam turbines can reach efficiencies of 40 percent. About 90 percent of newly installed electrical generating capacity is driven by steam turbines due to high efficiency and large capacity. The impulse turbine is the most basic modern turbine. An impulse stage consists of one or more stationary steam nozzles mounted in the casing and a rotating wheel with curved blades mounted along its periphery. Steam expanding through the nozzles exchanges pressure and temperature for great speed. Directing this steam against curved blades transfers momentum to the wheel. The Branca engine was an early impulse turbine. Reaction turbines exploit Newton's third law: "For every action there is an equal and opposite reaction." While Hero's aeolipile lacks blades like a modern turbine, recoil from steam did rotate the sphere, making it a reaction turbine.

The allure of producing rotary motion without intermediary crankshafts and connecting rods attracted efforts of engineers to little success until the 1880s, when two entirely different designs appeared in England and Sweden. In 1884 Charles Algernon Parsons patented the first practical reaction turbine. Unlike Hero's turbine, which had two jets, Parson's turbine had blades mounted around the periphery of fourteen pairs of rotating disks mounted on a single shaft. The area between blades was designed similarly to nozzles; steam passing through these spaces expanded and accelerated, imparting thrust to the rotating disk. Succeeding blade sets became larger, permitting steam to expand farther in each stage and is analogous to compounding in reciprocating engines. Fixed blades between the disks smoothly redirected steam to the following rotating blades. Large turbine sets can employ many more stages of expansion than reciprocating steam engines, since each stage adds only a small amount of friction, while additional cylinders soon add enough friction to offset any efficiency gains. The final stages in these machines can often effectively extract energy from steam at pressures below atmospheric. This aggressive use of compounding leads to significantly more efficient operation than for reciprocating steam engines.

Parsons adopted this multistage design to reduce turbine operating speeds to a useful level. His initial turbine was developed to produce electricity onboard ships and had an output of about 7 kW. In 1888 he designed the first steam turbine generating unit for public utility service. By the time of his death in 1931, his company manufactured turbines generating more than 50,000 kW.

In 1894 the success of his electrical generating plants led Parsons to undertake development of the first marine turbine, with the goal of building the fastest ship afloat. The prototype vessel *Turbinia* was 100 feet long. Initial trials with a single turbine directly connected to a propeller were unsatisfactory, as the turbine speed was too high for efficient operation. In 1896 the single turbine was replaced by three separate turbines, with steam flowing from the high-pressure turbine element to the intermediate element and then the low-pressure turbine element. Each turbine shaft was fitted with three propellers and she made 34½ knots, a new world's record.

In 1888 Carl G. P. de Laval invented his impulse turbine, which was based on the Pelton water wheel. Steam was accelerated to nearly the speed of sound through a nozzle or a series of nozzles. Directing the steam toward curved vanes mounted upon a disk resulted in high rotating speeds. Conceptually this is identical to Branca's turbine, but August Camille Edmond Rateau of Paris learned that wasp-waisted nozzles greatly enhanced velocity, making turbines feasible. Noticing that excessive rotating speeds were a weakness of de Laval turbines, Rateau patented a multistage impulse turbine in 1896. His machine was essentially a series of de Laval turbines. Breaking expansion into stages resulted in smaller pressure drops across each set of vanes and lower velocity for the same mechanical output. Over the years many variants of turbine arrangements have been tried and adopted. Efficiency has more than doubled since the turn of the twentieth century, and output has risen enormously. The first Parsons turbines of 1884 were perhaps 10 hp. By 1900, when the demand for electric power was growing tremendously, the largest steam turbine-generator unit produced about 1,250 kW, but advances over the next twenty years increased capacity to more than 60,000 kW. This output far exceeded what even the largest steam engines could provide, and paved the way for central generating plants with millions of turbine horsepower to be the principal electric power generators of the twentieth century.

The steady march of improvements in performance of steam turbines can be largely attributed to exploiting higher steam pressures and steam temperatures. Steam generator pressures and temperatures increased from about 175 psi and 440°F (227°C) in 1903 to 450 psi and 650°F (343°C) in 1921 and 1,200 psi and 950°F (510°C) by 1940. Since the late 1950s steam pressures of more than 3,500 psi and 1,150°F (621°C) have become available and at least one unit has reached 5,300 psi and 1,210°F (654°C). Nationwide, average plant efficiency climbed steadily, with an average of 29 percent in the late 1930s, 35.5 percent in the late 1940s, and 38.5 percent in the late 1950s. Efficiency improved little in the following decades as boiler and turbine operating temperatures and pressures had reached the limits of available material strength and durability.

By the turn of the twentieth century, reciprocating steam engines had become common items in everyday life. Transportation was dominated by steam-powered ships, locomotives, and boats. Steam was the preferred method of driving the earliest automobiles, with more steamers on U.S. roads than internal-combustion engines until about 1906. Portable and self-propelled steam engines provided power to drive agricultural and lumbering machinery on site. Most factories, shops, and mills had a steam engine that drove machinery by way of overhead shafts and leather belts. The sound of the mill steam whistle signaling the shift change set the routine in many communities. Within a few years this personal relationship with steam receded. Internal-combustion engines largely supplanted small steam engines where self-contained power was necessary, while electric motors became dominant in stationary applications.

Except for a few countries that still use steam engine locomotives, reciprocating steam engines have passed into history. They are used only rarely, more often than not as a hobby. Steam turbines, on the other hand, have continually increased in importance, as they are used to generate most of our electricity. All nuclear power plants generate steam to drive turbines, and some geothermal installations use the earth's heat similarly. Although steam turbines are still common on oceangoing ships, they are being replaced by large diesel and gas turbine installations due to high intrinsic efficiency, low first cost, and more economical labor and maintenance requirements.

Even when large diesels and gas turbines take over as prime movers, steam turbines still often have a vital role. A process called bottom cycling or combined cycling can team the steam turbine with an internal-combustion engine, which is inherently efficient due to its high temperature operation. Exhaust gases from these operating cycles are often sufficiently hot enough to operate a boiler and steam turbine. Since the steam system is driven by heat that would otherwise be discarded, overall system efficiency soars. Such bottom cycling plants, which are gaining favor in new generating facilities in the United States, are now 50 percent efficient and are likely to close in on 60 percent in the near future.

Kenneth A. Helmick

See also: Black, Joseph; Engines; Gasoline Engines; Newcomen, Thomas; Parsons, Charles Algernon; Rankine, William John Macquorn; Savery, Thomas; Trevithick, Richard; Turbines, Steam; Watt, James.

BIBLIOGRAPHY

Bentley, J. (1953). *Old Time Steam Cars.* Greenwich, CT: Fawcett Publications.

Briggs, A. (1982). *The Power of Steam.* Chicago: University of Chicago Press.

Hunter, L. C. (1985). *A History of Industrial Power in the United States, Volume Two: Steam Power.* Charlottesville: University of Virginia Press.

Rose, J. (1887). *Modern Steam Engines.* Philadelphia: Henry Carey Baird. Reprint, Bradley, IL: Lindsay, 1993.

Storer, J. D. (1969). *A Simple History of the Steam Engine.* London: John Baker.

Watkins, G. (1979). *The Steam Engine in History.* Derbyshire, Eng.: Moorland.

STEAM TURBINES

See: Turbines, Steam

STEPHENSON, GEORGE (1781–1848)

Many successful people are inspired by their parents and George Stephenson was no exception. In modern day vernacular, his father had one of the "coolest jobs in town" in the small village of Wylam, on the River Tyne, eight miles west of Newcastle, England, where George was born. Robert Stephenson raised his family in a primitive dwelling while serving as the pumping engine fireman at the local colliery, keeping the mines free of water. The company's coal wagon passed very close to the family's front door, helping George make up for his lack of formal education with up close and very personal attentiveness.

When George was eight yeas old, the mine's pumping engine broke down and work temporarily ceased, forcing his father to move the family in search of work. With the impact of engines permanently etched in his mind, the young boy began to make clay models in great detail, simulating his father's machines. Unknowingly, he was destined to study engines for the rest of his life. Defying parental wishes to take up farming, at age thirteen he joined his brother James as a picker in another mine. Working his way up the ladder at the Black Callerton mine, George began driving an engine and by seventeen he earned the position of assistant engine fireman—higher than any post his father had ever attained.

He was eventually promoted to engineman—or plugman—which required intricate engine knowledge and skills. He quickly learned to devote all his free time to the study and care of his engines, tearing down and rebuilding them to better understand their inner workings.

He became interested in stories about James Watt and Matthew Boulton but faced a significant hurdle: at the age of eighteen, he was still unable to read. He began reading lessons with a young schoolmaster, Robin Cowens, and then, to further his knowledge of engineering, was tutored in math by Andrew Robertson, a Scottish mathematician.

George earned the prestigious position of brakesman at the Dolly Pit colliery, responsible for the engine used for hauling coal out of the mines. He was married in 1802. During that year he made an unsuccessful attempt to build a perpetual motion machine. But he did gain respect in the community as an accomplished clock-mender.

In 1808, his career breakthrough came at the Killingworth High Pit, where he and two other brakesman contracted with management to work on a royalty basis, rather than a fixed weekly wage. The risk was offset by his ability to create mechanical engineering efficiencies, considered one of Stephenson's greatest strengths. Opportunity knocked in the form of a pumping engine built by John Smeaton that had been a complete failure during a period of twelve months. Approaching the investors, Stephenson made an audacious claim that he could have the engine running within a week. In fact, he did so in only three days of breakdown and reconstruction, acquiring great local renown and the true launching of his career as a self-taught "doctor of engines." Proving his flexibility, he drained an entire quarry with a unique miniature pump that he invented.

George continued his adult education with John Wigham, focusing on the areas of math and physics. In addition, he befriended an engineer, John Steele, who had apprenticed under Richard Trevithick in the building of a successful locomotive engine during 1803 to 1804.

Side view of George Stephenson's locomotive, "The Rocket." (Corbis Corporation)

Stephenson applied theories developed by others, including Ralph Allen's work on self-acting inclines, in which empty coal wagons were pulled up a track, powered by full wagons moving downhill. Other inventions included a winding-engine and new design for pumping engines, before turning his attention to the self-propelled steam engine.

In the early nineteenth century, hardly anyone imagined that passenger transportation would evolve, concentrating instead on the need to transport coal. Most of the inventiveness applied before Stephenson was to reduce track friction, allowing horses to pull more load. There were even experiments in the seventeenth century by Sir Humphrey Mackworth to build a wind-powered land-ship, which were disappointing due to the inevitable limits of nature. The first successful steam locomotive was built by French engineer Cugnot in 1763. Designed for war, its top speed was a paltry two and one-half miles per hour, with endurance of only fifteen minutes. There were numerous failures for the rest of the century, hampered by persistent attempts to operate them on roads.

It is noteworthy that William Murdock achieved significant progress in the 1780s, despite discouraging remarks by James Watt and his partner, Matthew Bolton, who thought that steam locomotion was impractical. This chiding among rivals was common and it seemed that personal credibility was sometimes as important as genius. When Trevithick did his brilliant work with locomotives, his eccentricities and irascibility deterred his acceptance, as did his unbusinesslike and unprofessional nature. Stephenson's predecessors who did earn credibility, however, were William Hedley and aide, Timothy Hackworth, at coal mines in the Newcastle area.

George Stephenson was not truly the inventor of railways or steam locomotives. But, like Thomas Newcomen and his work with the first practical atmospheric steam engine, George and his son, Robert, were the first to use concentrated imagination to turn existing work and theory into fully functional locomotive engines on rails. The Stephensons made the greatest contribution toward the building of the British railway system.

George Stephenson's first engine, named the Blucher, took ten months to complete, and first operated on the Killingworth Railway on July 25, 1814. His most notable achievement was building the Liverpool to Manchester railway, that featured thirty miles of track and sixty-three bridges. The most famous locomotive ever built—"The Rocket"—was actually designed and made by his son, Robert, although its most serious fault was reconfigured by George, who approved all concepts of design and manufacture.

Despite their growing reputation, there was great resistance at times to the Stephensons' locomotive building, most notably among farmers. Although the noise that scared animals was largely reduced, there was an economic concern to farmers—if steam trains and carriages succeeded, there would be less demand for horses and less demand to grow oats to feed them. But technological progress prevailed, even if it was often settled in court.

In addition to George's acclaimed work with locomotives, he and his son Robert designed numerous contraptions such as a scarecrow with wind-aided arms, a sundial, an oil light that burned underwater, and an automatic cradle rocker. While George Stephenson derived immediate satisfaction from his mechanical accomplishments, there is some evidence that he resented or disdained the public acclaim through his rejection of invitations to join the Royal Society or to accept Knighthood. Despite his brilliance, his relative late education kept him from feeling socially accepted. This may have started with a grudge as early as 1815, when he lost out in a patent dispute with a more privileged rival over a safety lamp that George invented.

Nonetheless, Stephenson realized that he had a great impact on society and finished his career as a traveling consultant in engineering. He was known to dine with kings, such as the King of Belgium, and richly earned his permanent place in the annals of industrial history.

Dennis R. Diehl

See also: Steam Engines; Turbines, Steam.

BIBLIOGRAPHY

Rolt, L.T.C. and Allen, J. S. (1997). *The Steam Engine of Thomas Newcomen.* Ashbourne, UK: Landmark Publishing.

Rowland, J. (1954). *George Stephenson: Creator of Britain's Railways.* Long Acre, London, Odhams Press Limited.11

STIRLING, ROBERT (1790–1878)

Robert Stirling was born on October 25, 1790, in the parish of Methvin, Perthshire, Scotland. Stirling, the son of a farmer, received a classical education followed by studies in divinity at the universities of Edinburgh and Glasgow. The Church of Scotland, after examining his knowledge of Hebrew, Greek, divinity, and church history, licensed Stirling to preach the Gospel in March 1816. Shortly after, Stirling accepted the patronage of the Duke of Portland, who proposed Stirling as a suitable candidate for the vacant post of assistant minister at Kilmarnock. Stirling was ordained on September 19, 1816.

Just eight days after his ordination, Stirling applied for a patent for "Improvements for diminishing the consumption of fuel." He had evidently carried out his experiments while he was a student at Edinburgh, but why a student of divinity should have developed an interest in fuel economy, and gone to the expense of applying for a patent to protect his invention, is something of a mystery. Stirling's patent was in several parts. The first part was for a method of diminishing the quantity of fuel used in furnaces for melting glass, while still producing the high degree of heat required. The second was for a heat exchanger suitable for use in breweries and factories where economies could be made by transferring heat from one fluid or vapor to another. The third was for an engine for moving machinery.

Stirling's engine consisted of an externally heated, hot-air engine fitted with what he termed an "economiser." The economiser absorbed and released heat to and from the enclosed air, which was shifted alternately between hot and cold spaces, with the resulting pressure changes communicated to a power piston. Stirling was the first to design an engine with a working cycle that depended on recovering waste heat for efficient operation.

Stirling's patent was based on practical experiments and a working knowledge of radiant heat. Stirling may have been influenced by the works of John Leslie, Professor of Mathematics at Edinburgh, who in 1804 published the results of his investigations into the nature of radiant heat. The general layout of Stirling's engine embodies Leslie's experiments with spaced radiation screens to reduce the transmission of heat. In a second work published in 1813, Leslie argued that heat was transmitted as a series of vibrations, and if heat were applied to one end of a cylinder of metal, then a regular descending graduation of temperature would soon be established along the whole length. Stirling seems to have perceived that here was a process that might be made reversible. If heat were transmitted in a series of vibrating layers, and if the heat source were itself moved along the layers, heat could be discharged into bands of heat at decreasing temperatures. That is to say, as the heated mass touched each layer, the temperature of the heated mass would fall. If the mass of air were to be passed back over the bands of heat, then it would regain its former temperature. This was how Stirling described the workings of his economiser, although he took care to dispel any idea of perpetual motion by pointing out the losses.

Stirling's transfer to Kilmarnock did not put a stop to his experimenting. Here he meet Thomas Morton, an inventor and manufacturer who supplied him with a workshop. Morton also built telescopes and observatories and was influential in developing Stirling's interest in constructing optical and scientific instruments. In 1818 Stirling erected an engine of his own design to drain water from a local stone quarry. This engine performed its duty well until, as a result of carelessness of the engine-man, it overheated and was rendered useless.

James Stirling, who was ten years younger than his brother Robert, initially studied for the church, taking a classical education at Edinburgh and Saint Andrews University. The pull of mechanics proved stronger than religion, and he took up an apprenticeship with Claud Girdwood and Co., of Glasgow, machine makers and iron founders. In 1824 he suggested his brother use compressed air for his engine rather than air at atmospheric pressure, and by 1827 James was influential enough to have built an experimental twenty horse-powerengine fitted with a sheet iron economiser, in which too much faith was placed, resulting in the engine failing for want of effective cooling. This engine embodied a number of improvements in the economiser, which were patented in 1827 in the joint names of Robert and James Stirling. James left Girdwood's in 1830 to become manager of the Dundee Foundry, Dundee.

At Dundee the brothers continued experimenting, directing their efforts into improving the efficiency of both cooling apparatus and economiser. The results of these experiments were patented in 1840 (the year Robert was awarded a Doctor of Divinity), and in 1843 a forty-five horsepower engine was set up to power the whole of the foundry, which it did until 1847. The trial engine successfully vindicated the Stirling brothers' claims for improved fuel efficiency, but ultimately failed as a practical engine because the available material—cast iron—could not withstand the high temperatures required for efficient operation. The development of the Stirling engine was effectively halted until improvements in metallurgy were made.

In 1837 James married Susan Hunter, the daughter of a Saint Andrews Professor; they had no children. James left Dundee in 1845 to become a successful consulting engineer in Edinburgh, where he died at the age of seventy-five. In 1819 Robert had married Jane Rankine, the daughter of a Kilmarnock merchant. They had two daughters and five sons. Robert had moved to the parish of Galston in 1823, and he remained a respected minister until his death at the age of eighty-eight, although some of his parishioners felt he devoted too much time to his mechanical pursuits.

Such was Robert Stirling's interest in science and engineering that of his sons, four became leading railway engineers in Britain and South America, and one became a church minister. His son-in-law became the general manager of an iron works. Stirling never sought to benefit from his invention; indeed, he did little to prevent infringements of his patents. Stirling's 1816 patent foresaw all possible applications of what came to be called the regenerative process, a term coined by John Ericsson, who never accepted the priority of Stirling's patent. However, Ernst Warner von Siemens, who profitably developed the regenerative furnace for smelting metals, did acknowledge Stirling as the originator.

Robert Sier

See also: Ericsson, John; Siemens, Ernst Warner von; Stirling Engines.

BIBLIOGRAPHY

Sier. R. (1987). *A History of Hot Air and Caloric Engines.* London: Argus Books.

Sier. R. (1995). *Rev. Robert Stirling D.D.* Chelmsford: L. A. Mair.

STIRLING ENGINES

The principle that makes Stirling engines possible is quite simple. When air is heated it expands, and when it is cooled it contracts. Stirling engines work by cyclically heating and cooling air (or perhaps another gas such as helium) inside a leak tight container and using the pressure changes to drive a piston. The heating and cooling process works like this: One part of the engine is kept hot while another part is kept cold. A mechanism then moves the air back and forth between the hot side and the cold side. When the air is moved to the hot side, it expands and pushes up on the piston, and when the air is moved back to the cold side, it contracts and pulls down on the piston.

While Stirling engines are conceptually quite simple, understanding how any particular engine design works is often quite difficult because there are hundreds of different mechanical configurations that can achieve the Stirling cycle. Figure 1 shows a schematic of a transparent educational demonstration engine that runs on the top of a cup of hot coffee. This engine uses a piece of foam similar to what would be used as a filter for a window air conditioning unit to "displace" the air between the hot side and the cold side. This foam displacer is carefully mounted so it does not touch the walls of the cylinder. Figure 2 shows how this particular engine achieves the Stirling cycle. In this engine, the air flows through and around the displacer from the hot side then back to the cold side, producing a power pulse during both the hot and cold portion of the cycle. Stirling engines can be mechanically quite simple since they have no valves, and no sparkplugs. This can result in extremely high reliability as there are fewer parts to fail.

It is worthwhile to compare Stirling engines to other more familiar engines and note their similarities as well as their differences. Stirling engines are a type of heat engine. They turn heat into mechanical work and in this sense they perform the same function as other well known heat engines such as gasoline, diesel, and steam engines. Like steam engines, Stirling engines are external combustion engines, since the heat is supplied to the engine from a source outside the cylinder instead of being supplied by a fuel burning inside the cylinder. Because the heat in a Stirling engine comes from outside of the engine, Stirling engines can be designed that will run on any heat source from fossil fuel heat, to geothermal heat, to sunshine. Unlike steam engines, Stirling engines do not use a boiler that might explode if not carefully monitored.

When operating on sunshine, or geothermal heat, Stirling engines obviously produce no pollution at all, but they can be exceedingly low emissions engines even when burning gasoline, diesel, or home heating oil. Unlike gasoline or diesel engines that have many thousands of start stop cycles of combustion each minute, burners in Stirling engines burn fuel continuously. It's much easier to make a continuous combustion engine burn very cleanly than one that has to start and stop. An excellent demonstration of this principle is to strike a match, let it burn for a few seconds, then blow it out. Most of the smoke is produced during the starting and stopping phases of combustion.

A BRIEF HISTORY

In the early days of the industrial revolution, steam engine explosions were a real problem. Metal fatigue

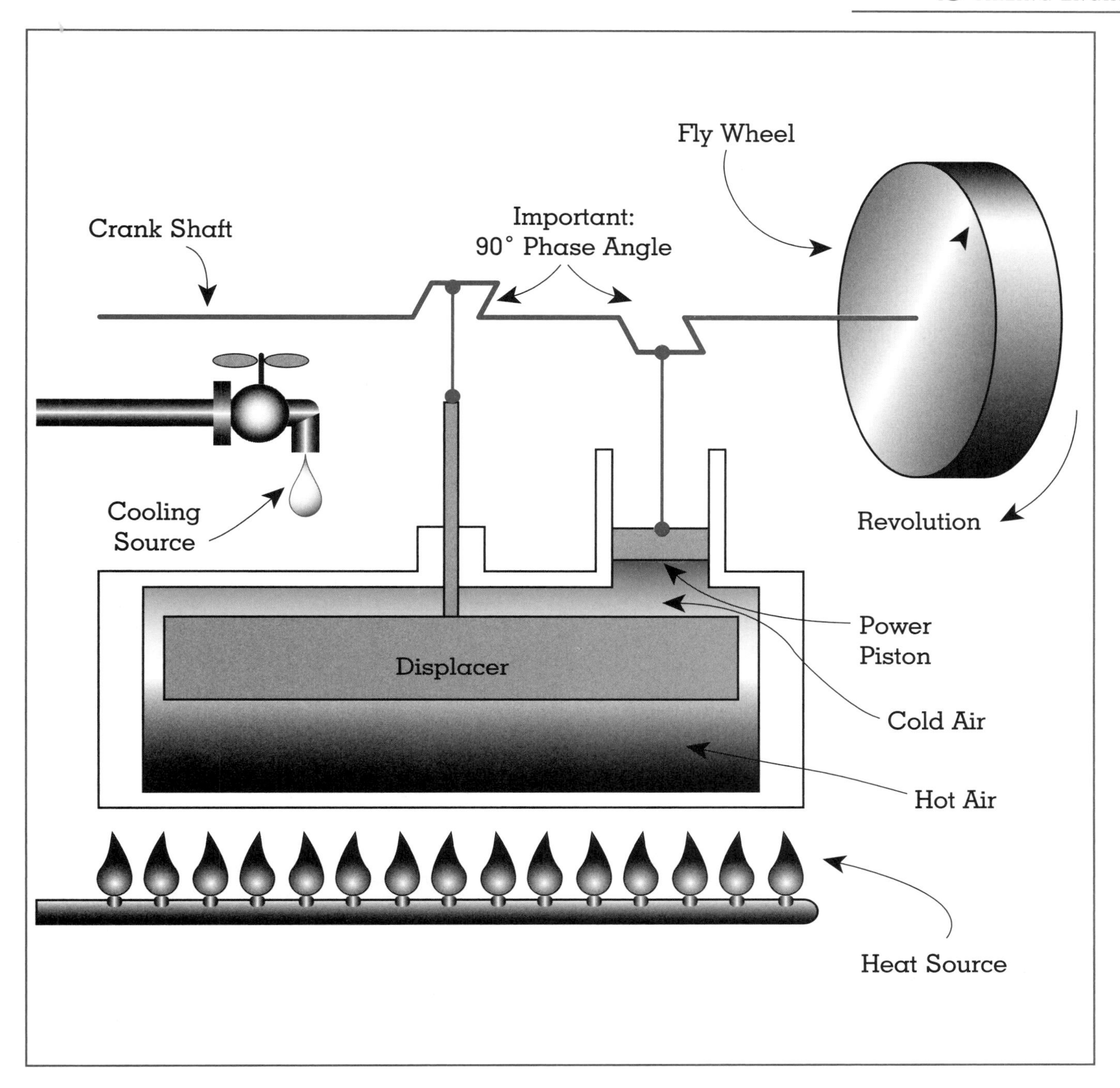

Figure 1.
Key to Stirling Engine. The air flows both through and around the porous displacer. The displacer looks like a piston but it is not.

was not well understood, and the steam engines of the day would often explode, killing and injuring people nearby. In 1816 the Reverend Robert Stirling, a minister of the Church of Scotland, invented what he called "A New Type of Hot Air Engine with Economiser" as a safe and economical alternative to steam. His engines couldn't explode, used less fuel, and put out more power than the steam engines of the day.

The engines designed by Stirling and those who followed him were very innovative engines, but there was a problem with the material that was used to build them. In a Stirling engine, the hot side of the

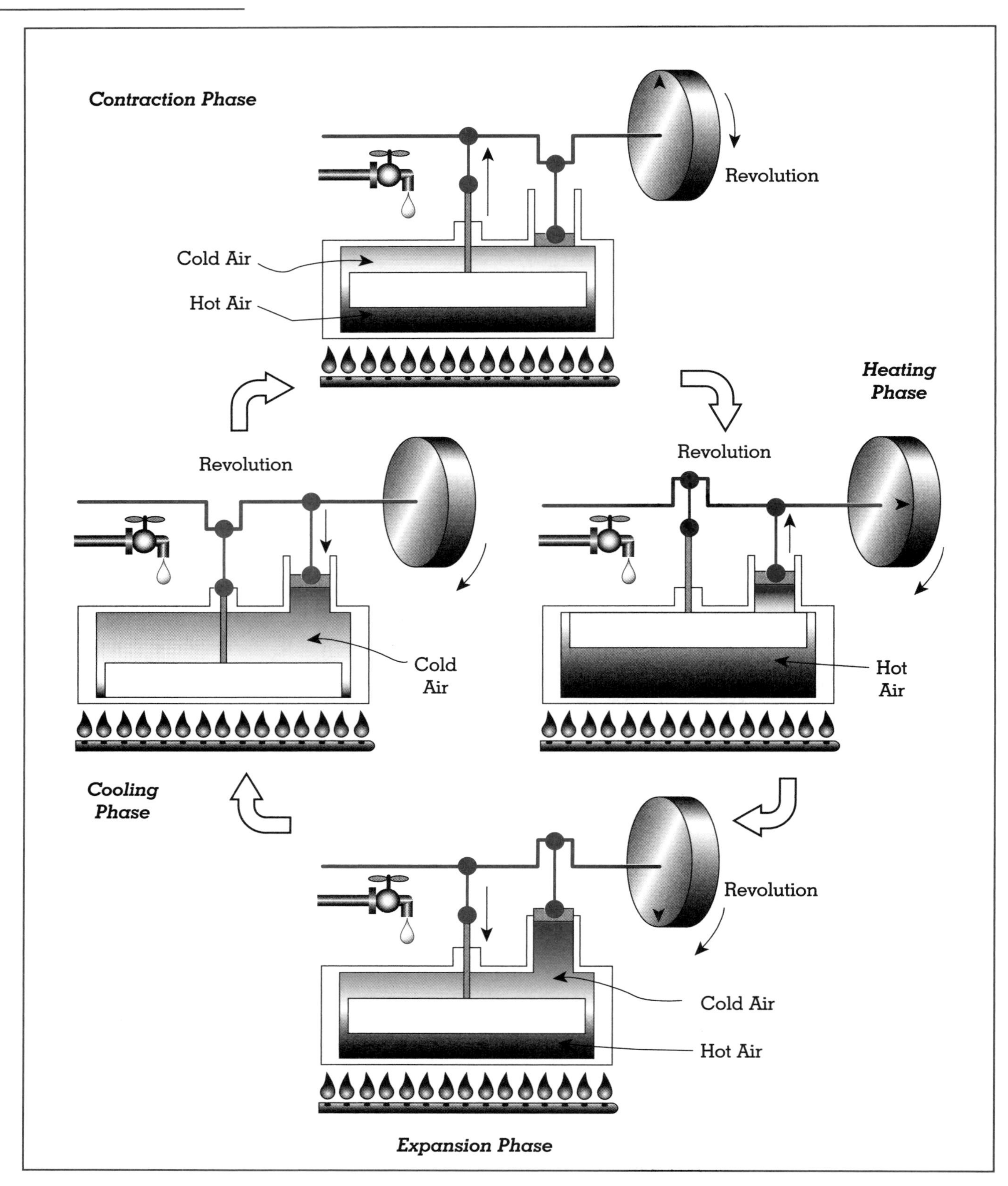

Figure 2.
Four phases of the Stirling engine power cycle.

engine heats up to the average temperature of the flame used to heat it and remains at that temperature. There is no time for the cylinder head to cool off briefly between power pulses. When Stirling built his first engines, cast iron was the only readily available material, and when the hot side of a cast iron Stirling engine was heated to almost red hot, it would oxidize fairly quickly. The result was that quite often a hole would burn through the hot side causing the engine to quit. In spite of the difficulties with materials, tens of thousands of Stirling engines were used to power water pumps, run small machines, and turn fans, from the time of their invention up until about 1915.

As electricity became more widely available in the early 1900s, and as gasoline became readily available as a fuel for automobiles, electric motors and gasoline engines began to replace Stirling engines.

REGENERATION

Robert Stirling's most important invention was probably a feature of his engines that he called an "economiser." Stirling realized that heat engines usually get their power from the force of an expanding gas that pushes up on a piston. The steam engines that he observed dumped all of their waste heat into the environment through their exhaust and the heat was lost forever. Stirling engines changed all that. Robert Stirling invented what he called an "economiser" that saved some heat from one cycle and used it again to preheat the air for the next cycle.

It worked like this: After the hot air had expanded and pushed the piston as far as the connecting rod would allow, the air still had quite a bit of heat energy left in it. Stirling's engines stored some of this waste heat by making the air flow through economiser tubes that absorbed some of the heat from the air. This precooled air was then moved to the cold part of the engine where it cooled very quickly and as it cooled it contracted, pulling down on the piston. Next the air was mechanically moved back through the preheating economiser tubes to the hot side of the engine where it was heated even further, expanding and pushing up on the piston. This type of heat storage is used in many industrial processes and today is called "regeneration." Stirling engines do not have to have regenerators to work, but well designed engines will run faster and put out more power if they have a regenerator.

CONTINUED INTEREST

In spite of the fact that the world offers many competing sources of power there are some very good reasons why interest in Stirling engines has remained strong among scientists, engineers, and public policy makers. Stirling engines can be made to run on any heat source. Every imaginable heat source from fossil fuel heat to solar energy heat can and has been used to power a Stirling engine.

Stirling engines also have the maximum theoretical possible efficiency because their power cycle (their theoretical pressure volume diagram) matches the Carnot cycle. The Carnot cycle, first described by the French physicist Sadi Carnot, determines the maximum theoretical efficiency of any heat engine operating between a hot and a cold reservoir. The Carnot efficiency formula is

$$(T(hot)-T(cold))/T(hot)$$

T(hot) is the temperature on the hot side of the engine. T(cold) is the temperature on the cold side of the engine. These temperatures must be measured in absolute degrees (Kelvin or Rankine).

STIRLING APPLICATIONS

Stirling engines make sense in applications that take advantage of their best features while avoiding their drawbacks. Unfortunately, there have been some extremely dedicated research efforts that apparently overlooked the critical importance of matching the right technology to the right application.

In the 1970s and 1980s a huge amount of research was done on Stirling engines for automobiles by companies such as General Motors, Ford, and Philips Electronics. The difficulty was that Stirling engines have several intrinsic characteristics that make building a good automobile Stirling engine quite difficult. Stirling engines like to run at a constant power setting, which is perfect for pumping water, but is a real challenge for the stop and go driving of an automobile.

Automobile engines need to be able to change power levels very quickly as a driver accelerates from a stop to highway speed. It is easy to design a Stirling engine power control mechanism that will change power levels efficiently, by simply turning up or down the burner. But this is a relatively slow method of changing power levels and probably is not a good way to add the power necessary to accelerate across an intersection. It's also easy to design a simple Stirling engine control device that can change power levels quickly but allows the engine to continue to consume fuel at the full power rate even while producing low amounts of power. However it seems to be quite difficult to design a power control mechanism that can change power levels both quickly and efficiently. A few research Stirling engines have done this, but they all used very complex mechanical methods for achieving their goal.

Stirling engines do not develop power immediately after the heat source is turned on. It can take a minute or longer for the hot side of the engine to get up to operating temperature and make full power available. Automobile drivers are used to having full power available almost instantly after they start their engines.

In spite of these difficulties, there are some automobile Stirling applications that make sense. Hybrid electric cars that include both batteries and a Stirling engine generator would probably be an extremely effective power system. The batteries would give the car the instant acceleration that drivers are used to, while a silent and clean running Stirling engine would give drivers the freedom to make long trips away from battery charging stations. On long trips, the hybrid car could burn either gasoline or diesel, depending on which fuel was cheaper.

To generate electricity for homes and businesses, research Stirling generators fueled by either solar energy or natural gas have been tested. They run on solar power when the sun is shining and automatically convert to clean burning natural gas at night or when the weather is cloudy.

There are no explosions inside Stirling engines, so they can be designed to be extremely quiet. The Swedish defense contractor Kockums has produced Stirling engine powered submarines for the Swedish navy that are said to be the quietest submarines in the world.

Aircraft engines operate in an environment that gets increasingly colder as the aircraft climbs to altitude, so Stirling aircraft engines, unlike any other type of aircraft engine may derive some performance benefit from climbing to altitude. The communities near airports would benefit from the extremely quiet operation that is possible. Stirling engines make sense where these conditions are met:

1. There is a premium on quiet.
2. There is a very good cooling source available.
3. Relatively slow revolutions are desired.
4. Multiple fuel capacity is desired.
5. The engine can run at a constant power output.
6. The engine does not need to change power levels quickly.
7. A warm-up period of several minutes is acceptable.

LOW TEMPERATURE DIFFERENCE ENGINES

In 1983, Ivo Kolin, a professor at the University of Zagreb in Croatia, demonstrated the first Stirling engine that would run on a heat source cooler than boiling water. After he published his work, James Senft, a mathematics professor at the University of Wisconsin, River Falls built improved engines that would run on increasingly small temperature differences, culminating in an elegant and delicate Stirling engine that would run on a temperature difference smaller than 1°F.

These delicate engines provide value as educational tools, but they immediately inspire curiosity into the possibility of generating power from one of the many sources of low temperature waste heat (less than 100°C) that are available. A quick look at the Carnot formula shows that an engine operating with a hot side at 100°C and a cold side at 23°C will have a maximum Carnot efficiency of [((373 K–296 K)/373 K) × 100] about 21 percent. If an engine could be built that achieved 25 percent of the possible 21 percent Carnot efficiency it would have about 5 percent overall Carnot efficiency.

That figure seems quite low until one realizes that calculating Carnot efficiency for an engine that uses a free heat source might not make much sense. For this type of engine it would probably be more worthwhile to first consider what types of engines can be built, then use dollars per watt as the appropriate figure of merit.

Stirling engines that run on low temperature differences tend to be rather large for the amount of power they put out. However, this may not be a significant drawback since these engines can be largely manufactured from lightweight and cheap materials such as plastics. These engines could be used for applications such as irrigation and remote water pumping.

CRYOCOOLERS

It isn't immediately obvious, but Stirling engines are a reversible device. If one end is heated while the other end is cooled, they will produce mechanical work. But if mechanical work is input into the engine by connecting an electric motor to the power output shaft, one end will get hot and the other end will get cold. In a correctly designed Stirling cooler, the cold end will get extremely cold. Stirling coolers have been built for research use that will cool to below 10 K. Cigarette pack sized Stirling coolers have been produced in large numbers for cooling infrared chips down to 80 K. These micro Stirling coolers have been used in high end night vision devices, antiaircraft missile tracking systems, and even some satellite infrared cameras.

Brent H. Van Arsdell

See also: Cogneration Technolgies; Diesel Cycle Engines; Gasoline Engines; Steam Engines; Stiring, Robert.

BIBLIOGRAPHY

Senft, J. R. (1993). *An Introduction to Stirling Engines.* River Falls, WI: Moriya Press.

Senft, J. R. (1996). *An Introduction to Low Temperature Differential Stirling Engines.* River Falls, WI: Moriya Press.

Walker, G. (1980). *Stirling Engines.* Oxford: Oxford University Press.

Walker, G.; Fauvel, O. R.; Reader, G.; Bingham, E. R. (1994). *The Stirling Alternative.* Yverdon, Switzerland: Gordon and Breach Science Publishers.

West, C. (1986). *Principles and Applications of Stirling Engines.* New York: Van Nostrand Reinhold.

STORAGE

Energy storage is having energy in reserve for future needs. While it takes millions of years to create huge energy stores in the form of coal, oil, natural gas, and uranium, humans are able to quickly manufacture energy stores such as batteries, and to build dams to store the gravitational energy of water. Although batteries and dams are very useful and important storage systems, the amount stored is very small compared to natural stores. These natural energy stores are large,

Solar panels on a house roof in Soldier's Grove, Wisconsin. (Corbis Corporation)

but finite, and it is important to realize that they will not last forever.

The food a person eats is a personal energy store banking on photosynthesis. In photosynthesis, gaseous carbon dioxide, liquid water, and solar energy interact to produce solid carbohydrates, such as sugar and starch, and gaseous oxygen.

$$\text{carbon dioxide} + \text{oxygen} + \text{solar energy} \rightarrow \text{carbohydrates} + \text{oxygen}$$

Solar energy for photosynthesis is converted and stored in the molecular bonds of carbohydrate molecules. When the process is reversed by combining carbohydrates with oxygen, stored energy is released. We literally see and feel the results when wood is burned to produce heat. A flame is not produced when carbohydrates and oxygen interact in the human body, but the process is similar and appropriately called "burning." The energy we derive from food for performing our daily activities comes from burning carbohydrates.

About 80 percent of the electric energy used in the United States is derived from stored energy in coal. The stored energy has its origin in photosynthesis. Coal is the end product of the accumulation of plant matter in an oxygen-deficient environment where burning is thwarted. Formation takes millions of years. Proven reserves of coal in the United States are upwards of 500 billion tons, a reserve so great that even if coal continues to be burned at a rate of over one billion tons per year, the reserves will last for hundreds of years.

Stores of petroleum, from which gasoline is derived, are vital to industry and the transportation sector. Stores of natural gas are important to industry and the residential and commercial sectors. Both petroleum and natural gas result from the decay of animal and plant life in an oxygen-deficient environment. Solar radiation is the source of the energy stored in the molecules making up all the fossil fuels (i.e., coal, petroleum, and natural gas). The energy is released when petroleum and natural gas are burned. Economic considerations figure prominently in determining proven reserves of petroleum and natural gas. To be proven, a reserve must provide an economically competitive product. There may be a substantial quantity of an energy product in the ground, but if it cannot be recovered and sold at a profit, it is not proven. There are substantive quantities of energy stored as oil in shale deposits in the western United States, but, to date, extracting them are not economical. Proven reserves of oil and natural gas for the United States are around thirty-five billion barrels and 250 trillion cubic feet, respectively. Consumption rates of about three billion barrels per year and twenty trillion cubic feet per year suggest that they will be depleted in the early part of the twenty-first century. But economic conditions can change and new proven reserves will emerge.

An early Energizer watch battery, held against shirt cuff button for size comparison. (Corbis Corporation)

For several decades, the United States has relied on foreign sources of petroleum energy. The availability of petroleum from foreign sources is subject to political instability and changes in the economies in these countries. To guard against losses in petroleum supplies, the United States has established the Strategic Petroleum Reserve. The reserve is an emergency supply of crude oil stored in huge underground salt caverns along the coastline of the Gulf of Mexico. Upwards of one billion barrels of petroleum are stored in six sites along the coasts of Texas and Louisiana.

A battery stores electric energy. Although the concentration of energy is small compared, for example to gasoline, we see a myriad of uses of batteries in radios, cellular phones, flashlights, computers, watches, and so on. The public's demand for these portable products is ever increasing, and scientists strive to develop lighter and better batteries.

An electric power plant generates huge amounts of electric energy, but it does this only on demand by consumers because there is no economical way to store electric energy. Electric power plants are massive units that are difficult to shut down and expensive to start up. At night when consumer demand is low, electric utilities are willing to reduce the cost of electric energy in order to keep the plants operating. During the night, it is often economical to use electricity to run pumps that move water to an elevated position. Then when demand for electricity is high, the water is released to run turbogenerators at a lower level. Such a system is called a pumped storage unit. There are fourteen of these in the United States, each having an electrical output of more than 240 megawatts, which is comparable to a large electric power plant.

A nuclear power plant converts energy stored in uranium and plutonium. Uranium occurs naturally; plutonium is made through nuclear transmutations. The nucleus of a uranium atom consists of protons and neutrons that are bound together by a nuclear force. The stored energy in a uranium nucleus is associated with nuclear forces holding the nucleus together. Energy is released in a nuclear reactor by nuclear (fission) reactions initiated by interaction of neutrons with uranium nuclei. Typically, a single reaction produces two other nuclei and three neutrons. Schematically

$$n + U \rightarrow X + Y + 3n.$$

Importantly, the neutrons and protons in the reacting products (n + U) are rearranged into lower energy configurations in the reaction products (X + Y + 3n), and stored energy is released. A single nuclear fission reaction releases upwards of 10,000,000 times more energy than a chemical reaction in the burning of coal. Whereas a coal-burning electric power plant producing 1,000 megawatts of electric power must burn some 10,000 tons of coal per day, a nuclear power plant producing comparable power will need an initial loading of around a hundred tons of uranium of which about twenty-five tons are changed each year.

Water freezes and ice melts at 0°C (32°F). Melting requires 334 kilojoules (kJ) of energy for each kilogram (kg) of ice turned to water. The 334 kJ of energy are stored in the one kg of water. The 334 kj of energy must be removed from the one kg of water in order to convert the water back to ice. If heat is added to water at temperatures between the freezing point (0°C) and boiling point (100°C), the temperature of the water increases 1°C for every 4200 J added to a kilogram of water. The 4,200 kJ/kg °C is called "the specific heat" of water. The specific heat is a property of all materials. Importantly, the specific heat is substantially larger for water than for nearly all other substances in nature. Heat transferred to a substance to increase the temperature is called "sensible heat." The energy added stays within the substance making the substance a reservoir of thermal energy. Because of its high specific heat, water is a superb coolant for an automobile engine or steam turbine. Fortunately, water is cheap and usually available. In a climate where the coolant in an automobile engine might freeze, antifreezes keep this from happening by lowering the freezing point of the coolant.

Solar energy systems are very popular among many for environmental reasons, and people are even willing to pay more for this "clean" energy. The biggest problem with these sources is that the time when energy is available does not necessarily match when it is needed. One way that solar system designers address this problem is by incorporating a thermal storage system. A solar-heated house or commercial building must have a thermal energy store to provide heat at night or on days when clouds block solar radiation. Solar heating systems using roof-top collectors often use water or rocks to store thermal energy. Some solar houses employ large, south-facing windows to maximize the amount of solar energy entering the house. The solar energy may warm a brick or concrete wall or floor so that it becomes a thermal energy store that provides heat when needed at night.

Joseph Priest

See also: Biological Energy Use, Ecosystem Functioning of; Conservation of Energy; Flywheels; Storage Technology.

BIBLIOGRAPHY

"Energy Conversion and Storage," *NREL: Center for Basic Sciences home page*. Golden, CO: National Renewable Energy Laboratory, Center for Basic Sciences. <http://www.nrel.gov/basic_sciences/>.

Priest, J. (2000). *Energy: Principles, Problems, Alternatives*, 5th ed. Dubuque, IA: Kendall/Hunt Publishing.

STORAGE TECHNOLOGY

MOTIVATION

Electric utilities are expected to provide uninterrupted service. To fulfill this expectation, utilities use a range of energy storage technology so that electricity can be produced in the most economical and efficient manner. Gasoline refined from crude oil is drawn from an underground storage tank when a driver replenishes the fuel in an automobile. Imagine the difficulty meeting the demands of the drivers if the crude oil had to be refined at the time of the replenishment. The task is imposing because consumer demand varies throughout the day and throughout the year. The same is true for utilities providing electricity to consumers.

A large electric power plant produces electricity on consumer demand, and the utility must be prepared to meet demands that vary throughout the day, the week, and the year. The demand is divided roughly into three parts termed baseload, part time, and peak. Base load demand is met with large units having slow response. Usually these works are fueled by fossil fuels, such as coal and natural gas, or uranium. In some cases the electricity may come from a large hydroelectric plant. Generation of baseload electricity is the most economical of the three types. If demand were constant, only baseload electricity would be needed and the cost of electricity to consumers would be minimized. The demand labeled part time is generally provided by the large plants used for baseload. Because of the difficulty of having to start and stop large systems, electricity produced in the part time mode is some two to three times more expensive than baseload electricity. Peak demands requiring fast response are met with oil- and gas-fired turbo-generators, hydroelectricity, pumped-storage hydroelectric systems, purchases from other utilities and, in a few cases, wind-powered generators and solar-powered systems. Peak electricity may cost three to four times more than baseload electricity. Apart from the generating capacity needed to meet peak demands, the distribution system of transmission lines and associated equipment must be able to handle the electric power. Economically, it is in the interest of both the electric utility and the consumer to make an effort to "flatten out" the demand curve. To do this, a utility sometimes encourages consumers to do chores requiring electricity at night by offering a cheaper night rate. It is also in the economic interest of both consumers and utilities if the utilities can produce electric energy during the off-peak times, store the energy in some form, and then convert the stored energy to electric energy during times of greater need. In addition to using pumped storage hydroelectric systems or flywheels for this scheme, there are serious efforts for developing energy storage technology using compressed air, superconducting magnets, and batteries.

EVALUATING ENERGY STORAGE SYSTEMS: ENERGY DENSITY

Every source of energy has a certain mass and occupies a certain amount of space, and both mass and volume are important practical considerations. Clearly, some energy storage systems will be rejected simply because their physical size and cost are impractical. Accordingly, developers of energy storage systems need to know the energy density, measured in kilowatt-hours per kilogram (kWh/kg) or in kilowatt-hours per cubic meter (kWh/m^3). Determining the energy densities for a pumped hydroelectric storage system, which is a practical scheme in many areas of the United States, is straightforward and provides a basis of comparison for other energy storage technologies.

A pumped hydroelectric system makes use of the potential energy of an elevated mass of water. The potential energy (E) of a mass (m) of water elevated a height (H) is $E = mgH$, where g is the acceleration due to gravity. Dividing both sides of this equation by the mass produces $E/m = gH$, which is the energy density in kWh/kg. For water elevated a height of 100 meters, the energy density is about 1000 kWh/kg.

To measure the energy density in kWh/m^3, we relate the mass (m) to mass density (ρ) and volume (V) by $m = \rho V$. The potential energy may then be written as $E = \rho VgH$. Dividing both sides of $E = rVgH$ by the volume produces $E/V = \rho gH$, which is the energy density in kWh/m^3. For H = 100 m the energy density of water is about 0.3 kWh/m^3. Because pumped storage technology is workable, the energy density provides a benchmark for evaluating other energy storage systems.

COMPRESSED AIR ENERGY STORAGE (CAES)

Compressed air can be stored in a container until it is released to turn the blades of a fan or turbine. To determine the energy density of the compressed air, one has to make thermodynamic assumptions about the process. For reasonable assumptions, the energy density is about 2 kWh/m^3, or about 10 times the energy density of water in a pumped storage hydroelectric system. Compressed air energy storage is very attractive from energy density considerations. On the other hand, the volume required for significant amounts of energy is quite large. For example, storing a million kilowatt-hours of energy requires a volume of about 500,000 cubic meters. For comparison, a box 50 meters wide and 100 meters long (roughly the size of the playing area of a football field) and 100 meters high has a volume of 500,000 cubic meters. At this time, there are two operational compressed air energy storage facilities in the world, and both use natural underground reservoirs for the compressed air.

The compressed air energy storage unit operated by the Alabama Electric Cooperative in McIntosh, Alabama utilizes a 19-million cubic foot underground cavern. Air is released during peak periods, heated with natural gas, and expanded through a turbogenerator to produce 110 megawatts of electric power. There is sufficient energy stored to deliver power for 26 hours. A compressed air energy storage system in Huntoff, Germany employs 300,000 cubic meters of space in an underground cavity in a natural salt deposit. Electric generators in the system produce 300 megawatts and there is sufficient energy stored to operate the generators for 2 hours. The Electric Power Research Institute (EPRI) has estimated that more than 85 percent of the United States has geological characteristics that could accommodate an underground CAES reservoir. If the disparity in costs between baseline and part-time/peak load increases, the use of more underground CAES reservoirs is likely.

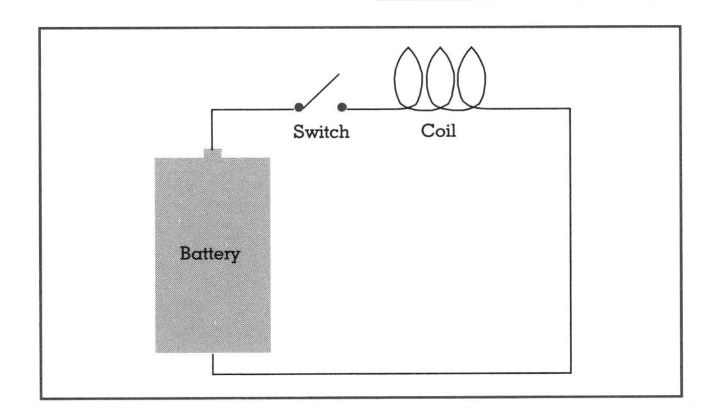

Figure 1.
An elementary electric circuit.

SUPERCONDUCTOR MAGNETIC ENERGY STORAGE (SMES)

Electric utilities can anticipate peak demands and be prepared to meet them. On the other hand, there are short-term fluctuations in the power levels that a utility cannot anticipate but must handle quickly. Superconducting magnetic energy storage (SMES) is a technology that has potential for meeting both anticipated peak demands and those requiring quick response.

Imagine an electric circuit consisting of a battery, a switch, and a coil of wire (Figure 1). There is no electric current in the coil when the switch is open. When the switch is closed, a current is initiated and a magnetic field is produced in the region around the wire. This is an example of the Oersted effect. As long as the magnetic field is changing, voltage is induced in the coil according to an example of Faraday's law. The induced voltage provides opposition to the current established by the battery, this opposition is the essence of Lenz's law. Eventually the current will stabilize at a constant value and a constant magnetic field is established. Electric forces provided by the battery do work to establish the magnetic field and this work gives rise to energy stored in the magnetic field. The energy can be recovered when the switch is opened and the current eventually falls to zero. The amount of energy stored in the magnetic field depends on the electric current and a property of the coil called the inductance. Inductance depends on the construction of the coil and the medium in which the coil is placed. Formally, the magnetic energy is given by $E = (1/2) LI^2$, where L

Paul Chu holds a superconductor computer chip capable of storing more information than conventional chips, 1987. (Corbis Corporation)

and I represent the inductance and the electric current. For any coil, it would appear that any amount of magnetic energy could be stored by producing a sufficiently large current. However, the wire from which the coil is made has electric resistance, and the combination of current and resistance produces heat. Accordingly, the battery must continually do work and expend energy in order to maintain the current. For ordinary materials, there is a practical limit to the current that can be maintained. Despite the practical limitations imposed by heat considerations, there are many practical applications utilizing energy stored in the magnetic field of a current-carrying coil.

The electrical resistance of most conductors, metals in particular, decreases as the temperature of the conductor decreases. For some pure metals and compounds of the metals, the resistance decreases with temperature as usual, but at some critical temperature the resistance drops identically to zero. The resistance remains zero as long as the material is maintained at a temperature below the critical temperature. Such a material is termed a superconductor and the zero resistance state is termed superconductivity. Mercury, first observed to be a superconductor by Heike Kamerlingh Onnes in 1908, has a critical temperature of 4.2 K (room temperature is approximately 300 K). Since Onnes's discovery several other superconductors have been identified but until 1986 the material with the highest critical temperature, 23 K, was a compound of niobium and germanium (Nb_3Ge). In spite of requiring a very low temperature, several very important technologies involving large magnetic fields evolved, including magnetic resonance imaging for medical diagnostic purposes, and superconducting magnets for particle accelerators and nuclear fusion research. Technology for magnetic energy storage never advanced significantly because of the economics involved in producing the low temperatures required for the superconducting state of existing materials.

The optimism about inexpensive superconductivity was stimulated by two notable discoveries. In 1986 Alex Muller and Georg Bednorz discovered a new class of superconductors, ceramic in form, having a critical temperature of 35 K. In 1987 Paul Chu produced a compound that became superconducting at 94 K. While 94 K is still a very low temperature, it is easily and inexpensively attainable with liquid nitrogen (77 K). Materials with critical temperatures as high as 135 K have been found and there is speculation that someday a material may be discovered that is superconducting at room temperature. Because the critical temperatures of these superconductors are considerably higher than that of their metallic counterparts, the phenomenon has been labeled "high-temperature superconductivity" or HTS.

Although no heat is produced by current in a superconductor, the current cannot be made arbitrarily large because the superconducting state vanishes for a well-defined critical current. Because magnetic energy is proportional to the square of the current, research on HTS for magnetic energy storage focuses on developing materials with high critical current. The technology has proceeded to the development stage. For example, a superconducting magnetic energy storage system for quick response to voltage fluctuations has been field tested at Carolina Power and Light. Energy is drawn from the magnetic storage system and fed to the electrical system during moments when power drops because of faults in the electrical system.

BATTERY STORAGE

When a battery, such as the lead-acid type used in automobiles, is charged with an electric generator, an electric current stimulates chemical reactions that store energy. When a device, such as a light bulb, is connected to the battery's terminals, the chemical reactions reverse and produce an electric current in the bulb. The energy stored by a battery is relatively low. For example, a lead-acid battery has an energy density of about 0.05 kWh/kg. Nevertheless, batteries are extremely useful and they play an increasingly important role in a variety of energy storage applications for electric utilities.

In the electric power industry, batteries are used in conjunction with other ways of generating electricity. For example, a home may employ photovoltaic cells or a wind turbine for electricity. However, photovoltaic cells cannot operate without sunlight and wind turbines cannot produce electricity without wind. Batteries charged by photovoltaic cells or wind turbines in off-use times or by the electric utility at hand can provide electricity when the local system is inoperative. The electric power delivered in this mode tends to be in the range of 1 to 4 kilowatts. Batteries have been used for this purpose since the 1970s.

Electricity is delivered to a home or industry at a nominal voltage and frequency. For a home, the voltage is nominally 230 volts, which is reduced to 115 volts for most appliances, and the frequency is 60 hertz. For a variety of reasons, either (or both) the voltage and frequency may change. For example, if the demand during a peak period is excessive, the voltage level may drop. Voltage fluctuations are fairly common and sometimes abrupt. Brief fluctuations sometimes cause computers, for example, to "crash." Avoiding system crashes is important because they are costly and reduce productivity. Although the technology of quality power storage systems that can quickly correct voltage fluctuations is not widespread, there is interest in them and development is proceeding. These systems would likely be designed for a commercial establishment, an industrial complex, or a village, and would have power capability between 30 and 100 kilowatts.

A typical large coal-burning or nuclear-fueled electric power plant is capable of delivering about 1,000 megawatts of electric power, and an average home during peak usage needs about 3 kilowatts of electric power. Some 330,000 homes could be served if all the power from a large plant went to homes during peak usage. Because the power involved is significant, changes in demand during peak periods can be significant, and a utility must be prepared to meet them. One way of doing this is with storage batteries. The technology is complicated by having to change the alternating current (a.c.) voltage produced by a generator to direct current (d.c.) for charging the batteries and then changing d.c. voltage from the batteries to a.c. to meet peak demands. Nevertheless, the technology exists. The power delivered by these battery storage systems tend to be on the order of megawatts. In 1997, more than 70 megawatts of capacity was installed in ten states across the United States.

The low energy density of batteries is a major deterrent to more widespread use in the electric utility sector. Although progress in developing better batteries has been slow and discouraging, research continues. The status of battery development at the time of this writing is presented in Table 1. A battery subscribing to the long-term goal of the United States Advanced Battery Consortium would still have an energy density about 70 times less than that of gasoline (about 15 kWh/kg).

Type	*Energy* (kWh/kg)	*Power* (kW/kg)
Lead acid	0.050	0.200
Nickel cadmium	0.053	0.160
Nickel metal hydride	0.070	0.175
Sodium sulfur	0.085	0.115
Lithium iron sulfide	0.095	0.104
Zinc bromine	0.075	0.045
Long-term goal of the U.S. Advanced Battery Consortium	0.2	0.4

Table 1.
Status of Battery Technology in 1999

COOL THERMAL ENERGY STORAGE

Electric power usage peaks in summer in most areas mainly because of demand for air conditioning. If consumers could satisfy their air conditioning needs without operating energy intensive conventional systems, then the electric utility could reduce the peak demand. This can be accomplished using electricity for conventional refrigeration in non-peak periods, usually at night, to remove heat from a storage medium such as water or materials that undergo a phase change around room temperature. Then, during peak demand periods, the storage medium is tapped for air conditioning and other cooling purposes. During the 1970s and early 1980s, utilities extended cost incentives to customers for installing thermal storage systems. The financial incentive was based on the cost of constructing new power plants. Offering a rebate to customers who shifted peak demand was viewed as a very cost-effective alternative to acquiring new generating capacity, and thermal storage was seen as one of the best ways to shift demand.

There are over 1,500 cool thermal energy sytems in operation in the United States. Eighty seven percent of all systems employ ice storage because it requires smaller tanks and can be installed in a large range of configurations. It is a well-developed technology and a number of systems have been installed. For example, an ice-storage plant located near the Palmer House Hotel in Chicago cools several commercial buildings in the Loop area.

FUTURE GROWTH

Better energy storage technology benefits everyone. For consumers, batteries that are lighter and longer lasting and that recharge more rapidly will continue to accelerate the growth in cordless technology, technology that offers consumers ever greater mobility. For the utilities, more and better storage options limit the need to invest in peak capacity generation, and thereby help lower the electric rates of customers. And for industry, whose rates are discounted if demand can be shifted away from peak periods, there is a great financial incentive to install storage systems. With so many benefits from research and development, the significant storage technology advances in the twentieth century should continue well into the twenty-first century.

Joseph Priest

See also: Batteries; Capacitors and Ultracapacitors; Flywheels; Hydroelectric Energy.

BIBLIOGRAPHY

Electric Power Research Institute. (1987). *Commercial Cool Storage Design Guide*. Washington: Hemisphere Pub. Corp.

Garg, H. P.; Mullick, S. C.; and Bhargava, A. K. (1985). *Solar Thermal Energy Storage*. Boston: D. Reidel Publishing Col.

Ter-Gazarian, A. (1994). *Energy Storage for Power Systems*. Stevenage, Harts., U.K.: Peter Perigrinus Ltd.

STRATEGIC PETROLEUM RESERVE

See: Reserves and Resources

SUBSIDIES AND ENERGY COSTS

Subsidy is a transfer of money from government to an individual or a firm to stimulate undertaking a particular activity. Subsidies and subsidy-like programs are major parts of energy policies. However, such energy grants are only one example of government aid and not the most important examples. The basic economic principle that people respond favorably to financial incentives implies that such transfers stimulate the action that is aided.

More broadly, subsidy can comprise any government policy that favors an action. Such assistance includes tax provisions that lessen the burdens on some activities, and regulations designed to promote an activity. Government ownership with regular deficits, whether or not consciously planned, is another subsidy mechanism. Particularly in the United States, where the government is a major landowner, access to government property at below-market prices is another possible form of subsidy. An additional source of subvention is regulation causing the affected firms to favor (cross-subsidize) some customers.

Conversely, in principle but not in practice, completely nongovernmental subsidy could arise in energy. Private subsidies do occur, as with parental support or patronage of the arts. These illustrations suggest that the incentives to private aid are limited. More critically, only governments have the taxing, spending, and regulating powers to facilitate extensive subsidies.

THE NATURE AND EFFECTS OF SUBSIDIES

Implementation can differ widely. Aid may be freely and unconditionally available to anyone undertaking the favored activity, or restrictions may be imposed.

The awards may have an upper limit, be dependent on the income or other characteristics of the recipient, or require the beneficiary to make a contribution from its own resources. Subsidies may be awarded by holding contests to select beneficiaries.

The impact of a subsidy depends upon many factors. These include first the magnitude of the gap between costs and what can be earned in an unregulated market. To attain any goal, the subsidy must fully cover that gap. The more of an activity the government wants to encourage, the more aid is needed. The effectiveness of an outlay depends upon the elasticity (responsiveness to incentives) of production and consumption. In principle, everything from a thriving new industry to total failure can emerge.

THE ECONOMICS OF SUBSIDIES

Economic theory stresses efficiency—encouraging placing resources in their most valuable uses. An extensive theory of market failure shows when markets might fail to lead resources to their most valuable uses and when subsidies could be desirable. The failures most relevant to energy subsidies are (1) publicness, (2) cost-recovery problems, and (3) imperfect knowledge. Publicness is the simultaneous consumption of a product (or the absorption of damages) by all of society. Cost-recovery problems stem from the existence of economies of scale exist such that firms cannot price efficiently without losing money. Imperfect knowledge means that market participants may lack the knowledge needed to make the most efficient choice.

Economists differ widely in their views on the practical importance of these market failures, on what remedies are appropriate, and whether governments can satisfactorily remedy the problem. One clear principle is that the problem be identified as precisely as possible and the assistance be targeted as closely as possible to the problem. For example, if information is inadequate, providing knowledge about options would be preferable to aiding a specific option. The three justifications for subsidies set a *minimum* test of aid that many programs fail to pass.

Further problems arise in moving from principles to practice. To implement an assistance program, quantitative estimates of the magnitude of market failure must exist. Accurate measurement is a formidable problem. Determination of impacts is often difficult. Disputes usually arise about the importance of subsidy to market outcomes. Moreover, with complex tax systems, it is difficult to determine what comprises equal treatment of all activities and what is a favorable (tax-subsidy) measure.

Economists also recognize that another concern is about equity, the fairness of the distribution of income. Economists differ as much as everyone else about the proper criteria of equity. However, economic principles create considerable agreement about how best to attain any given equity goal. A primary principle is that the aid should go to people rather than specific commodities.

Several considerations are involved. First is a preference for consumer sovereignty, allowing recipients to choose how to spend. Second, being a consumer or a producer of a particular commodity is a poor indicator of whether assistance is deserved. Everyone from the poorest residents (but not the homeless) to the richest Americans are residential electricity consumers. Third, commodity-specific policies tend more to aid producers in the industry than consumers. Politicians often dispute these appraisals and institute subsidies on equity grounds.

THE PRACTICE OF ASSISTANCE

Energy subsidies are widely attacked from many different viewpoints about public policy. Thus, among economists, both severe critics of the oil companies, such as Blair, and those such as Adelman and McAvoy who stress defective government policy criticize the subsidies. Environmentalists and other supporters of active government often feel that too many programs aid the unworthy.

Others, however, suggest reorientation to more appropriate realms. The valid point is that uncorrected market failure may prevail. However, often the arguments for subsidies are misstated. Some suggest that the existence of inappropriate subsidies elsewhere justifies aid to those excluded. The preferable reaction to bad existing subsidies is their removal. Past waste cannot be recovered.

PRESERVING ENERGY INDUSTRIES

A major element of energy policy in leading industrial countries such as the United States, Germany, Britain, Japan, and France has been provision of substantial aid to existing energy industries. Part of the aid provided by governments went to meet commitments to provide pensions to retired workers. Subsidy-like measures such as trade restrictions, regulations, and forced purchase plans also were critical parts of the assistance.

Programs were instituted to preserve coal production in Western Europe and Japan. Every involved country (Britain, Germany, France, Spain, Belgium, and Japan) transferred taxpayers' money to the coal industry. The programs started in the period between the end of World War II and the resurgence of the oil supply in the late 1950s.

Germany initially had a particularly complex program. Coal sales to the steel industry were largely subsidized by direct government grants. A tax on electricity consumption partially subsidized electric-power coal use. Electricity producers were forced to contract to buy German coal and cover in higher production costs the substantial portion of their costs that the electricity tax failed to repay. In the late 1990s, a court decision invalidating the tax and the expiration of the purchase contract forced a shift to using general tax revenues to provide aid.

Every one of these programs has contracted. Costs mounted, forcing cutbacks in how much coal output was maintained. Belgium allowed its coal industry to die. Britain cut off aid, ended government ownership, and left the industry to survive on its own. France and Japan are committed to ending subvention. Germany still is willing only periodically to reduce the output target goals used to orient assistance. The nominal defenses relate to alleged national security and sovereignty issues. However, critics contend that the goal is an unwise retardation of transfer of employment from the industry.

Reliance on forcing consumers to subsidize is hardly unique to German coal. U.S. "public utility" regulation long has required unequal pricing among consumers. Historically, emphasis was on rates that caused industrial and commercial users to subsidize residential users. This has encouraged industrial customers to secure rate reforms. These are equity measures that are subject to the criticisms stated above.

STIMULATING ALTERNATIVE ENERGY

Since the energy shocks of the 1970s, other efforts to subsidize energy have arisen. One often-used justification is inadequate foresight. For example, encouraging specific forms of nonutility generation of electricity was part of 1978 U.S. federal energy legislation. That legislation gave individual states discretion about evaluating the economic attractiveness of the types of nonutility generation encompassed. These included various special sources, such as solar and wind. The most important category by far was "cogeneration." This was a new name for a long-established (but, at the time, waning) practice of simultaneously producing electricity and steam in industrial plants and at times selling some of the electricity to utilities. States anxious to promote such alternatives issued predictions of future costs upon which to base current decisions. Typically, the forecasts exceeded the prices that actually prevailed. Thus, governments acting to offset allegedly defective private insight provided forecasts to the private sector that turned out to be inaccurate.

OIL AND GAS PRICING

Complex price-control plans have been devised. Examples include cost-based prices of electricity and the multitiered pricing systems associated with U.S. oil and gas regulation. The U.S. government and many other governments are major producers of electricity. At a minimum, this involves subsidies from the failure to charge prices that fully recover costs. Beyond this, much of the electricity is from waterpower and is sold at cost (or, more precisely, a measure of costs based on economically defective regulatory accounting techniques) rather than market clearing prices. The profits that would result from such market prices thus are transferred to those lucky enough to be charged the lower price. Since the power goes into the transmission network and loses its identity, the actual users cannot be identified. Arbitrary rules are devised to determine which of those drawing from the grid are favored with low prices. Historically, access was based on the ownership of the purchasing utilities. Government and cooperative-owned utilities and thus all their customers were favored. Ultimately, the conflict of this rule with the traditional preferences for households was recognized. The rules for the Pacific Northwest, where the largest portion of federal waterpower is produced, were changed in the late 1970s so that household users, whatever the ownership of their distribution utility, received the low price.

The oil and gas price-control systems were complex devices that subdivided each fuel into numerous categories based on such considerations as the time of discovery and the difficulty of extraction. Each category was subject to different rules about initial prices; allowable price adjustments; and, in the case of gas, the duration of control. This system had the potential, realized in gas, for low-cost producers to subsidize high-cost producers. A further complication was that the operative law required charging households lower gas prices than industrial users. The result was the signing of numerous contracts at unsustainably high prices.

Further government regulation that totally changed how natural gas was distributed resolved this difficulty. Historically, pipeline companies were required to act as purchasers and resellers of gas. The new approach of the 1980s limited pipelines to selling transmission services. Customers of the pipelines purchase the gas directly from producers. By exiting from gas purchasing, the pipelines could similarly back out of unattractive contracts.

DEPLETION ALLOWANCES

Those concerned with proper pricing of energy often also allude to a special provision of the U.S. Internal Revenue Service code that gives tax favors called depletion allowances to mineral production. In particular, producers can reduce their taxable income by a percentage of total gross income (that differs both among minerals and over time for individual minerals), but the deduction is capped at 50 percent of net income without the allowance. In 1963, Stephen L. McDonald presented the first extensive analysis of these provisions and found no reason why they should exist. His updates and others have supported this conclusion. The assistance to large companies in the oil and gas sector was eliminated and aid to smaller producers was lowered in the legislation in 1975.

ENERGY RESEARCH

Market failure arguments exist for government aid to some forms of research. Since the results of research are so broadly disseminated, it may be prohibitively expensive for private investors to charge the users of their discoveries. Thus the inventors are undercompensated, and underexpenditure arises unless the government provides aid. In addition, some theorists contend (and others deny) that governments are more farsighted than private firms.

Nuclear power long has been a major element in government energy assistance. Some critics of nuclear power argue that this aid was wasted because no nuclear technologies came close to recovering the investments made and did not generate a surplus that repaid government outlays. A less severe assertion is that only the initial work on light-water reactors (the dominant technology) was justified. Some argue that even aid to that technology was extended past the time at which the private sector was ready to undertake the needed further steps.

Several alternative technologies that were heavily supported failed to become commercially viable. The most obvious case was the fast breeder reactor. Such reactors are designed to produce more fissionable material from nonfissionable uranium than is consumed. The effort was justified by fears of uranium exhaustion made moot by massive discoveries in Australia and Canada. Prior to these discoveries extensive programs to develop breeder reactors were government-supported. In addition, several different conventional reactor technologies were aided. The main ongoing nuclear effort is research to develop a means to effect controlled fusion of atoms.

The greatest success in new fossil fuel technology has a distant government-aid basis. The combustion turbine is a stationary adaptation of the jet engine. The combustion turbine's development and improvement were aided by government military aircraft programs. However, turbine manufacturers independently developed the electric-power version.

Since the decline of the U.S. coal industry from 1946 to 1960, many proposals emerged (and received government funding) to promote coal use. One type of effort sought to transform coal into a more attractive fuel. The options ranged from producing a better solid fuel to synthesizing natural gas and petroleum products. Efforts also were directed at facilitating coal use. These included design of better coal-using devices (such as a longtime dream of a viable coal-using locomotive technology). Radically new methods of generating electricity such as fuel cells and magnetohydrodynamics were advocated. Such technologies might be desirable whatever the fuel choice but also might particularly encourage coal use. Other suggestions involve noncoal alternatives such as petroleum in very heavy crude oil, tar sands, and oil shale. Once again, fears of depletion of oil and gas are used as reasons for the aid. Despite extensive government aid, these technologies failed to succeed.

South Africa massively invested in an oil and gas substitution program involving synthesis from coal; the effort was at least partially motivated by international attempts to curtail South African access to conventional sources of oil. A U.S. effort in the Carter administration of the late 1970s produced only one modest coal gassification venture that continued operation. The latest drives to find replacements for the internal-combustion engine for automobiles are behind the ambitious schedule established in 1992 U.S. energy legislation.

Government efforts to promote solar power and wind have failed to date to produce substantial results. World data for 1997 indicate that only 0.41 percent of consumption was geothermal, solar, or wind; 0.39 percent was geothermal and the others accounted for 0.02 percent. The 1999 U.S. percents are 0.06 geothermal, 0.04 solar, and 0.02 wind.

It is widely agreed among economists that market failures in energy exist. However, the record in practice has produced widespread criticism of implementaion of energy subsidies.

Richard L. Gordon

See also: Economic Externalities; Energy Economics; Governmental Intervention in Energy Markets; Industry and Business, Energy Use as a Factor of Production in; Market Imperfections; Regulation and Rates for Electricity; Supply and Demand and Energy Prices; Taxation of Energy; True Energy Costs; Utility Planning.

BIBLIOGRAPHY

Adelman, M. A. (1993). *The Economics of Petroleum Supply: Papers by M. A. Adelman, 1962–1993.* Cambridge, MA: MIT Press.

Bator, F. (1958). "The Anatomy of Market Failure." *Quarterly Journal of Economics* 72(3):351–79.

Blair, J. M. (1976). *The Control of Oil.* New York: Pantheon Books.

Bradley, R. L., Jr. (1995). *Oil, Gas & Government: The U. S. Experience.* Lanham, MD: Rowman & Littlefield.

Cohen, L., and Noll, R. (1991). *The Technology Pork Barrel.* Washington, DC: Brookings Institution.

Gordon, R. L. (1970). *The Evolution of Energy Policy in Western Europe: the Reluctant Retreat from Coal.* New York: Praeger

Gordon, R. L. (1982). *Reforming the Regulation of Electric Utilities.* Lexington, MA: Lexington Books.

Gordon, R. L. (1987). *World Coal: Economics, Policies and Prospects.* Cambridge, Eng.: Cambridge University Press.

Heilemann, U. and Hillebrand, B. (1992). "The German Coal Market after 1992." *Energy Journal* 13(3):141–56.

International Energy Agency. (annual since 1978). *Energy Policies and Programmes of IEA Countries.* Paris: Organisation for Economic Co-operation and Development.

MacAvoy, P. W. (1962). *Price Formation in Natural Gas Fields.* New Haven, CT: Yale University Press.

McDonald, S. L. (1963). *Federal Tax Treatment of Income from Oil and Gas.* Washington, DC: Brookings Institution.

SUPERCONDUCTING ELECTRONICS

See: Electricity

SUPERSONIC TRANSPORT

See: Aircraft

SUPPLY AND DEMAND AND ENERGY PRICES

The quantity of energy supplied is the flow of energy brought onto the market, and the quantity of energy demanded is the amount of energy purchased for a particular period of time. Quantity can be measured in terms of the number of kilowatt hours produced by an electric generator in a day, the number of barrels of oil or cubic feet of gas brought to the market in a month, or the number of tons of coal produced and sold in a year. Primary energy takes the form of fossil fuels or electricity from primary, sources including hydro, nuclear, solar, geothermal, and biomass, while secondary energy is electricity generated from fossil fuels.

DATA SOURCES

Data on the quantity of energy supplied, called energy production, are available from a variety of government, trade association, and international sources. Some of the better sources include the U.S. Energy Information Administration, the American Petroleum Institute, the International Energy Agency, and the United Nations. Secondary energy quantity is reported as net or gross. Net energy is the amount of energy produced and gross is the amount of primary energy required to produce it.

Quantities of energy demanded and supplied are reported in a bewildering variety of ways. The three typical units of measurement are energy content, volume and weight. Energy content includes British thermal units in the United States or kilocalories and kilojoules in the rest of the world, as well as coal or oil equivalents. Kilowatts are the most common units of measurement for electricity and are sometimes applied to other energy sources as well. Weights are most often expressed as short tons, long tons or metric tonnes. Volumes are expressed as barrels, cubic feet, gallons or liters.

DETERMINANTS OF ENERGY SUPPLY

The quantity of energy supplied is a function of the economic and technical variables that influence the cost of bringing the energy source to market and the price that a supplier receives for the energy source in the market. For example, in a competitive market the quantity of coal (Q_s) supplied at the mine mouth is a function of the price received for the coal (P); the price of the capital necessary to produce the coal, such as drag lines, cutting tools, and loaders (P_k); the price of labor, which includes wages, salaries, and indirect labor costs (P_l); the price of using land or any other natural resource or other factor of production (P_n); and technical variables that could include the technologies available and the geology of the deposits (T). An increasing T in this context represents better technology or more favorable geology.

Prices of other related goods also influence quantity supplied. For example, uranium production may produce vanadium as a byproduct. Thus, uranium and vanadium are complementary goods or are goods produced together. When the price of vanadium increases, uranium production becomes more profitable and will increase. Gas found with oil is called associated gas and is considered complementary to oil in production. If the price of oil goes up, drillers may look harder for oil and find more gas to produce as well.

Alternatively, goods may be substitutes in production. If the price of other minerals increase, coal producers may look for other minerals and produce less coal. If gas is nonassociated, it is found without oil and is a substitute for oil in supply rather than a complement. If the price of natural gas, a substitute good (P_s) or a good that could be produced instead of oil, increases, drillers may spend less time looking for oil and more time looking for and producing nonassociated gas, thus decreasing oil production.

Governments often interfere in energy markets and their policies may influence quantity supplied. For example, environmental regulations (E_r) that require less pollution or more safety when producing fuels decrease the quantity of fuel supplied. Such regulations in the United States include the removal of sulfur from fuels, the addition of oxygenates to gasolines in some areas of the country, and more safety in coal mines. Additional environmental regulations increase cost and should decrease quantity supplied. Aggregate market supply also depends on the number of suppliers (#S) in the industry.

We can write a general supply function as follows:

$$Q_s = f(+ P + P_c - P_s - P_l - P_k - P_n + T - E_r + \#S)$$

The sign before each variable indicates how the variable influences quantity supplied. A minus sign indicates they are inversely related and a positive sign indicates they are directly related. Thus, in the example of coal, raising the price of coal is likely to increase quantity supplied, whereas raising the price of labor is likely to decrease the quantity supplied.

For nonrenewable energy sources such as fossil fuels, expectations about the future price and interest rates influence the current quantity supplied. Expectations of higher future prices should cause less production today and more production tomorrow.

ELASTICITY OF SUPPLY

A measure of how responsive quantity supplied is to a variable (say price) is called the elasticity of supply with respect to that variable. Elasticity of supply is the percentage change in quantity divided by the percentage change in the variable in question or

$$\frac{\%\text{ change } Q_5}{\%\text{ change P}}$$

If the supply price elasticity of oil is 1.27, it follows that if the price of oil increases by 1 percent, the quantity of oil supplied increases by 1.27 percent. A cross elasticity of supply indicates how quantity produced is related to another price. For example, if the cross elasticity of oil supply with respect to the price of gas is 0.15, then if the price of gas increases 1 percent, the quantity of oil produced goes up 0.15 percent. Because energy production is capital-intensive, supply price elasticities are larger or more elastic in the long run than in the short run. The long run is the time it takes for producers to totally adjust to changing circumstances and allows for totally changing the capital stock. In contrast, in the short run capital stock is fixed and total adjustment does not take place. Often the short run is considered a year or less, but the exact length of time depends on the context.

Information about supply elasticities would be highly useful for those involved in energy markets, but unfortunately little is available. Carol Dahl and T. Duggan (1996) surveyed studies that use simple models to estimate energy supply or elasticities. They found estimates for the various fossil fuels and uranium in the United States and concluded that studies estimating these elasticities using reserve costs are the most promising. Such studies yielded a U.S. gas supply own-price elasticity of 0.41, a uranium supply own-price elasticity from 0.74 to 3.08, an Appalachia coal supply own-price elasticity of 0.41 to 7.90, and a U.S. oil supply own-price elasticity of 1.27. Even less is known about cross-price elasticities. Dahl and Duggan (1998) surveyed oil and gas exploration models that include cross-price elasticities for oil and gas but did not find strong statistical results from any of the models.

DETERMINANTS OF ENERGY DEMAND

Energy demand is a derived demand. Consumers and businesses demand energy not for itself but for the services that the energy can provide. A consumer may want energy for lighting, space conditioning in the form of heat in the winters and cooling in the summer, and energy to run vehicles and appliances.

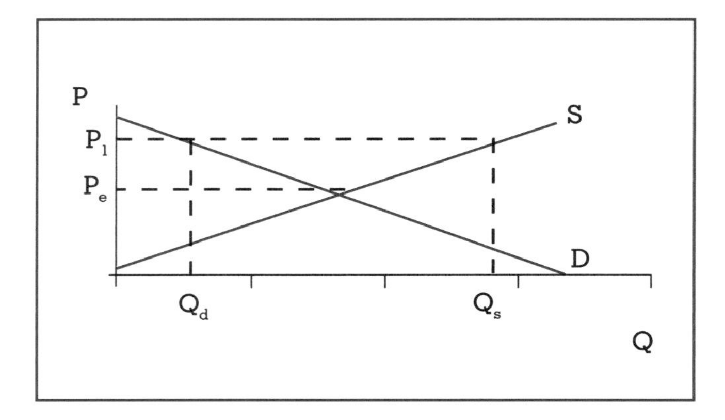

Figure 1.
Energy demand and supply.

Businesses often have these same needs and also need energy to run motors and for process heat.

For consumers, quantity demanded of energy (Q_{cd}) is a function of the price of energy (P), the price of other related goods, disposable income (Y), and other variables (O) such as personal preferences, lifestyle, weather, and demographic variables and, if it is aggregate demand, the number of consumers (#C). Take for example the quantity of electricity demanded by a household. If the price of electricity increases consumers may use less electricity. If the price of natural gas, a substitute for electricity in consumption (P_s), decreases, that may cause consumers to shift away from electric water heaters, clothes driers and furnaces to ones that use natural gas, thus increasing the quantity of natural gas demanded. If the price of electric appliances (P_c) increases, or decreases quantity of electricity demanded. consumers may buy less appliances and, hence, use less electricity. Increasing disposable income is likely to cause consumers to buy larger homes and more appliances increasing the quantity of electricity consumed. Interestingly, the effect of an increase in income does not have to be positive. For example, in the past as income increased, homes that heated with coal switched to cleaner fuels such as fuel oil or gas. In the developing world, kerosene is used for lighting, but as households become richer they switch to electricity. In these contexts coal and kerosene are inferior goods and their consumption decreases as income increases. We can write a general consumer energy demand function as follows:

$$Q_{cd} = f(-P + P_c - P_s \pm Y \pm O + \#C)$$

Again the signs before the variables indicate how the variables influence quantity. The sign before income (Y) is ±, since the sign would be + for a normal good and – for an inferior good. The sign before other variables is also ±, since the sign depends on what the other variable is. For example, colder weather would raise the demand for natural gas, but an aging population, which drives less than a younger one, would decrease the demand for gasoline.

For businesses the demand for energy is the demand for a factor of production. Its demand depends on the price of the energy demanded (P) as well as the price of its output (P_o), technology (T) and prices of other factors of production—land, labor, and capital—that might be substitutes (P_s) or complements (P_c) in consumption. Environmental policy (E_p) might also affect the demand for fuel. If this is aggregate business demand for energy the number of businesses is also relevant.

We can write a general business energy demand function as follows:

$$Q_d = f(+ P + P_c - P_s \pm T \pm E_p + \#B)$$

The sign on technology (T) and environmental policy (E_p) are uncertain and depend on the particular technology and policy. For example, environmental regulations requiring lower sulfur emissions favor gas over coal, while new technologies that make oil cheaper to use increase oil demand at the expense of gas and coal.

ELASTICITY OF DEMAND

The responsiveness of energy consumption to a variable can be represented by an elasticity, as was the case for energy production. Again the elasticity is measured as the percentage change in quantity over the percentage change in the variable. The demand elasticity is negative, since raising price lowers quantity demanded, and if it is less than –1, the demand is called elastic. Lowering price when demand is elastic means that quantity demanded increases by a larger percent than price falls, thus energy expenditures go up. Alternatively, raising price lowers expenditures. If the demand elasticity is between 0 and –1, then it is called inelastic. Now lowering price means that quantity increases by a smaller percent than price falls. Thus energy expenditures go down. Alternatively, raising price lowers expenditures.

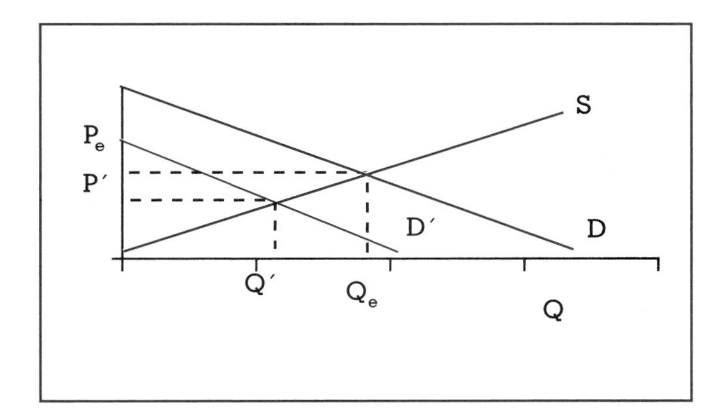

Figure 2.
Shift in demand in response to Asian crisis.

INCOME ELASTICITIES

Income elasticities are positive for normal and noninferior goods, since raising income increases total consumption of these goods. Goods with income elasticities greater than 1 are said to have income elastic demand. Suppose jet fuel has an income elasticity of 3. The in income goes up by 1 percent, the quantity of jet fuel demanded goes up by 3 percent and a larger share of income is spent on jet fuel. Such goods with income elastic demand are called luxuries and they become a larger share of spending as a person gets richer. If the income elasticity is between 0 and 1, it is called inelastic. As income increases, a smaller share of income is spent on goods with inelastic demand. Often necessities such are fuel for heating are income ineleastic.

A cross-price elasticity indicated how the quantity of one good changes when the price of another related good changes. The sign of the cross-price elasticity tells us whether goods are substitutes or complements. Take the case of a cement producer who needs a great deal of heat to produce cement. The producer's cross-price elasticity of demand for natural gas with respect to the price of the substitute good, coal, is the percentage change in natural gas demand divided by the percentage change in coal price. This elasticity for substitute goods will be positive, since an increase in the price of coal will cause an increase in the quantity of natural gas demanded. Alternatively, the cross-price elasticity of demand for complementary goods is negative. If the price of gas furnaces (a complement to gas) goes down, the cement producer may buy more gas

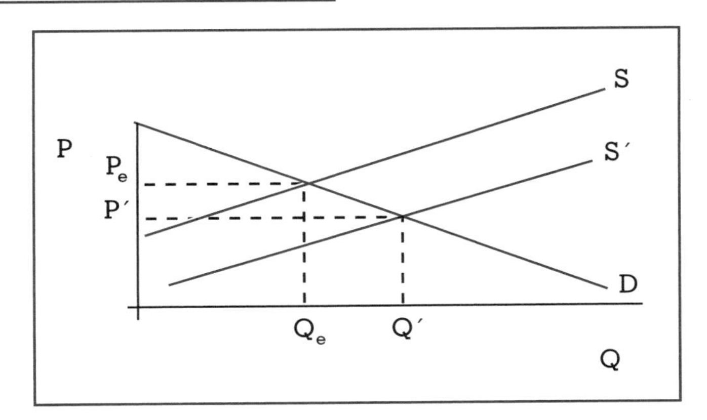

Figure 3.
Shift in supply in response to improved technology.

furnaces, instead of coal furnaces, and use more natural gas.

ESTIMATED ELASTICITIES

More statistical work has been done estimating demand elasticities. Dahl (1993a) surveyed this work for the United States. She found considerable variation in ownprice and income elasticities across studies with the most consistency for studies of residential demand and for gasoline. More products seem to be price and income inelastic and short run elasticities are more certain than long run elasticity. Short run price elasticities for a year are probably between 0 and –0.5 for energy products. Dahl (1995) surveyed transportation demand studies and concluded that the price elasticity of demand for gasoline in the United States is –0.6 and the income elasticity is just below 1. Dahl (1992, 1993b, 1994) looked at energy demand elasticities in developing countries and found that oil and price elasticities of demand are inelastic and near –0.3, while income elasticities of demand are elastic and near 1.3. Dahl (1995b) surveyed natural gas's own, cross and income elasticities in industrial countries but was not able to come to a conclusion about the magnitude of these elasticities. Studies are also often inconsistent about whether coal, oil, and electricity are substitutes or complements to natural gas.

ENERGY PRICES

Elasticities tell us how responsive are quantities demanded and supplied. In a competitive market where many buyers and sellers are competing with one another, the interactions of supply and demand determine the price in energy markets. To see how, we simplify demand and supply by holding all variables constant except price and quantity and graph both functions in Figure 1. Equilibrium in this market is at P_e where quantity demanded equals quantity supplied. If price is at P_1 quantity demanded is Q_d and quantity supplied is at Q_s. There is excess quantity supplied and there would be pressure on prices downward until quantity demanded equaled quantity supplied. Thus interactions of demand and supply determine price.

CHANGE IN DEMAND

If one of the variables held constant in the demand curve were to change, it would shift the whole demand curve, called a change in demand. For example, suppose Figure 2 represents the world market for crude oil. The Asian crisis beginning in 1997 reduced Asian income, which in turn reduced Asia's demand for oil. This decrease in demand lowered price moving along the supply curve, called a decrease in quantity supplied.

CHANGE IN SUPPLY

The 1990s have seen technical changes in finding oil such as 3D seismic and horizontal drilling that have reduced costs. These technical changes shifted the supply of oil as shown in Figure 3. This shift is called a change in supply and is the result a change in the a variable other than the own price. The increase in supply lowers price causing a movement along the demand curve called a change in quantity demanded.

MARKET SETS PRICES

Thus, as economic and political events occur along with changes in demography, preferences and technology, shifting demand and supply interact to form prices in competitive energy markets. The above discussion assumes competitive markets where consumers and producers compete to buy and to sell products, and they have to take market price as given. This is probably the case most often for buyers of energy products. However, in the case of production, sometime market power exists. For

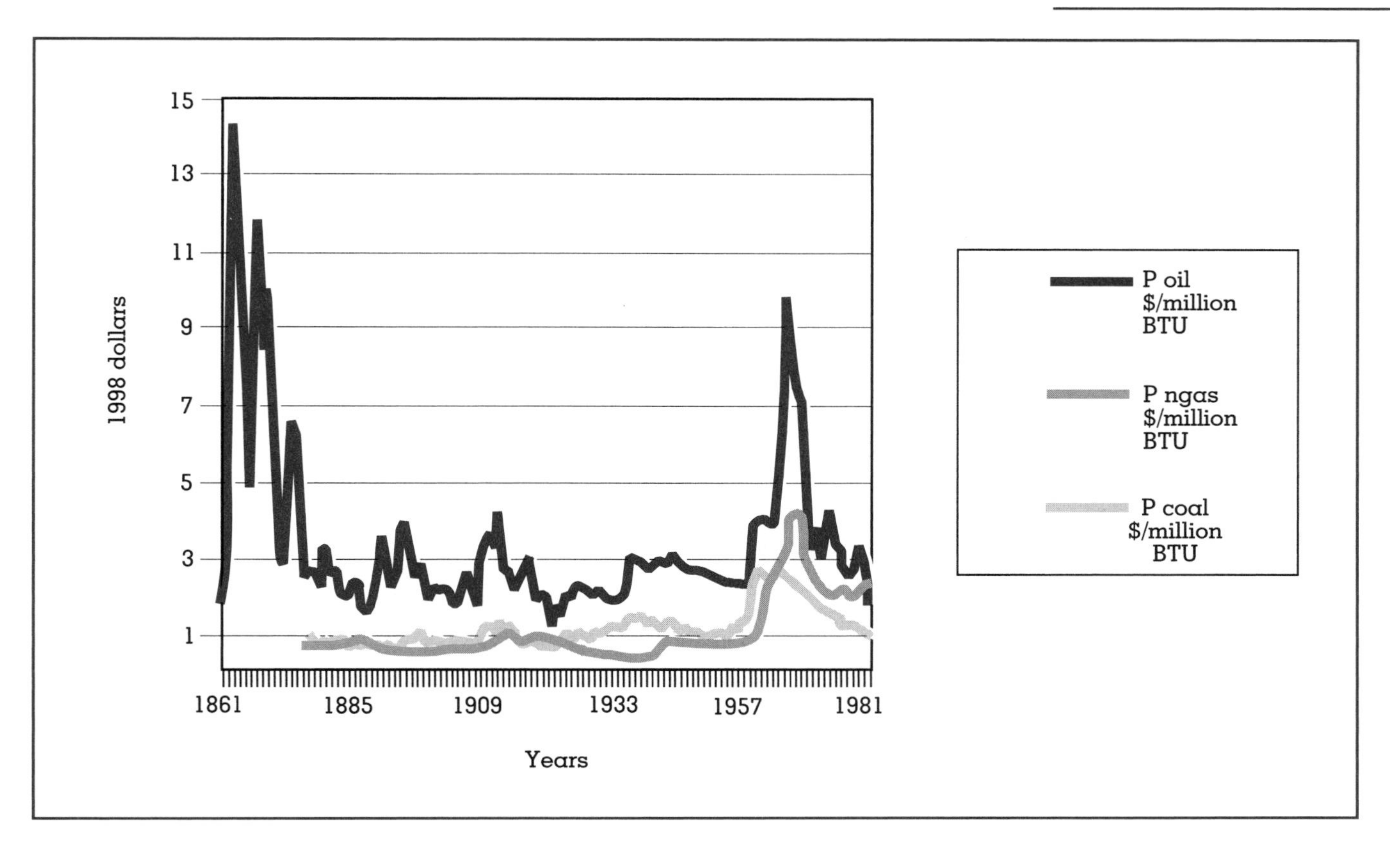

Figure 4.
U.S. prices for coal, oil, and natural gas, 1861–1981.

SOURCE: U.S. Department of Commerce, Bureau or the Census. (1975); U.S. Energy Information Administration, Department of Energy. (1998).

NOTE: Prices have been converted into Btus assuming 5,800,000 Btus per barrel of oil, 3,412,000 Btus per 1000 kWh, 1,000,000 Btus per 1000 mcf of natural gas, and 22,500,000 Btus per short ton of coal. A Btu is about ¼ of a food calorie or a kilocalorie, 1,000 Btus contain the energy content of a candy bar.

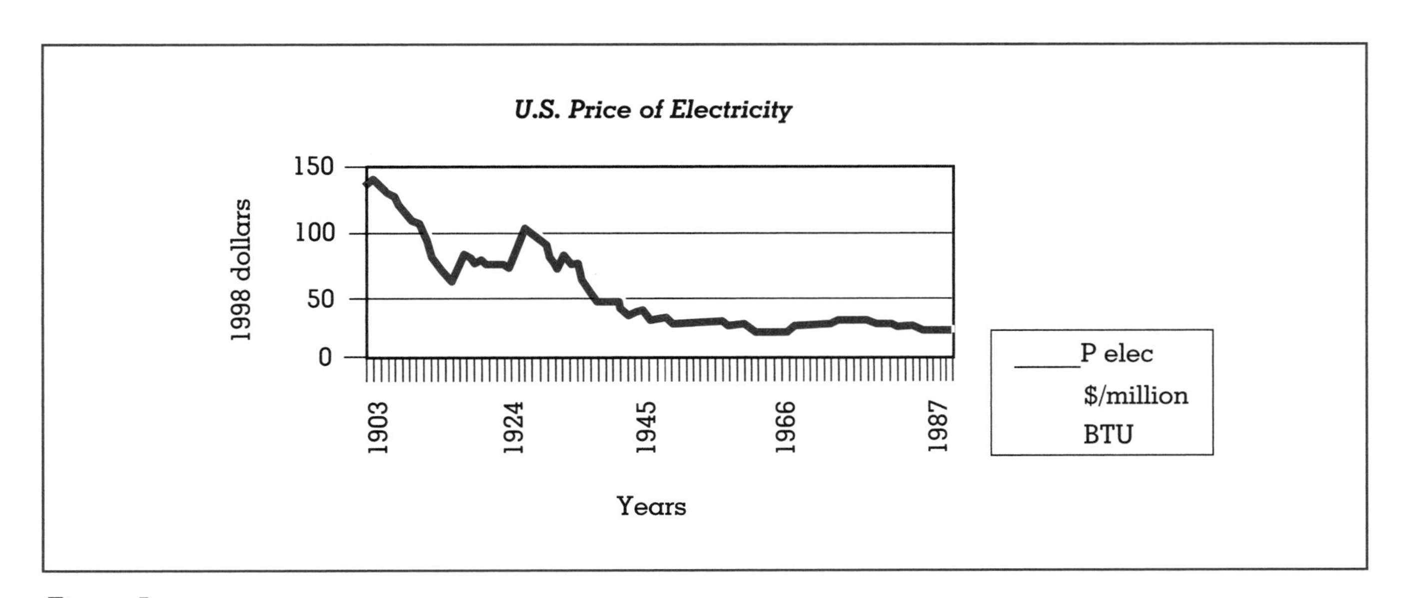

Figure 5.
U.S. Prices for Electricity, 1913–2000.

SOURCE: U.S. Department of Commerce, Bureau or the Census. (1975); U.S. Energy Information Administration, Department of Energy. (1998).

example, in the oil market, first Rockfeller, then the large multinational oil companies, and state regulatory commissions such as the Texas Railroad Commission and then OPEC have exercised pricing power. In such a case, the producer tries to influence price or quantity in order to receive higher prices and earn excess profits.

FOSSIL FUEL PRICES

Figure 4 gives an historical overview of how fossil fuel prices have changed in response to market changes. Price availability varies from series to series and is reported for the period 1861–1998 for the well head price of oil (Poil), 1880–1998 for the fob mine price of bituminous coal, and 1922-1998 for the well head price of natural gas. Figure 4 shows the volatility of oil prices as cartels formed and raised prices and then lost control as higher profits encouraged entry by other. Since the oil market is a global market these prices would be similar to those on the world market, as would prices for coal. The coal market has been reasonably competitive, and from the figure, we can see its price has been more stable than oil. However, coal price has been increased by oil price changes since oil is a substitute for coal in under-the-boiler and heating uses.

Sometimes governments interfere with markets by setting price controls. In the United States wellhead price controls were set on natural gas sold into interstate markets beginning in the early 1950s. These controls which were not completely removed until the early 1990s and sometimes caused natural gas shortages. Prices, which had been reasonably stable, became quite volatile as the price controls were increasingly relaxed throughout the 1980s and into the 1990s.

For electricity, economies of scale have existed making it more economical for one firm to produce and distribute electricity for a given market. One firm would be able to monopolize the market and earn excess profits. This has led governments to regulate electricity in the United States and to produce electricity in most of the rest of the world throughout much of the twentieth century. In such a case, electricity price is set not by the market but by the government. In Figure 5 we can see the evolution of constant dollar electricity prices in the United States from the early 1900s to the present, based on the average real consumer price of electricity (Plec). The falling real price reflects the cost reductions in producing and distributing electricity.

As the size of markets have increased and the optimal size of electricity generation units have decreased more electricity markets are being privatized and restructured to allow more competition into the markets and less government control over pricing.

Carol Dahl

See also: Subsidies and Energy Costs.

BIBLIOGRAPHY

Apostolakis, B. E. (1990). "Interfuel and Energy-Capital Complementarity in Manufacturing Industries." *Applied Energy* 35:83–107.

Dahl, C. A. (1992). "A Survey of Energy Demand Elasticities for the Developing World." *Journal of Energy and Development* 18(I):1–48.

Dahl, C. A. (1993a). "A Survey of Energy Demand Elasticities in Support of the Development of the NEMS" Prepared for the United States Department of Energy, Contract De-AP01-93EI23499 October 19.

Dahl, C. A. (1993b). "A Survey of Oil Demand Elasticities for Developing Countries." *OPEC Review*. 17(4):399–419.

Dahl, C. A. (1994). "A Survey of Oil Product Demand Elasticities for Developing Countries," *OPEC Review* 18(1):47-87.

Dahl, C. A. (1995a). "Demand for Transportation Fuels: A Survey of Demand Elasticities and Their Components" *Journal of Energy Literature* 1(2):3-27.

Dahl, C. A. (1995b). "A Survey of Natural Gas Demand Elasticities: Implications for Natural Gas Substitution in New Zealand." Draft. CSM/IFP Petroleum Economics and Management. Golden, CO: Colorado School of Mines, Division of Economics and Business.

Dahl, C. A., and Duggan, T. (1996). "U. S. Energy Product Supply Elasticities: A Survey and Application to the U.S. Oil Market," *Resources and Energy Economics* 18:243–263.

Dahl, C. A., and Duggan, T. (1998). "Survey of Price Elasticities from Economic Exploration Models of U.S. Oil and Gas Supply," *Journal of Energy Finance and Development* 3(2):129–169.

U.S. Department of Commerce, Bureau or the Census. (1975). *Historical Statistics of the United States: Colonial Times to 1970*. Washington, DC: U.S. Government Printing Office.

U.S. Energy Information Administration/Department of Energy. (1998). *Annual Energy Review*. Washington, DC: U.S. Government Printing Office.

Welsh, H. E., and Gormley, J. C. (1995). "Sources of Information on the Oil Industry." In *The New Global Oil Market: Understanding Energy Issues in the World Economy*, ed. S. Shojai. Westport, CT: Praeger.

SUSTAINABLE DEVELOPMENT AND ENERGY

Sustainable development is defined as "development that meets the needs of the present without compromising the ability of future generations to meet their own needs" (Brundtland, 1987). Applied to energy, it is the ability of a society to continue present energy production and consumption patterns many generations into the future, with a focus on the relationship of available energy resource to the rate of resource exhaustion. A sustainable energy market requires the quantity, quality, and utility of energy to improve over time—energy becomes more available, more affordable, more usable and reliable, and cleaner over time (Bradley and Simon, 2000).

Development and population growth adversely impact the sustainability future by accelerating energy resource exhaustion and environmental degradation. For example, the greater the population growth, the greater the desire for development by the growing population, the greater the desire for short-term exploitation of energy resources for the development (less concern for future generations), and the greater the environmental degradation from the greater development and energy use.

All fossil fuels are considered unsustainable because someday they will reach a point of depletion when it becomes uneconomic to produce. Petroleum is the least sustainable because it is the most finite fossil fuel. Although levels of production are expected to begin declining no later than 2030 (U.S. production peaked in 1970), the U.S. and world reserves could be further expanded by technological advances that continue to improve discovery rates and individual well productivity. The extraction of oils found in shales (exceeds three trillion barrels of oil equivalent worldwide) and sands (reserves of at least two trillion barrels worldwide) could also significantly increase reserves. The reserves of natural gas are comparable to that of oil, but natural gas is considered a more sustainable resource since consumption rates are lower and it burns cleaner than petroleum products (more environmentally sustainable).

Coal is the least expensive and most sustainable fossil fuel energy source (world reserves of 1.2 to 1.8 trillion short tons, or five times the world's oil reserves), but many feel it is the least sustainable from an environmental perspective. Coal combustion emits far more of the pollutants sulfur dioxide, carbon monoxide, nitrogen oxides and particulates, and is the fossil fuel that emits the greatest amount of carbon dioxide—the greenhouse gas suspected to be responsible for global warming. However, advances in technology have made the conversion of all fossil fuels to energy more efficient and environmentally sustainable: fossil-fuel availability has been increasing even though consumption continues to increase; efficiency improvements have saved trillions of dollars, and vehicle and power plant emissions in 2000 were only a fraction of what they were in 1970.

Nuclear energy is more sustainable than the fossil fuels. If uranium is used in fast breeder reactors designed to produce large quantities of plutonium fuel as they produce electricity, the world resources could produce approximately 200 times the total global energy used in 1997. However, concerns about the danger of nuclear power, the high cost of building and maintaining plants, and the environmental dilemma of what to do with the nuclear wastes that result is the reason many consider it less sustainable.

Renewable energy is the most sustainable energy because the sources, such as the sun, wind and water are inexhaustible and their environmental impact minimal. But the problem with renewable resources is that most sources produce intermittently (when the sun is shining, wind is blowing). Except for hydroelectricity, which produces about 18 percent of the world's electricity, all the other renewable energy sources combined produce less than 2 percent. Advances in renewable energy technology, coupled with continued efficiency improvements in energy-using products, may be part of the energy sustainability solution, but the demand for energy from the world's ever-growing population is too great for it to be the only solution.

Over 80 percent of the world's energy consumption comes from nonrenewable sources that cannot be sustained indefinitely under current practices. If technological advances continue to make conventional energy resources plentiful and affordable for many years to come, the transition to more sustainable energy sources can be smooth and minimally disruptive.

John Zumerchik

See also: Efficiency of Energy Use; Emission Control, Power Plants; Emission Control, Vehicle; Oil and Gas, Exploration for; Oil and Gas, Production of.

BIBLIOGRAPHY

Bartlett, A. A. (1997–1998). "Reflections on Sustainability, Population Growth, and the Environment." *Renewable Resources Journal* 15(4):6–23.

Bartlett, A. A. (2000). "An analysis of U.S. and World Oil Production Patterns Using Hubbert-Style Curves." *Mathematical Geology* 32(1):1–17.

Bradley, R., and Simon, J. (2000). *The Triumph of Energy Sustainability.* Washington, DC: American Legislative Exchange Council.

Brundtland, G. H. (1987). *Our Common Future, World Commission on Environment and Development.* New York: Oxford University Press.

SYNTHETIC FUELS

Synthetic fuels are usually thought of as liquid fuel substitutes for gasoline and diesel fuel made from petroleum sources. In broad context, the source of these synthetics can be any feedstock containing the combustible elements carbon or hydrogen. These include coal, oil shale, peat, biomass, tar sands, and natural gas. Water could be included here because it can be decomposed to produce hydrogen gas, which has the highest heating value per unit weight of any material (excluding nuclear reactions). The conversion of natural gas is treated in a separate article, so this article will emphasize solid feedstocks.

From a practical standpoint, coal, because of its abundance, has received the most attention as a source for synthetic fuels. As early as 1807, a coal-gas system was used to light the streets of London, and until the 1930s, when less expensive and safer natural gas started to flow through newly constructed pipelines, gas piped to homes in the Eastern United States was derived from coal. Kerosene, originally a byproduct from the coking of coal for metallurgical applications, can be considered the first synthetic liquid fuel made in quantity. But once crude oil became cheap and abundant, there was little serious research on synthetic liquid fuels in the industrial world until the Energy Crisis of 1973. The main exceptions to this generalization are the important work on coal conversion in Germany, cut off from oil imports during the two World Wars, and the Sasol Process in South Africa, which produces a synthetic, waxy "crude oil" from indigenous coal deposits.

The Sasol Synfuel refinery in Secunda, South Africa, produces synthetic fuels for consumers. (Corbis Corporation)

After 1973 the United States invested heavily in synthetic fuel research and development, hoping synthetics could serve as economical substitutes for crude oil. However, coal conversion is not profitable unless the price of crude oil is over $50 per barrel, which is why the processes developed were mothballed when world crude oil prices fell in the 1980s.

CONVERSION

The major chemical difference between natural gas, crude oil, and coal is their hydrogen-to-carbon ratios. Coal is carbon-rich and hydrogen-poor, so to produce a synthetic liquid or gas from coal requires an increase in the hydrogen-to-carbon ratio. Coal's ratio of about 0.8 has to be raised to 1.4 to 1.8 for a liquid, and to over 3 to produce a synthetic gaseous fuel. Natural gas (chiefly methane) has a ratio of 4. This can be done by either adding hydrogen or rejecting carbon.

Addition of hydrogen can involve reacting pulverized coal with hydrogen-rich liquids. More commonly, pressurized hydrogen gas is reacted with coal (hydrogenation) in the presence of a catalyst. The latter scheme also removes many of the noxious sulfurous and nitrogenous impurities in coal by converting them to gaseous hydrogen sulfide and ammonia. Carbon removal entails pyrolysis in the absence of air (coking) to produce varying amounts of gases, liquids, and char, depending on the reaction time-temperature-pressure conditions employed.

Synthetic Fuel Liquids via Gas Intermediates

Liquids can also be synthesized via an indirect scheme where the coal is first gasified in an intermediate step. The coal is pulverized and reacted with steam to produce "water gas," an equimolar mixture of carbon monoxide and hydrogen:

$$C + H_2O \rightarrow CO + H_2 \quad (1)$$

The carbon monoxide can then be further reacted with steam and/or hydrogen in the water gas shift reaction:

$$CO + H_2O \rightarrow CO_2 + H_2 \quad (2)$$

$$CO + 3H_2 \dashrightarrow CH_4 + H_2O \quad (3)$$

Combining Equation 1 and Equation 2, one can generate hydrogen and carbon dioxide:

$$C + 2H_2O \rightarrow CO_2 + 2H_2 \quad (4)$$

And adding reactions from Equation 1 and Equation 3 yields just methane:

$$C + 2H_2 \rightarrow CH_4 \quad (5)$$

These operations carry energy penalties, and the heat of combustion released when burning the methane, hydrogen, or carbon monoxide produced is less than the energy that would have been released had the coal been burned directly. To produce heavier liquids, the equimolar mixture of hydrogen and carbon monoxide (water gas, also known as synthesis gas) is the preferred feedstock. Badische Anilin pioneered the synthesis of methanol at high temperature and pressure after World War I:

$$CO + 2H_2 \rightarrow CH_3OH \quad (6)$$

Shortly thereafter, Fischer and Tropsch discovered an iron catalyst that would convert synthesis gas to a mixture of oxygenated hydrocarbons (alcohols, acids, aldehydes, and ketones) at atmospheric pressure. In the next decade Ruhrchemie developed new cobalt catalysts that could produce a mixture of hydrocarbon liquids and paraffin wax from Fischer Tropsch liquids at moderate pressure. This, plus direct hydrogenation of coal, was the basis of German synfuel capacity during World War II. Finally, in 1955, the South African Coal, Oil and Gas Co. further improved the technology and commercialized the Sasol Process as the basis for South Africa's fuel and chemicals industry. Favorable economics were possible because government mandates and subsidies dictated the use of local coal resources rather than spending scarce hard currency to import foreign crude oil.

Methanol is an excellent, high-octane motor fuel, but it was not cost-competitive with gasoline made from cheap, abundant, crude oil. It also has a lower specific energy (miles per gallon) because it is already partially oxidized. In the late 1970s, Mobil Oil announced its "Methanol to Gasoline" Process, which efficiently converts methanol to C2 to C10 hydrocarbons via a synthetic zsm-5 zeolite catalyst. This coal-methanol-gasoline technology will probably enjoy use as crude oil prices rise in the future. Since natural gas (largely methane or CH_4) is readily converted into methanol via partial oxidation with pure oxygen, this also offers an alternate to low-temperature liquefaction or pipelining as a way to utilize natural gas deposits that are in remote locations.

There is a parallel technology for the partial oxidation of methane to make ethylene and other olefins. These can then be polymerized, alkylated, and hydrotreated as need be to make hydrocarbon fuels in the gasoline and diesel fuel boiling range. Methane can also be partially oxidized to produce oxygenates, such as methyl tert-butyl ether (MBTE), that are used in reformulated gasoline as blending agents.

Direct Liquefaction of Coal

In comparison to the capital-intensive, multistep gaseous route from coal to gasoline, the direct liquefaction of coal in a single processing step is attractive. While the removal of carbon from coal to generate liquids richer in hydrogen sounds simple, there is not enough hydrogen in coal to yield much useable liquid fuel. Further, the process must have a ready market, such as a nearby smelter, to utilize the major coke fraction. Addition of hydrogen during pyrolysis (hydropyrolysis) increases liquid yields somewhat, but the direct hydrogenation of coal has always seemed a more attractive route. Nevertheless, the Office of Coal Research of the U.S. Deptartment of the Interior sponsored research on this scheme for many years. As a result, the Char-Oil-Energy Development (COED) process was developed by the FMC Corporation in the period 1965–1975. In the COED process, 50 to 60 percent of the coal feed is rejected as char, and finding a market for char is problematic. Working on subbituminous western coals, the Oil Shale Corporation developed a similar scheme (Toscoal Process), which also produced 50 percent solid char.

High-pressure hydrogenation of coal was patented by Bergius in 1918, but the liquids were of low quality. However, by the mid 1930s, plants using on the order of 500 tons per day of low-rank coals and coal pitch had been built in England and Germany. Toward the end of World War II, Germany had 12 large "coal refineries" producing over 100,000 barrels a day of motor and aviation fuels for the war effort. After the war, in spite of the abundance of crude oil in the United States, pilot plant work on coal hydrogenation was continued in by the U.S. Bureau of Mines. This work was further developed by Ashland Oil into the "H-coal" process, and a pilot plant was built in 1980. The hydrocarbon liquids produced were rather high in nitrogen and oxygen, but all heavy materials were recycled to extinction so that no pitch or char was produced.

In the 1920s, I. G. Farben discovered that certain hydrogen-rich solvents, such as tetralin, could dissolve heavy coal components. The heavy extracted liquids could more conveniently be upgraded by hydrogenation, and the extracted residue was converted to metallurgical coke and carbon electrodes. After the war a low-level research effort was undertaken in the United States on solvent-refined coal processes, with primary emphasis on producing boiler fuel. Yields of useful liquids were gradually improved by inserting additional processing steps such as hydrocracking some of the heavier fractions.

Spurred by the OPEC oil crisis, the U.S. Department of Energy encouraged major oil companies to increase efforts to produce liquid fuels from coal, tar sands and shale oil deposits. A key development was the Exxon Donor Solvent Coal Liquefaction Process (EDS). It involves reacting a coal slurry with hydrogenated recycle solvent and hydrogen. The donor solvent transfers some of its hydrogen to the coal, is distilled from the reaction products, hydrogenated, and then recycled to the process. The light ends from distillation are steam-reformed to make process hydrogen. The heavy vacuum bottoms are Flexicoked, and the coke is gasified to provide process fuel gas. Except for some residual carbon in the gasifier ash, very little of value in the coal is wasted.

Work has also continued on the solvent-refined coal + hydrocracking concept (the NTSL, or non-integrated, two-stage liquefaction process), and a pilot plant was operated by Amoco, DOE and the Electric Power Research Institute (EPRI) from 1974 to 1992.

In addition to coal, there have been extensive post-oil-crisis studies aimed at utilizing the extensive western oil shale deposits. Oil shale utilization has many problems common to coal conversion and, in addition, extensive inorganic residues must be disposed of. The products are also high in nitrogen, and this increases refining costs. Tar sands constitute a final hydrocarbon reserve of interest. Here the problems are more tractable. The sands can be extracted with hot water to produce a material similar to a very viscous crude oil. It is refined as such, and thus Great Canadian Oil Sands, Ltd. has been able to justify operation of several very large tar sands refineries at Cold Lake, Alberta. The major process involves fluid coking of the heavy bottoms from distillation.

ECONOMIC AND ENVIRONMENTAL OUTLOOK

The coal conversion efficiency to synthetic, pipeline-quality natural gas or liquid crude oil is in the 60 to 70 percent range. This means that only 60 to 70 percent of the latent heat energy in the coal can be obtained by burning the product of the conversion. However, for the lower Btu per cubic foot products of water gas and coke oven gas, conversion efficiencies can reach over 95 percent.

The reason for the poor conversion efficiency to synthetic fuels is the high energy cost in liberating hydrogen from water (thermal dissociation, electrolysis) and, when distillation does not involve water, the partial combustion needed to produce the gas (CO). The price of crude oil would have to rise to around $40 to $50 a barrel, or need government subsidies of $20 to $30 a barrel, for liquid synthetic fuel to be competitive. Because the conversion to gaseous fuels is less complex and costly than liquid fuels, the subsidy or rise in natural gas prices would not have to be as dramatic.

Besides the economic feasibility problem, synthetic fuels face significant environmental hurdles. During direct liquefaction, heavy, high-boiling polyaromatics organics are produced. Scientists are trying to eliminate these carcinogenic fractions by recycling them through the liquefaction process. Production by the indirect method is less problematic because it tends to produce fewer toxic chain hydrocarbons. There is also a significant release of solid, liquid, and gaseous residual waste that comes from the boilers, heaters and incinerators, or as part of the processing stream.

The process will adversely affect air quality by releasing nitrogen oxides, sulfur oxides, carbon monoxides and other particulates into the atmosphere. Better control of the conversion conditions and better control of emissions can make the process cleaner, yet technology cannot do anything to curb carbon emissions. Since much of the carbon in coal is converted to carbon dioxide in the synthesis process, and is not part of the synthetic fuel itself, the amount of carbon dioxide that will be released to the environment during combustion is 50 to 100 percent more than coal, and around three times more than natural gas.

Since most systems use tremendous amounts of water, the production of synthetic fuels will have a detrimental effect on water quality as well. It will require major technological advances to more effectively handle waste streams—waste-water treatment systems, sulfur recover systems and cooling towers—to make synthetic fuels an acceptable option from an environmental perspective. This emission control technology will be expensive, only adding to the economic disadvantages the synthetic fuel market already faces.

As crude oil reserves dwindle, the marketplace will either transition to the electrifying of the transportation system (electric and fuel-cell vehicles and electric railways), with the electricity being produced by coal, natural gas, nuclear and renewables, or see the development of an industry to produce liquid fuel substitutes from coal, oil shale, and tar sands. It might also turn out to be a combination of both. The transition will vary by nation and will be dictated strongly by the fuels available, the economic and technological efficiencies of competitive systems, the relative environmental impacts of each technology, and the role government takes in the marketplace.

John Zumerchik
Herman Bieber

See also: Hydrogen; Natural Gas, Processing and Conversion of.

REFERENCES

American Petroleum Institute. (1987). *Liquid Fuels from Natural Gas.* API 34–5250.

Burgt, M. J. v. d., et al. (1988). *Methane Conversion.* New York: Elsevier Science, Inc.

Crow, M. (1988). *Synthetic Fuel Technology Development in the United States.* New York: Praeger Publishing.

Haar, L., et al. (1981). *Advances in Coal Utilization.* Chicago: Institute of Gas Technology.

Hinman, G. W. (1991). "Unconventional Petroleum Resources." In *The Energy Sourcebook*, ed. R. Howes and A. Fainberg. New York: American Institute of Physics.

Lee, S. (1996). *Applied Technology Series: Alternative Fuels.* New York: Taylor & Francis.

Storch, H. H., et al. (1951). *The Fischer Tropsch and Related Synthesis.* New York: John Wiley & Sons.

Supp, E. (1990). *How to Produce Methanol from Coal.* New York: Springer Verlag.

T

TAXATION OF ENERGY

In the United States, the federal government does not impose an energy tax or a general sales tax that is broadly applicable to energy. However, excise taxes are imposed on certain fuels, and there are a number of income tax provisions specific to the energy sector. There are three separate categories of taxes and fees that affect energy use: (1) excise taxes/fees that primarily affect energy demand; (2) income tax provisions that primarily affect energy supply by operating on the after-tax rate of return on investment; and (3) income tax provisions that primarily affect the demand for specific energy sources.

EXCISE TAXES

Excise taxes placed on specific energy sources tend to reduce the demand for these energy sources in both the short and the long run. The federal government imposes excise taxes on almost all petroleum products and coal (see Table 1). The federal government also imposes excise taxes on many transportation uses of methanol, ethanol, natural gas, and propane and imposes a fee on electricity produced from nuclear power plants.

By far the most substantial energy excise taxes imposed by the federal government are those imposed on motor fuels. Gasoline generally is taxed at a rate of 18.4 cents per gallon, and diesel fuel is taxed at a rate of 24.4 cents per gallon. Commercial aviation fuels are taxed at 4.4 cents per gallon, noncommercial aviation gasoline is taxed at 19.4 cents per gallon, and noncommercial aviation jet fuel is taxed at 21.9 cents per gallon. Diesel fuel used by commercial cargo vessels on inland or intracoastal waterways is taxed at 24.4 cents per gallon. Certain users of motor fuels are exempt from the excise tax or pay a reduced rate. For example, state and local governments and motor fuels for off-road use (such as by farmers) are exempt from gasoline and diesel taxes. Reduced rates of tax are paid on gasoline blended with ethanol (gasohol) and by providers of intercity bus service.

Receipts from these taxes are allocated to various trust funds, the largest recipients being the Highway Trust Fund (gasoline and diesel fuel taxes) and the Airport and Airway Trust Fund (aviation fuels taxes). Congress appropriates monies from these funds for surface and air transportation capital projects and operating subsidies. Each of the above tax rates includes 0.1 cent per gallon dedicated to the Leaking Underground Storage Tank Trust Fund. Monies from this fund are appropriated by Congress for remediation of environmental damage at underground fuel storage facilities.

Coal from underground mines is assessed an excise tax of $1.10 per ton, and coal from surface mines is assessed a tax of $0.55 per ton, but in each case, a constraint is added that states that the tax cannot exceed 4.4 percent of the sales price. The receipts from these taxes are allocated to the Black Lung Disability Trust Fund. Monies from this trust fund are used to pay for health benefits to coal miners.

Producers of electricity from nuclear power plants are assessed a fee of 0.1 cent per kilowatt-hour to pay for future storage of spent nuclear fuel at a federal facility. Receipts from this fee are allocated to the Nuclear Waste Trust Fund and are appropriated by Congress to cover the costs of developing and constructing a permanent storage facility.

Each of these excise taxes raises the price of the fuel for uses subject to the tax and might be expected to reduce the demand for petroleum and coal. However, to some extent the excise tax receipts col-

Item	$ Billions
Highway Trust Fund Fuel Excise Taxes	
Gasoline	21.7
Diesel fuel	7.7
Other highway fuels	1.4
Total	30.8
Noncommercial aviation fuels excise tax	0.2
Commercial aviation fuels excise tax	0.6
Inland waterways trust fund excise tax	0.1
Leaking underground storage tank trust fund excise tax	0.2
Motororboat gasoline and special fuels excise taxes	0.2
Black lung trust fund coal excise tax	0.6

Table 1
Estimated 1999 Federal Energy-Related Excise Tax Receipts
SOURCE: Joint Committee on Taxation, U.S. Congress, 1998.

lected are dedicated to trust funds to finance expenditures that directly benefit those paying the tax. To the extent that these taxes are benefit taxes, they may have no independent effect on the demand for the taxed fuel because those paying the taxes are receiving commensurate services. For example, the receipts from aviation fuels taxes generally finance airport expansion and safety projects, providing a wide set of benefits to the flying public. But the tax and the benefit of the tax are not always strongly coincident. For example, in the case of motor fuels taxes, a portion of the Highway Trust Fund is used to support mass transit projects, which are not likely to directly benefit highway users except to the extent that the projects relieve traffic congestion. In the case of electricity, the effect of the fee on overall demand is not clear, because the electricity generation industry is transitioning from a regulated to a more competitive posture and because about 20 percent of electricity is generated from nuclear power plants. Given that the proceeds of this fee are to be spent on disposal of nuclear waste, a cost of production, the fee may make the cost of electricity production more fully reflect social costs. In the case of the Black Lung Disability Trust Fund, the supported benefits are for workers, not users, and hence the tax on coal is expected to operate solely to reduce demand. This tax also is related to the cost of production such that it makes the cost of coal reflect better the social costs.

In a different context, purchasers of automobiles that do not meet specific fuel economy standards must pay a "gas guzzler" excise tax. This tax presumably decreases the demand for such vehicles and should indirectly affect demand for gasoline.

INCOME TAXATION AND ENERGY SUPPLY

Various provisions in the federal income tax treat energy producers more or less favorably than other businesses. By changing the after-tax rate of return on investments in the energy sector, the Tax Code may alter the long-run supply of specific types of energy.

In general, the income of all participants in the energy sector is subject to income tax of one form or another. Two notable exceptions arise in the generation and sale of electricity. Governmental agencies (such as the Tennessee Valley Authority, the Bonneville Power Administration, and municipally owned power companies) account for approximately 14 percent of the electricity sold in the United States. Such public power producers and suppliers are not subject to federal income tax. Electricity providers organized as cooperatives are common in rural areas, accounting for about 8 percent of electricity sales. Income from sale of electricity is not subject to income tax if the income is paid to co-op members as a patronage refund.

To compute taxable income, a taxpayer is permitted to subtract from revenues those expenses that are necessary to create current sales. Expenses related to future sales are required to be capitalized and recovered over time. Distinctions between these two categories—current and suture sales—are sometimes hard to make. In some cases the Tax Code permits energy producers to claim currently, or on an accelerated basis, expenditures that are related to future production and sales. This lowers income for current tax purposes, thereby increasing the after-tax return to investments in the energy sector compared to other investments.

The Tax Code, for example, provides special rules for the treatment of intangible drilling costs, or IDCs. These include expenditures made for wages, fuel, materials, and the like necessary for drilling wells and preparing wells for the production of oil and natural gas. IDCs also may include expenditures for the construction of derricks and tanks, for gathering pipelines, and for roads. Certain taxpayers (generally those without major refining operations) may elect to expense, rather than capitalize, IDCs incurred for domestic properties. Such acceleration of cost recovery may provide preferential treatment for investment in oil and gas production.

The Tax Code also provides special tax advantages through "percentage depletion" allowances. To determine taxable income, the Tax Code permits the owner of a mineral reserve a deduction in recognition of the fact that the mineral reserve is depleted as the mineral is extracted. Some owners of oil and natural gas properties are permitted to compute this deduction as a percentage of gross income from the property (percentage depletion is not available to owners of producing properties who also have substantial refining capacity). Generally, eligible taxpayers may deduct annually 15 percent of gross income, up to 100 percent of the net income from the property (reaching this limit means that income from the property is effectively tax-exempt). Because percentage depletion is computed without regard to the taxpayer's basis in the depletable property, the cumulative depletion deductions may exceed the amount expended by the taxpayer to develop or acquire the property. In such circumstances, an investment in elgible oil and gas properties is tax-favored compared to other investments.

There also are circumstances where income tax provisions reduce the after-tax rate of return on such investments below those available in other sectors. For example, certain types of electricity-generating property (such as those used in distributed power applications) may have useful lives that are substantially shorter than the tax lives permitted for cost recovery. In such instances, the after-tax rate of return will be negatively affected by these specific provisions.

The Tax Code allows credits against income tax liability for various types of specific energy production, including electricity produced from wind or "closed loop" biomass (essentially agricultural operations devoted to electricity generation) and oil and natural gas from "nonconventional" sources (such as natural gas from coal seams, and oil from shale). Credits also are allowed for investments in specific forms of energy production, including facilities to produce geothermal or solar power, and expenditures associated with enhanced oil recovery technologies. Credits against income tax liability directly reduce taxes paid and therefore reduce the cost of production and increase the return on investment. Accordingly, these credits would be expected to increase production of energy from the specified sources.

Item	$ Billions
Expensing of exploration and development costs	2.5
Excess of percentage depletion of fuels over cost depletion	5.0
Tax credit for production of non-conventional fuels	7.1
Tax credit for enhanced oil recovery costs	0.3
Tax credit for electricity production from wind and closed-loop biomass	0.4
Tax credit for investments in solar and geothermal energy	0.3
Exclusion of energy conservation subsidies	0.2

Table 2
Estimates of Selected Energy-Related Federal Tax Expenditures, FY 1999–2003
SOURCE: Joint Committee on Taxation, U.S. Congress, 1999.

In addition to specific deductions and credits, the Tax Code permits state and local governments to issue bonds on which the interest is exempt from federal income tax. This provision means that states and local governments can borrow at interest rates below those paid by private corporations. Municipally owned electricity providers often can issue tax-exempt debt; the lower interest rate may have the effect of increasing the provision of electricity by these entities.

To provide a sense of the value of some of these deviations from neutral taxation, Table 2 reports cumulative estimates over five years of selected energy-related "tax expenditures." The Congressional Budget and Impoundment Control Act defines a "tax expenditure" as a revenue loss attributable to a special inclusion, exemption, or deduction from gross income or a special credit, preferential tax rate, or deferral of tax.

INCOME TAXATION AND ENERGY DEMAND

The Tax Code contains various provisions that may affect the demand for energy from specific sources. In the transportation area, purchasers of electric vehicles may claim a credit against their income tax liability for a portion of the purchase price. Taxpayers who purchase or retrofit a vehicle that runs on alternative fuels (such as natural gas, propane, methanol, or ethanol) may claim an accelerated deduction for their expenditure. Taxpayers who commute to work by automobile and whose employer provides parking may have a portion of the value of their parking space included in taxable income (if the value of the parking space exceeds a specified amount). Alternatively, employers may provide their employees with mass transit passes and not include the full value of the benefit in the employee's taxable income. Each of these provisions is expected to reduce the demand for gasoline to some extent.

Taxpayers who receive incentive payments for installing energy-efficient equipment, such as a high-efficiency furnace, do not have to include the payment in their taxable income. This tax-favored treatment may make consumers more likely to purchase energy-saving equipment, thereby reducing demand for energy.

As mentioned above, state and local governments can borrow at relatively low interest rates by issuing tax-exempt debt, and this finance technique is used to a great extent in highway and road construction. By reducing the cost of road construction, tax-exempt debt may increase the amount of construction undertaken, thereby increasing the demand for fuels.

SUMMARY

Tax policy (including excise taxes and income tax provisions) operates in several different ways to affect energy supply and demand, for energy in general and for specific fuels. There is no consensus about the net overall effect on supply and demand, in part because these provisions operate in different sectors, in different markets, and sometimes in different directions.

Thomas A Barthold
Mark J. Mazur

See also: Capital Investment Decisions; Energy Economics; Environmental Economics; Government Agencies; Government Intervention in Energy Markets; Industry and Business, Operating Decisions and; Oil and Gas, Drilling for; Regulation and Rates for Elec-tricity; Subsidies and Energy Costs.

BIBLIOGRAPHY

Committee on the Budget, U.S. Senate. (1998). *Tax Expenditures: Compendium of Background Material on Individual Provisions*, S. Prt. 105-70. Washington, DC: U.S. Government Printing Office.

Joint Committee on Taxation, U.S. Congress. (1998). *Estimates of Federal Tax Expenditures for Fiscal Years 1999–2003*. Washington, DC: U.S. Government Printing Office.

Joint Committee on Taxation, U.S. Congress. (1999). *Schedule of Present Federal Excise Taxes (as of January 1, 1999)*. Washington, DC: U.S. Government Printing Office.

U.S. Office of Management and Budget. (1999). *Budget of the United States Government, Fiscal Year 2000*, Washington, DC: U.S. Government Printing Office.

TEMPERATURE REGULATION

See: Energy Management Control Systems

TENNESSEE VALLEY AUTHORITY

See: Government Agencies

TESLA, NIKOLA (1856–1943)

Nikola Tesla was born in 1856 in Smilja, Croatia, to parents of Serbian heritage. The region of his birth was at the time part of the Austro-Hungarian Empire. His scientific and engineering aptitude were obvious at an early age, evidenced by stories of inventions consisting of simple mechanical devices he worked on during his childhood. For this reason his parents were persuaded to allow him to pursue an education in engineering rather than becoming a Serbian Orthodox priest as they had hoped. While Tesla studied at Graz Polytechnic and at the University of Prague, he developed an interest in electrical power transmission.

Moving into the commercial world, he accepted a position with a new telephone company in Budapest, where he developed a telephone repeater system, which formed the basis for the modern loudspeaker. Tesla was promoted within the company to a position in Paris. His potential was now clear to the manager of the plant, who recommended that he move to the United States to work with Thomas Edison, then considered the premier engineering genius of the age.

Tesla arrived in New York in 1884 and was hired by Edison. Edison understood Tesla's ability but remained unconvinced by his new employee's insistence on the use of alternating current for electrical power transmission. Nevertheless, Tesla accepted his assignment to work on improving Edison's direct-current method, which was then in use. Working long hours Tesla increased the system output and asked Edison for a $50,000 bonus, which Tesla understood he was to receive. Edison refused, claiming that the bonus had only been a joke. Tesla quit Edison's employ, and thereafter relations were strained between the two men.

Tesla was awarded patents on the alternating-current motor. In the single-phase AC motor, two circuits passing current are set up diagonally opposite with respect to a circular armature. With the currents ninety degrees out of phase, the armature rotates. In the polyphase AC motor, three or more circuits, each having a different phase separation, are employed. George Westinghouse, Jr., is reported to have bought all of Tesla's related patents for $1 million plus royalties, although the exact amount remains in dispute. The two men collaborated very close by. At the World's Colombian Exposition in Chicago in 1893, Tesla's system of alternating-current power transmission was successfully employed. The Westinghouse Company later won the contract to harness the hydroelectric power of Niagara Falls, producing a viable, long-distance, electrical distribution system. When financial difficulties beset the Westinghouse Company between 1891 and 1893, Tesla agreed to give up the royalties to which he was entitled.

Tesla now devoted his energies to work in his own laboratory in New York City. His inventions were numerous. He experimented with the transmission of signals using electromagnetic waves at the same time as Marconi was developing radio. At Madison Square Garden Tesla demonstrated remote control of mechanical objects. His Tesla Coil was capable of generating extremely high voltages. It consisted of three circuits. AC power was applied to the first circuit and, by means of transformers, the voltage was stepped up in the second and third circuits, both of which possessed spark gaps. The third circuit also included a capacitor tunable for resonance, thus allowing extremely large voltages to be developed.

Nikola Tesla. (Corbis Corporation)

In 1891 Tesla became a U.S. citizen. His naturalization papers remained among his most prized possessions.

Tesla accepted an offer of land and free electricity in Colorado Springs by the local electrical company to continue his research. In 1899, he conducted experiments he considered to be of extreme importance in the conduction of electricity through the earth without the use of wires. He reported that he was able, by means of this principle, to illuminate electric lightbulbs twenty-six miles from the power source. This transmission mechanism, which Tesla explained as taking place by means of the resonant frequency of the earth, has yet to be adequately verified to the satisfaction of the scientific community. In his Colorado Springs laboratory Tesla also performed experiments to simulate lightning that were successful enough to produce a power outage and along with it a withdrawal of the local company's offer of free electricity.

Tesla left Colorado Springs in 1900 and continued his research in a new laboratory on Long Island, which was opened in 1901. However, his efforts were beset by difficulties, not least of which were financial problems, which caused closure of this facility.

His flow of innovative ideas continued unabated. These included improvements for turbines, methods for communication with life on other worlds; and an idea characterized by the press as a "death ray," which may be interpreted as a precursor to the modern laser.

In 1915 it was falsely reported that Edison and Tesla had been jointly awarded the Nobel Prize in physics and that Tesla refused the honor because of his differences with Edison. The circumstances surrounding this news remain cloudy. Nevertheless, Tesla deeply felt the hurt of not receiving this recognition. In 1917 he was persuaded to accept the Edison Medal from the American Institute of Electrical Engineers as an acknowledgment of his pioneering contributions. Among the many honors bestowed on Tesla, perhaps the most important was having an electrical unit named after him, the tesla being the unit for magnetic flux density.

Toward the end of Tesla's life, unflattering articles were written about him, and there were innuendoes that he was involved in the occult. Money problems were never far away, and he moved his residence from one hotel to another, each one cheaper than the one before. His circle of friends contracted. Feeding the pigeons that lived close to his hotel became very important to him, and he developed an almost spiritual bond with them.

Tesla died in 1943. His funeral service was held in the Cathedral of St. John the Divine. As during his lifetime, controversy was not far away. The Serb and Croat mourners sat on opposite sides of the cathedral.

Tesla remains a fascinating man because of his personal life and his engineering genius. However, from a technical point of view Tesla is most remembered for his contributions to the use of alternating-current power transmission.

James D. Allan

See also: Edison, Thomas Alva; Electricity; Electricity, History of; Electric Power, Generation of; Electric Power Transmission and Distribution Systems; Lighting.

BIBLIOGRAPHY

Corson, D. R., and Lorain P. (1962). *Introduction to Electromagnetic Fields and Waves.* San Francisco: W. H. Freeman.

Hunt, I., and Draper, W. W. (1964). *Lightning in His Hand: The Life Story of Nikola Tesla.* Denver: Sage Books.

Peat, D. (1983). *In Search of Nikola Tesla.* Bath, Eng: Ashgrove Press.

Seifer, M. J. (1996). *The Life and Times of Nikola Tesla.* Secaucus, NJ: Carol Publishing Group.

Walters, H. B. (1961). *Nikola Tesla: Giant of Electricity.* New York: Thomas Y. Crowell.

THERMAL ENERGY

Thermal energy is the sum of all the random kinetic energies of the molecules in a substance, that is, the energy in their motions. The higher the temperature, the greater the thermal energy. On the Kelvin temperature scale, thermal energy is directly proportional to temperature.

All matter is composed of molecules or, in some cases, just atoms. In gases and liquids, molecules are relatively free to move around. In a solid they are not so free to move around, but they can vibrate. Although molecules are microscopic, they do have some mass. Combining mass and speed gives them kinetic energy. Depending on the substance, the particles may interact with each other or their surroundings, in which case they would have potential energy along with kinetic energy. Summing the kinetic energy and potential energy of all the molecules gives the total energy of the substance. This energy is called internal energy because it is internal to the confines of the substance.

In a substance where the molecules may move around, the motion is random. No molecule has a definite speed or kinetic energy, but a molecule has a definite average kinetic energy that depends on the temperature. On the Kelvin temperature scale, the average kinetic energy is directly proportional to the Kelvin temperature. The sum of all the random kinetic energies of the molecules is called thermal energy, therefore, thermal energy is directly proportional to the Kelvin temperature.

Citizen Watch unveils the Eco-Drive Thermo, a wristwatch powered by the heat of the human wearing it. (Corbis Corporation)

When one puts a warm hand in contact with cold water, the hand cools and the thermal energy in the hand decreases. The water warms and the thermal energy of the water increases. The exchange of energy stops when both the hand and water come to the same temperature. While in transit, the energy is called *heat.* When two objects at the same temperature are in contact, no heat flows between either. Accordingly, there is no change in the thermal energy of either, no matter how much thermal energy in either one.

Gasoline-powered engines, diesel engines, steam turbines, and gas turbines are examples of heat engines. All heat engines work on a cyclic principle of extracting thermal energy from some source, converting some of this energy to useful work, and rejecting the remaining energy to something at a lower temperature. In an automobile engine, the ignition of a gasoline vapor-air mixture produces a gas at a temperature several hundred degrees above room temperature. The pressure of the gas forces a piston downward, doing work. The gas cools and is ejected out the exhaust at a temperature significantly lower than at the time of ignition. A heat engine converting thermal energy to work cannot function unless there is a temperature difference between the source and exhaust. The larger the temperature difference, the greater the efficiency of the engine. Usually, the lower temperature is that of the engine's surroundings and the ignition temperature is significantly higher.

If it were it is practical to have a temperature lower than what exists naturally in our environment, a heat engine could be built in which this temperature was the exhaust temperature and the temperature of the environment was the higher temperature. Heat engines extracting thermal energy from the surface water of an ocean, and rejecting thermal energy to the cooler sub-surface water, have been proposed. They would not be very efficient because the temperature difference would be small and they would not be easy to construct. The attraction is related to the huge amount of thermal energy in the oceans, which cover roughly two-thirds of the earth's surface.

Joseph Priest

BIBLIOGRAPHY

Hobson, A. (1995). *Physics: Concepts and Connections.* Englewood Cliffs, NJ: Prentice-Hall.

Serway, R. A. (1998). *Principles of Physics.* Fort Worth, TX: Saunders College Publishing.

THERMODYNAMICS

Thermodynamics, which comes from the Greek word meaning "heat motion" is the science that studies the transfer of energy among different bodies. When a typical power plant produces electricity, water is heated to produce steam by burning fuel. The steam is allowed to expand and perform work, which turns an electric generator that produces electrical energy. The efficiency of this process of energy production—how much electrical energy can be produced for burning a given amount of fuel—is determined by the laws of thermodynamics. These same laws are equally important for the heat engines that propel millions of planes, trains and automobiles around the world. In fact, nearly all energy that powers our energy intensive economy involves, in some manner, the laws of thermodynamics.

Scientific descriptions of nature can be categorized as microscopic or macroscopic. A microscopic description involves a discussion of the system using the properties of, and laws governing, atoms and molecules, while in a macroscopic description, one focuses on the large scale, properties of the system. Thermodynamics is a special macroscopic theory that does not involve use of underlying microscopic laws and is, therefore, valid regardless of the form of these laws. A thermodynamic system (or just system) is a specified macroscopic system, that is, a certain specified collection of particles. The matter not included in the system is called the surroundings, and the system plus the surroundings is called the universe. In this context, the universe is not the astronomical universe of galaxies, but the system plus the rest of the matter that can interact with the system, in a measurable manner. As an example of a thermodynamic system consider a gas inside a container (can). The gas could be the system and the walls of the can plus the immediate surroundings would be the universe. Or the gas plus the walls of the can could be the system. The definition of the system is arbitrary and is up to us to define. Failure to have in mind a clear definition of the system often leads to errors in reasoning using thermodynamics principles.

Thermodynamic variables are macroscopic variables associated with the system, such as the system's energy E, volume V, temperature T, pressure P, among others. The thermodynamic variables for a system are measurable properties of the system, and their values define the state of the system. A system is in a thermodynamic equilibrium state if the values of its thermodynamic variables do not change in time. For the gas in the can system, in equilibrium, no change in the values of the temperature, pressure, volume, or any other thermodynamic variable of the gas would be observed. If the system is in an equilibrium state, then one says the system is in a particular thermodynamic state. Thermodynamic processes occur when the system changes from an initial equilibrium state to another final equilibrium state; the thermodynamic variables may undergo changes during this process. If the initial equilibrium state is labeled by i and the final equilibrium state by f then the process $i \rightarrow f$ corresponds to the energy, temperature and pressure changing from E_i, T_i, P_i to the values E_f, T_f, P_f. An important concept in thermodynamics is the idea of a reversible process. In a reversible process, the system undergoes changes where the system is at all times very close to an equilibrium state; the changes in the thermodynamic variables take place very slowly and can also be reversed by very small changes in the agents causing the process. A reversible process can be realized by changing the external conditions so slowly that the system can adjust gradually to the changes. In our can of gas a very slow compression of the system would be a reversible process. A process that is not reversible is called irreversible; a very rapid compression of the gas that would set up pressure waves and gas flows would be an irreversible process. All real processes that occur in nature are irreversible to a certain extent, but it is possible to carry out processes that are reversible for all practical purposes, and to calculate what would happen in a truly reversible process. Various types of thermodynamic processes are named depending on conditions placed on the system. For example, an isothermal process, such as the compression of a gas held at constant temperature, is a process that takes place at constant temperature while an isobaric process, such as the expansion of a gas held at constant pressure, is a process that takes place at constant pressure.

A process in a system may be associated with work W being done on the system. Work is defined the same as in mechanics—as a force times the distance through which the force acts. For example, in our can of gas it may be stirred, thereby doing work on the gas and changing, say, its temperature. In a compression or expansion of the gas, work is also done on the system. If energy is exchanged between the system and the surroundings due solely to a temperature difference, then this energy is called heat. A thermally insulated system is a system that maintains its thermodynamic state regardless of the temperature of its surroundings. A system enclosed in thermally insulated walls like the walls of a very good thermos bottle cannot sense the temperature of the surroundings. Processes that occur in thermally insulated systems are called adiabatic processes. For an insulated system, the only way to change its energy is by the performance of work; in this case the work is called adiabatic work W_a.

The first law of thermodynamics has two parts. The first part associates the change in the energy of the system in adiabatic processes to the adiabatic work done on the system:

$$DE = E_f - E_i = W_a \quad \text{(adiabatic processes).} \tag{1a}$$

In Equation 1a the symbol DE is used for the change in energy for the process i→f; the final value of the system E_f minus the initial energy of the system E_i. Thus, if the system undergoes an adiabatic compression, the work done on the system would give the change in the energy of the system using Equation 1a. Starting from a given state i, one can perform adiabatic work on the system and define the energy at other states f using Equation 1a written in the form $E_f = E_i+W_a$. In words, Equation 1a can be described by saying that energy is conserved for adiabatic processes; if work is done on the system then this work shows up in the energy of the system. Energy is neither created nor destroyed, but is transferred from the surroundings to the system, or vice versa through the work as agent.

The second part of the first law of thermodynamics arises when the requirement that the process be adiabatic is dropped; recall that this means the system is not insulated, and processes can be caused by heating and cooling. In a general process (the only assumption is that matter is not added or removed from the system), if an amount of work W is done on the system and the energy changes by DE then the heat supplied to the system Q is defined by

$$DE = E_f-E_i = W+Q \quad \text{(general processes).} \qquad (1b)$$

Notice that work is defined from mechanics (force times distance) and can be used to define the energy difference between any two states in Equation 1a. This energy difference and the work are then used to define the heat Q in Equation 1b. In words, Equation 1b says that energy is conserved and heat must be included as a form of energy; or the final energy of the system is the initial energy plus the work done on the system plus the energy added to the system by heating.

Note the following sign conventions for W and Q: the work W in Equation 1a and 1b is positive if it increases the energy of the system, and the heat Q is positive if it increases the energy of the system. If W is positive, then the surroundings do work on the system, for example, compressing the gas, but if W is negative, the system does work on the surroundings, expanding the gas. If Q is positive, the surroundings give energy to the system by means of a temperature difference, but if Q is negative, the energy passes from the system to the surroundings.

In summary, the first law of thermodynamics, Equations 1a and 1b, states that energy is conserved and the energy associated with heat must be included as a form of energy. No process i→f is possible if it violates the first law of thermodynamics; energy is always conserved in our world as dictated by Equation 1b. If Equation 1b is applied to an adiabatic process, then because Q = 0 the first part, Equation 1a is recovered, but one still needs both parts of the first law to define the quantities.

The second law of thermodynamics further restricts the types of processes that are possible in nature. The second law is particularly important in discussions of energy since it contains the theoretical limiting value for the efficiency of devices used to produce work from heat for our use.

The second law of thermodynamics also consists of two parts. The first part is used to *define* a new thermodynamic variable called entropy, denoted by S. Entropy is the measure of a system's energy that is unavailable for work.The first part of the second law says that if a reversible process i→f takes place in a system, then the entropy change of the system can be found by adding up the heat added to the system divided by the absolute temperature of the system when each small amount of heat is added:

$$DS = S_f-S_i = SUM(dQ/T) \quad \text{(reversible process).} \qquad (2a)$$

dQ is a small amount of energy added *reversibly* to the system as heat when the temperature of the system is at an absolute temperature of T. SUM means to add the individual dQ/T terms over the entire reversible process from i→f. The reversible way of adding heat to a system is to put the system in contact, through a heat conducting wall (diathermal wall), with the surroundings that are at a very slightly different temperature. The absolute temperature scale is defined so that the freezing point of water is T = 273 Kelvin or 273 K. The zero of temperature on this scale is an unattainable temperature referred to as absolute zero, although temperatures within one-millionth of a Kelvin and lower of absolute zero have been produced (Note at absolute zero all motion in the system does not cease). Notice that Equations 1a and 2a have a certain similarity since they both are used to define a relevant thermodynamic variable: energy E for Equation 1a and entropy S for Equation 2a. Equation 2a allows the definition of the entropy S

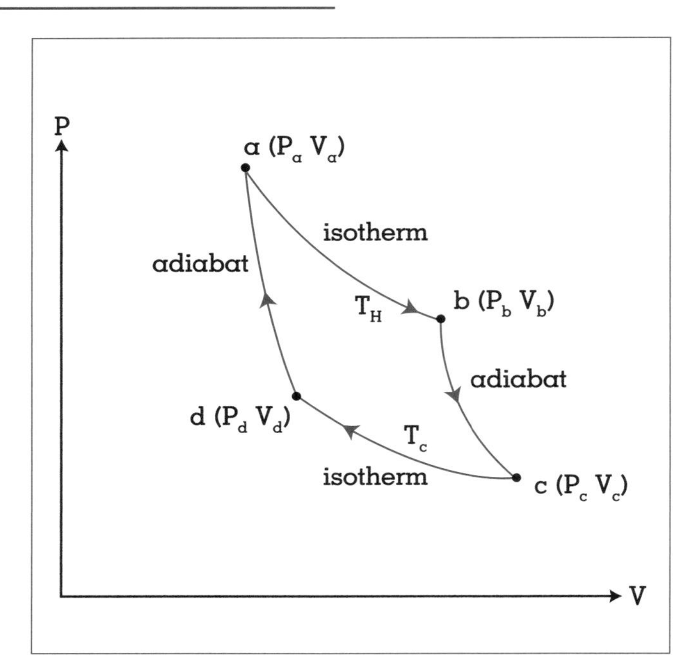

Figure 1.
Carnot cycle for a gas.

at every equilibrium state of the system by starting with the entropy at a particular state and carrying out reversible processes to other states.

The second part of the second law states that where system undergoes an adiabatic process (system surrounded by insulating walls), i→f and the process is reversible, the entropy is not changed, while when the adiabatic process is not reversible the entropy must increase:

$$DS = S_f - S_i = 0$$
(process i→f is adiabatic and reversible),

$$DS = S_f - S_i > 0$$
(process i→f is adiabatic but not reversible). (2b)

Thus, in adiabatic processes the entropy of a system must always increase or remain constant. In words, the second law of thermodynamics states that the entropy of a system that undergoes an adiabatic process can never decrease. Notice that for the system plus the surroundings, that is, the universe, all processes are adiabatic since there are no surroundings, hence in the universe the entropy can never decrease. Thus, the first law deals with the conservation of energy in any type of process, while the second law states that in adiabatic processes the entropy of the system can never decrease.

A heat engine is a device to use the flow of heat energy from a higher temperature to a lower temperature to perform useful work, often mechanical or electrical work. A steam engine heats steam and uses this heated steam and a cooler surroundings to perform mechanical work such as lifting a weight or propelling a steamship. A certain amount of heat Q is converted into a certain amount of work W per cycle of the engine.

In order to investigate heat engines, first focus on a particular cyclic process called a Carnot cycle. Figure 1 shows a Carnot cycle for our gas in the can, and shows the changes in the pressure P and the volume V for the gas during the cycle. The Carnot cycle is the process a→b→c→d→a: from a→b the system at temperature T_H undergoes an isothermal expansion from volume V_a to V_b, from b→c the system undergoes an adiabatic expansion from volume V_b to V_c while the temperature falls to the lower value T_C. Next, the system undergoes an isothermal compression c→d. Finally, the system completes the cycle by undergoing an adiabatic compression d→a, after which its state, a, is precisely the same as before the process was started.

All of the processes in the Carnot cycle are reversible, which is indicated by drawing the lines passing through equilibrium states on the diagram in Figure 1. During the isothermal expansion a→b the system will do work and absorb the energy Q_H from the surroundings that are at the temperature T_H. During the process c→d the system will have work performed on it and liberate energy Q_C to surroundings at lower temperature T_C. The reversible heat exchanges are arranged by having the system in contact with the surroundings, which are held at constant temperature during the isothermal expansion and compression. The surroundings function as a (thermal) reservoir at the temperature T_H and T_C. A reservoir is able to absorb or liberate as much energy as required without a change in its temperature. An oven held at temperature T_H functions as a reservoir. During the adiabatic expansion, the system exchanges no heat, but work is done. Note that in the Carnot cycle only two reservoirs exchange energy during the isothermal process. In other reveisible cycle one needs a series of reservoirs to exchange energy. The laws of thermodynamics may be applied to the Carnot cycle and lead to important results.

Applying the first law of thermodynamics to the Carnot cycle gives

$$DE = E_f - E_i = 0 = Q_H - Q_C - W, \quad (3)$$

or, solving this for the work done during the cycle in terms of the heat absorbed per cycle and the heat liberated per cycle:

$$W = Q_H - Q_C. \quad (4)$$

The system does work on the surroundings during the two expansions, and work is done on the system in the two compressions in the Carnot cycle. The net work done in the cycle is work done on the surroundings by the system and is W in Equations 3 and 4. Note that $E_f = E_i$ in the complete cycle a→a since i = f = a, and the total process is cyclic. It is easy to imagine carrying out the Carnot cycle on our can of gas example. In Equations 3 and 4, the proper signs have been used so that the quantities Q_H, Q_C, and W are all positive numbers; the system absorbs energy Q_H from the high temperature reservoir, rejects or liberates energy Q_C at the low temperature reservoir, the system performs work W on the surroundings, and the system is finally returned to the starting equilibrium state. Equation 4 is the application of the first law of thermodynamics to the Carnot engine.

Because the gas in the Carnot cycle starts and ends at the same state, the system's entropy does not change during a cycle. Now apply the second law to the universe for the case of the Carnot cycle. Because the processes are reversible, the entropy of the universe does not change by Equation 2b. This can be written:

$$DS = 0 = Q_C/T_C - Q_H/T_H \quad \text{(Carnot cycle)}, \quad (5)$$

where the first part of the second law, Equation 2a, is used to determine the entropy change of the two reservoirs in the Carnot cycle. Solving Equation 5 for the ratio of the heat absorbed to the heat liberated, one finds the important Carnot relation connecting the energy Q_H/Q_C ratio to absolute temperature ratio:

$$Q_H/Q_C = T_H/T_C \quad \text{(Carnot cycle)}. \quad (6)$$

This relation is used to define the absolute temperature scale in terms of energy exchanged with reservoirs in a Carnot cycle. The relations for the Carnot cycle, given by Equations 4 and 6 are general and valid for any system that is taken through a Carnot cycle and not just for our can of gas example.

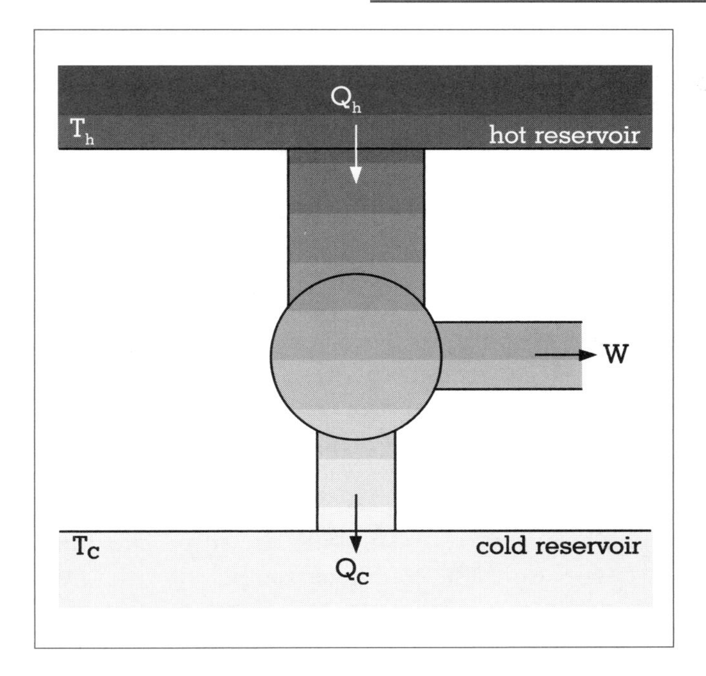

Figure 2.
Schematic of a heat engine.

Figure 1 shows the thermodynamic processes associated with the Carnot cycle. Since these thermodynamic processes produce work, this is also called a Carnot engine. Figure 2 shows another more general picture of a heat engine in thermodynamics. This engine could be a Carnot engine or some other engine. For practical purposes, an efficiency, e, of an engine is defined as the work obtained in one cycle divided by the heat absorbed; this is the what can be sold, W, divided by what must be paid for to produce W, namely Q_H:

$$e = W/Q_H \quad \text{(efficiency of a heat engine)}. \quad (7)$$

The efficiency of the engine would be greatest if it could be arranged that W = QH, QC = 0, giving an efficiency of 1 or 100 percent efficient. In this case, all the heat absorbed at the high temperature is converted into work. This is perfectly valid from the point of view of the first law of thermodynamics, because energy is conserved. However, from the second law,

Equation 5, $DS = 0 = -Q_H/T_H$, which is impossible since $Q_H = W$ is not zero and this implies the entropy of the universe would decrease.

Thus, the second law leads to the conclusion that energy cannot be absorbed from one reservoir and converted completely into work with no other changes in the universe. This is an alternate form of the second law of thermodynamics called the Kelvin-Plank form of the second law. This result is of great practical importance for discussions of energy, since without this result the conversion of heat completely into work would be possible. A steamship could sail on the ocean, absorbing energy from the ocean, and power itself with no need to carry fuel. The fact that heat must be rejected at a lower temperature means that the steamship must carry fuel to produce a higher temperature than the surroundings. The second law of thermodynamics forbids this so-called perpetual motion machine of the second kind, which is an imaginary device that converts heat completely into work with no other change in the universe.

An estimate of the efficiency of a heat engine working between two temperatures T_H and T_C can be obtained by assuming the Carnot cycle is used. By combining the results from applying the first and second laws to the Carnot cycle, the Carnot efficiency e_c may be written:

$$e_c = 1-Q_C/Q_H = 1-T_C/T_H \quad \text{(Carnot engine efficiency).} \qquad (8)$$

This remarkable result shows that the efficiency of a Carnot engine is simply related to the ratio of the two absolute temperatures used in the cycle. In normal applications in a power plant, the cold temperature is around room temperature $T = 300$ K while the hot temperature in a power plant is around $T = 600$ K, and thus has an efficiency of 0.5, or 50 percent. This is approximately the maximum efficiency of a typical power plant. The heated steam in a power plant is used to drive a turbine and some such arrangement is used in most heat engines. A Carnot engine operating between 600 K and 300 K must be inefficient, only approximately 50 percent of the heat being converted to work, or the second law of thermodynamics would be violated. The actual efficiency of heat engines must be lower than the Carnot efficiency because they use different thermodynamic cycles and the processes are not reversible.

The first and second laws of thermodynamics can also be used to determine whether other processes are possible. Consider the movement of energy from a cold temperature reservoir to a hot temperature reservoir with no other changes in the universe. Such spontaneous cooling processes would violate the second law. From the first law, for such a process $Q_H = Q_C$, the heat lost by the low-temperature reservoir is gained by the high-temperature reservoir; again, the signs are chosen so that both Q_H and Q_C are positive. The change of entropy of the universe in this case is just the change in entropy of the two reservoirs, which, by Equation 2a is $DS = Q_H/T_H-Q_C/T_C = -Q_C(1/T_C-1/T_H)$ which is less than zero since Q_C and $(1/T_C-1/T_H)$ are positive. This means that the entropy of the universe would decrease in such a process. Thus, if heat flows from a cold reservoir to a hot reservoir, with no other changes in the universe, it violates the second law, which says the entropy of the universe can never decrease).

This leads to what is called the Clausius form of the second law of thermodynamics. No processes are possible whose only result is the removal of energy from one reservoir and its absorption by another reservoir at a higher temperature. On the other hand, if energy flows from the hot reservoir to the cold reservoir with no other changes in the universe, then the same arguments can be used to show that the entropy increases, or remains constant for reversible processes. Therefore, such energy flows, which are very familiar, are in agreement with the laws of thermodynamics.

Since the Earth plus the Sun form an approximately isolated system, the entropy of this system is increasing by the second law, while the energy is constant by the first law. The system (Earth plus Sun) is not in an equilibrium state since energy is being transferred from the Sun to the Earth and the Earth is radiating energy into space. The balance of energy absorbed by the Earth (gain minus loss) controls the Earth's temperature. Manmade processes occurring on Earth (burning, automobiles, energy production, etc.) lead to a continual increase in entropy and to an increase in atmospheric pollution. This increase in atmospheric pollution changes the net energy absorbed by the Earth and hence the Earth's temperature. This is the greenhouse effect or global warming. Thus, there is a connection between the entropy increase of the Earth-Sun system and global warming. The main worry is the tremendous increase in atmospheric pollution by the increasing population of the Earth. It is not known how this will end but a "heat-death" of the Earth like that which exists on Venus is a disturbing possibility. Because of the complexity of the process it is very difficult to calculate and predict if this will happen. Unfortunately, past a certain point, such processes could lead to an irreversible increase of the temperature of the Earth, a runaway greenhouse effect, even if we stopped our atmospheric pollution.

The Carnot cycle is reversible so a Carnot refrigerator or air conditioner which removes heat from the cold temperature reservoir is in essence a heat engine operated in reverse. In Figure 3 is shown a schematic diagram of a general refrigerator, where Q_C is absorbed from the low temperature reservoir, work W is done on the system, and heat Q_H is delivered to the high temperature reservoir per cycle. The signs have again been chosen so that Q_C, Q_H, and W are all positive numbers. The first law applied to the refrigerator cycle gives:

$$W = Q_H - Q_C \quad \text{(refrigerator cycle)}, \qquad (9)$$

which has the same form as for an engine, Equation 5, since it is just conservation of energy for a cycle. The quantity of interest for a refrigerator is the coefficient of performance k:

$$k = Q_C/W \quad \text{(coefficient of performance)}, \qquad (10)$$

which is what is valuable, Q_C, divided by what must be paid for, W. If it could be arranged so that $W = 0$ this would give the best value for k since cooling without cost would be realized. However, this violates the Clausius form of the second law because energy would be transferred from the cold reservoir to the hot reservoir with no other change in the universe. Therefore, the coefficient of performance of any refrigerator cycle must be finite. The application of the second law to the Carnot refrigerator cycle gives the same fundamental Carnot cycle result relating the temperature ratio to the heat ratio in Equation 6. Equations 6, 9, and 10 can be combined to obtain an expression for the value of k for the Carnot refrigerator:

$$k_c = Q_C/(Q_H - Q_C) = T_C/(T_H - T_C) \quad \text{(Carnot cycle)}. \qquad (11)$$

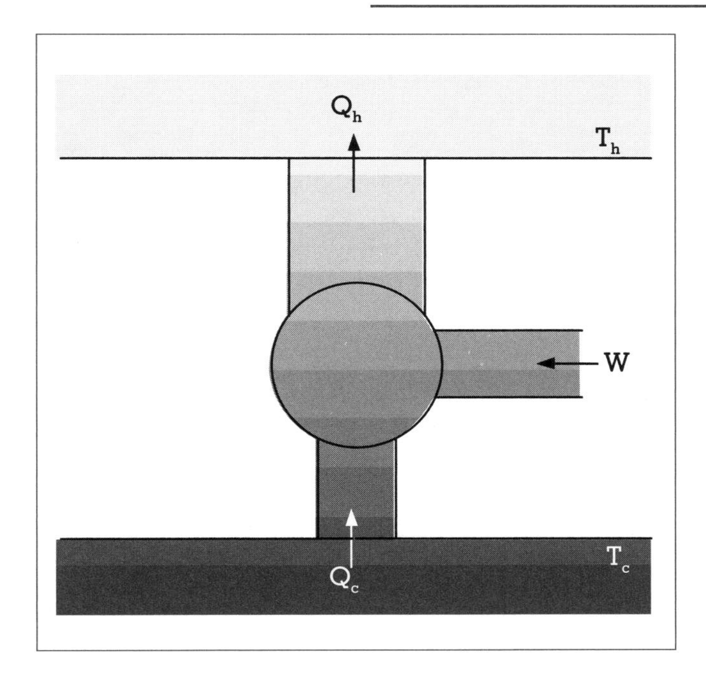

Figure 3.
Schematic of a refrigerator.

In a typical refrigerator, the cold temperature is around 273 K and the compressor raises the temperature of the gas on the high-temperature side by around 100 K so the coefficient of performance is approximately $k = 273\ K/100\ K = 2.7$. This means that the amount of heat removed is 2.7 times the amount of work done during one cycle. A heat pump cools the inside of the house in the summer and cools the outside of the house in the winter. It can be discussed as a refrigerator in both cases, but the cold temperature reservoir changes from inside to outside the house as summer changes to winter. This is accomplished by reversing the flow of the working substance in the heat pump. In the winter, as the outside cold temperature falls, the coefficient of performance falls according to Equation 11, and the heat pump produces less heating for the same amount of work. Often the heat pump is supplemented with resistive heating. As long as the coefficient of performance stays above 1, the heat pump is more efficient than simple resistive heating, which produces the same amount of heat as work. One way to make a heat pump have a larger coefficient of performance is to bury the heat exchange unit underground so that the cold temperature is higher than the ambient air temperature. With this change the heat pump can function efficiently even in cold climates.

No heat engine operating between two temperature reservoirs at temperatures T_C and T_H can be more efficient than a Carnot engine operating between these same two temperatures. To show this, first note that the engine can operate in a reversible or irreversible manner. If the engine is reversible, then it has the same efficiency as the Carnot engine since only two temperatures are involved. If the engine is irreversible, then the second law Equations 2a and 2b give

$$DS = Q_C/T_C - Q_H/T_H > 0$$

or

$$Q_C/Q_H > T_C/T_H.$$

If this is used in the definition of the efficiency of an engine,

$$e = 1 - Q_C/Q_H$$

and, compared to the efficiency of the Carnot cycle given in Equation 8, the efficiency of the engine must be less than the Carnot efficiency. This means that the work done in an irreversible engine is less for the same amount of heat absorbed from the high temperature reservoir. In the same way it follows that no refrigerator operating between two temperatures can have a greater coefficient of performance than a Carnot refrigerator. The laws of thermodynamics are of central importance for discussions of energy in our world.

John R. Ray

BIBLIOGRAPHY

Callen, H. B. (1985). *Thermodynamics: An Introduction to Thermostatics.* New York: John Wiley and Sons.

Fermi, E. (1957). *Thermodynamics.* New York: Dover Publications.

Pippard, A. B. (1966). *Elements of Classical Thermo-dynamics.* Cambridge, UK: Cambridge University Press.

Zemansky, M. W., and Dittman, R. H. (1997) *Heat and Thermodynamics*, 7th ed. New York: McGraw Hill.

THERMOSTATS

See: Energy Management Control Systems

THOMPSON, BENJAMIN (COUNT RUMFORD) (1753–1814)

PERSONAL LIFE AND CAREER

Benjamin Thompson, born in Woburn, Massachusetts, in 1753, acquired a deep interest in books and scientific instruments as a youth, and matured into a person of great charm and intellect. In 1772 he received a position as a teacher in Concord, New Hampshire, where he soon met and rather quickly married Sarah Walker Rolfe, a wealthy widow fourteen years his senior, who had been charmed by his brilliant mind and dashing manners. Overnight his status changed to that of a country gentleman, managing his wife's estate and helping John Wentworth, the British governor of New Hampshire, with some agricultural experiments.

To reward Thompson, in 1773 Wentworth commissioned him a major in the New Hampshire militia. There Thompson acted as an informant for the British, so arousing the rancor of his fellow colonists that they planned to tar and feather him. He quietly departed for Boston, leaving his wife and baby daughter behind. He never saw his wife again, and only saw his daughter briefly many years later when he was living in France.

In Boston Thompson worked briefly for General Gage, the highest-ranking British officer in the Massachusetts Bay Colony. When the American Revolution began in 1776, his loyalty to the American cause was clearly suspect, and he left for England. Upon arriving in London his obvious talents led to rapid advancement in his career, and eventually he was appointed Undersecretary of State for the colonies. He also had a brief military career with the British Army in America, and retired in 1784 with the rank of Colonel.

In 1784 this freelance diplomat joined the court of Karl Theodor, Elector of Bavaria, and rapidly rose to become head of the ineffectual Bavarian Army. For his contributions to building up Bavaria's defensive strength, in 1793 he was made a Count of the Holy Roman Empire and took the name "Count Rumford," since that was the original name of the

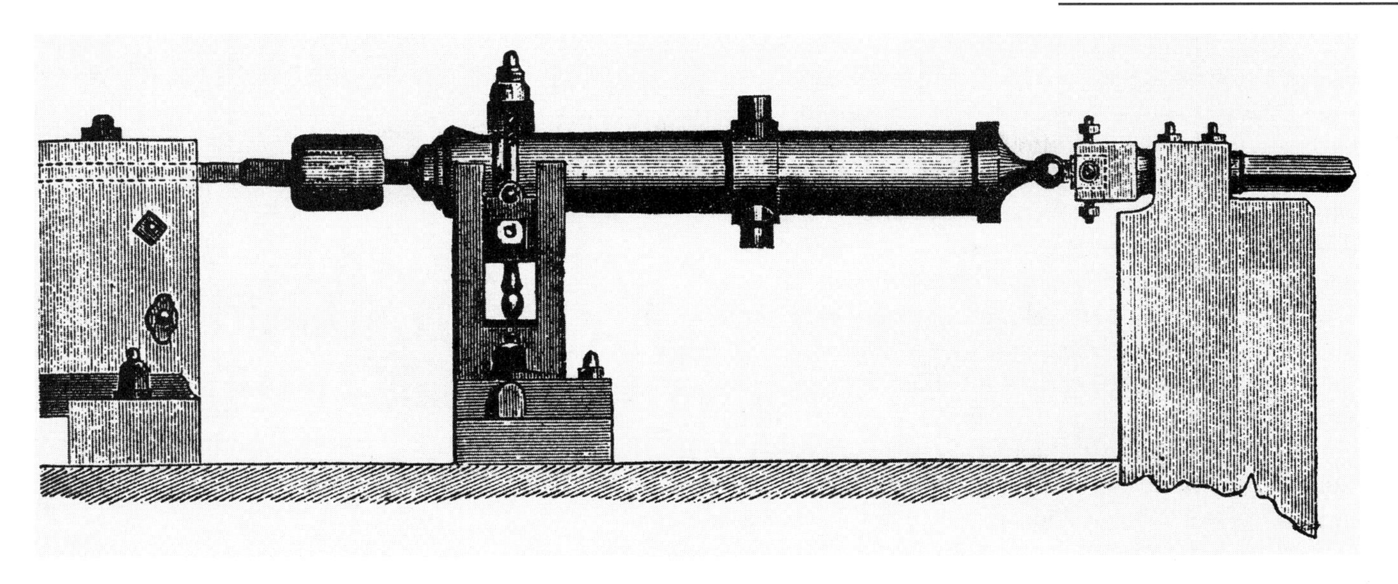

Benjamin Thompson's 1798 invention for producing heat based on his idea that friction generates heat by increasing the motion of molecules. (Corbis-Bettmann)

New Hampshire town in which he had first taught school.

After returning briefly to London in 1800, Rumford was on the move again, this time to Paris, where in 1805 he married the widow of the great French chemist, Antoine Lavoisier. Rumsford's first wife had passed away by this point. His marriage to Madame Lavoisier was, however, a very stormy affair and two years later, after having ensured a handsome lifetime annuity for himself, he and his wife separated. Rumford then retired to Auteuil, outside Paris, where for the rest of his life he worked energetically and with much success on applied physics and technology.

During his life Count Rumford did much good with the funds he obtained by marrying wealthy widows, and by dazzling the leaders of three European nations with his scientific accomplishments. He established sizable prizes for outstanding scientific research, to be awarded by the American Academy of Arts and Science in Boston, and the Royal Society of London. He designed the lovely English Gardens in Munich, and supervised their construction. Finally he provided the funds to start the Royal Institution in London, which later attained great scientific prestige under the direction of Humphry Davy and Michael Faraday.

RUMFORD'S CONTRIBUTIONS TO SCIENCE

During the eighteenth century, the kinetic theory of heat had gradually lost favor and been replaced by the conception of heat as an indestructible fluid, to which Lavoisier had given the name "caloric." In 1798 Rumford, as Minister of War for the Elector of Bavaria, performed pivotal experiments in negating the existence of caloric. While watching the boring of a cannon barrel at the Munich military arsenal, Rumford was struck by the large amount of heat produced in the process. As long as the mechanical boring continued, heat continued to appear. This was hard to explain on the basis of the prevalent caloric theory. It appeared that continued boring was able to produce an inexaustible amount of caloric, which the limited amount of metal in the cannon barrel could not possibly contain.

Rumford decided to try a more controlled experiment. He placed a brass gun barrel in a wooden box containing about nineteen pounds of cold water, and used a team of horses to rotate a blunt steel borer inside the barrel. After 2.5 hours, the water boiled! As Rumford described it, "It would be difficult to describe the surprise and astonishment expressed on the countenances of the bystanders on seeing so large a quantity of cold water heated, and actually made to boil, without any fire."

Rumford suggested that anything that an isolated body can supply *without limitation* could not possibly be a material fluid. The only thing that could be communicated in this fashion was *motion*, in this case the motion of the steel borer that first produced *heat* in the form of molecular motion of the cannon molecules, which was then passed on to the water as random motion of its molecules. Therefore, according to Rumford heat was a form of random molecular motion. By this and other experiments Rumford had indeed demonstrated that a caloric fluid did not exist, but he had not yet seen the intimate connection that exists between heat, work and energy. That connection was to come later with the research of Sadi Carnot, Julius Robert Mayer, James Joule, and Hermann von Helmholtz.

In addition to this groundbreaking research on work and heat, Rumford made an extraordinary number of important contributions to applied science. He studied the insulating properties of the cloth and fur used in army uniforms, and the nutritive value of various foods and liquids. He decided that thick soup and coffee were the best sources of strength for any army in battle; and introduced the potato as a food into central Europe. He designed a large number of drip-type coffee makers and introduced the first kitchen range, the double boiler, and the pressure boiler found in modern kitchens. Rumford also designed better chimneys and steam-heating systems for houses. His continued interest in the scientific principles behind such devices made Rumford one of the world's first great applied physicists.

Rumford died suddenly at Auteuil in August 1814, leaving his entire estate to Harvard College to endow a professorship in his name. While considered a great scientist and a charming man by the best and brightest of his contemporaries, Rumford was at heart a soldier-of-fortune, and could be arrogant, obnoxious, and cruel to those he considered beneath him. For these reasons, he is less highly respected today as a man than he is as a scientist.

Joseph F. Mulligan

BIBLIOGRAPHY

Brown, S. C. (1952). "Count Rumford's Concept of Heat." *American Journal of Physics* 20:331–334.

Brown, S. C. (1953). "Count Rumford—Physicist and Technologist." *Proceedings of the American Academy of Arts and Sciences* 82:266–289.

Brown, S. C. (1962). *Count Rumford, Physicist Extra-ordinary.* Garden City, NY: Anchor Books.

Brown, S. C. (1976). "Thompson, Benjamin (Count Rumford)." In *Dictionary of Scientific Biography,* ed. C. C. Gillispie, Vol. 13, pp. 350–352. New York: Scribner.

Brown, S. C. (1979). *Benjamin Thompson, Count Rumford.* Cambridge, MA: MIT Press.

Ellis, G. E. (1871). *Memoir of Sir Benjamin Thompson, Count Rumford.* Boston: American Academy of Arts and Sciences.

Wilson, M. (1960). "Count Rumford," *Scientific American* 203 (October):158–168.

THOMSON, JOSEPH JOHN (1856–1940)

The British physicist, famous for his discovery of the electron, was born in Cheetham, near Manchester, on December 18, 1856. He first entered Owens College (later Manchester University) at the early age of fourteen. In 1876 Thomson won a scholarship in Mathematics to Trinity College, Cambridge, and remained a member of the College for the rest of his life. He became a Fellow in 1880, Lecturer in 1883 and Master in 1918, a position he held with great flair until his death on August 30, 1940.

Thomson met Rose Paget in 1889 when, as one of the first women to be allowed to conduct advanced work at Cambridge, she attended some of his lectures. They married on January 22, 1890, and had two children: a son, George, and a daughter, Joan. The marriage was a long and happy one.

Many were surprised when, at the end of 1884, Thomson was appointed Cavendish Professor of Experimental Physics in succession to Lord Rayleigh. Thomson was not yet twenty-eight, and he had almost no experience in experimentation. Nevertheless he immediately began work on the conduction of electricity through gases and single-mindedly pursued this topic until he resigned the Chair in 1919. Under Thomson's inspiration and guidance the Cavendish Laboratory became the world's foremost research institution. Seven of those who started or continued their careers there went on to win Nobel Prizes, including his own son.

J.J., as Thomson was commonly called, received the 1906 Nobel Prize in Physics in "recognition of the great merits of his theoretical and experimental investigations of the conduction of electricity by gases". He was knighted in 1908, received the Order of Merit in 1912 and was successively President of the Physical Society, the Royal Society and the Institute of Physics.

The demonstration of the independent existence of the negative electron of small mass was a watershed in the long quest to understand the nature of electricity. Many scientists had contributed to this search in the 300 years prior to Thomson's discovery: Gilbert, Franklin, Coulomb, Poisson, Galvani, Volta, Davy, Oersted, Ampère, Ohm, Faraday, Maxwell, and others. In 1874 the Irish physicist George Johnstone Stoney pointed out that, on the basis of Faraday's law of electrolysis, there exists an absolute unit of electricity, associated with each chemical bond or valency. Hermann von Helmholtz, independently, drew similar conclusions, declaring that electricity, both positive and negative, was divided into elementary portions that behaved like atoms of electricity. Stoney later suggested the name "electron" for this elementary charge.

The phenomena exhibited by the electric discharge in rarefied gases had long been known to German and British physicists. Many British physicists held that cathode rays were particles of matter, similar in size to ordinary molecules, projected from the negative pole. In contrast, most German physicists maintained that cathode rays were analogous to electric waves, getting strong support from Heinrich Hertz who in 1892 showed that the rays could pass through thin sheets of metal. That fact was difficult to reconcile with the molecule-sized particle interpretation.

Thomson was convinced that whenever a gas conducted electricity, some of its molecules split up, and it was these particles that carried electricity. Originally, he thought that the molecule was split into atoms. It was not until 1897 he realized the decomposition to be quite different from ordinary atomic dissociation. At the beginning of that year Thomson performed some experiments to test his particle theory.

First he verified that cathode rays carried a negative charge of electricity and measured their deflection in magnetic and electrostatic fields. He concluded that cathode rays were charges of negative electricity carried by particles of matter. Thomson found that the ratio of the mass, m, of each of these particles to the charge, e, carried by it was independent of the gas in the discharge tube, and its value was of the order of 1/1,000 of the smallest value known at that time, namely the value for the hydrogen ion in electrolysis of solutions. He then devised a method for direct measurements of e as well as m/e, thus allowing the mass of the particles to be determined. The measurements showed that e carried the same

Joseph John Thomson. (Library of Congress)

charge as the hydrogen ion. Thus m was of the order of 1/1000 of the mass of the hydrogen atom, the smallest mass known at that time. This numerical result was perfectly adequate for the interpretation adopted and, if not at first very accurate, was soon improved by later experiments. Thomson concluded that the negative charge carrier, its mass and charge being invariable, must represent a fundamental concept of electricity, or indeed of matter in any state. With regard to its size he estimated what is now called the "classical electron radius" to be 10^{-15} m.

Thus, the search for the nature of electricity led to the discovery of the electron and the proof that it is a constituent of all atoms. These achievements gave scientists the first definite line of attack on the constitution of atoms and the structure of matter. The electron was the first of the many fundamental particles later proposed. Though Thomson was the undisputed discoverer of the electron, there were others in the hunt who came close to the prize, notably the French physicist Jean Perrin, the German physicists Emil Wiechert and Walter Kaufmann, and the Dutch physicist Pieter Zeeman. The latter had, in 1896, calculated the e/m

ratio for a vibrating "ion"—in other words a bound electron—emitting light. His result was similar to Thomson's corresponding value for a free electron.

Thomson's achievement not only produced explanations for many historically puzzling observations but also opened up new fields of science. Of these the richest is surely the electronic structure of matter, a field of unceasing development that has produced, among other things, the silicon chip, the computer and the technology that provides near-instantaneous global intercommunication and access to an unprecedented amount of information.

Leif Gerward
Christopher Cousins

BIBLIOGRAPHY

Gerward, L., and Cousins, C. (1997). "The Discovery of the Electron: A Centenary." *Physics Education* 32:219–225.

Random House Webster's Dictionary of Scientists. (1997). New York: Random House.

Thomson, J. J. (1897). "Cathode Rays." *Philosophical Magazine* 44:293–316.

Thomson, J. J. (1899). "On the Masses of the Ions in Gases at Low Pressures." *Philosophical Magazine* 48:547–567.

THOMSON, WILLIAM (LORD KELVIN) (1824–1907)

In the 1850s the Glasgow professor of natural philosophy (physics), William Thomson, and his colleague in engineering science, Macquorn Rankine, revolutionized the traditional language of mechanics with new terms such as "actual" ("kinetic" from 1862) and "potential" energy. Rankine also constructed a new "science of thermodynamics" by which engineers could evaluate the imperfections of heat engines. By the end of the decade, Thomson and Rankine had been joined by like-minded scientific reformers, most notably the Scottish natural philosophers James Clerk Maxwell, Peter Guthrie Tait, and Balfour Stewart, and the telegraph engineer Fleeming Jenkin. As a group, these physicists and engineers created a new "science of energy" intended to account for everything from the smallest particle to the largest heavenly body.

The fourth child of James and Margaret Thomson, William was born in 1824 in Belfast, then Ireland's leading industrial center, where his father taught mathematics in the politically radical Belfast Academical Institution. His mother came from a Glasgow commercial family, but died when William was just six. Encouraged throughout his early years by his father (mathematics professor at Glasgow University from 1832 until his death from cholera in 1849), Thomson received the best education then available in Britain for a future mathematical physicist. He moved easily from the broad philosophical education of Glasgow University to the intensive mathematical training offered by Cambridge University, where he came second in the mathematics examination (the "Mathematics Tripos") in 1845. Having spent some weeks in Paris acquiring experimental skills in 1846 at the age of twenty-two, Thomson was elected to the Glasgow chair of natural philosophy which turned out to be a post that he held for fifty-three years.

Thomson married Margaret Crum in 1852. She died, childless, in 1870 after a very long illness that rendered her unable to walk. William's second marriage was to Frances Blandy in 1874. Again, there were no children. She outlived him.

Through the engineering influence of his older brother James, William became increasingly committed in the late 1840s to Sadi Carnot's theory of the motive power of heat. Since its somewhat obscure publication in 1824, Carnot's theory had become better known through its analytical reformulation ten years later by the French engineer Emile Clapeyron. The theory explained the action of heat engines by analogy to waterwheels. Just as a "fall" of water drove a waterwheel, so the "fall" of heat between the high temperature of the boiler and the low temperature of the condenser drove a steam engine. This representation gave Thomson the means of formulating in 1848 an 'absolute' temperature scale (later named the "Kelvin scale" in his honor), which correlated temperature difference with work done, thereby making the scale independent of any specific working substance such as mercury or air.

In the same period, James Prescott Joule, son of a Manchester brewery owner, had been carrying out a series of experiments to determine the relationship between work done and heat produced. In the wake of Michael Faraday's electrical researches, electromagnetic engines appeared as a possible future rival to steam power, but in practice their performance failed to match the economy of the best steam engines. Attempting to explain this discrepancy, Joule's early research located the resistances to useful work in various parts of the electrical circuit including the battery. He concluded that, in order to account for all the gains and losses in a circuit, there had to be more than a mere transfer of heat from one part to another; that is, there had to be mutual conversion of heat into work according to a mechanical equivalent of heat. His investigations presupposed that no work (or its equivalent) could be truly lost in nature—only God could create or annihilate the basic building blocks of the universe.

Having met Joule for the first time at the 1847 meeting of the British Association for the Advancement of Science in Oxford, Thomson initially accepted that Joule's experiments had shown that work converted into heat. Committed to Carnot's theory of the production of work from a fall of heat, however, he could not accept the converse proposition that work had been converted into heat could simply be recovered as useful work. Therefore, he could not agree to Joule's claim for mutual convertibility. By 1848 he had appropriated from the lectures of the late Thomas Young (reprinted in the mid-1840s) the term "energy" as a synonym for *vis viva* (the term in use at the time, traditionally measured as mv^2) and its equivalent terms such as work, but as yet the term appeared only in a footnote.

Prompted by the competing investigations of Rankine and the German physicist Rudolf Clausius, Thomson finally reconciled the theories of Carnot and Joule in 1850–51. For the production of motive power, a "thermo-dynamic engine" (Thomson's new name for a heat engine) required both the transfer of heat from high to low temperature and the conversion of an amount of heat exactly equivalent to the work done. His long-delayed acceptance of Joule's proposition rested on a resolution of the problem of the irrecoverability of work lost as heat. He now claimed that work "is lost to man irrecoverably though not lost in the material world." Like Joule, he believed that God alone could create or destroy energy (that is, energy was conserved in total quantity) but, following Carnot and Clapeyron, he also held that human beings could only utilize and direct transformations of energy from higher to lower states, for example in waterwheels or heat engines. Failure to do so resulted in irrecoverable losses of useful work.

William Thomson (Lord Kelvin). (Corbis Corporation)

Thomson's "On a Universal Tendency in Nature to the Dissipation of Mechanical Energy" (1852) took the new "energy" perspective to a wide audience. In this short paper for the *Philosophical Magazine*, the term "energy" achieved public prominence for the first time, and the dual principles of conservation and dissipation of energy were made explicit: "As it is most certain that Creative Power alone can either call into existence or annihilate mechanical energy, the 'waste' referred to cannot be annihilation, but must be some transformation of energy." Two years later Thomson told the Liverpool meeting of the British Association that Joule's discovery of the conversion of work into heat by fluid friction—the experimental foundation of the new energy physics—had "led to the greatest reform that physical science has experienced since the days of Newton."

Through the British Association, Thomson and his associates offered a powerful rival reform program to that of metropolitan scientific naturalists (including T. H. Huxley and John Tyndall) who promoted a professionalized science, free from the perceived shackles of Christianity and grounded on

Darwinian evolution and the doctrine of energy conservation. In critical response to Charles Darwin's demand for a much longer time for evolution by natural selection, and in opposition to Charles Lyell's uniformitarian geology upon which Darwin's claims were grounded, Thomson deployed Fourier's conduction law (now a special case of energy dissipation) to make order-of-magnitude estimates for the earth's age. The limited time-scale of about 100 million years (later reduced) appeared to make evolution by natural selection untenable. But the new cosmogony (theory of the origin of the universe) was itself evolutionary, offering little or no comfort to strict biblical literalists in Victorian Britain.

Thomson also examined the principal source of all useful work on earth. Arguing that the sun's energy was too great to be supplied by chemical means or by a mere molten mass cooling, he at first suggested that the sun's heat was provided by vast quantities of meteors orbiting around the sun, but inside the earth's orbit. Retarded in their orbits by an ethereal medium, the meteors would progressively spiral toward the sun's surface in a cosmic vortex. As the meteors vaporized by friction, they would generate immense quantities of heat. In the early 1860s, however, he adopted Hermann Helmholtz's version of the sun's heat, whereby contraction of the body of the sun released heat over long periods. Either way, the sun's energy was finite and calculable, making possible his estimates of its limited past and future duration.

The most celebrated textual embodiment of the "science of energy" was Thomson and Tait's *Treatise on Natural Philosophy* (1867). Originally intending to treat all branches of natural philosophy, Thomson and Tait in fact produced only the first volume of the *Treatise*. Taking statics to be derivative from dynamics, they reinterpreted Newton's third law (action-reaction) as conservation of energy, with action viewed as rate of working. Fundamental to the new energy physics was the move to make extremum (maximum or minimum) conditions, rather than point forces, the theoretical foundation of dynamics. The tendency of an entire system to move from one place to another in the most economical way would determine the forces and motions of the various parts of the system. Variational principles (especially least action) thus played a central role in the new dynamics.

Throughout the 1860s Thomson and his associates (especially Jenkin, Maxwell, and Balfour Stewart) played a leading role, both in shaping the design of electrical measuring apparatus and in promoting the adoption of an absolute system of physical measurement such that all the units (including electrical resistance) of the system should bear a definite relation to the unit of work, "the great connecting link between all physical measurements." These researches were conducted in the aftermath of the failure of a number of deep-sea telegraph projects, most notably the Transatlantic Cable of 1858. They provided the scientific foundation for a dramatic expansion in British telegraphic communication around the globe in the remaining decades of the nineteenth century, when British Imperial power reached its zenith.

Elevation to the peerage of the United Kingdom in 1892 brought William Thomson the title Baron Kelvin of Largs (usually abbreviated to Lord Kelvin). He was the first British scientist to be thus honored and took the title Kelvin from the tributary of the River Clyde that flowed close to the University of Glasgow. By the closing decades of his long life, when most of his associates had passed away, Kelvin was very much the elder statesman of British science. But the science of energy had been taken up by a younger generation of physical scientists and transformed into quite different modes of scientific understanding, ranging from the "energetics" of the German physical chemist Wilhelm Ostwald (denying atoms in favor of energy) to the "radioactive" physics of Ernest Rutherford (offering new possibilities for estimating the ages of earth and sun). Resisting many of the consequences of these new conceptions, Kelvin published right up to his death at the age of eighty-three, and found at last a resting place in Westminster Abbey, not far from his hero, Sir Isaac Newton.

Crosbie Smith

See also: Carnot, Nicolas Leonard Sadi; Faraday, Michael; Fourier, Jean Baptiste Joseph; Helmholtz, Hermann von; Joule, James Prescott; Maxwell, James Clerk; Rankine, William John Macquorn.

BIBLIOGRAPHY

Smith, C., and Wise, N. (1989). *Energy and Empire. A Biographical Study of Lord Kelvin.* Cambridge: Cambridge University Press.

Smith, C. (1998). *The Science of Energy. A Cultural History of Energy Physics in Victorian Britain.* Chicago: University of Chicago Press.

Thompson, S. P. (1910). *The Life of William Thomson, Baron Kelvin of Largs*, 2 vols. London: Macmillan and Co.

Thomson, W. (1872). *Reprint of Papers on Electrostatics and Magnetism.* London: Macmillan and Co.
Thomson, W. (1882–1911). *Mathematical and Physical Papers*, 6 vols. Cambridge: Cambridge University Press.
Thomson, W. (1891-1894). *Popular Lectures and Addresses*, 3 vols. London: Macmillan and Co.

THREE MILE ISLAND

See: Environmental Problems and Energy Use; Nuclear Energy, Historical Evolution of the Use of

TIDAL ENERGY

See: Ocean Energy

TIRES

Scottish inventor John Boyd Dunlop, displays an invention he patented in 1888—the bicycle with pneumatic tires. (Corbis-Bettmann)

HISTORICAL DEVELOPMENT

The first commercially successful pneumatic tire was developed in 1888 in Belfast by the Scottish veterinarian John Boyd Dunlop primarily to improve the riding comfort of bicycles. Dunlop also showed, albeit qualitatively, that his air-inflated "pneumatic" took less effort to rotate than did the solid rubber tires in use at that time. His qualitative tests were the first known rolling resistance experiments on pneumatic tires. Due to this significant reduction in rolling loss, many professional cyclists in Britain and Ireland adopted air-inflated tires for their bicycles by the early 1890s. Pneumatics for the nascent automobile industry soon followed.

Tires, like everything that rolls, encounter resistance. The resistance encountered by the tire rolling across a surface is a major factor in determining the amount of energy needed to move vehicles. Since Dunlop's original efforts, a considerable number of tire design improvements have been made that have tended to cause a decrease in tire power consumption. For example: separate plies of cotton cord were introduced in the early 1900s to replace Dunlop's square-woven flax (as early tires failed from fabric fatigue before wearing out); in the late 1950s the fuel-efficient radial-ply construction was commercialized in Europe to replace the bias-ply tire; this change also improved vehicle handling and increased mileage. (The radial tire features one or more layers of reinforcing cords or plies disposed perpendicular to the two beads plus a steel belt in the tread region, while the bias construction is built-up with an even number of plies arrayed at alternating, opposing angles between beads without a belt.) There has been a trend toward using larger-diameter tires, which are more rolling-efficient for comfort, appearance and safety reasons as vehicles were downsized in the United States during the 1980s. Elimination of the inner tube by tubeless tires, the use of fewer but stronger cord plies, and the production of more dimensionally uniform tires have each made small but measurable reductions in tire energy loss, although each of these changes was instituted primarily for other reasons.

By the 1990s, automobile tires in the industrialized portions of the world were often taken for granted by the motoring public because of the high level of performance they routinely provide in operation. This casual attitude toward tires by a large segment of the driving public can be partially explained by the sheer number of units manufactured and sold each year. In 1996 about one billion tires of all sizes and types were marketed worldwide with a value of more than $70 billion; this includes 731 million passenger car tires and 231 million truck tires with agricultural, earthmover, aircraft, and motorcycle tires constituting the remainder. Three regions of the world dominate tire production and sales: North America, Western Europe and Japan. Among them, these mature markets are responsible for more than three-quarters of all passenger car tires and one-half of all truck tires. Bicycle tires are now largely produced in less-developed countries, so an accurate number is hard to assess.

The pneumatic tire has the geometry of a thin-walled toroidal shell. It consists of as many as fifty different materials, including natural rubber and a variety of synthetic elastomers, plus carbon black of various types, tire cord, bead wire, and many chemical compounding ingredients, such as sulfur and zinc oxide. These constituent materials are combined in different proportions to form the key components of the composite tire structure. The compliant tread of a passenger car tire, for example, provides road grip; the sidewall protects the internal cords from curb abrasion; in turn, the cords, prestressed by inflation pressure, reinforce the rubber matrix and carry the majority of applied loads; finally, the two circumferential bundles of bead wire anchor the pressurized torus securely to the rim of the wheel.

However, it is the inelastic properties of the cord and rubber components of the tire that are responsible for the heat buildup and energy dissipation that occur during each tire revolution. This loss of energy results in a drag force that impedes tire rotation. The cyclic energy dissipation of tire materials is a mixed blessing for the tire development engineer. It is required in the contact patch between tire and road to produce frictional forces that accelerate, brake, and/or corner the vehicle, but it must be minimized throughout the tire in the free-rolling condition so as not to adversely impact fuel economy.

Depending on the specific car model and service conditions, tires can consume 10 to 15 percent of the total energy in the fuel tank, and in this respect the power loss of the four tires is comparable in magnitude to the aerodynamic drag opposing the forward motion of the vehicle during urban driving at about 45 miles per hour.

	Rolling Resistance Coefficient
Radial passenger tire	.008 – .015
Bias passenger tire	.016 – .025
Radial truck tire	.005 – .007
Bias truck tire	.008 – .010
Railroad wheel (steel)	.002 – .003

Table 1.
Rolling Resistance Coefficient of Various Tire Types

Rolling resistance coefficient is defined as the nondimensional ratio of drag force retarding tire rotation to wheel load—with lower being better. For most passenger car tires freely rolling on smooth, hard surfaces, this coefficient varies between 0.01 and 0.02, but may increase to 0.03 or higher at increased speeds, at very high loads, at low-inflation pressures, and/or on soft surfaces.

Typical ranges for the rolling resistance coefficients of passenger car and truck tires in normal service (vs. a steel wheel) are given in Table 1.

Truck tires are operated at about four times the inflation pressure of passenger car tires, which principally accounts for truck tires' lower rolling resistance coefficients.

Power consumption is the product of drag force and speed—and four tires on a typical American sedan consume approximately 10 horsepower of engine output at 65 miles per hour.

The radial passenger tire introduced in North America during the 1970s had a 20 to 25 percent improvement in rolling resistance compared to the bias ply tire then in use. This was a major improvement that resulted in a 4 to 5 percent gain in vehicle fuel economy under steady-state driving conditions—that is, an approximately 5:1 ratio between rolling resistance reduction and fuel economy improvement. By the 1990s, evolutionary advances in tire design and materials continued with the general use of higher operating pressures, but an 8 to 10 percent improvement in rolling resistance now translates into only a 1 percent gain in fuel economy.

Trade-offs between rolling resistance (and therefore fuel economy) and safety occur in tire design and usage just as with vehicles. For example, the least hysteretic polymers and tread compound formulas that lower rolling resistance tend to compromise wet grip and tread wear. Also, worn-out tires with little or no remaining tread depth are much more fuel-efficient than new tires, but require greater distances to stop when braking on wet surfaces.

ENERGY FOR MANUFACTURING

The energy required to produce a tire is only about 10 percent of that consumed while in use on an automobile overcoming rolling resistance during 40,000 miles of service. The majority of the manufacturing energy expended, 60 to 70 percent, is for the mold during the vulcanization process.

At the end of its useful life, the tire, with its hydrocarbon-based constituents, is a valuable source of energy, with a higher energy value than coal. By the mid-1990s, approximately 80 percent of the tires worn out annually in the United States were being recycled, recovered, or reused in some fashion, with three-fourths of these serving as fuel for energy production in boilers and cement kilns.

Joseph D. Walter

See also: Automobile Performance; Efficiency of Energy Use; Transportation, Evolution of Energy Use and.

BIBLIOGRAPHY

Mullineux, N. (1997). *The World Tyre Industry*. London: *The Economist* Intelligence Unit.

Schuring, D. J. (1980). "The Rolling Loss of Pneumatic Tires." *Rubber Chemistry and Technology* 53:600–727.

Tabor, D. (1994). "The Rolling and Skidding of Automobile Tyres." *Physics Education* 29:301–306.

Walter, J. D. (1974). "Energy Losses in Tires." *Tire Science and Technology* 2:235–260.

TOKAMAK

See: Nuclear Energy

TOWNES, CHARLES HARD (1915–)

Charles Hard Townes had a long, distinguished career as a physicist and educator with years of service to the military and academic research communities. One particular line of work will ensure that his name be remembered by history—his contributions to the early development of the laser. The laser is considered by many historians to be one of the most important technological achievements of the twentieth century. As a source of energy, the laser is extremely inefficient—it devours electricity, converting only a fraction into the laser beam. Yet, due to the special properties of laser versus normal light, this technology is extraordinarily useful. A 100-watt light bulb, for example, can barely light a living room, but a pulsed, 100-watt laser can be used to cut or drill holes in metal.

Born in Greenville, South Carolina, Charles Townes early in his life showed the characteristics of both a highly intelligent individual and a scientist. As a boy, Townes studied continuously, covering daunting subjects such as Greek, Latin, and Old English. He also read every issue of Popular Mechanics. Not limiting himself as a book scholar, however, Townes plunged into the hands-on work of technical subjects. Townes's father owned a shop that he rented to a clock and watch dealer. Townes reveled in disassembling the old, broken clocks. In high school, he took shop and mechanical drawing and developed a special interest in electricity and radio. He even attempted (unsuccessfully) to build his own crystal radio.

At the age of sixteen, Townes entered Furman University and received two bachelor's degrees (modern languages and physics) in 1935. He continued his education, receiving a master's at Duke University in 1937 and a doctorate at Cal Tech in 1939. In the summer of 1939, Bell Labs hired him. Numerous lines of research were being undertaken simultaneously at Bell Labs. Most of the work done by Townes initially dealt with basic research and the transmission of telephone and television signals. Worldwide political events, however, soon changed this emphasis.

In September of 1939, Germany invaded Poland and started what would become World War II. Although the United States was still politically neu-

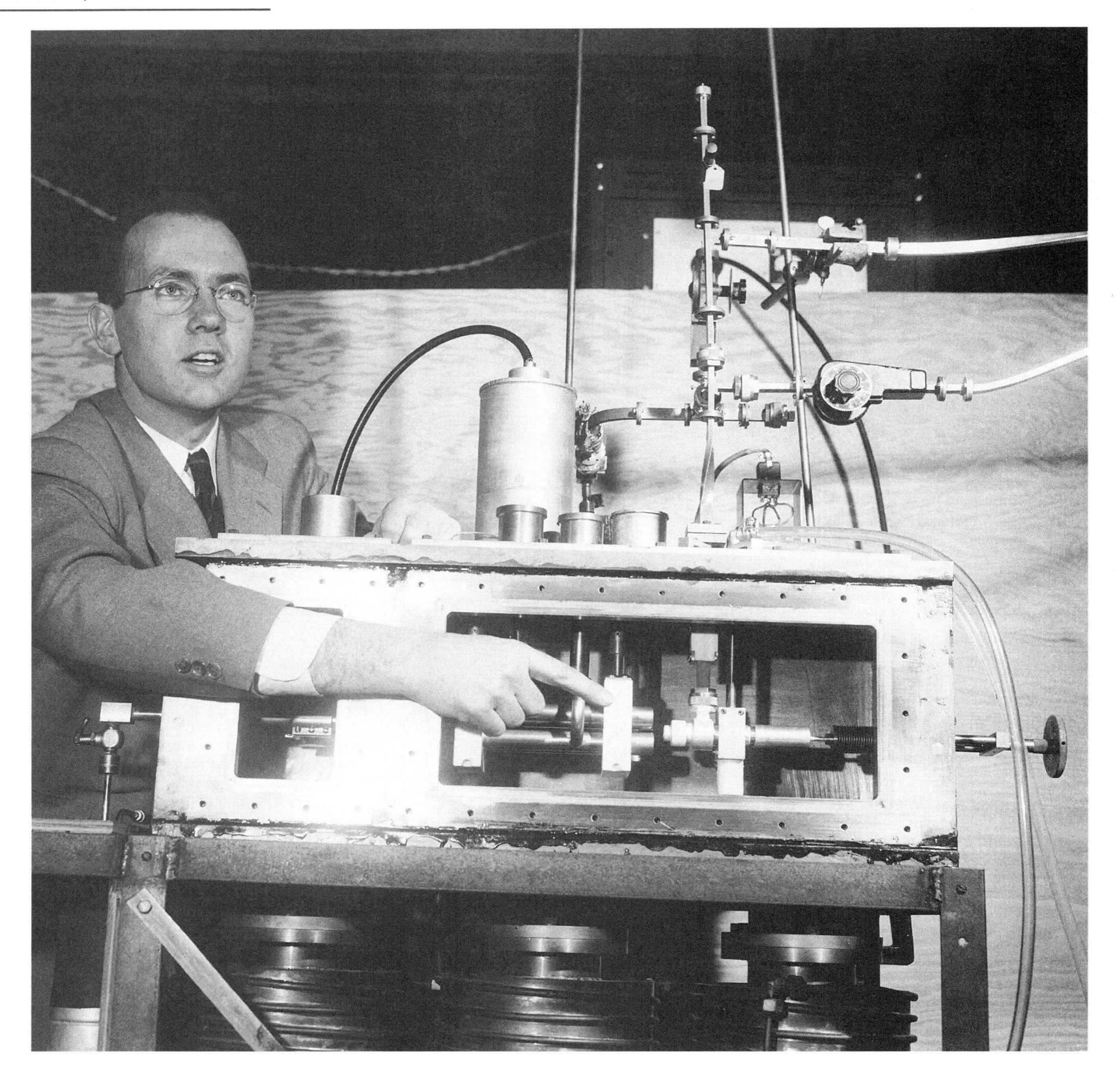

Charles Hard Townes operates an atomic clock. (Corbis-Bettmann)

tral at the time, in March, 1941 Townes was reassigned to work on radar bombing. Bell Labs had done some research on anti-aircraft aiming systems, and Townes and his colleagues used this knowledge to develop a way to replace the bombsight with radar targeting. Although this advance was not put into practice during the war, it helped forge the future of high-tech warfare and was influential in Townes's later civilian endeavors.

Just as the intellectual atmosphere of Bell Labs shaped Townes's career, the cosmopolitan diversity of New York City molded his personal life. He enjoyed the many theaters, museums, and restaurants there. He signed up for voice and music theory lessons at

Juilliard. And most significantly, he met Frances Brown, an activity director at the International House at Columbia University, whom he married in May 1941. The couple had four daughters. His work at Bell Labs had one downside—it often took him away from his family for trips to Florida where the radar bombing system was tested and re-tested.

In 1948, however, Townes returned to the academic world, accepting a professorship at Columbia University. There, during the early 1950s, he speculated that stimulated emission (an ambitious theory first proposed by Albert Einstein in 1917) could generate microwaves, but he also knew a population inversion was necessary. Population inversion occurs when energy is introduced, causing the majority of a material's atoms to reach an excited level rather than their normal "ground level." Edward Purcell and Robert Pound, researchers at Harvard University, had demonstrated population inversion in 1950 using a crystal composed of lithium fluoride. Using this concept as well as his own radar-related work on amplifying a signal, Townes came up with the idea for a closed cavity in which radiation would bounce back and forth, exciting more and more molecules with each pass. With that, Townes had the basic concept of the laser.

Over the next few years, Townes, with the help of graduate students Herbert Zeiger and James Gordon, calculated the precise size of the necessary cavity and built a device that used ammonia as an active medium. Success came in 1954 with the completion of the first maser, an acronym for microwave amplification by stimulated emission of radiation.

In 1957, Townes developed the equation that showed that this same process could also obtain much smaller wavelengths (in the infrared and visible light range). Townes collaborated with Arthur Schawlow, a research assistant in his laboratory from 1949 to 1951, who then moved on to become a physicist at Bell Labs where Townes was still doing consulting work. When he was a postdoctoral fellow at Columbia, Schawlow met Aurelia, Townes' younger sister, who had come there to study singing. Soon the two married.

In 1957, this team of brothers-in-law started working together on Townes's idea for an optical maser. They found atoms that they felt had the most potential, based on transitional probabilities and lifetimes. However, there was still one major problem: In the visible light portion of the electromagnetic spectrum, atoms don't remain in an excited state as long as microwaves. A new cavity seemed the most appropriate solution. In 1958, Schawlow proposed using a long, slim tube as the resonating cavity to produce a very narrow beam of radiation. With that breakthrough, Townes and Schawlow were well on their way to developing a laser. They authored a paper on the subject and submitted a patent on behalf of Bell Labs.

Others, however, were reaching similar ideas at the same time. R. Gordon Gould, a graduate student at Columbia, had come to the same conclusions. In November 1957 he wrote up his notes on a possible laser and had them notarized. Theodore Maiman, a physicist who had been working with ruby masers at Hughes Aircraft Company in Malibu, California, learned of laser research in September 1959 at a conference on quantum electronics that Townes had organized. Afterwards, he began his research on a laser using a pink ruby as an active medium.

Townes's participation in the race for the first laser lessened in the fall of 1959. He was offered and accepted a position as the director of research at the Institute for Defense Analysis in Washington, D.C. With cold war tensions throughout the world, and science playing an increasingly prominent role in government thinking and funding, Townes felt obligated and honored to serve his country in this role. He still met with and directed graduate students at Columbia on Saturdays, but his active role was dramatically reduced.

Schawlow continued working on his laser at Bell Labs. He had rejected ruby as an active medium because he felt it would not reach population inversion. By pumping the ruby with the light from a photographer's flash lamp, however, Maiman succeeded, created the world's first laser in June 1960.

Townes's academic life continued. He served as provost of MIT from 1961 to 1966. In 1964, Townes received the Nobel Prize in physics "for work in quantum electronics leading to construction of oscillators and amplifiers based on the maser-laser principle." He was named university professor at the University of California-Berkeley in 1967. There he worked for more than 20 years in astrophysics. Ironically, this field is one of many that were transformed by the laser, and Townes often used lasers in his subsequent research.

The career of Charles H. Townes and the development of the laser exemplify the technological revolution of the twentieth century. Following the second world war, the United States experienced a golden age of science. Basic research flourished with unprecedented government funding, and it provided the underlying principles for thousands of future devices.

Pioneers such as Charles Hard Townes helped steer the course for this era in the history of energy, and in doing so forged a path for scientists to come.

Karl J. Hejlik

See also: Lasers.

BIBLIOGRAPHY

Bromberg, J. L. (1992). "Amazing Light." *Invention and Technology*. Spring: 18–26.

Bromberg, J. L. (1991). *The Laser in America: 1950–1970*. Cambridge, MA: MIT Press.

Hecht, J. (1992). *Laser Pioneers*. Boston, MA: Academic Press.

Perry, T. S. (1991). "Charles H. Townes: Masers, Lasers & More." *IEEE Spectrum* 28:32.

National Academy of Sciences. (1987). *Lasers: Invention to Application*. Washington, DC: National Academy Press.

TRAFFIC FLOW MANAGEMENT

Americans living in the fifty most congested cities spend an average of thirty-three hours each year stuck in traffic. Congestion causes much more than driver aggravation: air quality suffers, vehicle idling and stop-and-go traffic reduce fuel economy by as much as 30 percent, and we lose billions of dollars in productivity. These are the consequences as the automobile does what it is designed to do—transport a highly mobile population. Continued suburban expansion, reduction in household size, increase in number of workers per household, and general changes in lifestyle have all contributed to increased travel demand and greater congestion.

Even without congestion, from the perspective of capital utilization and energy consumption, automobile and roadway use is inefficient. First, the majority of personal transportation energy is consumed in moving personal vehicles that contain only one occupant and drive one or two hours a day. Second, the transportation infrastructure usually operates below capacity. States expend tremendous resources building highways to accommodate peak period demands (7 A.M. to 9 A.M. and 3 P.M. to 6 P.M.). Most of these lanes are not needed for the rest of the day. Rush hour demand still exceeds capacity in many places, resulting in disruption of traffic flow, stop-and-go driving conditions, a drop in freeway throughput, increased fuel consumption, increased vehicle emissions, and wasted time. Nevertheless, the personal vehicle remains the preferred means of transportation in much of the world given the personal freedom and economic opportunity (access to jobs) that it affords.

Using capital and energy more efficiently is a common goal of government, business, and the individual. State agencies have responded to congestion delay by building more capacity and/or increasing the efficiency of existing capacity. At the local level, however, there is growing opposition to the detrimental noise, air quality, and space infringements of building more and bigger highways. Federal environmental and transportation regulations have shifted from capacity expansion to a focus on congestion reduction and air quality improvement. Because mobility enhances regional economic development and productivity, shifting travel demand out of the peak period, or efficiently improving transportation supply (without necessarily adding new major facilities) still provides tremendous public benefits. Recent strategies have focused on behavior modification (transportation demand management strategies) and targeted improvements to the transportation infrastructure (transportation supply improvement strategies).

TRANSPORTATION DEMAND MANAGEMENT STRATEGIES

The objective of demand management strategies is to encourage or require drivers to reduce the frequency and length of automobile trips, to share rides, or to use alternative modes of transportation. When peak period automobile trips are shifted out of the peak, or into alternative transportation (such as carpools, mass transit, bicycling, or walking) congestion declines and the remaining automobile users benefit from improved travel times. Demand management measures include no-drive days, employer-based trip reduction programs, parking management, park and ride programs, alternative work schedules, transit fare subsidies, and public awareness programs. Demand management measures may or may not require intensive capital investment, but are usually characterized by ongoing operating costs.

Researchers have been able to identify strategies (and specific incentives and disincentives) that can

Cars in a traffic jam on the Hollywood Freeway in California. (Corbis Corporation)

change travel behavior at individual facilities. Effective strategies can be tailored for individual businesses and activities as a function of employment classification, land use, and transit service availability. The fact that many strategies are effective at reducing travel demand at individual facilities is not controversial. But it is difficult to implement regional demand management programs that are acceptable to the public without micromanagement by the government. Such regional demand reduction strategies are typically implemented in the form of regulatory mandates, economic incentives, and education campaigns.

Regulatory Mandates and Employer-Based Trip Reduction

Businesses, operating in a market economy, have little incentive to implement demand management strategies on their own. They pay for the costs of shipment delays associated with moving goods and services to the marketplace (these costs are incorporated into the selling price of the goods and services). However, the vast majority of society's congestion cost is external to marketplace price-setting. Companies sometimes implement measures to reduce employee travel to their facility when there's insufficient parking. Normally, they provide convenient automobile access because it makes it easier to attract and keep employees.

Regulatory mandates require, either directly or indirectly, that specific segments of the population change their trip-making behavior. Examples of direct regulatory mandates include: automobile bans in downtown areas, restrictions on motor vehicle idling time (i.e., heavy-duty vehicles), restricted access to airport terminals for certain types of vehicles, odd/even day gasoline rationing at retail filling stations (based on license plate numbers), restricted hours for goods delivery in urban areas, and peak hour restrictions on truck usage in downtown areas. Direct mandates have proven extremely unpopular. Although developing and rapidly growing industrial countries (e.g., Singapore, Mexico, and China) do implement such measures, none are implemented on a sustained and widespread basis in the United States.

Implementation of indirect regulatory mandates has been more common than direct mandates, primarily in the form of trip reduction ordinances implemented by local governments. These ordinances require employers to increase the average vehicle occupancy of employee vehicles during commute periods, but usually allow employers great flexibility in developing their strategies.

During the late 1980s and early 1990s, employers implemented trip reduction measures in many urban areas. Regulatory agencies developed commute vehicle occupancy goals that would result in fewer vehicle trips to the facility during the morning peak. Employers could offer their employees incentives (e.g., cash rebates for carpool participants) or impose disincentives (e.g., parking fees for employees who drove alone)to achieve these ridership goals.

The largest and most prominent experience with trip reduction ordinances in the United States was Regulation XV, adopted December 11, 1987, in the South Coast (Los Angeles) area. The regulation required employers of a hundred or more individuals to prepare and implement trip reduction plans between the home and worksite. Employers developed their own incentive programs to encourage workers to rideshare or use alternative transportation. Facilities reported progress annually, and they adjusted their plans each year until they achieved their ridership goal.

The tripmaking aspects of specific measures implemented by employers under Regulation XV were widely variable. Trip reductions depended on such local factors as employer size and location, employment and site characteristics, location of labor pool, and socioeconomic composition. A 1991 study of seventy-six facilities in the Regulation XV program found no apparent correlation between the number of incentives offered and the improvement in ridership levels. The quality, not the quantity of incentives offered was the driving force for behavioral change. Two factors had a significant effect on ridesharing: (1) use of parking incentives and disincentives coupled with transit pass or commute subsidies; and (2) management commitment coupled with the presence of an on-site transportation coordinator. A program to guarantee rides home for emergencies and last minute work/personal schedule changes was necessary but insufficient condition to encourage ridesharing.

Over the entire Los Angeles region, employer-based demand management strategies were slow to evolve. Employers were required only to develop "approvable" plans to achieve specified ridership goals with no penalty for failure to achieve the goals. A detailed study of 1,110 work sites found that the implementation of Regulation XV reduced vehicle use to participating facilities by about 5 percent during the first year of the program. The most instructive finding of this study is that the primary improvements in ridership came from increased use of carpools. All other changes were trivial: a slight increase in vanpools and compressed workweeks, a slight decrease in bicycling/walking and telecommuting, and no change in transit use. This finding suggested that reduced vehicle use can be achieved with little or no institutional change, because carpools do not require the same level of organizational effort and financial support as many other options.

Even if successfully implemented, the overall travel implications of programs similar to Regulation XV would be modest. Commute trips represent about 25 percent of daily trips, and commute trips to facilities with a hundred or more employees represent approximately 40 perent of commute trips in the Los Angeles area. Even if commute trips to affected facilities are reduced between 5 percent and 20 percent, employer-based trip-reduction strategies may yield total daily trip reductions of between 0.5 percent and 2 percent (although primarily achieved during peak periods). The costs of such initiatives are substantial. Employers often hire rideshare coordinators and provide incentives, and regulators must monitor and enforce the program. In a survey of more than 400 Los Angeles area facilities the typical cost of placing employees in carpools or transit through personalized ridesharing assistance ranged from $7.72 per employee in large firms (~10,000+ employees) to $33.91 per employee in small firms (~100 employees).

A few years after Regulation XV was implemented, when medium-sized businesses (100–250 employees) came under the regulatory requirements, the business community began exerting significantly increased political pressure on the South Coast Air Quality Management District (SCAQMD) to repeal the regulation, and took their case directly to Congress. On December 23, 1995, Congress amended Section 182(d)(1)(b) of the Clean Air Act. The new language allowed air quality planning agencies to opt out of the required employer-based programs.

Overnight, employee commute programs around the nation became voluntary, disappearing entirely from air quality management plans.

The SCAQMD undertook an eighteen-month study in which the agency encouraged voluntary rideshare efforts for medium-sized facilities. However, the exemption of facilities from employer-based trip reduction and ineffective implementation of voluntary programs yielded an increase in pollutant emissions. Despite the failure of voluntary measures, the California State Legislature permanently exempted medium-sized businesses from the Los Angeles regulations. The largest air pollution control agency with the worst air quality in the nation could not retain their commute program for medium-sized employers over public objection.

Public Information and Education

Recent behavioral shifts, such as the overall decrease in the number of smokers and the increase in residential recycling activity, suggest that ongoing media campaigns coupled with formal education programs can effectively influence human behavior. Many of California's local air pollution control districts have implemented education programs as a means of increasing public awareness of how travel behavior affects air quality. The California Air Resources Board prepared a variety of information packets for government decision-makers as well as the public. Numerous states and local air pollution control agencies have followed suit across the United States.

Education campaigns implemented in conjunction with regulatory mandates can make employer-based trip reduction strategies more efficient at both the facility and regional levels. The SCAQMD implemented an education program to support their employer trip reduction program. District staff members advised corporate representatives on cost-effective strategies implemented by other companies in the region. They recommended compressed work weeks, in-house rideshare matching, subsidized transit passes, carpool/vanpool subsidies, preferential carpool parking, flexible hours, telecommuting, bicycle lockers and showers, and company award/prize programs. Agency staff also recommended guaranteed ride home programs as a necessary supplement to successful carpooling strategies. Of the sixty-five employers that received advice and provided final cost information, about 88 percent (fifty-seven) reported a significant decrease in program implementation costs as a direct result of switching to the new options recommended by SCAQMD in 1997. The average annual cost per worksite declined from $35,616 to $16,043 (an average of $19,573 per worksite, or $70 per employee).

Programs aimed at the younger generation through grade school may achieve positive results over time. Children's programs are likely to yield a future generation that is educated about the economic, environmental, and social costs associated with transportation. Given the historic failures of regional travel programs and public resistance of pricing strategies, public awareness campaigns are becoming a major focus of U.S. regulatory agencies. It remains to be seen if these investments will prove cost-effective.

Economic Incentives

Economic incentives in transportation include monetary incentives or disincentives to the transportation consumer (i.e., vehicle operator or passenger) as encouragement to change travel behavior. Economists have long argued that monetary signals serve as the most economically efficient method to achieve changes in transportation demand. They argue that consumers will consume goods and services most efficiently when they are required to pay the full cost of the goods and services. Provided costs are not set too high, economic incentives should achieve behavioral change more efficiently than prescriptive rules. Economic incentives including congestion pricing and gasoline tax increases have received strong support from businesses, environmentalists, and local press editorials in San Francisco.

Transportation economics literature argues persuasively for the implementation of congestion pricing to improve the efficiency of the current transportation system. Area licensing schemes are also an option, where vehicle owners purchase stickers allowing the car to enter the most congested areas of the city during peak hours. Singapore, Hong Kong, and Oslo and Bergen in Norway provide examples of area license schemes.

The Environmental Defense Fund and Regional Institute of Southern California sponsored comprehensive modeling studies of pricing on transportation behavior for the Los Angeles area. The findings are as follows:

- Regional congestion pricing of $0.15 per mile may yield a vehicle miles traveled (VMT) reduction of about 5.0 percent and trip reduction of 3.8 percent;

- Regional $3.00 per day parking charges may yield a VMT reduction of about 1.5 percent and a trip reduction of 1.8 percent;
- Regional $0.60 per hour nonemployee parking charges may yield a VMT reduction of about 3.5 percent and trip reduction of 4.3 percent;
- Mileage and smog-based registration fees averaging $110 per vehicle per year may yield a VMT reduction of about 0.4 percent and trip reduction of 0.7 percent.

Because of academic interest in congestion pricing theory, Congress established the Congestion Pricing Pilot Program in the Intermodal Surface Transportation Efficiency Act (ISTEA) of 1991 to fund congestion pricing studies and demonstration programs. Due to a lack of political support for congestion pricing on public roads, no significant congestion pricing programs have been implemented in the United States.

High occupancy toll (HOT) lanes are proving to be a potentially viable alternative to congestion pricing. HOT lanes allow single occupant vehicles to access new high occupancy vehicle lanes or facilities by paying a toll. HOT lane facilities are now operating in San Diego, California; Riverside, California; and Houston, Texas. Because the public perceives HOT lanes as providing new capacity and appears more accepting of tolls on these facilities, additional investigation into the consumer acceptance and economic benefits of HOT lanes will continue.

Gasoline Taxes. Studies in 1994 indicated that a $2.00 increase per gallon in gasoline tax (raising United States prices from about $1.00 to about $3.00 per gallon) could yield a VMT reduction of about 8.1 percent and trip reduction of 7.6 percent. Determining the long-term effects of higher gasoline prices can be difficult. Fuel costs are a small component of the total cost of owning and operating an automobile. Research indicates that gasoline demand is relatively inelastic over the short-term and somewhat more elastic over the long-term. Significant increases in fuel price can affect short-term automobile use. However, when fuel prices rise significantly, demand for new fuel-efficient vehicles also increases. Individuals who purchase more fuel efficient vehicles can retain many of the trips and VMT without experiencing significant increases in total operating cost. Fuel price, VMT demand, and fuel intensity are interlinked, and the cumulative effect yields the change in net travel demand and net fuel consumption.

Parking Pricing. In the United States, paying for parking is the exception rather than the rule; 90 to 95 percent of auto commuters pay nothing for parking. Nationwide, employers provide 85 million free parking spaces to commuters, with a fair market value of nearly $36 billion a year. Even in the central business district of Los Angeles, where parking fees are more common than in most areas, of the 172,000 office workers, more than 54,000 drivers park at their employer's expense.

Despite the fact that most employees pay nothing for parking, it is important to remember that there is no such thing as free parking. Even if employers do not pay an outside vendor and instead provide their own lots for employee parking, these lots must be constructed, gated, monitored, and maintained. Companies providing parking to employees at no charge pay an opportunity cost for not putting their property to more productive uses. The land could be developed and used in producing more company income or could be sold for development by others. Employers simply pass on the real and opportunity costs of parking to the consumers of the goods and services provided by the company. These consumers benefit little, if at all, from the parking provided to employees. Failure to charge employees for parking constitutes an inefficient pricing structure.

Travelers are highly sensitive to parking charges because the charges represent a large change in their out-of-pocket costs. Parking costs is one of the three most frequently cited factors (along with convenience and time saved) in the carpool decision. This responsiveness to parking prices is economically rational because motorists treat the vehicle purchase and annual insurance payments as sunk costs with respect to daily travel decisions, leaving parking costs as a large percentage of out-of-pocket expenses. For instance, typical commute trips to the Los Angeles core business district cost less than $2.00 per day in gasoline costs. Adding the fair market value of parking to gasoline costs can increase out-of-pocket expenses to roughly $6.00 per day. Studies show that free parking is a greater incentive to driving alone than providing free gasoline.

Various case studies lend support to the finding that parking prices are significant in affecting trip-generation and mode choice. An early study of employer-paid parking effects on mode choice, conducted in the late 1960s in the central business district of Los Angeles, examined county workers receiving employer-paid parking and federal employees paying for their own parking. The CARB (1987) study found that only 40 percent of the employees subject to parking fees drove to work alone, while 72 percent of similar employees that were not subject to parking fees drove alone. The availability of transit and alternative modes, and the amount of available free parking, influence the effectiveness of parking pricing. An analysis of thirteen employers in 1991 found that when they instituted paid parking programs the number of trips at the worksite was reduced by 20 percent. Other studies in the Los Angeles area indicate that between 19 percent and 81 percent fewer employees drive to work alone when they are required to pay for their own parking.

Employer-paid parking is changing in California because of a law requiring employers to "cash out" free parking. The 1992 California law, and subsequent regulations, require employers who provide free or subsidized offsite parking to their employees to provide a cash equivalent to those employees who do not use the subsidized parking (California Health and Safety Code 43845). The program applies to employers of more than fifty persons in areas that do not meet the state ambient air quality standards (between 3 percent and 13% of the medium and large employers in the Los Angeles area).

The California Air Resources Board (CARB) sponsored research in 1997 to evaluate eight case studies of firms that complied with California's parking cash-out. The number of drive-alone trips to these facilities dropped 17 percent after cashing out. The number of carpool participants increased by 64 percent; the number of transit riders increased by 50 percent; the number of workers arriving on foot or bicycle increased by 39 percent; and total commute trip vehicle miles of travel dropped by 12 percent. These findings are a revelation because this significant shift in travel behavior resulted from a regional policy.

Most businesses in Los Angeles have hesitated to implement employee parking pricing, despite the fact that increased parking fees can increase vehicle occupancy much more efficiently that other strategies. Most employees who receive free parking view this subsidy as a right, or fringe benefit of employment.

TRANSPORTATION SUPPLY IMPROVEMENT STRATEGIES

Transportation supply improvement strategies change the physical infrastructure or operating characteristics to improve traffic flow and decrease stop and go movements. They often require intensive capital investment and comprise bottleneck relief, capacity expansion, and construction improvements. Regional transportation plans usually include integrating high-occupancy vehicle (HOV) systems (differentiated from construction of HOV lanes). Many areas are also refocusing efforts on traditional technology-oriented means that reduce traffic congestion: signal timing optimization, rapid incident response, ramp metering, and applications of intelligent transportation system technology.

HOV Systems

Serious congestion delay often exists in the same urbanized areas that fail to meet federal air quality standards. In these areas, capacity expansion projects are usually difficult to implement. The emissions from the increased travel demand that follows corridor expansion can create additional air quality problems. One effective means of expanding capacity while restricting a corridor's travel demand and emissions growth is the addition of HOV lanes. Only carpools and transit vehicles can use these HOV lanes. The efficiency of carpool lanes as a system improvement is a function of the characteristics of the overall HOV system. Vehicles need to remain in free-flowing HOV lanes until they reach the exit for their destination or the time benefits associated with the carpooling activity are limited. Forcing carpools out of dedicated lanes, into congested lanes, and back into dedicated lanes is an impediment to carpool formation. HOV systems that ensure faster travel through congested regions are more successful in attracting users. Many urban areas are integrating HOV bypass facilities, and HOV onramps/offramps into interconnected systems.

Signal-Timing Optimization

Traffic signal-timing improvement is the most widespread congestion management practice in the United States. During the late 1970s and early 1980s, many cities and municipalities began focusing on improving signal timing as a means to reduce fuel consumption. Traffic engineers program traffic signal green, yellow, and red times at a local traffic control box located at the intersection. Signal-timing improvements can range from a simple change of a timing plan (such as increasing green time on one leg of an intersection during a peak period), to complex computer-controlled signal coordination along an

Carpool lanes, like this one on a California freeway near the Westwood neighborhood in Los Angeles, were designed to encourage more people per car. (Corbis-Bettmann)

entire transportation corridor. By linking control boxes at consecutive intersections and coordinating the green lights along a traffic corridor, vehicles moving in platoons can pass through consecutive intersections without stopping. In general, signal-timing programs reduce the number of stops, reduce hours of stop delay, and increase overall system efficiency. Consumers benefit from reduced congestion, increased safety, reduced stress, and improved response times for emergency vehicles.

Transportation engineers model signal-timing improvements using simulation programs and system optimization routines. In optimization programming, timing plans will be changed to purposely delay some vehicles, or platoons of vehicles, whenever such a delay has the potential of reducing congestion delay for many other vehicles on the road.

Numerous studies agree that signal-timing optimization programs are cost-effective. In 1986 California had an average first-year fuel use reduction of 8.6 percent from signal-timing programs. In 1980 the Federal Highway Administration's National Signal Timing Optimization Project examined the effectiveness of signal-timing improvements in eleven U.S. cities. The study found that these cities had reduced vehicle delay by more than 15,500 vehicle hours per year, stops per intersection by more than 455,000 per year, and fuel consumption by more than 10,500 gallons per year. Signal-timing optimization also has the potential to significantly decrease vehicle emissions in urban areas. Many cities do not place signal-timing improvement near the top of their annual resource priority list for transportation funding. Future programs that employ real-time signal controls (traffic responsive control strategies) will become widespread once program resources are made available by states and municipalities for the necessary technology and expertise to implement these systems.

Rapid Incident Response Programs

Roadside assistance programs detect and rapidly respond to crashes and breakdowns to clear the vehicles from the roadway as quickly as possible. Crashes often result in lane blockages that reduce capacity until the crash is completely removed from the system. Even a vehicle stalled on the shoulder of the road can create significant delays. After the incident is cleared, additional time is required for the congestion to clear. Reducing the duration of lane blockages by even a few minutes each can save thousands of hours of total congestion delay each year.

Many urban areas are now using video detection systems to monitor traffic flow. The systems, such as Autoscope, detect changes in video pixels to estimate traffic flow and vehicle speeds on freeways and major arterials. Significant differences in average speeds from one monitor to another indicate the presence of an accident between video monitoring locations. With early detection of an incident, rapid accident response teams (either roving the system or staged in specific locations along the system) can be dispatched immediately. Cellular telephones will also play an increasing role in incident detection programs as the cellular network density increases.

A recent study of the Penn-Lincoln Parkway indicated that the annual program investment of $220,000.00 for their roadside assistance program reduces congestion delay by more than 547,000 hours per year (and a benefit-cost ratio of 30:1).

Ramp Metering

Major urban areas in the United States have begun implementing ramp metering to optimize performance of major freeway systems. A ramp meter brings vehicles on the onramps entering the freeway to a complete stop during peak periods and releases vehicles to the freeway one or two at a time. Metering prevents queues of vehicles from entering the traffic stream and disrupting freeway traffic flow. Ramp meters delay the onset of congested stop-and-go activity, allowing freeways to continue to operate at efficient flow rates for extended periods. Ramp metering delays vehicles arriving at the ramps, but the net congestion savings on the freeway more than offsets the ramp delay. Although emissions are significantly lower on the freeway systems when metering maintains smooth traffic flow, the net emissions benefits of such systems are not clear at this time. The hard accelerations associated with vehicles leaving the stop bar on the ramp result in significantly elevated emissions levels. New ramp metering research projects are trying to determine the systemwide tradeoffs in emissions.

Intelligent Transportation Systems

A variety of intelligent transportation systems (ITS) with the potential to improve traffic flow and relieve congestion are currently being explored. ITS technologies range from simple delivery of traffic information into the vehicle (helping drivers optimize route choice decisions) to complete vehicle automation. Although many of the proposed ITS strategies require development of new roadway and electronic infrastructure systems, some new technologies will evolve over the near term. Drivers can expect to access, via cellular Internet connections, detailed transportation system performance data. Electronic in-vehicle maps and systems data will allow users to optimize travel decisions and avoid getting lost. New radar systems will transmit warnings from roadside systems to vehicles, giving construction zones a new level of protection. Onboard collision avoidance systems may also prevent a significant number of rear-end crashes that cause congestion.

RELATIVE EFFECTIVENESS OF DEMAND AND SUPPLY STRATEGIES

Due to political changes regarding transportation control measures (TCMs), the CARB shifted from advocating TCMs in the late 1980s to evaluating their cost effectiveness by the mid 1990s. In 1995 they studied the cost effectiveness of twenty emissions reduction strategies funded by California vehicle registration fees. The CARB analyses indicate that signal timing, purchases of new alternative fuel vehicles, and construction of bicycle facilities can be cost effective compared to many new stationary source control measures. A videophone probation interview system designed to substitute for office trips resulted in an agency cost savings, indicating that there are still technology projects that can simultaneously attain emissions reductions and save public dollars. Of the twenty measures, eleven tried to change travel demand, and nine tried to improve traffic flow or to shift drivers to new vehicles or alternative fuels. Only six of the eleven demand management measures proved cost-effective. The technology-oriented supply improvement strategies (such as signal timing) and fuel shifts fared better, with eight of the nine measures proving cost-effective.

MODELING THE EMISSIONS IMPACTS OF TRAFFIC FLOW IMPROVEMENTS

Experts believe that current emissions models overestimate emissions at low operating speeds. They predict emissions solely as a function of average speed. However, emissions are a strong function of speed/acceleration operating conditions. Modal emission rate models demonstrate that emissions in grams/second are fairly consistent when the vehicle is operating at low to moderate speed with low acceleration rates. Under these conditions, the gram/mile emissions rates are a function of the constant gram/mile rate and the amount of time the vehicle spends on the road segment in question (which is a function of average speed). Emissions can skyrocket under high-speed conditions and conditions of moderate speed with high acceleration and deceleration rates. The operating condition corresponding to the lowest emissions and reasonably efficient fuel consumption is smooth traffic flow, probably between 20 to 35 mph, with little acceleration and deceleration.

Evaluation of traffic flow improvement projects using the current emission rate modeling regime is bound to underestimate the air quality benefits. Because emissions are activity-specific, it is important to develop methods that can estimate the effect of traffic flow improvements on trip emissions in terms of changes in speed/acceleration profiles. The Georgia Institute of Technology and University of California at Riverside are currently developing emissions models that can assess the benefits of traffic flow improvements. The application of these modal emissions models is expected to show that improved signal timing may have a pronounced impact on improving emissions.

CONCLUSION

Evidence suggests that demand management initiatives have had relatively small impacts on travel behavior and fuel consumption in the United States. Direct agency intervention at the regional level has not worked as intended. During the 1990s, economic incentives designed to internalize the personal and social costs of the automobile seemed to be the most logical and promising ways of achieving changes in travel behavior. As consumers internalize the true costs of owning and operating the personal automobile, individual tripmaking decisions become more rational, increasing system efficiency. Strategies such as congestion pricing, emission fees, and even pay-as-you drive automobile insurance received a great deal of attention in state legislatures and in the popular press. Before the public widely accepts pricing strategies in the United States, regulators need to address a variety of equity dilemmas. In an era of cheap and abundant energy, it is likely that such pricing arguments will focus on potential air quality and congestion benefits rather than on energy consumption benefits.

The most successful transportation-related economic incentive to date has been the parking cash-out program in California. Limited-scale implementation has been very successful. Regional parking pricing is likely to be a viable travel demand strategy, but will be difficult to adopt. Tax codes that allow employers to provide parking as a tax-exempt employee benefit would need to change. Employers and employees need to feel that the costs of parking are real. By implementing pricing strategies that the public will support, and simultaneously continuing public education campaigns explaining the energy and environmental problems associated with owning and operating an automobile, gradual acceptance of more widespread incentives might be achieved.

Despite the implementation and performance limitations of demand management strategies, metropolitan areas have been able to significantly improve traffic flow by implementing of transportation system improvement strategies. Lane expansion projects that provide bottleneck relief have become the focus of new roadway construction. Traffic signal timing optimization, rapid incident response systems, and implementation of HOV systems and high-occupancy toll lanes have significantly enhanced system capacity and reduced congestion. These strategies are capital intensive, but the returns on investment in congestion relief and fuel and emissions savings have been significant. The public seems much more willing to expend resources to enhance capacity than to endure less personal mobility imposed by demand management strategies.

Because of continued increases in vehicle ownership, tripmaking, and mileage, many researchers question the extent to which system improvement projects can mitigate congestion growth. As congestion worsens in major urban areas, there will be a greater focus on demand management strategies. In addition, land use strategies are getting more attention. Transit trips typically account for only 3 percent of the total trips made in the United States. Only in cases where transit is available, where origin and destination land use characteristics match desired tripmaking characteristics, and where transit costs (transit time, wait time, fares, and opportunity) are lower than automobile costs, do Americans opt to commute by transit. Proper design of land use, at densities that support transit operation and provide more opportunity to walk to common destinations, can provide significant long-term changes in travel patterns that minimize congestion, fuel consumption, and motor vehicle emissions. Because land use decisions come under local jurisdiction, federal agencies are now developing guidance documents for state and local agencies.

Randall Guensler

See also: Air Pollution; Emission Control, Vehicle; Energy Management Control Systems; Government Intervention in Energy Markets.

BIBLIOGRAPHY

Adler, K.; Grant, M.; and Schroeer, W. (1998). *Emissions Reduction Potential of the Congestion Mitigation and Air Quality Improvement Program: A Preliminary Assessment.* Transportation Research Record 1641. Transportation Research Board. Washington, DC: National Research Council.

Bay Area Economic Forum. (1990). *Congestion Pricing—The Concept.* Oakland, CA: Author.

California Air Resources Board. (1991). *Employer-Based Trip Reduction: A Reasonably Available Transportation Control Measure.* Executive Office. Sacramento, CA: Author.

California Air Resources Board. (1995). *Evaluation of Selected Projects Funded by Motor Vehicle Registration Fees, Revised.* Technical Support Document. Sacramento, CA: Author.

Cambridge Systematics, Inc. (1991). *Transportation Control Measure Information Documents.* Prepared for the U.S. Environmental Protection Agency Office of Mobile Sources. Cambridge, MA: Author.

Cambridge Systematics, Inc.; COMSIS Corporation; K. T. Analytics; Deakin, E.; Harvey, G.; and Skabardonis, A. (1990). *Transportation Control Measure Information* (December Draft). Prepared for the U.S. Environmental Protection Agency, Office of Mobile Sources. Ann Arbor, MI: Author.

Chatterjee, A.; Wholley, T.; Guensler, R.; Hartgen, D.; Margiotta, R.; Miller, T.; Philpot, J.; and Stopher, P. (1997). *Improving Transportation Data for Mobile Source Emissions Estimates* (NCHRP Report 394). Transportation Research Board. Washington, DC: National Research Council.

COMSIS Corporation. (1994). *Implementing Effective Employer-Based Travel Demand Management Programs.* Institute of Transportation Engineers. Washington, DC: Author.

Dahl, C., and Sterner, T. (1991). "Analyzing Gasoline Demand Elasticities: A Survey." *Energy Economics* 13(3):203–210.

Deakin, E; Skabardonis, A; and May, A. (1986). *Traffic Signal Timing as a Transportation System Management Measure: The California Experience.* Transportation Research Record 1081. Transportation Research Board. Washingto, DC: National Research Council.

Donnell, E.; Patten, M.; and Mason J. (2000). *Evaluating a Roadside Assistance Program: Penn-Lincoln Parkway Service Patrol.* Transportation Research Record 1683. Transportation Research Board. Washington, DC: National Research Council.

Ferguson, E. (1989). *An Evaluation of Employer Ridesharing Programs in Southern California.* Atlanta, GA: Georgia Institute of Technology.

Giuliano, G. (1992). "Transportation Demand Management: Promise or Panacea?" *APA Journal* (Summer):327–334.

Giuliano, G.; Hwang, K.; Perrine, D.; and Wachs, M. (1991). *Preliminary Evaluation of Regulation XV of the South Coast Air Quality Management District.* Working Paper No. 60. The University of California Transportation Center. Berkeley, CA: University of California.

Gordon, D., and Levenson, L. (1989). *DRIVE+: A Proposal for California to Use Consumer Fees and Rebates to Reduce New Motor Vehicle Emissions and Fuel Consumption.* Berkeley, CA: Lawrence Berkeley Laboratory, Applied Science Division.

Guensler, R. (1998). *Increasing Vehicle Occupancy in the United States. L'Avenir Des Deplacements en Ville (The Future of Urban Travel).* Lyon, France: Laboratoire d'Economie des Transports.

Guensler, R., and Sperling, D. (1994). "Congestion Pricing and Motor Vehicle Emissions: An Initial Review." In *Curbing Gridlock: Peak Period Fees to Relieve Traffic Congestion*, Vol. 2. Washington, DC: National Academy Press.

Hallmark, S; Bachman, W.; and Guensler, R. (2000). *Assessing the Impacts of Improved Signal Timing as a Transportation Control Measure Using an Activity-Specific Modeling Approach.* Transportation Research Record. Transportation Research Board. Washington, DC: National Research Council.

Hartgen, D., and Casey, M. (1990). "Using the Media to Encourage Changes in Travel Behavior" (890423). *Proceedings of the 1990 Annual Meeting of the Transportation Research Board.* Washington, DC: National Research Council.

Harvey, G., and Deakin, E. (1990). *Mobility and Emissions in the San Francisco Bay Area.* Prepared for the Metropolitan Transportation Commission. Oakland, CA: Author.

Harvey, G. (1994). "Transportation Pricing and Travel Behavior." In *Curbing Gridlock: Peak Period Fees to Relieve Traffic Congestion*, Vol. 2. Washington, DC: National Academy Press.

Haug International and LDA Consulting. (1998). *SB836 Evaluation of Rule 2202 Voluntary Replacement Measures*, Final Report. Presented to the SB 836 Oversight Committee. Los Angeles, CA: Haug International.

Hirst, E. (1978). "Transportation Energy Conservation Policies." In *Energy II: Use Conservation and Supply*, eds. P. H. Abelson and A. L. Hammond. Washington, DC: American Association for the Advancement of Science.

Kain, J. (1994). "The Impacts of Congestion Pricing on Transit and Carpool Demand and Supply." In *Curbing Gridlock: Peak Period Fees to Relieve Traffic Congestion*, Vol. 2. Washington, DC: National Academy Press.

Knapp, K.; Rao, K.; Crawford, J.; and Krammes, R. (1994). *The Use and Evaluation of Transportation Control Measures*. Research Report 1279-6. Texas Transportation Institute. Texas: Federal Highway Administration and Texas Department of Transportation.

Meyer, M. (1997). *A Toolbox for Alleviating Traffic Congestion and Enhancing Mobility*. Washington, DC: Institute of Transportation Engineers.

Mohring, H., and Anderson, D. (1996). "Congestion Costs and Congestion Pricing." In *Buying Time Symposium*. Hubert H. Humphrey Institute of Public Affairs. Minneapolis, MN: University of Minnesota.

Public's Capital, The. (1992). "The Rising Tide of User Fees." In *Governing* (April). Washington, DC: Author.

Robinson, F. (1996). "Selected Findings from the Twin Cities Congestion Pricing Study." In *Buying Time Symposium*. Hubert H. Humphrey Institute of Public Affairs. Minneapolis, MN: University of Minnesota.

Schreffler, E.; Costa, T.; and Moyer, C. (1996). *Evaluating Travel and Air Quality Cost-Effectiveness of Transportation Demand Management Projects*. Transportation Research Record 1520. Transportation Research Board. Washington, DC: National Research Council.

Shoup, D. (1997). "Evaluating the Effects of Cashing Out Employer-Paid Parking: Eight Case Studies." *Transport Policy* 4(4).

Shoup, D., and Breinholt M. (1997). "Employer-Paid Parking: A Nationwide Survey of Employers' Parking Subsidy Policies." In *The Full Costs and Benefits of Transportation*, eds. D. L. Greene, D. Jones, and M. A. Delucchi. Berlin: Springer.

Sierra Club. (1990). *Heading the Wrong Way: Redirecting California's Transportation Policies*. Sacramento, CA: Author.

South Coast Air Quality Management District. (1998). *Rule 2202—On-Road Motor Vehicle Mitigation Options Status Report*. Diamond Bar, CA: Author.

Spock, L. (1998). *Tolling Practices for Highway Facilities*. Synthesis of Highway Practice 262. National Cooperative Highway Research Program (NCHRP). Transportation Research Board, National Research Council. Washington, DC: National Academy Press.

Ullberg, C. (1991). "Parking Policy and Transportation Demand Management." Proceedings of the 84th Annual Meeting of the Air and Waste Management Association. Pittsburgh, PA.

U.S. Department of Transportation. (1997). *1995 National Personal Transportation Survey Data Files* (FHWA-PL-97-034). Federal Highway Administration. Washington, DC: Author.

U.S. Environmental Protection Agency. (1998). *Assessing the Emissions and Fuel Consumption Impacts of Intelligent Transportation Systems (ITS)*. Report Number 231-R-98-007. Washington DC: Author.

U.S. Environmental Protection Agency and U.S. Department of Transportation. (1998). *It All Adds Up to Cleaner Air*. Pilot Site Resource Kit. USEPA, Office of Mobile Sources, Transportation Air Quality Center. Ann Arbor, MI: Author.

Wachs, M. (1991). "Transportation Demand Management: Policy Implications of Recent Behavioral Research." *Journal of Planning Literature* 5(4)333–341.

Wilson, R., and Shoup, D. (1990). "Parking Subsidies and Travel Choices: Assessing the Evidence." *Transportation* 17:141–157.

Winston, C., and Shirley, C. (1998). *Alternate Route: Toward Efficient Urban Transportation*. Washington, DC: Brookings Institution Press.

Winters, P.; Cleland, F.; Pietrzyk, M.; Burris, M; and Perez, R. (2000). *Predicting Changes in Average Vehicle Ridership on the Basis of Employer Trip Reduction Plans*. Transportation Research Record 1682. Transportation Research Board. Washington, DC: National Research Council.

Zavattero, D.; Ward, J.; and Rice, D. (2000). *Analysis of Transportation Management Strategies for the 2020 Regional Transportation Plan*. Transportation Research Record 1682. Transportation Research Board. Washington, DC: National Research Council.

TRAINS

See: Freight Movement; Mass Transit

TRANSFORMERS

A transformer is an electrical component used to connect one alternating current (ac) circuit to another through the process of electromagnetic induction. The input current travels through a conductor (the primary) wound around a conductive core. The current traveling through the primary windings creates an alternating magnetic field in the core. An output conductor (the secondary) is wound around the same core so that the magnetic

Power transformer station in South Carolina. (Corbis-Bettmann)

field cuts through the secondary windings, inducing the output electrical current. For most transformers, the primary and secondary windings never come into direct electrical contact with each other. Instead, the transfer of energy from primary to secondary is accomplished solely through electromagnetic induction.

Transformers were developed through a series of scientific discoveries in the nineteenth century. Most notably, Michael Faraday showed in 1831 that a variable magnetic field could be used to create a current, thus pioneering the concept of electromagnetic induction. It was not until the 1880s that Nikola Tesla was able to use this principle to bolster his patents for a universal ac distribution network.

The majority of power transformers change the voltage from one side of the transformer to the other. The change in voltage is directly related to the number of turns each conductor (primary and secondary) makes around the transformer's core. For example, if the primary makes ten turns around the core, and the secondary makes five turns around the core, the secondary voltage will be half of the primary voltage. This type of transformer would be called a step-down transformer, since it steps down the voltage. On the contrary, if the number of turns in the primary is less than the number of turns in the secondary, the transformer will be a step-up transformer.

The power transformer also must maintain a balance of power from one side to the other. Power is the product of voltage and current. Therefore, neglecting any internal losses, if a transformer steps up the voltage by a given factor, it will also step down the current by the same factor. This has great application in the generation and transmission of electricity. It is difficult to generate extremely high voltages of electricity. Generating plants produce electricity at a relatively low voltage and high current, and step the voltage up to transmission levels through a transformer. This has the added effect of reducing transmission losses, since as the voltage increases, the current decreases, thereby reducing resistive voltage drops.

Iron (or steel) is frequently used as a transformer core because it provides adequate magnetic flux at a relatively low reluctance. In other words, iron cores, however, require additional design considerations to prevent excessive power losses and heat dissipation. The changing magnetic field inside the core creates current not only in the secondary windings of the transformer but also within the core itself. Using a laminated core constructed from small laminated plates stacked together and insulated from each other reduces these losses.

Other types of transformers include instrument, radio-frequency, wide-band, narrow-band, and electronic transformers. Each of these transformers operates similarly and is used in specific applications best suited for the transformer's design characteristics.

Brian F. Thumm

See also: Electric Motor Systems; Electric Power, Generation of; electric Power Transmission and Distribution Systems; Faraday, Michael; Magnetism and Magnets; Tesla, Nikola.

BIBLIOGRAPHY

Faraday, M. (1965). *Experimental Researches in Electricity.* New York: Dover Publications.

Fink, D. G., and Beaty, H. W., eds. (1993). *Standard Handbook for Electrical Engineers*, 13th ed. New York: McGraw-Hill.

TRANSIT

See: Mass Transit

TRANSMISSIONS

See: Drivetrains

TRANSPORTATION, EVOLUTION OF ENERGY USE AND

Transportation is energy in motion. Transportation, in a fundamental sense, is the application of energy to move goods and people over geographic distances. Freight transportation may be regarded as part of complex logistical and/or distribution systems that carry needed materials to production facilities and finished goods to customers. Transportation of passengers can serve long-distance travelers, daily commuters, and vacationers, among others. Special systems accommodate the requirements people have for mobility throughout the day. A wide variety of specific technologies and management systems are involved, and from the mid-nineteenth century, a large portion of the world's energy supply has been devoted to transportation.

When sailing vessels predominated at sea and when horses or draft animals provided basic land transport, energy demands by transport systems were small. With the advent of increasingly reliable steam power for river, lake, and ocean vessels beginning in the late 1820s, together with the spread in the 1830s of steam railways running long distances in Europe and North America, demand rose sharply in the industrialized countries for wood and coal as fuel. Coal became the fuel of choice for ocean transport as steam gradually displaced sail from the mid- to the late nineteenth century, since far more Btus per volume of fuel could be stored in the limited space of fuel bunkers.

Nearly all locomotives in Britain, and many on the Continent, used coal as well. In North America, wood—because of its plentiful supply—was the principal fuel for locomotives and for western and southern riverboats through the late 1860s. Locomotives of some railroads in Pennsylvania burned anthracite—"hard" coal. Most of the earliest coal production in the United States, starting in the 1820s, was anthracite, which was slow-burning but produced very little smoke. That virtue made anthracite the preferred type of coal for home heating and for industries within large urban centers such as Philadelphia and New York, to which the costs of transporting wood were higher than to smaller cities and towns closer to their wood supplies.

Electric propulsion street cars. (Corbis Corporation)

Railway use of wood as fuel in the United States peaked at almost 3.8 million cords per year in the 1860s and fell slowly thereafter. From the 1850s, the use of coal—preponderantly the more common type, called bituminous or "soft" coal—increased on railways and on waterways, especially in the East and Midwest as coal production in those regions accelerated. (A ton of good bituminous equaled the heating value of 1¾ to 2 cords of seasoned hardwood.) Two major factors influenced the broad conversion to coal: First, as the quantity mined increased, prices fell—from $3.00 per ton in the 1850s to about $1.00 per ton to high-volume purchasers just ten years later. Second, the country's voracious appetite for wood—for domestic and industrial fuel, for charcoal used in ironmaking, and for construction—had denuded vast stretches of the eastern forests by 1880. In that year U.S. railroads consumed 9.5 million tons of coal, which accounted for about 90 percent of their fuel supply.

Electricity, generated primarily from coal, became widely used for transport within and near cities, as trolleycar systems proliferated after the late 1880s. The enabling developments were, first, successful forms of large-scale electric generation from steam-driven dynamos and, second, a practical method of electricity distribution to the railcars. Most such cars took their power from an overhead wire, with electrical grounding through the track. In many parts of the world, trolley systems became a fixture of urban transportation in large and medium-size cities. Trolley companies usually owned their own power-generation stations and often provided a given city with its first electrical distribution network. Trolley companies were therefore eager to sell electricity and often sponsored efforts to spread the use of electric lighting systems and other electric appliances. Elevated urban railways, powered at first by tiny, coal-fired steam locomotives in the 1880s and converted later to electricity, also helped ease urban congestion in New York and Chicago. A related development by 1900 in a few of the world's major cities was the electric-powered subway.

Especially in North America, a direct effect of the electric trolley—from about 1890 through the 1920s—was the rapid growth of suburbs. In some cases, trolley systems that extended beyond city boundaries accelerated the expansion of existing suburbs. In many other cases, new suburbs sprang up, facilitated by the cheap mobility offered by newly built trolley lines. Long-distance, electric interurban rail systems also grew at a fast rate in the United States after 1890. Throughout New England, the East, and the Midwest, and in many places in the South and West, travelers and longer-distance commuters could use electric railcars connecting small towns with medium-size and large cities, over distances of up to a hundred miles or so.

Since the mid-twentieth century, petroleum has been the predominant fuel stock for energy used in transportation on land, water, and in the air. The shift began slowly. Oil-fired boilers for ships were tried in the 1860s. A railway in Russia fired more than a hundred of its steam locomotives on oil in the 1880s. By the first decade of the twentieth century—and although coal was still the favored fuel for steam transportation throughout the world—several thousand steam locomotives in the western part of the United States as well as a few ocean vessels were fired by residual oil from a rapidly growing number of refineries or, in some cases, by light petroleum.

Regional scarcities of coal initially drove these uses. As petroleum became more abundant and as its price fell, oil became more attractive. In firing boilers, fuel oil possessed only a slight advantage over good-quality coal in Btus per unit volume. But liquid fuels were much easier to handle and store than coal. Competitive pressures kept the prices per Btu of residual oil and coal quite close.

The transition to oil for the boilers of oceangoing steamships was well along by the 1920s. A coincidental development in the early part of the century was the steam-turbine engine for ships, which mostly displaced multicylindered steam piston engines in marine applications; the turbine is more energy-efficient at constant speeds maintained over long periods of time. A decreasing number of older riverboats, towboats, lake boats, and freighters continued to use coal through the 1950s (and a rare few such steam vessels were still in use in the 1980s). As late as 1957, U.S. railroads purchased $48 million worth of bituminous coal for steam locomotive fuel—and $374 million of diesel fuel. By 1960, all use of steam locomotives on major lines in the United States and Canada had ended, and by 1970 the change to diesel locomotives was virtually complete worldwide, except in some parts of China, India, Africa, and Eastern Europe. Thus the use of coal as a transport fuel ended for the most part in the late twentieth century, except for electricity (generated from coal) used on a small percentage of the world's railway mileage.

The burgeoning growth in the use of petroleum-based fuels in transportation came with the widespread use of automobiles and trucks after 1910 and the beginning of air travel after World War I. In the 1890s, little gasoline was consumed. It was regarded as an explosively dangerous by-product of the refining process in the production of kerosene (a petroleum distillate) for lighting. But with the invention of practical automobiles in the 1890s in the United States and Europe, demand slowly grew for a high-energy fuel that could be used readily in small, internal-combustion engines. In such use, the high Btu per unit volume of gasoline was a distinct advantage. The story of the huge oil companies and gasoline distribution systems that arose in the United States and overseas in the early twentieth century is one driven entirely by the rise of the automobile.

During the first two decades of the twentieth century, several hundred firms in Britain attempted automotive production, and nearly 3,000 tried to do so in the United States. But there was one car that created much of the rising demand for gasoline: the Ford Model T. Between 1908 and 1927, 15 million Model Ts rolled out. Ford built half of the automobiles manufactured in the world in 1920. The Model T's low price and high reliability created a vast pool of willing purchasers. A result was to sweep away the economic incentives that might have supported a viable market for alternative fuels or motors for automobiles, such as electric or steam propulsion. Versus the electric, a gasoline engine provided much greater operating range; versus steam (such as the compact boiler and engine used in the temporarily popular Stanley steamer), a gasoline engine was much simpler and less expensive.

After 1910, gasoline-powered trucks became common. Their utility eventually replaced horse-drawn vehicles in farm-to-market and intracity use, though that replacement took decades. Even in the industrial world, horses provided a large but declining share of intracity freight transport through at least the 1930s and early 1940s. Highway buses became popular in the 1920s. Such buses quickly displaced the long-distance, electric interurban rail systems of the 1890s and early 1900s.

Rudolph Diesel first patented his engine in 1892, but it was not until 1910 that a successful application was made in a vessel. The first diesel engines suitable for use in trucks came in the early 1920s. Annual U.S. production of diesels did not exceed one million horsepower until 1935. Over the next two years that production doubled, largely because of railroad locomotive applications. Due to the high compression pressures a diesel engine must withstand in its cylinders, it must be built heavier than a gasoline engine of similar output. But the diesel's advantage is that it is the most energy-efficient internal-combustion engine yet developed. Marine diesel design advanced in the 1930s and during World War II, leading to greater use in larger transport vessels from the late 1940s. The demand for high-quality diesel fuel—similar to kerosene—thus expanded.

With the inception of the U.S. interstate highway system in the late 1950s, the modern long-distance highway tractor and its semitrailer evolved and became the predominant method of freight transportation for time-sensitive, high-value goods. Elsewhere in the world, different forms of the long-distance truck took over most freight transport. Today, in the United States and elsewhere, these trucks are almost universally diesel. In the United States in the 1990s, some 39 percent of intercity ton-miles of commercial freight were borne by rail, 28 percent by trucks, and 15 percent on lakes, rivers, and canals (the rest was carried by pipelines and a fraction of 1% by air). Diesel fuel powered nearly all this road, rail, and water transport.

An aspect of transportation that has a heavy bearing on energy consumption is horsepower. Transport can be measured in units of *work*—tons hauled per mile, for example—but energy consumption is more proportionately related to *power*, such as tons per mile *per unit of time*. For a given load carried over a given distance, higher speed requires more power and hence more energy. Thus there is a trade-off especially important in all forms of transportation: Fuel-saving strategies often involve reduced speeds, while increased transport capacities of systems per unit of time often necessitate higher speeds and therefore require higher fuel use per ton-mile or per seat-mile.

Aviation has been powered from its beginnings by gasoline. Only spark-ignited gasoline engines proved able to provide the extremely high power-to-weight ratios needed by aircraft engines, where weight is always at a premium. During World War II, German and British engineers developed the gas-turbine engine for aircraft (known commonly as the jet engine). Such engines ran best on a variation of kerosene. In the "turbojet" form of the gas-turbine, there is no propeller; in the "turboprop" form, the central turbine shaft of the engine turns a propeller through a gearcase.

In transport applications in the 1950s, the jet's favorable power-to-weight ratio and efficiency at high altitudes resulted in a travel revolution. (The first commercial jet, the de Havilland Comet, first flew with paying passengers in 1952; the pioneering Boeing 707 came a few years later.) Speed was only one factor. Flying at more than 30,000 feet—well above the turbulent weather systems ordinarily encountered by piston-engined aircraft, which operate most efficiently up to about 20,000 feet or below—the comfort level of jets allowed millions of people to accept these new planes as long-distance transport.

Turboprop types of aircraft proved more reliable and cheaper to operate than piston-engined airplanes for short hauls and more fuel-efficient than turbojets in such use. Overall airline patronage shot up dramatically after 1960. Since the 1980s, air express has boomed as well. Thus most aviation fuel used today is well-distilled kerosene, with a small portion of antifreeze for the extremely low ambient temperatures at high altitude. High-octane gasoline powers much of the "general aviation" sector of private aircraft, where piston engines are still the norm for planes of up to four or five passengers.

During the oil shortages of the 1970s and early 1980s, researchers in the United States and Europe investigated alternative fuels for transportation. Shale oil deposits were tapped, and other projects experimented with the conversion of various grades of coal to oil (a technology developed by Germany in World War II). Two U.S. railroads (the Burlington Northern and the Chessie System) seriously considered a new-technology steam locomotive that would burn coal in a gasification furnace, thus controlling emissions.

Test flight of the De Havilland Comet 1 jet transport prototype. (Corbis-Bettmann)

Meanwhile, diesel manufacturers experimented with modified engines burning blends of finely pulverized coal mixed in fuel oil. While these engines ran reasonably well, tiny amounts of noncarbon impurities in the pulverized coal inevitably built up within the engines' cylinders, causing rough running or damage. Stabilizing oil prices from 1982 through 1999 removed most incentives for further research.

From 1980 to 1998, total U.S. petroleum consumption ranged between 15 million and 19 million barrels daily, with consumption at the end of the 1990s about the same as in the peak of 1978. Use of petroleum for U.S. transportation was slightly less than 10 million barrels per day in 1980 and was about 12 million barrels in 1997, or some 60 percent of total domestic oil consumption. That oil consumption rate supported a growth in U.S. freight of about 20 percent from 1980 to the late 1990s, amounting in 1997 to a bit less than 15,000 ton-miles of freight transported per person per year.

Travel has also multiplied. The U.S. Departments of Commerce and Transportation surveyed long-distance trips in 1977 and 1995 (with a long-distance trip defined as a round-trip of at least a hundred miles each way). Between those years, long-distance trips per person per year by automobile grew by almost 90 percent; by air, such trips nearly trebled. Increased mobility is a central feature of modern life.

All modes of transport—by road, rail, water, and air—have increased engine fuel efficiency and have instituted other fuel-saving strategies since the oil shortages of the 1970s. New turbine-engine aircraft have doubled the average fuel efficiency of earlier jets. On railroads, a gallon of diesel fuel moved 235 ton-miles of freight in 1980; in the early 1990s, a gallon moved more than 360 ton-miles. Average automobile fuel mileages have improved. Redesigned hull shapes and large, higher-efficiency diesel engines have lowered fuel costs in maritime transport.

New vehicles, propulsion systems, and fuels are on the horizon: "hybrid" automobiles (combining an electric motor, batteries, and a gasoline or diesel engine), better electric cars, greater use of compressed natural gas (CNG) and propane for urban fleet vehicles such as taxis and delivery trucks, fuel cells (using gasoline, natural gas, or propane as the feedstock for hydrogen), methanol or ethanol to supplement gasoline, and other technologies are advancing. For such technologies and fuels successfully to compete with oil in the transportation market, the techniques to liquefy and compress natural gas would need to become cheaper, new types of storage batteries must be made practical, and the necessary infrastructure to support CNG-, fuel-cell-, and electric-powered cars and trucks would need to be developed. The huge worldwide investment in petroleum extraction, refining, and distribution makes economic change in fuel technologies difficult.

In Europe, Japan, and the United States, newer high-speed electric railroad trains are being advanced. Japan has shelved its development since the 1980s of magnetic-levitation trains, but German firms continue work on such systems for possible application in Germany (a "maglev" line has been planned between Berlin and Hamburg) and elsewhere. Engine development for aircraft has recently cut the number of turbojet engines required for a long-distance, high-capacity transport airplane from three or four to two, cutting fuel consumption per seat-mile, and further fuel efficiency increases can be expected in the years ahead.

William L. Withuhn

See also: Aircraft; Air Travel; Automobile Performance; Aviation Fuel; Diesel, Rudolph; Diesel Cycle Engines; Diesel Fuel; Electric Motor Systems; Fossil Fuels; Freight Movement; Fuel Cell Vehicles; Gasoline and Additives; Gasoline Engines; Hybrid Vehicles; Kerosene; Locomotive Technology; Magnetic Levitation; Mass Transit; Methanol; Petroleum Consumption; Propellers; Railway Passenger Service; Ships; Steam Engines; Tires; Traffic Flow Management.

BIBLIOGRAPHY

Association of American Railroads. (1983). *Railroad Facts.* Washington, DC: Author.

Babcock & Wilcox. (1928). *Marine Steam.* New York: Author.

Davies, R. E. G. (1972). *Airlines of the United States Since 1914.* London: Putnam.

Davies, R. E. G. (1964). *A History of the World's Airlines.* London: Oxford University Press.

Flink, J. J. (1970). *America Adopts the Automobile, 1895–1910,* Cambridge, MA: MIT Press.

Flink, J. J. (1988). *The Automobile Age.* Cambridge, MA: MIT Press.

Grosser, M. (1978). *Diesel, the Man and the Engine.* New York: Atheneum.

Johnston, P. F. (1983). *Steam and the Sea.* Salem, MA: Peabody Museum of Salem.

Middleton, W. D. (1961). *The Interurban Era.* Milwaukee: Kalmbach Publishing Co.

Middleton, W. D. (1967). *The Time of the Trolley.* Milwaukee: Kalmbach Publishing Co.

Nevins, A., and Hill, F. E. (1957). *Ford: Expansion and Challenge, 1915–1933.* New York: Scribner.

Nevins, A., and Hill, F. E. (1954). *Ford: the Times, the Man, and the Company.* New York: Scribner.

Sternberg, E. R. (1981). *A History of Motor Truck Development.* Warrendale, PA: Society of Automotive Engineers.

U.S. Department of Transportation, Bureau of Transportation Statistics. (1999). *Pocket Guide to Transportation.* Washington, DC: Author.

U.S. War Industries Board, General Bureau of Planning and Statistics. (1918). *The World's Steamship Fuel Stations.* Washington, DC: Author.

Wakefield, E. H. (1994). *History of the Electric Automobile.* Warrendale, PA: Society of Automotive Engineers.

White, J. H., Jr. (1997). *American Locomotives: an Engineering History, 1830–1880,* rev. and expanded ed. Baltimore: Johns Hopkins University Press.

White, J. H., Jr. (1993). *The American Railroad Freight Car.* Baltimore: Johns Hopkins University Press.

TREVITHICK, RICHARD (1771–1833)

The father of high pressure steam power was born at the village of Illogan in Cornwall, England, on April 13, 1771, the only boy after five older sisters. His father, also Richard, was a mine "captain," as mine managers were called. In 1760 Richard, Sr., had married Anne Teague of Redruth. Her family included several more mine captains.

While the village schoolmaster thought the younger Richard disobedient, slow, obstinate and spoiled, the positive sides of these qualities emerged in independent thought, technical planning, persistence, and the loyalty of his workers and friends. His uninspiring record as a student and lack of formal training in engineering mattered less as he began work in the mines. There his talent and interest in

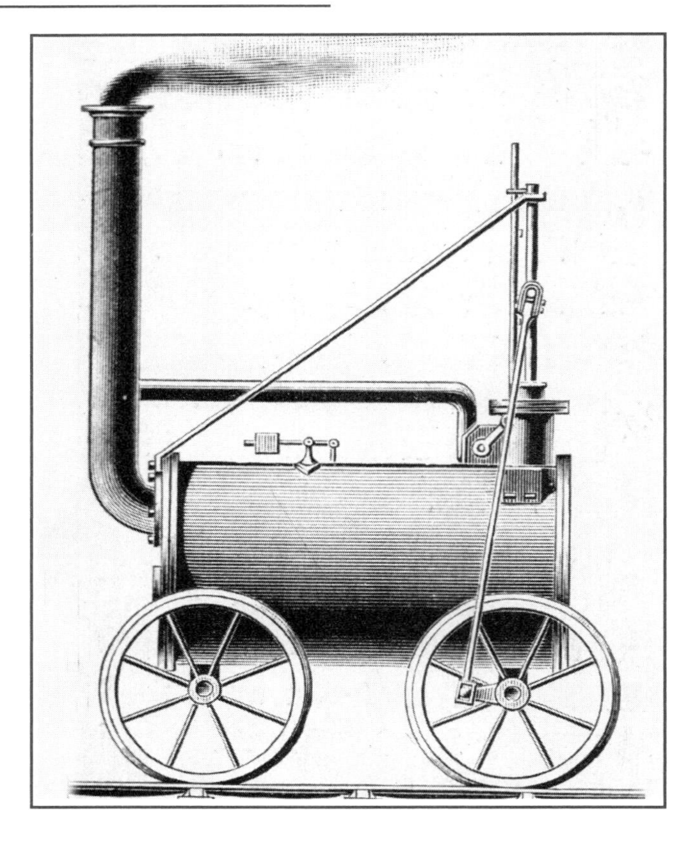

Richard Trevithick's locomotive around 1804. (Archive Photos, Inc.)

engineering gainied early notice. First employed as a mine engineer in 1790 at age nineteen, he was being called on as a consultant by 1792.

The steam-power systems in Trevithick's youth were massive but lightly loaded low-pressure engines. This technology was controlled by Boulton & Watt, whose business acumen had extended patents far beyond their normal expiration dates. The royalties typically took one-third of the savings in fuel over the Newcomen atmospheric engine. Another necessary evil was the expense of fuel, as coal was not locally mined. Cornish engineers worked incessantly to design and invent their way past these limitations.

Working around the Watt patents, he ensured safety while progressively raising the pressure of his systems to ten times atmospheric pressure, avoiding Watt's condenser altogether. This threat to the income of Boulton & Watt was met with a court injunction to stop the construction and operation of Trevithick's systems. Boulton & Watt had ruined other competitors, and their alarmist views of high pressure steam were largely due to lack of patent control in that emerging technology.

In 1797, Trevithick married Jane Harvey, lost his father, Richard, Sr., and made his first working models of high pressure engines. His wife was of the established engineering and foundry family of Hayle.

Trevithick's high-pressure engine models, including one with powered wheels, were the seeds for a line of improvements to steam engines and vehicles by many developers through the following decades. These departed from the preceding vehicle concepts and constructs of Murdock or Cugnot. These "puffer" engines made better use of the expansive properties of steam since they worked over a wider pressure differential. Using no condenser, Trevithick directed the exhaust up the chimney, inducing draft to increase the firing rate proportional to engine load, avoiding Watt's patents.

The first successful steam road vehicle was demonstrated climbing a hill in Camborne, Cornwall, on Christmas Eve, 1801. A new carriage, fitted with a proper coach body, was demonstrated in London in 1803. Eventually, a backer appeared with confidence and purchased a share in the developments. Samuel Homfray had Trevithick construct the first successful railway locomotive for his Penydarren Ironworks in South Wales. It was first run on February 13, 1804. Eight days later it won Homfray a wager of five hundred guineas. Homfray's existing cast-iron, horse-drawn rails were not yet suitable for continuous use. A similar locomotive was built at Newcastle-on-Tyne in 1805 to Trevithick's design. Again, the wooden rails were too light for the service. Trevithick's final showing of his locomotive concept was on a circular track at Gower Street in London where "Catch-me-who-can" gave rides for a shilling, but did not attract investment. Future developers would found an industry on Trevithick's technical generosity.

Trevithick's engineering reputation was established by the thirty high-pressure engines he built for winding, pumping and iron-rolling in Cornwall, Wales and Shropshire. His patent work and correspondence with members of the Royal Society might have brought him wider notice, but he was on a technical collision course with Watt and the engineering establishment, and conservative investors stayed away. Trevithick was left to financial alliances with indecisive or unscrupulous businessmen. While a near-fatal bout of typhus in 1811 kept him from work for months, his partner Richard Dickinson led him into bankruptcy in 1811.

Trevithick's high-pressure steam engines attracted attention, but his mechanical improvements enabling the boiler to withstand ten atmospheres of pressure were even more significant to power plant economy and practicality. He doubled the boiler efficiency. His wrought-iron boiler fired through an internal flue, the "Cornish" boiler became known worldwide. He applied the high-pressure engine to an iron-rolling mill (1805), a self-propelled barge using paddle-wheels (1805), a steam dredge (1806) and to powering a threshing machine (1812).

Some of Trevithick's other novel accomplishments and inventions include steam quarry drilling, applications for his superior iron tank construction, a "recoil engine" like Hero's Aeolipile, mining engines and mills in Peru, a one-piece cast-brass carbine for Bolivar's army, ship-raising by flotation, being the first European to cross the Isthmus of Nicaragua (from necessity), recoil-actuated gun-loading, iron ships, hydraulic dockside cranes, methods to drain Holland's lowlands, mechanical refrigeration, water-jet-propelled ships, and a portable heat storage heater.

Throughout his lifetime, Trevithick continued to measure his personal success in terms of his technical success in maintaining the fine balance of economy, utility and safety. At the end of his life he wrote, "I have been branded with folly and madness for attempting what the world calls impossibilities, and even from the great engineer, the late Mr. James Watt, who said...I deserved hanging for bringing into use the high-pressure engine. This so far has been my reward from the public; but should this be all, I shall be satisfied by the great secret pleasure and laudable pride that I feel in my own breast from having been the instrument of bringing forward and maturing new principles and new arrangements of boundless value to my country. . . . the great honour . . . far exceeds riches."

After becoming ill during a consulting stint at Dartford, Kent, England, Trevithick died, April 22, 1833, and was buried in a pauper's grave.

Continual disappointment in his business affairs kept him in relative obscurity even as his developments were reshaping the industrial world. Today, even the engineering community rarely notes his successes as the inventor of the first self-propelled road carriage we could call an automobile, and the railway locomotive.

Karl A. Petersen

See also: Watt, James.

BIBLIOGRAPHY

Boulton, I. W. Marginal notes, Science Museum copy of: Fletcher, William (1891). *Steam on Common Roads.*

Hodge, J. (1973). *Richard Trevithick: An Illustrated Life of Richard Trevithick.* Aylesbury: Shire Publications Ltd.

Rolt, L. T. C. (1960). *The Cornish Giant, The Story of Richard Trevithick, Father of the Steam Locomotive.* London: Lutterworth Press.

Trevithick, F. C. E. (1872). *Life of Richard Trevithick, with an Account of His Inventions.* London: E. & F.N. Spon.

TRIBOLOGY

The word tribology is derived from the Greek word, "tribos," which means "rubbing." Tribology covers the science of friction, lubrication and wear. Virtually any machine ever invented consists of components that must slide or roll relative to other components. Resistance to sliding and rolling caused by friction results in significant loss of energy. In fact, it has been estimated that as much as 33 percent of energy produced worldwide is used to overcome friction.

Friction and wear have many undesirable effects on individuals and society that are not widely known. Transportation vehicles have hundreds of moving parts that are subject to friction. Auto-mobiles generally wear out after about 100,000 miles of operation. Largely because of innovations to reduce friction and improve lubricants, the average lifetime (14 years in 2000) and miles of operation continue to increase. Since it requires as much energy to produce an automobile as it does to operate it for 100,000 miles, extending the life of an automobile would save considerable energy. The same applies to the steam and natural gas turbines that supply electricity and the home appliances that consume that electricity.

Friction and wear is even more important for national security. Downtime of military hardware as parts wear out and lower output power of military engines due to high friction can contribute to decreased effectiveness of the military and increases the costs of keeping highly specialized equipment in operation.

Wear of medical devices and biomaterials can affect quality of life. Wear of tooth fillings, artificial joints and heart valves can be inconvenient, costly (more frequent replacement) or even life-threatening (premature breakdowns). Wear of components can also cause accidents. Worn brakes and tires can cause automobile accidents, worn electrical cords can result in electrocution and fires and worn out seals can lead to radiation leaks at nuclear power plants.

RESEARCH

In the last few decades of the twentieth century the field of tribology has undergone a great surge in interest from industry and the scientific community. This is likely due to greater interest in conserving energy and natural resources. By lowering friction we conserve energy, and by lowering wear we increase product life, which conserves raw materials and the energy needed to turn raw materials into useful technology. However, there are many challenges to be overcome in the study of tribology.

Tribology is a surface phenomenon, and surfaces are extremely complex. As seen in Figure 1, when surfaces are sufficiently magnified we can see they are not flat, but consist of a series of peaks and valleys. In addition to being geometrically complex, surfaces are also chemically complex. Surfaces typically react with oxygen in the air to form a thin oxide film on the original surface. A thin gaseous film then generally forms on top of the oxide film. These films, which are usually so thin that they are transparent to the eye, significantly affect friction and wear.

Besides surface topography, the hardness of the sliding materials, thickness and properties of oxide films, temperature, and type of lubricant all affect tribology. These factors overlap many fields of study, including physics, chemistry, material science and mechanical engineering. A complete understanding of tribology will require scientists and engineers from these diverse disciplines to work together. Traditionally, interdisciplinary research in tribology has not been done due to researchers being unwilling to cross the boundaries of their own discipline.

Surfaces are not easily observed during sliding. Tribologists study surfaces before and after sliding to

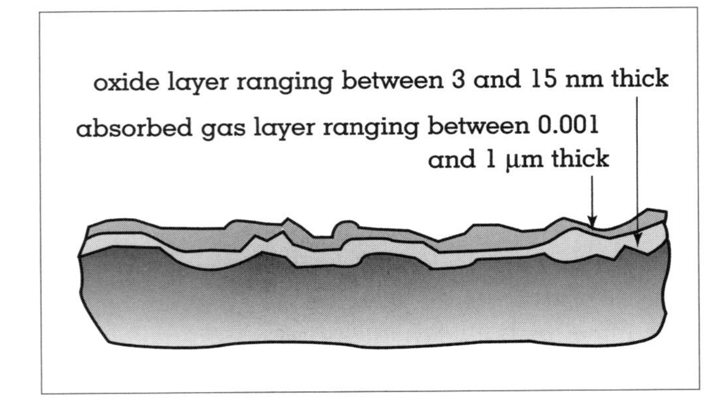

Figure 1.
Roughness of a "smooth" surface and surface films.

determine chemical and geometric changes that occurred, and to infer how these changes came about and how they affected friction and wear. Because of the microscopic nature of tribology research, and the difficulty in duplicating real world conditions, it is not easy to observe and determine the cause of wear.

HISTORY OF LUBRICANTS

The most common method of minimizing friction and wear is through lubrication. The first recorded use of a lubricant was in ancient Egypt in 2400 B.C.E. Records show that they would pour a lubricant in front of a sledge being used to move a stone statue, which weighed tens of tons. The lubricant used was probably either water or animal fat. Various lubricants derived from animals and vegetables were used for thousands of years. Some examples of lubricants derived from animals are sperm oil, whale oil, and lard oil. Sperm oil was taken from a cavity in the head of sperm whales, whale oil from whale blubber, and lard from pig fat. Examples of vegetable lubricants are olive oil, groundnut oil and castor oil.

Lubricants began to receive significantly more attention from the industrial and scientific community in the mid-1800s with the introduction of mineral oils as lubricants. These proved to be effective lubricants; the demand for their use in machinery led to the development of many oil companies.

One common application for oils is the lubrication of automotive engines. When automotive oils were first introduced, their viscosities were classified as light, medium, medium heavy, heavy and extra heavy. This method of classification was too subjective and led to some engines being improperly lubricated. To remedy this, the Society of Automotive Engineers in 1912 introduced a quantitative numbering system for automotive viscosity. These numbers range from zero to fifty with lower numbers indicating lower viscosities. A general trend observed in oils is that as temperature increases, viscosity decreases. If viscosity becomes too small, the oil will be ineffective; if it is too high, the oil will not flow properly. Multigrade oils are now available that behave as lower viscosity oil at lower temperatures and higher viscosity oil at higher temperatures. For instance, a 10W-30 oil would have the viscosity of a 10-grade oil at low temperatures and the viscosity of a 30-grade oil at high temperatures. The use of multigrade oils minimizes change in viscosity with temperature.

In addition to viscosity grade, automotive oils also have a quality designation based on the API Engine Service classification system. This system was developed through the combined efforts of the American Society of Testing Materials (ASTM), the Society of Automotive Engineers (SAE), the American Petroleum Institute (API), automotive manufacturers and the United States military. API ratings describe the oil's ability to provide resistance to factors such as wear and corrosion. The original API rating was termed SA. This type of oil is suitable only for engines operated in very mild conditions. Oils with an API rating of SB are of higher quality than SA; SC has a higher API rating than SB, and so on.

In addition to liquid lubricants, solid lubricants such as graphite and grease have also been used for several centuries. Graphite was mined in the sixteenth century in Cumberland, England. It was originally called "black-lead" and was used in writing instruments. It is still used in "lead" pencils. There are numerous reports of graphite being used to lubricate machinery and silk looms in the nineteenth century. A common modern application for graphite is in the lubrication of door locks.

A specification for grease was given in 1812 by Henry Hardacre as follows: "one hundred weight of plumbago (graphite) to four hundred weight of pork lard or beef suet, mutton suet, tallow, oil, goose grease or any other kind of grease or greasy substance, but pork lard is the best, which must be well mixed together, so as to appear to the sight to be only one substance." Grease is used in many industrial applications including lubrication of gears and bearings.

CONSERVATION OF LUBRICANTS AND ENERGY

Most lubricating oils and greases use mineral oil as their base component. However, oil is a natural resource with a finite supply. Thus, it is imperative that measures such as recycling and extending the life of lubricants be taken to conserve the world's supply of oil and energy. There are approximately 2.5 billion gallons of waste oil generated each year in the United States. Only 83 million gallons are refined and reused, while one billion gallons are burned as fuel and another billion gallons or more are released into the environment. In response to these statistics, the United States Congress encouraged recycling of used oil through the Energy Policy and Conservation Act of 1975 and Resource Conservation and Recovery Act of 1976. The United States Environmental Protection Agency has stated that used oil is an environmental hazard, making the need for recycling even more critical.

Extending the interval between oil changes in automobiles and machinery is another way to conserve lubricants. Oil life is limited by depletion of oil additives, overheating, chemical contamination, and contamination of the oil by foreign particles. Using improved additives and filtering of particles, oil life can be considerably extended.

HISTORY OF FRICTION

DaVinci is well known for his famous paintings, such as the Mona Lisa, but his genius had amazing breadth. The scientific study of friction began with his research in the later part of the fifteenth century. He performed experiments in which he measured the force required to slide wood blocks over a wooden table. One of his most important findings was that the force F, required to produce sliding, increases in proportion with the applied load, L. DaVinci also defined what is today called the coefficient of friction, f, between two sliding objects as:

$$f = F/L$$

where F is friction force and L is normal load, as shown in Figure 2.

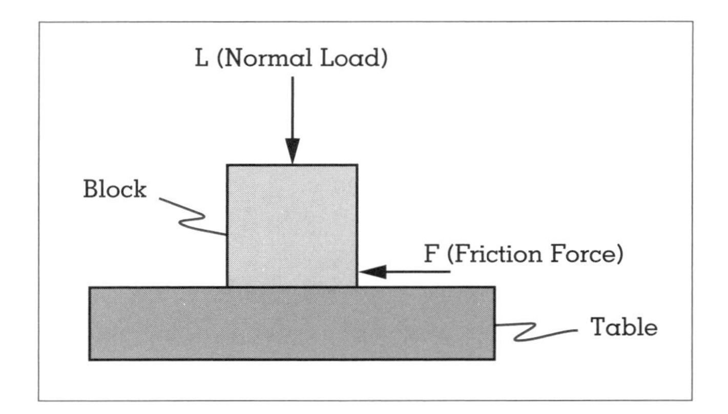

Figure 2.
The friction force between sliding objects resists movement. In this schematic, the block is being moved to the right.

DaVinci's experiments on friction also formed the basis for what today are called the first two laws of friction:

1. The force of friction, F, is directly proportional to the applied load, L.
2. The force of friction is independent of the apparent area of contact.

In general, a larger coefficient of friction between sliding bodies will require a larger force to produce sliding and, hence, more energy will be consumed. The most common method to reduce frictional energy losses is to lubricate the sliding surfaces. The reduction in coefficient of friction due to lubrication can be dramatic; typical values for dry sliding can range from 0.2 to 0.3, while typical values for lubricated sliding can range from 0.03 to 0.12.

The phenomenon of frictional heating is well known; for example, starting fires through the use of friction has been common since prehistoric times. However, frictional heating is generally a detrimental phenomenon. The temperature rise on sliding surfaces that occurs due to frictional heating generally produces a decrease in surface hardness, causing the surfaces to wear more easily, thus decreasing product life.

From an energy consumption point of view, we desire friction to be as small as possible. However, there are specific applications where friction should not be minimized. One common example is the brakes used in automobiles. If friction is too low, automobiles will not stop in a reasonable distance. However, if the friction is too high, a jerky ride will be produced. Other examples where high friction is desirable are shoe soles, nails, screws, and belt drives.

It is interesting to note that although friction has been studied for hundreds of years, there is no universal agreement on the fundamental mechanisms of how friction is produced. The two most popular theories are interlocking and adhesion. The interlocking theory states that as two rough surfaces slide over each other, the peaks on one surface "interlock" with peaks on the adjacent surface, producing resistance to sliding. The adhesion theory suggests that as the sliding surfaces contact each other, adhesive bonds form. These molecular bonds must be broken to continue sliding, which results in increased friction. The cost of not understanding the fundamentals of friction is high. Estimates have shown that as much as 0.5 percent of industrial countries' gross national products are wasted because we do not know enough about minimizing frictional losses during sliding.

FUTURE ADVANCES IN LUBRICATION

The transportation industries are being challenged with increasingly stringent government regulations that demand improved vehicle fuel economy and reduced emissions. Engine oils with greater ability to conserve energy are becoming more important as a way to lower engine friction and thus conserve natural resources. One specific approach under development is to produce engine oils with a friction-reducing additive such as molybdenum dialkyldithiocarbamate which has showed promise in developmental engine tests.

Another approach being used to conserve energy is to increase combustion temperatures in engines. This results in increased engine efficiency that produces energy savings. Combustion temperatures may eventually become so high that liquid lubricants break down and become ineffective. A possible approach for lubrication under these extreme temperatures is vapor phase lubrication, which refers to the deposition of lubricant from an atmosphere of vaporized lubricant. The vaporized lubricant is delivered in a carrier gas, such as nitrogen, to the component to be lubricated where it reacts chemically to form a solid lubricant film. These films are effective at much higher temperatures than liquid-based lubricants. Other approaches that can be used in components that operate in high operating temperatures are the use of gas bearings or magnetic bearings.

Gary C. Barber
Barbara Oakley

See also: Automobile Performance; Materials

BIBLIOGRAPHY

Booser, R. E. (1988). *Handbook of Lubrication, Theory and Practice of Tribology: Volume II.* Boca Raton: CRC Press.

Bowden, F. P., and Tabor, D. (1964). *The Friction and Lubrication of Solids, Part II.* Oxford: Clarenden Press.

Cameron, A. (1996). *Principles of Lubrication.* New York: John Wiley.

Czichos, H. (1978). *Tribology.* Amsterdam: Elsevier.

Dowson, D. (1979). *History of Tribology.* New York: Longman.

Fuller, D. D. (1984). *Theory and Practice of Lubrication for Engineers.* New York: John Wiley.

Groeneweg, M.; Hakim, N.; Barber, G. C.; and Klaus, E. (1991). "Vapor Delivered Lubrication of Diesel Engines—Cylinder Kit Rig Simulation." *Lubrication Engineering,* 47(12):1035.

Hamrock, B. J. (1994). *Fundamentals of Fluid Film Lubrication.* New York: McGraw-Hill.

Hutchings, I. M. (1992). *Tribology: Friction and Wear of Engineering Materials.* London: Edward Arnold.

Jost, H. P. (1975). "Economic Impact of Tribology." *Mechanical Engineering,* 97:26.

Ku, P. M. (1978). "Energy Conservation Through Tribology." *Tribology International* 11:153.

Ludema, K. C. (1996). *Friction, Wear, Lubrication: A Textbook in Tribology.* Boca Raton: CRC Press.

Peterson, M. B., and Winer, W. O., eds. (1980). *Wear Control Handbook.* New York: ASME.

Rabinowicz, E. (1995). *Friction and Wear of Materials.* New York: John Wiley.

Sarkar, A. D. (1980). *Friction and Wear.* New York: Academic Press.

Szeri, A. Z. (1980). *Tribology.* New York: McGraw-Hill.

Tung, S. C., and Tseregounis, S. I. (2000). "An Investigation of Tribological Characteristics of Energy-Conserving Engine Oils Using a Reciprocating Bench Test." SAE 2000-01-1781.

TRUE ENERGY COSTS

Market prices of energy often diverge from the true cost to society of consuming that energy. Two of the most common reasons for that divergence are external costs and subsidies, both of which make consumers think that energy is less expensive to society than it really is, and hence lead to more consumption of energy than would be economically optimal.

EXTERNAL AND INTERNAL COSTS

According to J. M. Griffin and H. B. Steele (1986), external costs exist when "the private calculation of costs differs from society's valuation of costs." Pollution represents an external cost because damages associated with it are borne by society as a whole, not just by the users of a particular fuel. Pollution causes external costs to the extent that the damages inflicted by the pollutant are not incorporated into the price of the fuel associated with the damages. External costs can be caused by air pollution, water pollution, toxic wastes, or any other damage to the environment not included in market prices for goods.

Some pollutants' external costs have been "internalized" because the resulting damage has already been incorporated into the price of energy by a tax on the energy source, or because an emissions-trading regime has been established to promote cost-effective control of that pollutant. The pollutants may still cause environmental damage, but these damages have been made internal to the economic calculations of consumers and firms by the taxes or emissions-trading schemes. They are thus no longer external costs.

For example, sulfur emissions from utility power plants in the United States are subject to an emissions cap and an allowance-trading system established under the Clean Air Act. An effective cap on annual sulfur dioxide emissions took effect in 2000, so no more than 8.95 million tons of SO_2 can be emitted annually. Utilities that want to build another coal plant must purchase sulfur emission allowances from others who do not need them. This system provides a market incentive for utilities to reduce their sulfur emissions as long as the cost of such reductions is less than the price of purchasing the allowances.

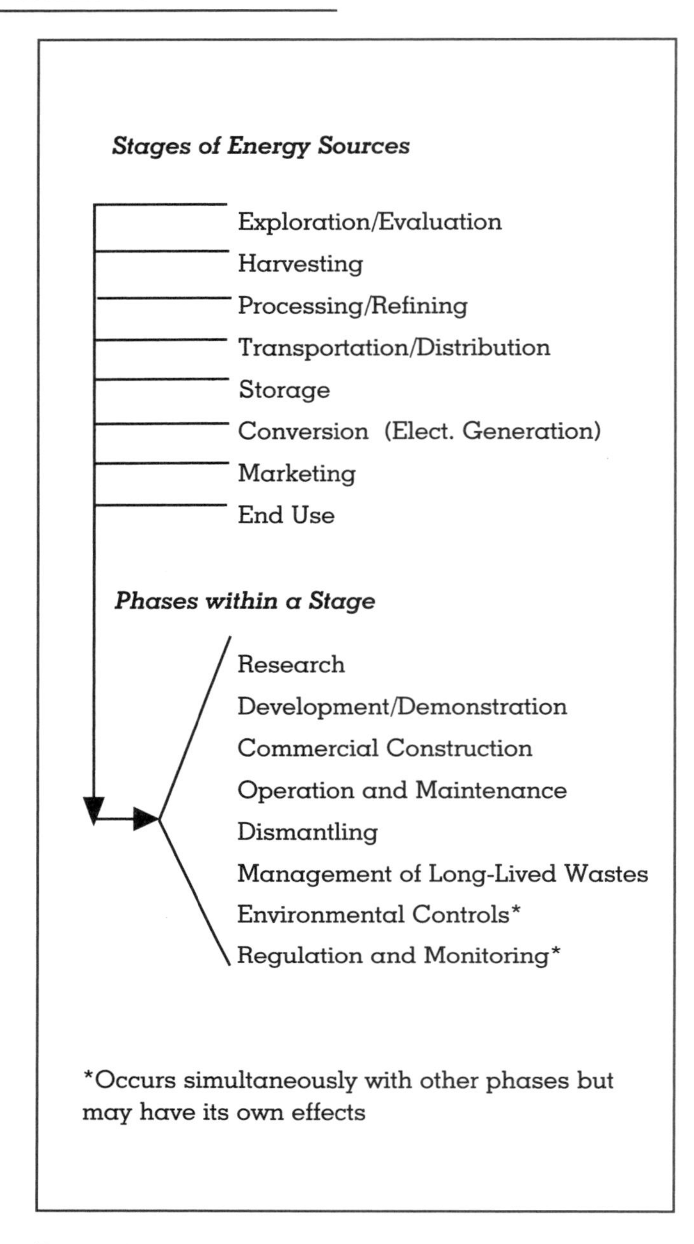

Figure 1.
Steps in energy production, processing, and use.
SOURCE: Holdren, 1981.

Subsidies

Subsidies represent an often hidden financial benefit that is given to particular energy sources by government institutions in the form of tax credits, research and development (R&D) funding, limits on liability for certain kinds of accidents, military spending to protect Middle East oil supply lines, below-market leasing fees for use of public lands, and other forms of direct and indirect government support for energy industries. Subsidies affect choices between different energy sources (which garner different levels of subsidy) and between energy-efficiency and energy supply choices (because some energy-efficiency options that would be cost-effective from society's point of view may not be adopted when energy prices are kept artificially low by the subsidies).

Subsidies can be created to reward important political constituencies, or they can promote the adoption of particular technologies. For example, an energy technology that has desirable social characteristics and low external costs might be a recipient of tax subsidies in the early stages of its development, so that it can gain a foothold against more established energy forms. This kind of support can be especially important for mass-produced technologies that have high costs when few units are being manufactured, but could achieve significant economies of scale in large volume production if the technology were widely adopted. The subsidy can provide the initial impetus that allows the technology to achieve lower costs and widespread acceptance. This rationale was the one used to justify the early subsidies of wind generation, which eventually led to the creation of an economically viable and internationally competitive U.S. wind industry.

Subsidies can also apply to the creation of new technology, through funding of research and development (R&D). The rationale for government subsidies for R&D (particularly long-term R&D) is well established in the economics literature. Companies will not fund the societally optimal level of basic R&D in new technologies, because many of the benefits of such research will flow to their competitors and to other parts of the economy (Mansfield 1982, pp. 454–455). Innovations are often easy to imitate, so which the innovators spend money on R&D, the followers can copy those innovations and avoid the risk and expense of such spending. R&D therefore has characteristics of a "public good."

Subsidies for relatively new energy technologies or fuels tend to be small in the aggregate and they tend not to have a measurable effect on the overall price of energy. When subsidies apply to widely used energy sources, however, they may create significant economic distortions, and the divergence from long-term economic efficiency may be substantial. Such distorting subsidies can sometimes exist for years because they are defended by powerful political constituencies.

EXTERNAL COSTS, SUBSIDIES, AND THE FUEL CYCLE

A comprehensive analysis of external costs and subsidies must treat each and every stage and phase in the process, which makes any such calculation inherently difficult. Uncertainties abound in these calculations, especially for external costs. As a result, estimates of total external costs and subsidies for different energy sectors vary widely.

External costs for fossil fuels are generally largest at the point of end use (combustion), though exploration (oil and gas drilling, mining), processing (refineries), and transportation (pipeline ruptures, tanker spills) can each contribute significant external costs in particular cases. For nuclear power, the accident risks associated with the conversion stage and the long-term issues surrounding disposal of spent fuel are the external costs that typically garner the most attention. There are also external costs from other stages of the nuclear fuel cycle, including the various effects of exploration, harvesting, and processing, as well as the risk of nuclear weapons proliferation from the spread of fissionable materials and related knowledge.

For nonfuel renewables such as hydroelectricity and wind power, external costs are most significant at the point of conversion (e.g., salmon migration blocked by dams, birds killed by wind turbine blades, noise and visual pollution from wind turbines). Construction of dams, particularly large ones, can also cause significant externalities by flooding large land areas, displacing people and wildlife, releasing methane from anaerobic decay of plant matter, and affecting local evapotranspiration rates. In contrast, generation of electricity using solar photovoltaics has few external costs, with the exception of small amounts of pollutant emissions from the manufacture and installation of the modules.

Most analyses of external costs have focused on electricity because of regulatory activities in that sector. One analysis of energy-related externalities in all sectors was conducted by Hohmeyer for Germany (Hohmeyer, 1988), but such comprehensive estimates are rare. For those analyses that have been done, estimates of external costs range from near zero (for photovoltaics and energy efficiency), to amounts that are significant relative to the market price of energy for some fossil fuels and nuclear power. The uncertainties in these calculations are typically quite large.

The most common subsidies for fossil fuels have been R&D and production incentives that have affected exploration and harvesting. R&D funding has also been important for fossil-fired, end-use technologies, such as furnaces that use a particular fossil fuel. For nuclear power, significant subsidies exist for processing/refining of fuel, R&D (which affects most stages), limitation of accident liability from operation of the plants, and management of long-lived wastes from the fuel cycle.

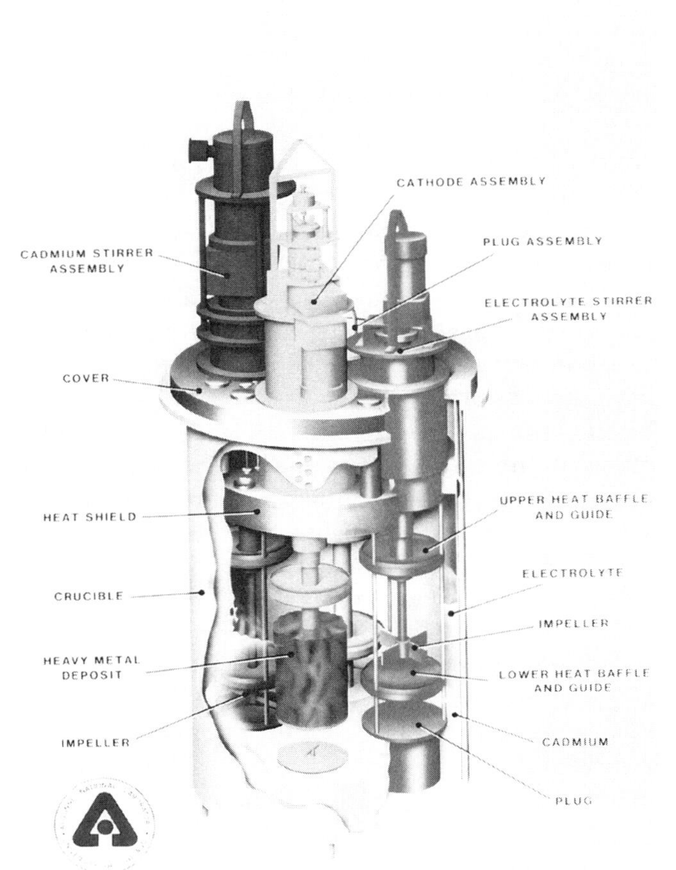

Cutaway diagram of an Integral Fast Refiner (IFR). It removes the short-lived by-products of nuclear fusion from the long-lived materials so the latter can be used to create new fuel. (Corbis-Bettmann)

A U.S. federal tax subsidy exists for wind generation and certain kinds of biomass-fired power plants built before December 31, 2001. Qualifying wind and biomass generators are paid 1.7 cents per kilowatt-hour generated over their first ten years of operation (the amount paid per kilowatt-hour increases over time at the rate of inflation). In the early days of wind generation there were subsidies in the United States for wind capacity installed, but these subsidies were phased out in the mid-1980s. R&D subsidies have been important for the development of new and more reliable renewable energy technologies. For energy-efficiency technologies, R&D funding and consumer rebates from electric and gas utilities have been the most important kinds of subsidies, but there has also been subsidization of installation of energy-efficiency measures in low-income housing. As the electric utility industry moves toward deregulation, states are increasingly relying on so-called "systems benefit charges" to fund subsidies for energy efficiency and renewable energy sources.

Unfortunately, the most recent estimates of energy subsidies in the United States date from the early 1990s and earlier, and such subsidies change constantly as the tax laws are modified and governmental priorities change. It is clear that total subsidies to the energy industries are in the billions to a few tens of billions of dollars each year, but the total is not known with precision.

ENERGY EFFICIENCY AND SUPPLY TECHNOLOGIES

External costs and subsidies for both energy efficiency and supply technologies must be included in any consistent comparison of the true energy costs of such technologies. Pollutant emissions from supply technologies are both direct (from the combustion of fossil fuels) and indirect (from the construction of supply equipment and the extraction, processing, and transportation of the fuel). Emissions from efficiency technologies are generally only of the indirect type. Increasing the efficiency of energy use almost always reduces emissions and other externalities.

THE TRUE COST OF ENERGY: AN EXAMPLE

While exact estimates of the magnitude of external costs and subsidies are highly dependent on particular situations, a hypothetical example can help explain how these two factors affect consumers' decisions for purchasing energy. In round numbers, if subsidies are about $20 billion for energy supply technologies, and total direct annual energy expenditures for the United States are about $550 billion, the combined cost of delivering that energy is $570 billion per year. This total represents an increase of about four percent over direct expenditures for energy.

Including external costs associated with energy supplies would increase the cost still further. Typical estimates for external costs for conventional energy supplies are in the range of 5 to 25 percent of the delivered price of fuel (for the sake of this example, we assume that these percentages are calculated relative to the price of fuel plus subsidies). If we choose ten percent for externalities in our example, we can calculate the "true energy price" that the consumer would see if subsidies and externalities were corrected in the market price. If the average price of fuel (without any taxes) is P, then the price of fuel correcting for subsidies is $P \times 1.04$, and the price of fuel correcting for both subsidies and externalities is $P \times 1.04 \times 1.10$. So in this example, the price of fuel would be about fourteen percent higher if subsidies and externalities were correctly included in the price.

If the particular energy source in question is already taxed (say at a five percent rate), then part of the external cost is already internalized. The true cost of fuel would remain the same ($P \times 1.04 \times 1.10$) but the size of the additional tax needed to correct for the externality would be smaller (about five percent instead of ten percent).

KEY POLICY ISSUES

New subsidies often outlive the public policy purpose that they were intended to address. The better-designed subsidies contain "sunset" provisions that require explicit action to reauthorize them after a certain time. Subsidy and externality policies are often interrelated. It may be politically difficult to tax an energy source with high external costs, but much easier to subsidize a competing energy source with low external costs. Such "second best" solutions are often implemented when political considerations block the preferred option.

Another important consideration is that "getting prices right" is not the end of the story. Many market imperfections and transaction costs affecting energy use will still remain after external costs are incorporated and subsidies that do not serve a legitimate public policy purpose are removed. These imperfections, such as imperfect information, asymmetric information, information costs, misplaced incentives, and bounded rationality, may be addressed by a variety of nonenergy-price policies, including efficiency standards, incentive programs, and information programs.

Jonathon G. Koomey

See also: Economic Externalities; Market Imperfections; Subsidies and Energy Costs.

BIBLIOGRAPHY

Bezdek, R. H., and Cone, B. W. (1980). "Federal Incentives for Energy Development." *Energy* 5(5):389–406.

Brannon, G. M. (1974). *Energy Taxes and Subsidies.* Cambridge, MA: Ballinger Publishing Co.

Golove, W. H., and Eto, J. H. (1996). *Market Barriers to Energy Efficiency: A Critical Reappraisal of the Rationale for Public Policies to Promote Energy Efficiency.* Berkeley, CA: Lawrence Berkeley Laboratory. LBL-38059.

Griffin, J. M., and Steele, H. B. (1986). *Energy Economics and Policy.* Orlando, FL: Academic Press College Division.

Heede, H. R. (1985). *A Preliminary Assessment of Federal Energy Subsidies in FY1984.* Washington, DC: Testimony submitted to the Subcommittee on Energy and Agricultural Taxation, Committee on Finance, United States Senate, June 21.

Hohmeyer, O. (1988). *Social Costs of Energy Consumption: External Effects of Electricity Generation in the Federal Republic of Germany.* Berlin: Springer-Verlag.

Holdren, J. P. (1981). "Chapter V. Energy and Human Environment: The Generation and Definition of Environmental Problems." In *The European Transition from Oil: Societal Impacts and Constraints on Energy Policy,* ed. by G. T. Goodman, L. A. Kristoferson, and J. M. Hollander. London: Academic Press.

Jaffe, A. B., and Stavins, R. N. (1994). "Energy-Efficiency Investments and Public Policy." *The Energy Journal* 15(2):43.

Koomey, J. (1990). *Energy Efficiency Choices in New Office Buildings: An Investigation of Market Failures and Corrective Policies.* PhD thesis, Energy and Resources Group, University of California, Berkeley. <http://enduse.lbl.gov/Projects/EfficiencyGap.html>.

Koomey, J.; Sanstad, A. H.; and Shown, L. J. (1996). "Energy-Efficient Lighting: Market Data, Market Imperfections, and Policy Success." *Contemporary Economic Policy* 14(3):98–111.

Koplow, D. N. (1993). *Federal Energy Subsidies: Energy, Environmental, and Fiscal Impacts.* Washington, DC: The Alliance to Save Energy.

Kosmo, M. (1987). *Money to Burn? The High Costs of Energy Subsidies.* World Resources Institute.

Krause, F.; Haites, E.; Howarth, R.; and Koomey, J. (1993). *Cutting Carbon Emissions—Burden or Benefit?: The Economics of Energy-Tax and Non-Price Policies.* El Cerrito, CA: International Project for Sustainable Energy Paths.

Kushler, M. (1998). *An Updated Status Report of Public Benefit Programs in an Evolving Electric Utility Industry.* Washington, DC: American Council for an Energy-Efficient Economy.

Levine, M. D.; Hirst, E.; Koomey, J. G.; McMahon, J. E.; and Sanstad, A. H. (1994). *Energy Efficiency, Market Failures, and Government Policy.* Berkeley, CA: Lawrence Berkeley Laboratory. LBL-35376.

Mansfield, E. (1982). *Microeconomics: Theory and Applications.* New York: W. W. Norton and Co.

Ottinger, R. L.; Wooley, D. R.; Robinson, N. A.; Hodas, D. R.; Babb, S. E.; Buchanan, S. C.; Chernick, P. L.; Caverhill, E.; Krupnick, A.; Harrington, W.; Radin, S.; and Fritsche, U. (1990). *Environmental Costs of Electricity.* New York: Oceana Publications, Inc., for the Pace University Center for Environmental and Legal Studies.

Sanstad, A. H., and Howarth, R. (1994). "'Normal' Markets, Market Imperfections, and Energy Efficiency." *Energy Policy* 22(10):826–832.

U.S. Department of Energy. (1980). *Selected Federal Tax and Non-Tax Subsidies for Energy Use and Production* (Energy Policy Study, Volume 6). Washington, DC: Energy Information Administration, U.S. Department of Energy. DOE/EIA-0201/6, AR/EA/80-01.

TURBINES, GAS

The aircraft gas turbine engine, developed more than sixty years ago, uses the principle of jet reaction and the turbine engine. The engine consists of three major elements: a compressor and a turbine expander, which are connected by a common shaft; and a combustor, located between the compressor and the turbine expander. The useful work of the engine is the difference between that produced by the turbine and that required by the compressor. For the simple cycle system shown in Figure 1, about two-thirds of all the power produced by the turbine is used to drive the compressor.

Jet reaction used in the first steam-powered engine, the aeolipile, is attributed to Hero of Alexandria around the time of Christ. In his concept, a closed spherical vessel, mounted on bearings, carried steam

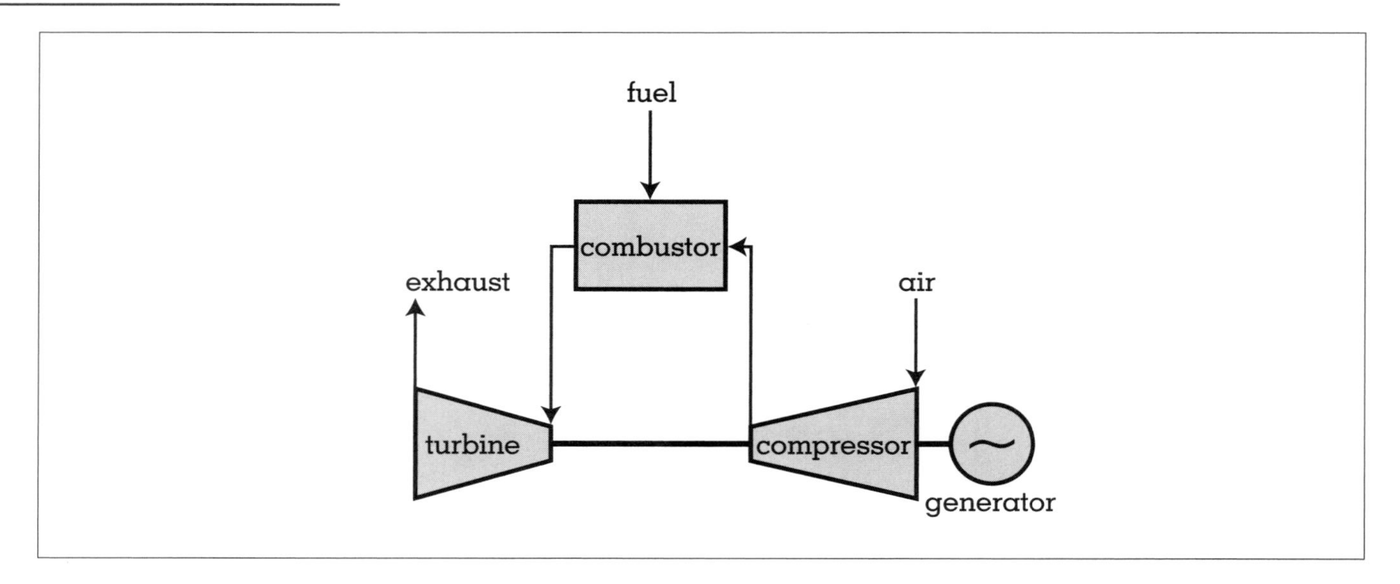

Figure 1.
Simple gas turbine cycle.

from a cauldron with one or more people discharging tangentially at the vessel's periphery, and was driven around by the reaction of steam jets. According to the literature, the first gas turbine power plant patent was awarded to John Barber, an Englishman, in 1791. Intended to operate on distilled coal, wood, or oil, it incorporated an air compressor driven through chains and gears by a turbine operated by the combustion gases. It was actually built but never worked.

There was a steady increase in the number of gas-turbine patents after Barber's disclosure. However, the attempts of the early inventors to reduce them to practice was entirely unsuccessful.

The early inventors and engineers were frustrated in their efforts to achieve a workable gas turbine engine because of the inadequate performance of the components and the available materials. However, gas turbine engine technology has advanced rapidly since the 1940s. Thrust and shaft horsepower (hp) has increased more than a hundredfold while fuel consumption has been cut by more than 50 percent. Any future advances will again depend upon improving the performance of components and finding better materials.

ORIGINS OF THE LAND-BASED GAS TURBINE

Perhaps the first approach to the modern conception of the gas-turbine power plant was that described in a patent issued to Franz Stolze in 1872. The arrangement of the Stolze plant, shown in Figure 2, consisted of a multistage axial-flow air compressor coupled directly to a multistage reaction turbine. The high-pressure air leaving the compressor was heated in an externally fired combustion chamber and thereafter expanded in the multistage turbine. The Stolze plant was tested in 1900 and 1904, but the unit was unsuccessful. The lack of success was due primarily to the inefficiency of the axial-flow compressor, which was based on aerodynamics, a science that was in its early stages of development. The lack of aerodynamic information also led Charles Algernon Parsons, the inventor of the reaction steam turbine, to abandon the development of the axial-flow compressor in 1908 after building approximately thirty compressors of that type with little success.

Several experimental gas-turbine power plants were built in France from 1903 to 1906. Those power plants operated on a cycle similar to that of a modern gas-turbine power plant. The most significant unit had a multistage centrifugal compressor that consisted essentially of a stationary casing containing a rotating impeller. Because of material limitations, the temperature of the gases entering these turbines was limited to 554°C (1,030°F). The thermal efficiency of the unit (work output divided by heat supplied) was less than 3 percent, but the unit is noteworthy, however, because

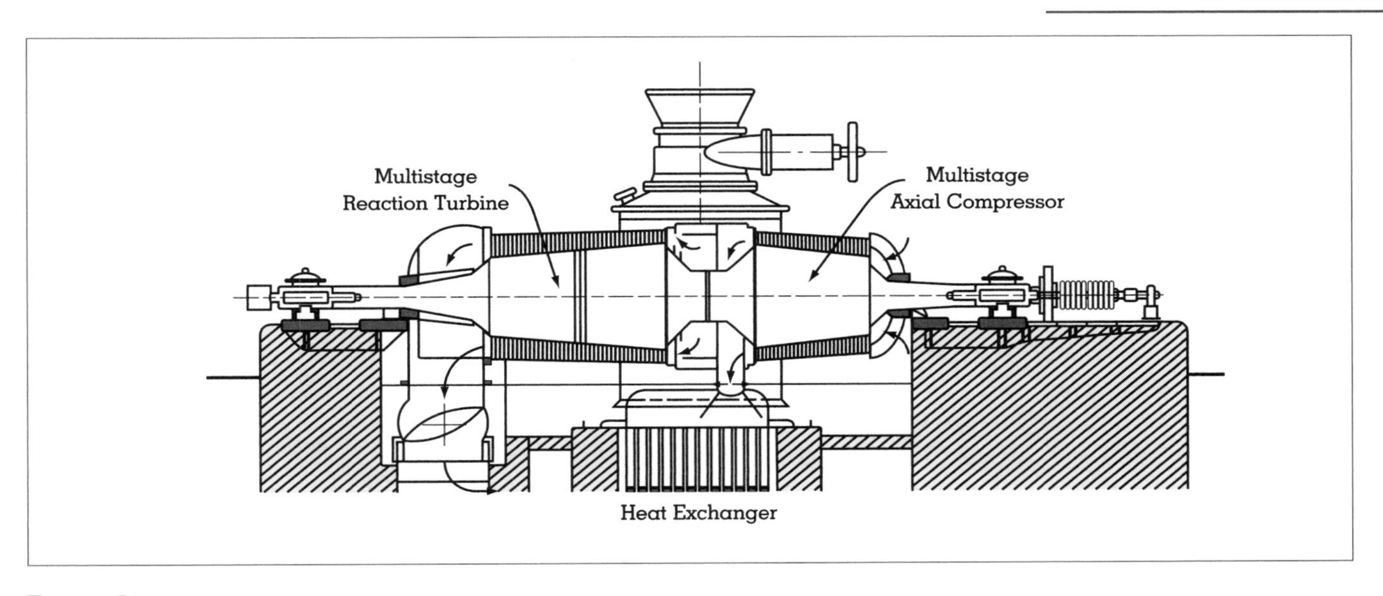

Figure 2.
The Stolze gas-turbine power plant.
SOURCE: Zipkin, 1964.

it was probably the first gas-turbine power plant to produce useful work. Its poor thermal efficiency was due to the low efficiencies of the compressors and the turbine, and also to the low turbine inlet temperature.

Brown Boveri is credited with building the first land-based gas turbine generating unit. Rated at 4 MW, it was installed in Switzerland in 1939. Leaders in the development of the aviation gas turbine during the Second World War included: Hans von Ohain (Germany), Frank Whittle (England), and Reinout Kroon (United States). Thus, combining the technology derived for aviation gas turbines with the experience of using turbines in the chemical industry, the birth of gas turbines for power generation started in the United States in 1945 with the development of a 2,000-hp gas turbine set that consisted of a compressor, twelve combustors, a turbine, and a single reduction gear. This turbine had a thermal efficiency of 18 percent. By 1949, land-based gas turbines in the United States had an output of 3.5 MW with a thermal efficiency of 26 percent at a firing temperature of 760°C (1,400°F).

COMPONENT DEVELOPMENT

Aircraft and land-based turbines have different performance criteria. Aircraft turbine engine performance is measured in terms of output, efficiency, and weight. The most significant parameter in establishing engine output (thrust or shaft hp) is turbine inlet temperature. Power output is extremely dependent on turbine inlet temperature. For example, an engine operating at 1,340°C (2,500°F) would produce more than over two and one-half times the output of an engine operating at 815°C (1,500°F). Engine efficiency, which is next in importance to engine output, is determined largely by overall engine or cycle pressure ratio. Increasing the pressure ratio from one to twenty reduces the fuel consumption by approximately 25 percent. Engine weight is affected most by turbine inlet temperatures (which acts to reduce the physical size of the engine) and by state-of-the-art materials technology that sets the temperature criteria.

During the Second World War, centrifugal compressors were used in early British and American fighter aircraft. As power requirements grew, it became clear that the axial flow compressor was more suitable for large engines because they can produce higher pressure ratios at higher efficiencies than centrifugal compressors, and a much larger flow rate is possible for a given frontal area. Today, the axial flow machine dominates the field for large power generation, and the centrifugal compressor is restricted to machines where the flow is too small to be used efficiently by axial blading.

For land-based gas turbines, the overall plant output, efficiency, emissions, and reliability are the important variables. In a gas turbine, the processes of compression, combustion, and expansion do not occur in a single component, as they do in a diesel engine. They occur in components that can be developed separately. Therefore, other technologies and components can be added as needed to the basic components, or entirely new components can be substituted.

Advanced two- and three-dimensional computer analysis methods are used today in the analyses of all critical components to verify aerodynamic, heat transfer, and mechanical performance. Additionally, the reduction of leakage paths in the compressor, as well as in the gas turbine expander, results in further plant efficiency improvements. At the compressor inlet, an advanced inlet flow design improves efficiency by reducing pressure loss. Rotor air cooler heat utilization and advanced blade and vane cooling are also used.

Several advanced turbine technologies may be applied to the gas turbine expander. These include the application of single crystal and ceramic components to increase material operating temperatures and reduce turbine cooling requirements, and active blade tip clearance control to minimize tip leakages and improve turbine efficiency. The objective of the latter scheme is to maintain large tip clearance at start-up and reduce them to minimum acceptable values when the engine has reached steady-state operating conditions.

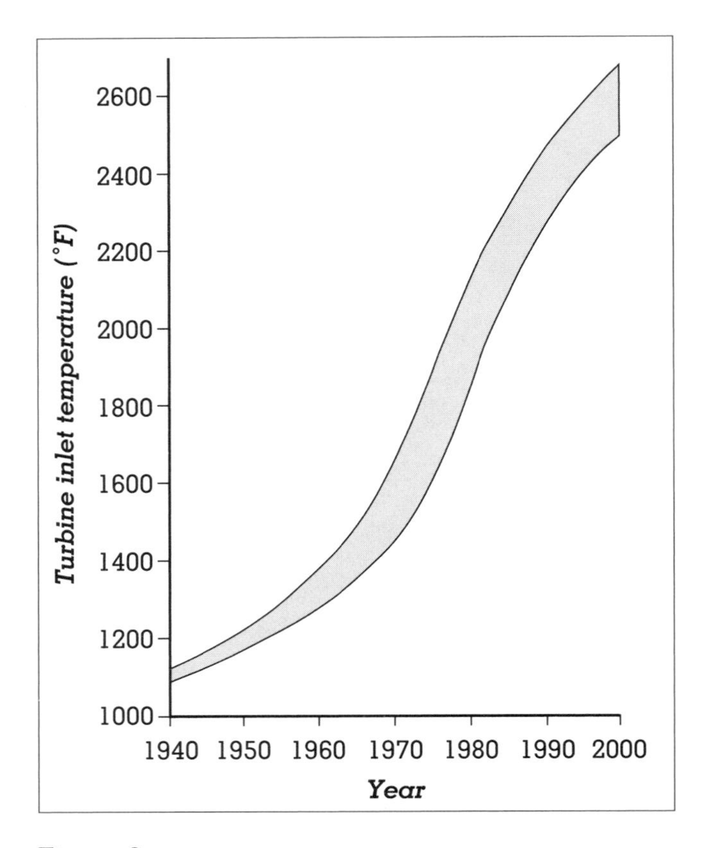

Figure 3.
Gas turbine inlet temperature trend.

ALTERNATIVE THERMAL CYCLES

As an alternative to raising firing temperature, overall power plant performance can be improved by modifications to the cycle. Combining a land-based simple cycle gas turbine with a steam turbine results in a combined cycle that is superior in performance to a simple gas turbine cycle or a simple cycle steam turbine considered separately. This is due to utilizing waste heat from the gas turbine cycle. By 1999, land-based sample cycle gas turbine efficiencies had improved from 18 percent to more than 42 percent, with the better combined cycles reaching 58 percent, and the ones in development likely to exceed 60 percent. Combined cycle efficiency improvements have followed the general advance in gas turbine technology reflected in the rising inlet temperature trend shown in Figure 3, which, in turn, was made possible by advances in components and materials.

Gas turbines can operate in open or closed cycles. In a simple cycle (also known as an open cycle), clean atmospheric air is continuously drawn into the compressor. Energy is added by the combustion of fuel with the air. Products of combustion are expanded through the turbine and exhausted to the atmosphere. In a closed cycle, the working fluid is continuously circulated through the compressor, turbine, and heat exchangers. The disadvantage of the closed cycle (also known as the indirect cycle), and the reason why there are only a few in operation, is the need for an external heating system. That is an expensive addition and lowers efficiency.

Most current technology gas turbine engines use air to cool the turbine vanes and rotors. This allows the turbine inlet temperature to be increased beyond the temperature at which the turbine material can be used without cooling, thus increasing the cycle efficiency and the power output. However, the cooling air itself is a detriment to cycle efficiency. By using closed-loop steam cooling, a new concept, the cooling air mixing loss mechanisms can be largely elimi-

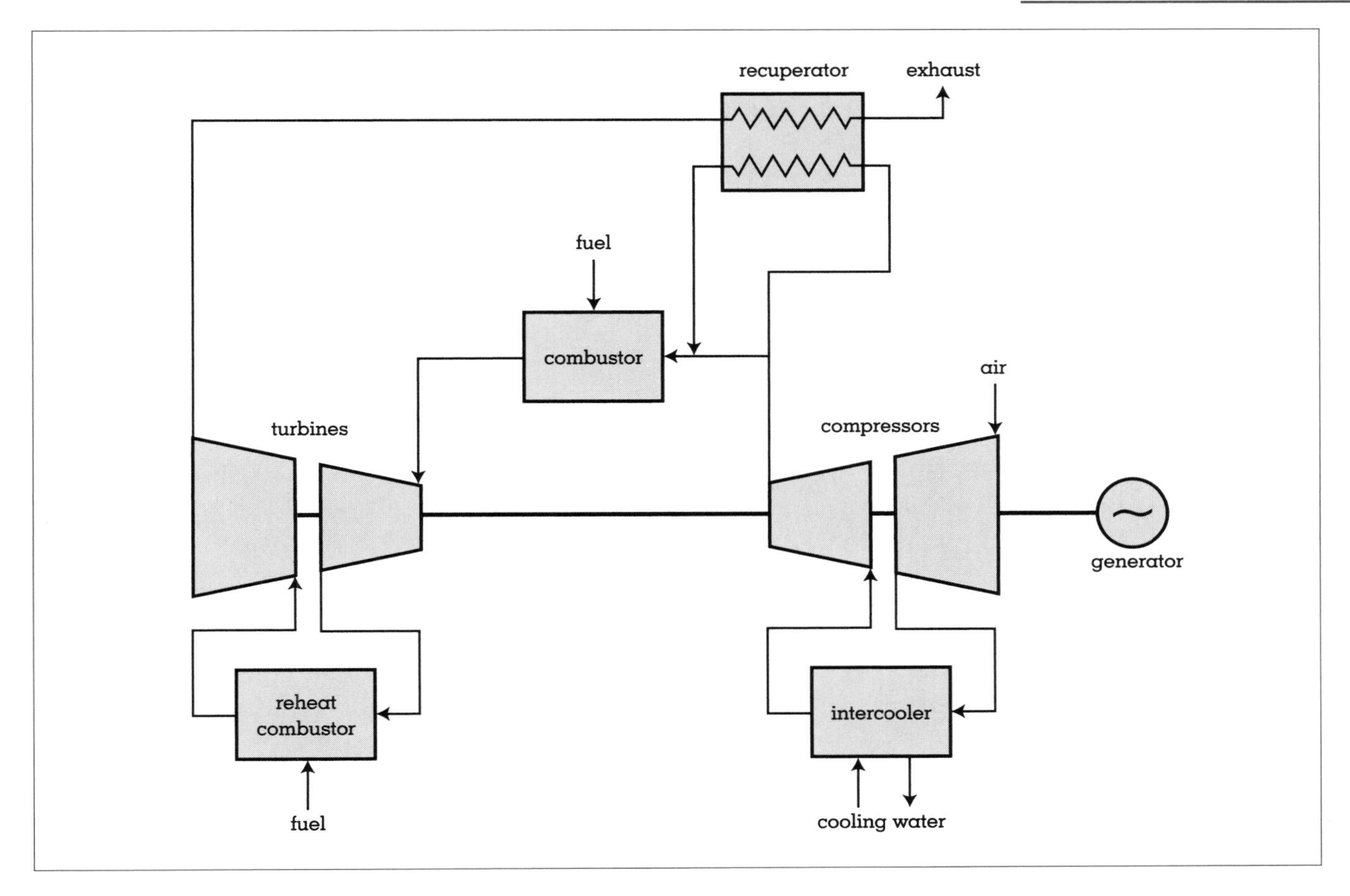

Figure 4.
Some gas turbine cycle options.
SOURCE: Maude and Kirchner, 1995.

nated, while still maintaining turbine material temperatures at an acceptable level. An additional benefit of closed-loop cooling is that more compressor delivery air is available for the lean premix combustion. The result is lower flame temperature and reduced oxides of nitrogen (NOx) emission. In combined cycles, the steam used for cooling the gas turbine hot parts is taken from the steam bottoming cycle, then returned to the bottoming cycle after it has absorbed heat in the closed-loop steam cooling system. For an advanced bottoming steam cycle, closed-loop steam cooling uses reheat steam from the exit of the high-pressure steam turbine to cool the gas turbine vane casing and rotor. The steam is passed through passageways within the vane and rotor assemblies and through the vanes and rotors themselves, then collected and sent back to the steam cycle intermediate-pressure steam turbine as hot reheat steam.

Several of the gas turbine cycle options discussed in this section (intercooling, recuperation, and reheat) are illustrated in Figure 4. These cycle options can be applied singly or in various combinations with other cycles to improve thermal efficiency. Other possible cycle concepts that are discussed include thermo-chemical recuperation, partial oxidation, use of a humid air turbine, and use of fuel cells.

The most typical arrangement for compressor intercooling involves removing the compressor air flow halfway through the compressor temperature rise, sending it through an air-to-water heat exchanger, and returning it to the compressor for further compression to combustor inlet pressure. The heat removed from the compressor air flow by the intercooler is rejected to the atmosphere because the heat is usually at too low a temperature to be of use to the cycle.

Another intercooling application is to spray water droplets into the compressor. As the air is compressed and increases in temperature, the water evaporates and absorbs heat. This results in a continuous cooling of the compressor. Note that for this concept the heat absorbed by the water is also rejected to the atmosphere. This water is never condensed by the cycle, but instead exhausted with the stack gases as low-pressure steam.

Compressor intercooling reduces the compressor work because it compresses the gas at a lower average temperature. The gas and steam turbines produce approximately the same output as in the nonintercooled case, so the overall cycle output is increased. However, since the compressor exit temperature is lowered, the amount of fuel that must be added to reach a given turbine inlet temperature is greater than that for the nonintercooled case. Intercooling adds output at approximately the simple cycle efficiency—the ratio of the amount of compressor work saved to the amount of extra fuel energy added is about equal to the simple cycle efficiency. Since combined cycle efficiencies are significantly greater than simple cycle efficiencies, the additional output at simple cycle efficiency for the intercooled case usually reduces the combined cycle net plant efficiency.

In recuperative cycles, turbine exhaust heat is recovered and returned to the gas turbine combustor, usually through a heat exchange between the turbine exhaust gases and the compressor exit airflow. The discharge from the compressor exit is piped to an exhaust gas-to-air heat exchanger located aft of the gas turbine. The air is then heated by the turbine exhaust and returned to the combustor. Since the resulting combustor air inlet temperature is increased above that of the nonrecuperated cycle, less fuel is required to heat the air to a given turbine inlet temperature. Because the turbine work and the compressor work are approximately the same as in the nonrecuperated cycle, the decrease in fuel flow results in an increase in thermal efficiency. For combined cycles the efficiency is also increased, because the gas turbine recovers the recuperated heat at the simple cycle efficiency, which is larger than the 30 to 35 percent thermal efficiency of a bottoming steam cycle, which recovers this heat in the nonrecuperated case. Since recuperative cycles return exhaust energy to the gas turbine, less energy is available to the steam cycle, and the resulting steam turbine output is lower than that of the baseline configuration. Even though the gas turbine output is approximately the same as in the baseline cycle (minus losses in the recuperation system), recuperative cycles carry a significant output penalty because of reduced steam turbine work, which is proportional to the amount of recuperation performed.

From a simple cycle standpoint, the combination of intercooling with recuperation eliminates the problem of the reduced combustor inlet temperature associated with intercooled cycles. The simple cycle then gets the benefit of the reduced compressor work and, at all but high pressure ratios, actually has a higher burner inlet temperature than the corresponding nonintercooled, nonrecuperated cycle. This results in a dramatic increase in the simple cycle efficiency.

Reheat gas turbines utilize a sequential combustion process in which the air is compressed, combusted, and expanded in a turbine to some pressure significantly greater than ambient, combusted again in a second combustor, and finally expanded by a second turbine to near ambient pressure. For a fixed turbine rotor inlet temperature limit, the simple cycle efficiency is increased for a reheat gas turbine compared to a nonreheat cycle operating at a pressure ratio corresponding to the second combustor's operating pressure. This is because the reheat cycle performs some of its combustion and expansion at a higher pressure ratio.

Thermochemical recuperation (TCR), also known as chemical recuperation, has been under evaluation for several years as a promising approach to increase power generation efficiencies. In a TCR power plant, a portion of the stack exhaust gas is removed from the stack, compressed, mixed with natural gas fuel, heated with exhaust heat from the gas turbine, and mixed with the air compressor discharge as it enters the combustor. As the mixture of natural gas and flue gas is heated by the gas turbine exhaust, a chemical reaction occurs between the methane in the fuel and the carbon dioxide and water in the flue gas. If this reaction occurs in the presence of a nickel-based catalyst, hydrogen and carbon monoxide are produced. For complete conversion of the methane, the effective fuel heating value is increased. Therefore, the natural gas/flue gas mixture absorbs heat thermally and chemically, resulting in a larger potential recuperation of exhaust energy than could be obtained by conventional recuperation, which recovers energy by heat alone. In fact, with full conversion of the natural gas fuel to hydrogen and carbon monoxide, up to twice the energy recuperated by the standard recuperative cycle may be recovered.

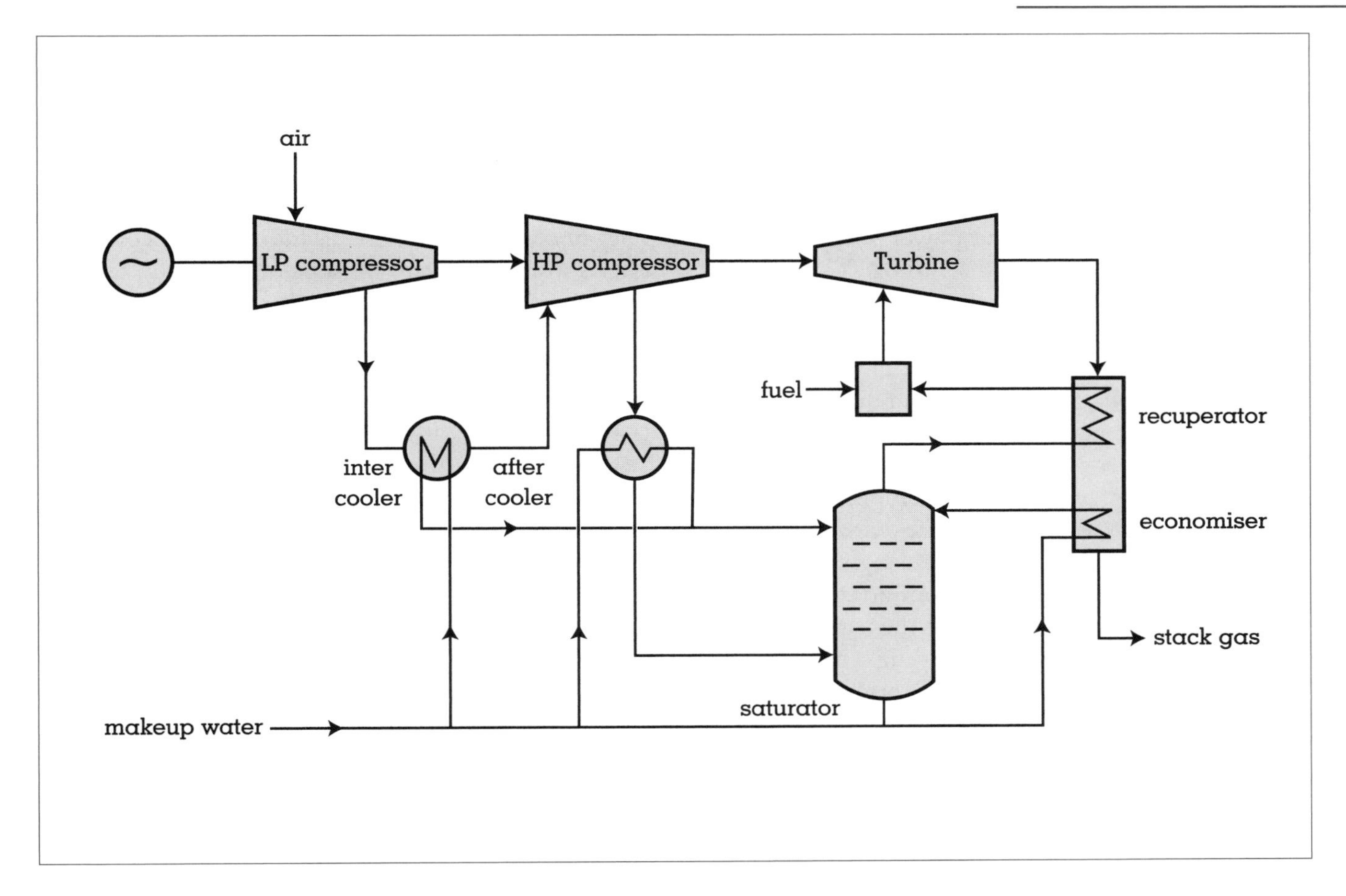

Figure 5.
HAT cycle flow diagram.
SOURCE: Morton and Rao, 1989.

Partial oxidation (PO) has also been proposed as a means to increase the performance of gas turbine power systems. PO is a commercial process used by the process industries to generate syngases from hydrocarbons, but under conditions differing from those in power generation applications. In this concept, a high-pressure, low-heating-value fuel gas is generated by partially combusting fuel with air. This fuel gas is expanded in a high-pressure turbine prior to being burned in a lower-pressure, conventional gas turbine. This process reduces the specific air requirements of the power system and increases the power output. PO has several potential advantages over conventional cycles or reheat cycles: lower specific air consumption and reduced air compressor work; increased power plant thermal efficiency; and potentially lower NOx emissions with improved combustion stability.

One difficulty associated with waste heat recovery systems is the satisfactory use of low grade heat to reduce stack losses. The humid air turbine (HAT), shown in Figure 5, uses a concept that should be considered on how to improve the use of low-grade heat. Warm waste is brought into contact with the compressor delivery air in a saturator tower to increase the moisture content and increase the total mass flow of the gases entering the turbine without significantly increasing the power demand from the compressor. Thus low-grade heat is made available for the direct production of useful power. Also, the high moisture content of the fuel gas helps to control NOx production during the combustion process.

Over a number of years, fuel cells have promised a new way to generate electricity and heat from fossil fuels using ion-exchange mechanisms. Fuel cells are categorized by operating temperature and the type of electrolyte they use. Each technology has its own operating characteristics such as size, power, density, and system configurations.

One concept, a solid oxide fuel cell (SOFC) supplied with natural gas and preheated oxidant air, produces direct-current electric power and an exhaust gas with a temperature of 850° to 925°C (1,560° to 1,700°F). In this hybrid system, the fuel cell contributes about 75 percent of the electric output, while the gas turbines contribute the remaining 25 percent. The SOFC exhaust can be expanded directly by a gas turbine with no need for an additional combustor. SOFC technology can be applied in a variety of ways to configure both power and combined heat and power (CHP) systems to operate with a range of electric generation efficiencies. An atmospheric pressure SOFC cycle, capable of economic efficiencies in the 45 to 50 percent range, can be the basis for a simple, reliable CHP system. Intergating with a gas turbine to form an atmospheric hybrid system, the simplicity of the atmospheric-pressure SOFC technology is retained, and moderately high efficiencies in the 50 to 55 percent range can be achieved in either power or CHP systems.

The pressurized hybrid cycle provides the basis for the high electric efficiency power system. Applying conventional gas turbine technology, power system efficiencies in the 55 to 60 percent range can be achieved. When the pressurized hybrid system is based on a more complex turbine cycle—such as one that is intercooled, reheated, and recuperated—electric efficiencies of 70 percent or higher are projected.

OTHER FUELS

The majority of today's turbines are fueled with natural gas or No. 2 distillate oil. Recently there has been increased interest in the burning of nonstandard liquid fuel oils or applications where fuel treatment is desirable. Gas turbines have been engineered to accommodate a wide spectrum of fuels. Over the years, units have been equipped to burn liquid fuels, including naphtha; various grades of distillate, crude oils, and residual oils; and blended, coal-derived liquids. Many of these nonstandard fuels require special provisions. For example, light fuels like naphtha require modifications to the fuel handling system to address high volatility and poor lubricity properties.

The need for heating, water washing, and the use of additives must be addressed when moving from the distillates toward the residuals. Fuel contaminants such as vanadium, sodium, potassium, and lead must be controlled to achieve acceptable turbine parts life. The same contaminants also can be introduced by the inlet air or by water/steam injection, and the combined effects from all sources must be considered.

The final decision as to which fuel type to use depends on several economic factors, including delivered price, cost of treatment, cost of modifying the fuel handling system, and increased maintenance costs associated with the grade of fuel. With careful attention paid to the fuel treatment process and the handling and operating practices at the installation, gas turbines can burn a wide range of liquid fuels. The ultimate decision on the burning of any fuel, including those fuel oils that require treatment, is generally an economic one.

Due to an estimated global recoverable reserve of more than 400 years of coal that can be used to generate electricity, future power generation systems must be designed to include coal. Today there are a number of gasification technologies available or being demonstrated at commercial capacity that are appropriate for power generation. Most of the development effort has been in the area of gasifiers that can be used in gasification combined cycle (GCC) applications. Other coal-fueled concepts under development include first- and second-generation pressurized fluidized bed (PFBC), and the indirect coal-fired case already discussed as a closed cycle.

Gasification is the reaction of fuel with oxygen, steam, and CO_2 to generate a fuel gas and low-residue carbon char. This char differentiates gasification from other technologies that burn residual char separately or apply char as a by-product. The gasification fuel gas is composed primarily of hydrogen and carbon oxides as well as methane, water vapor, and nitrogen. Generic classes of gasifiers that can be used in a GCC are moving bed, fluidized bed, and entrained flow.

GCC integrates two technologies: the manufacture of a clean-burning synthesis gas (syngas), and the use of that gas to produce electricity in a combined cycle power generation system. Primary inputs are hydrocarbon feeds (fuels), air, chemical additives, and water; primary GCC outputs are electricity (generated within the power block that contains a gas turbine, steam turbine, and a heat recovery steam generator), syngas, and sulfur by-products, waste, and flue gas. The flows

Workers assemble a 38-megawatt gas turbine at a General Electric plant in South Carolina. They build 100 turbines a year. (Corbis-Bettmann)

that integrate the subsystems include auxiliary power for an air separation unit and gasifier; air and nitrogen between an air separator and the gas turbine; heat from gasifier gas coolers to generate steam for the steam cycle; and steam to the gas turbine for NOx control.

Worldwide there are about thirty major current and planned GCC projects. Within the United States, the first GCC placed into commercial service, and operated from 1984 to 1989, was the 100-MW Cool Water coal gasification plant near Daggett, California.

Most commercial-size gasification projects have used oxygen rather than air as the oxidant for the gasifiers. Recent GCC evaluations have looked at using the gas turbine air compressor to supply air for the air separation unit. Typically this air stream is sent to a high-pressure air separator unit, which produces oxygen for gasification and high-pressure nitrogen for gas turbine NOx control. The diluent nitrogen lowers the flame temperature and therefore lowers the NOx.

Through Clean Coal Technology programs, the U.S. Department of Energy (DOE) is supporting several GCC demonstration projects that range in capacity from 100 to 262 MW. One of these plants is fueled by an advanced air-blown gasifier technology, and two of the projects will demonstrate energy-saving hot-gas cleanup systems for removal of sulfur and particulates. In Europe there are several commercial-size GCC demonstration plants in operation or under construction, ranging in size from 250 to 500 MW.

In a first-generation PFBC plant, the PFBC is used as the gas turbine combustor. For this application, the temperature to the gas turbine is limited to a bed temperature of about 870°C (1,600°F). This temperature level limits the effectiveness of this cycle as a coal-fired alternative.

In second-generation PFBC, a topping combustor is used to raise the turbine rotor inlet temperature to state-of-the-art levels. Pulverized coal is fed to a partial-gasifier unit that operates about 870° to 925°C (1,600° to 1,700°F) to produce a low heating value fuel gas and combustible char. The char is burned in the PFBC. The fuel gas, after filtration, is piped back to the gas turbine, along with the PFBC exhaust.

Fuel gas cleaning systems are being developed to fulfill two main functions: controlling the environmental emissions from the plant, and protecting downstream equipment from degraded performance. The fuel gas cleaning system also protects the gas turbine from corrosion, erosion, and deposition damage. Conventional GCC fuel gas cleaning systems, designated as cold gas cleaning, operate at temperatures of less than 315°C (600°F). Alternative technologies for fuel gas cleaning, which operate at considerably higher temperatures, are under development because of potential power plant performance benefits.

A biomass power generation industry is emerging that can provide substantial quantities of feedstock such as sugarcane residue (bagasse), sweet sorghum, rice straw, and switchgrass, as well as various wood byproducts. Today some of these residues are burned to generate power using a steam boiler and a steam turbine. However, operating problems result from ash agglomeration and deposition. Also, plant thermal efficiencies are relatively low, less than 30 percent, and in a number of plants less than 20 percent. The U.S. DOE has supported the development of biomass-fueled gasification systems that can be integrated into a combined cycle power plant and thereby obtain thermal efficiencies of greater than 40 percent.

Coal gasification is fuel flexible so that the process can use the most available feedstock at the best price. Gasifiers have successfully gasified heavy fuel oil and combinations of oil and waste gas. Other possible gasification feedstock includes petroleum coke, trash, used tires, and sewage sludge. Various combinations of feedstocks and coal have been successfully gasified.

Orimulsion is a relatively new fuel that is available for the gasification process. Orimulsion is an emulsified fuel, a mixture of natural bitumen (referred to as Orinoco-oil), water (about 30%), and a small quantity of surface active agents. Abundant Orinoco-oil reserves lie under the ground in the northern part of Venezuela.

ADVANCED GAS TURBINES

As discussed, the efficiency of a gas turbine cycle is limited by the ability of the combustion chamber and early turbine states to continuously operate at high temperatures. In 1992, the DOE started the Advanced Turbine Systems (ATS) Program that combined the resources of the government, major turbine manufactures, and universities to advance gas turbine technology and to develop systems for the twenty-first century. As pilot projects, two simple cycle industrial gas turbines are being developed for distributed generation and industrial and cogeneration markets, and two combined cycle gas turbines for use in large, baseload, central station, electric power generation markets.

DEVELOPMENT OF A HYDROGEN-FUELED GAS TURBINE

Looking to the future, the Japanese government is sponsoring the World Energy Network (WE-NET) Program through its New Energy and Industrial Technology Development Organization (NEDO). WE-NET is a twenty-eight-year global effort to define and implement technologies needed for hydrogen-based energy systems. A critical part of this effort is the development of a hydrogen-fueled gas turbine system to efficiently convert the chemical energy stored in hydrogen to electricity when hydrogen is combusted with pure oxygen. A steam cycle with reheat and recuperation was selected for the general reference system. Variations of this cycle have been examined to identify a reference system having maximum development feasibility while meeting the requirement of a minimum of 70.9 percent thermal efficiency. The strategy applied was to assess both a near-term and a long-term reference plant. The near-term plant requires moderate development based on extrapolation of current steam and gas turbine technology. In contrast, the long-term plant requires more extensive development for an additional high-pressure reheat turbine, has closed-loop steam cooling and extractive feedwater heating, and is more complex than the near-term plant.

OTHER GAS TURBINE APPLICATIONS

In addition to power generation and aircraft propulsion, gas turbine technology has been used for mechanical drive systems, gas and oil transmission pipelines, and naval propulsion. In natural gas pipelines, gas turbines provide mechanical pumping power, using the fluid being pumped as fuel. For marine applications, aero-derivative engines have been developed, yet a major disadvantage is its poor specific fuel consumption at part-load operation. For example, a naval vessel having a maximum speed of thirty knots and a cruise speed of fifteen knots loses considerable efficiency at cruising speed, since the cruise power will be only one-eighth of the maximum power (the power required being proportional to the cube of the speed). One alternative to reduce this concern is the use of a recuperator to heat the air going to the combustor with the heat from the gas turbine's exhaust. Another alternative to overcome high specific fuel consumption at part-load operation is to develop a shipboard combined cycle power plant consisting of gas turbines in conjunction with a steam turbine.

Gas turbines also have been considered for rail and road transportation. The Union Pacific successfully operated large freight trains from 1955 to 1975, several high-speed passenger trains and locomotives were built using aviation-type gas turbines. These, however, gave way to more economical diesels.

The first gas-turbine-propelled car (at 150 kW) was produced in the United Kingdom in 1950. For more than fifty years, significant efforts have been expended on automotive programs; however, diesel and gasoline engines continue to dominate. The major problem is still the poor part-load thermal efficiency of the gas turbine despite the use of a recuperated cycle with variable area nozzle guide vanes. Other problems include lack of sufficient high-temperature material development, and the relatively long acceleration time from idle to full load. For long-haul trucks, gas turbines were developed in the range of 200 to 300 kW. All used a low-pressure cycle ratio with a centrifugal compressor, turbine, and rotary heat exchanger.

A recent convergence of economic opportunities and technical issues has resulted in the emergence of a new class of gas turbine engines called microturbines. Designed for use in a recuperative cycle and pressure ratios of three to one to five to one, they can produce power in the range of 30 to 300 Kw. Initial work on this concept, which is primarily packaged today for cogeneration power generation, started in the late 1970s. In cogeneration applications that can effectively use the waste heat, overall system efficiencies can be greater than 80 percent. Manufacturers are exploring how microturbines can be integrated with fuel cells to create hybrid systems that could raise overall system efficiencies. Many issues, however, are still to be resolved with this approach, including cost and integration.

Another application of gas turbines is in a compressed air energy storage (CAES) system that allows excess base load power to be used as peaking power. It is similar to hydro-pumped storage, and the idea is to store energy during low demand periods by converting electrical power to potential energy. Rather than pumping water up a hill, CAES compresses and stores air in large underground caverns. When the power is needed, the air is allowed to expand through a series of heaters and turbines, and the released energy is then converted back to electricity. To increase output and efficiency, fuel is mixed with the air as it is released, the mixture is burned, and the energy released by combustion is available for conversion to electricity and heat recovery. This is similar to the operation of a standard gas turbine, except with CAES the compressor and turbines are separate machines that each run when most advantageous.

ROLE OF ADVANCED GAS TURBINE TECHNOLOGY

The electricity industry is in the midst of a transition from a vertically integrated and regulated monopoly to an entity in a competitive market where retail customers choose the suppliers of their electricity. The change started in 1978, when the Public Utility Regulatory Act (PURPA) made it possible for nonutility power generators to enter the wholesale market.

From various U.S. DOE sources, projections have been made that the worldwide annual energy consumption in 2020 could be 75 percent higher than it was in 1995. The combined use of fossil fuels is projected to grow faster from 1995 to 2020 than it did from 1970 to 1995. Natural gas is expected to account for 30 percent of world electricity by 2020, compared to 16 percent in 1996.

The power generation cycle of choice today and tomorrow is the combined cycle that is fueled with natural gas. Power generating technologies, regardless of the energy source, must maximize efficiency and address environmental concerns. To support the fuel mix and minimize environmental concerns, advanced coal combustion, fuel cells, biomass, compressed air energy storage, advanced turbine systems, and other technologies such as the development of a hydrogen-fueled cycle are under development.

Beyond the ATS program, the DOE is looking at several new initiatives to work on with industry. One, Vision 21, aims to virtually eliminate environmental concerns associated with coal and fossil systems while achieving 60 percent efficiency for coal-based plants, 75 percent efficiency for gas-based plants, and 85 percent for coproduction facilities. Two additional fossil cycles have been proposed that can achieve 60 percent efficiency. One incorporates a gasifier and solid oxide fuel into a combined cycle; the other adds a pyrolyzer with a pressurized fluidized bed combustor. Also under consideration is the development of a flexible midsize gas turbine. This initiative would reduce the gap between the utility-size turbines and industrial turbines that occurred during the DOE ATS program.

Ronald C. Bannister

See also: Cogeneration; Cogeneration Technologies; Locomotive Technology; Parsons, Charles Algernon; Storage; Storage Technology; Turbines, Steam; Turbines, Wind.

BIBLIOGRAPHY

Bannister, R. L.; Amos, D. J.; Scalzo, A. J.; and Datsko, S. C. (1992). "Development of an Indirect Coal-Fired High Performance Power Generation System." ASME Paper 92-GT-261.

Bannister, R. L.; Cheruvu, N. S.; Little, D. A.; and McQuiggan, G. (1995). "Development Requirements for an Advanced Gas Turbine System." *Transactions of the ASME, Journal of Engineering for Gas Turbines and Power* 117:724–733.

Bannister, R. L., and Horazak, D. A. (1998). "Preliminary Assessment of Advanced Gas Turbines for CVX." ASME Paper 98-GT-278.

Bannister, R. L.; Newby, R. A.; and Diehl, R. C. (1992). "Development of a Direct Coal-Fired Combined Cycle." *Mechanical Engineering* 114(12):64–70.

Bannister, R. L.; Newby, R. A.; and Yang, W.-C. (1999). "Final Report on the Development of a Hydrogen-Fueled Gas Turbine Cycle for Power Generation." *Transactions of the ASME, Journal of Engineering for Gas Turbines and Power* 121:38–45.

Briesch, M. S.; Bannister, R. L.; Diakunchak, I. S.; and Huber, D. J. (1995). "A Combined Cycle Designed to Achieve Greater Than 60 Percent Efficiency." *Transactions of the ASME, Journal of Engineering for Gas Turbine and Power* 117:734–741.

Cohen, H.; Rogers, G. F. C.; and Saravamuttoo, H. I. J. (1996). *Gas Turbine Theory*, 4th ed. Essex, Eng.: Longman House.

Diakunchak, I. S.; Bannister, R. L.; Huber, D. J.; and Roan, F. (1996). "Technology Development Programs for the Advanced Turbine Systems Engine." ASME Paper 96-GT-5.

Diakunchak, I. S.; Krush, M. P.; McQuiggan, G.; and Southall, L. R. (1999). "The Siemens Westinghouse Advanced Turbine Systems Program." ASME Paper 99-GT-245.

Horlock, J. H. (1992). *Combined Power Plants*. Oxford, Eng.: Pergamon Press.

Horlock, J. H. (1994). "Combined Power Plants—Past, Present and Future." *Transactions of the ASME, Journal of Engineering for Gas Turbines and Power* 117:608–616.

Horlock, J. H. (1997). *Cogeneration-Combined Heat and Power (CHP)*. Malabar, FL: Krieger.

Larson, E. D., and Williams, R. H. (1987). "Steam Injection Gas Turbines." *Transactions of the ASME, Journal of Engineering for Gas Turbines and Power* 106:731–736.

Layne, A. W., and Hoffman, P. A. (1999). "Gas Turbine Systems for the 21st Century." ASME Paper 99-GT-367.

Layne, A. W.; Samuelsen, S.; Williams, M. C.; and Hoffman, P. A. (1999). "Developmental Status of Hybrids." ASME Paper 99-GT-400.

Maude, C. W. (1993). "Advanced Power Generation: A Comparative Study of Design Options for Coal." *IEA Coal Research*. 55:15-32.

Maude, C. W., and Kirchner, A. T. (1995). "Gas Turbine Developments." *IEA Coal Research* 84:16-36.

Morton, T., and Rao, A. (1989). "Perspective for Advanced High Efficiency Cycles Using Gas Turbines." *EPRI Conference on Technologies for Producing Electricity in the Twenty-first Century* 7:1–7.

Newby, R. A., Bannister, R. L. (1994). "Advanced Hot Gas Cleaning System for Coal Gasification Processes." *Transactions of the ASME, Journal of Engineering for Gas Turbines and Power* 116:338–344.

Newby, R. A., and Bannister, R. L. (1998). "A Direct Coal-Fired Combustion Turbine Power System Based on Slagging Gasification with In-Site Gas Cleaning." *Transactions of the ASME, Journal of Engineering for Gas Turbine and Power* 120:450–454.

Newby, R. A.; Yang, W.-C.; and Bannister, R. L. "Use of Thermochemical Recuperation in Combustion Turbine Power Systems." ASME Paper 97-GT-44.

Richerson, D. W. (1997). "Ceramics for Turbine Engines." *Mechanical Engineering* 119(9):80–83.

Scalzo, A. J.; Bannister, R. L.; Decorso, M.; and Howard, G. S. (1996). Evolution of Westinghouse Heavy - Duty Power Generation and Industrial Combustion Turbines. *Transactions of the ASME, Journal of Engineering for Gas Turbines and Power* 118(2):361–330.

Scott, P. (1995). "Birth of the Jet Engine." *Mechanical Engineering* 117(1):66–71.

Touchton, G. (1996). "Gas Turbines: Leading Technology for Competitive Markets." *Global Gas Turbine News* 36(1):10–14.

Valenti, M. (1993). "Propelling Jet Turbines to New Users." *Mechanical Engineering* 115(3):68–72.

Valenti, M. (1995). "Breaking the Thermal Efficiency Barrier." *Mechanical Engineering* 117(7):86–89.

Valenti, M. (1997). "New Gas Turbine Designs Push the Envelope." *Mechanical Engineering* 119(8):58–61.

Watts, J. H. (1999). "Microturbines: A New Class of Gas Turbine Engine." *Global Gas Turbine News* 39:4–8.

Williams, M.; Vanderborgh, N.; and Appleby, J. (1997). "Progress Towards a Performance Code for Fuel Cell Power Systems." *ASME Joint Power Generation Conference* 32:423–427.

Wilson, D. G., and Korakianitis, T. (1998). *The Design of High-Efficiency Turbomachinery and Gas Turbines*, 2nd ed. Upper Saddle River, NJ: Prentice Hall.

Wilson, J. M., and Baumgartner, H. (1999). "A New Turbine for Natural Gas Pipelines." *Mechanical Engineering* 121(5):72–74.

Zipkin, M. A. (1977). "Evolution of the Aircraft Gas Turbine Engine." *Israel Journal of Technology* 15:44–48.

TURBINES, STEAM

EVOLUTION OF AN INDUSTRY

Since the turn of the twentieth century, the steam turbine has evolved from an experimental device to the major source of electrical generation. Practical steam turbine inventions coincided with the development of direct-current electric dynamos first used to power arc-lighting street systems. In the United States, the first central station to provide electrical lighting service was Thomas Edison's Pearl Street Station in New York City in 1882. Powered by 72 kW of steam engines, it served 1,284 16-candlepower dc lamps. This installation demonstrated the feasibility of central station electricity. Initially, Edison's system needed a large number of scattered power plants because it utilized direct current, and dc transmission was uneconomical over large distances. In 1885, George Westinghouse's Union Switch & Signal Co. acquired rights to manufacture and sell a European-design transformer, and the company then developed alternating-current distribution capability to utilize its transformers, which made longer-distance transmission of electricity practical. The Westinghouse Electric Co. was formed to exploit this device. By 1900 there were numerous dc and a few ac generating stations in the United States, all with reciprocating steam engines or hydraulic turbines as prime movers. However, the ac technology quickly became a primary factor in stimulating the development of power generation.

Although they were reliable, the early steam engines were huge, heavy devices that were not very efficient. Nearly all the companies in the electric equipment business seized the opportunity to develop the steam turbine as an alternative. In 1895, Westinghouse acquired rights to manufacture reaction turbines invented and patented in 1884 by the English inventor Charles Algernon Parsons. Allis-Chalmers also acquired rights to manufacture under Parsons' patents, so early machines of these two manufacturers were quite similar. In 1987, the General Electric Co. (founded by Edison) entered into an agreement with Charles Curtis to exploit his steam turbine patent.

The Curtis and the Parsons turbine designs are based on different fundamental principles of fluid flow. The Curtis turbine has an impulse design, where the steam expands through nozzles so that it reaches a high velocity. The high-velocity, low-pressure steam jet then impacts the blades of a spinning wheel. In a reaction turbine such as the Parsons design, the steam expands as it passes through both the fixed nozzles and the rotating blades. High pressure stages are impulse blades. The high-pressure drops quickly through these stages, thus reducing the stress on the high pressure turbine casing. The many subsequent stages may be either impulse or reaction designs.

STEAM TURBINE CYCLES

The basic function of a steam turbine is to efficiently convert the stored energy of high-pressure, high-temperature steam to useful work. This is accomplished by a controlled expansion of the steam through stages consisting of stationary nozzle vanes and rotating blades (also called buckets by one major manufacturer). The size and shape of the nozzle vanes and rotating blades are such as to properly control the pressure distribution and steam velocities throughout the turbine flow path. Blading improvements have increased turbine cycle efficiency by reducing profile losses, end-wall losses, secondary flow losses, and leakage losses. Use of tapered twisted designs for longer blades reduces losses on the innermost and outmost portions of the blades.

A complete turbine generator unit could consist of several turbine elements connected in tandem on a single shaft to drive a generator. To extract as much energy from the steam as possible, as it decreases in temperature and pressure in its passage through the machine, the typical arrangement could include a high-pressure (HP), an intermediate-pressure (IP), and one or more low-pressure (LP) elements, as illustrated in Figure 1.

The HP, IP, and LP turbines may be either single-flow or double-flow designs, depending upon the volume of steam utilized. In a single-flow turbine, the total volume of steam enters at one end and exhausts at the other end. The double flow is designed so the steam enters at the center and divides. Half flows in one direction, and half in the other direction into exhausts at each end of the turbine.

The basic steam cycle for a steam turbine installation is called a Rankine cycle (named after Scottish engineer and physicist William John Macquorn Rankine). This cycle consists of a compression of liquid water, heating and evaporation in the heat source (a steam boiler or nuclear reactor), expansion of the

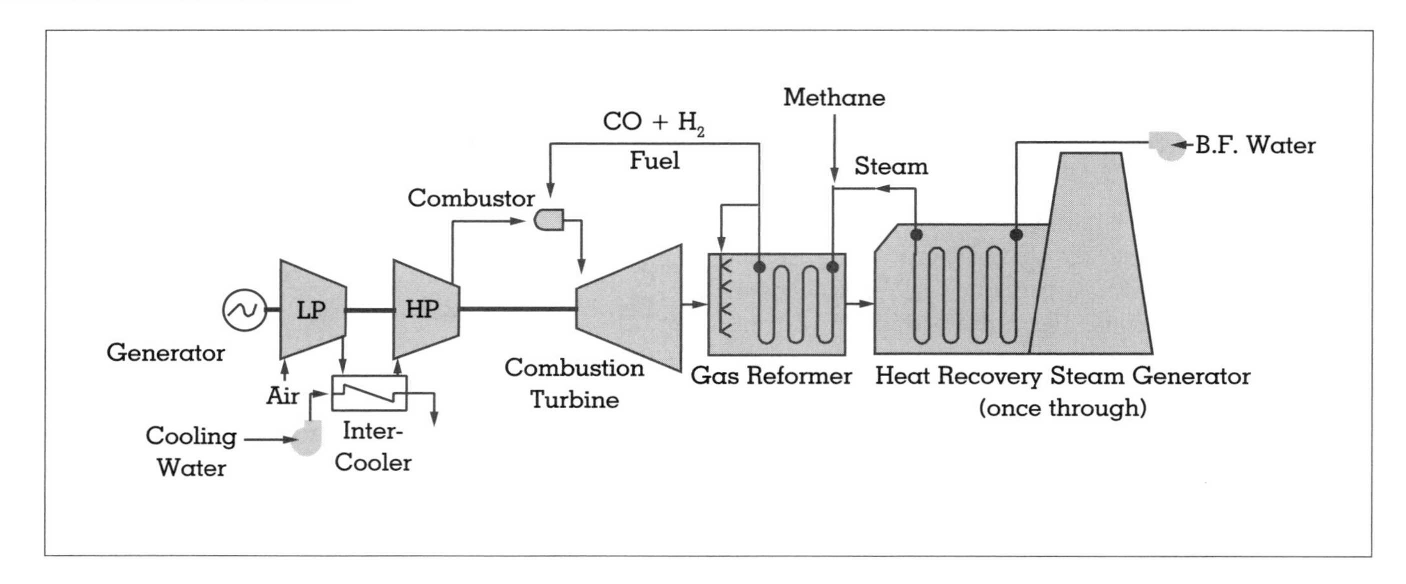

Figure 1.
Diagram of a complete turbine generator unit.

steam in the prime mover (a steam turbine), and condensation of the exhaust steam into a condenser. There is a continuous expansion of the steam, with no internal heat transfer and only one stage of heat addition. By increasing the pressure and/or temperature and decreasing the heat rejected by lowering exhaust temperature, cycle efficiency can be improved.

Weir patented a regenerative feedwater heating cycle in 1876. The regenerative Rankine cycle eliminates all or part of the external heating of the water to its boiling point. In this cycle, a small amount of expanded steam is extracted from a series of pressure zones during the expansion process of the cycle to heat water in a multiplicity of heat exchangers to a higher temperature. Theoretical and practical regenerative cycles reduce both the heat added to the cycle and the heat rejected from the cycle.

Reheat involves steam-to-steam heat exchange using steam at boiler discharge conditions. In the reheat cycle, after partially expanding through the turbine, steam returns to the reheater section of the boiler, where more heat is added. After leaving the reheater, the steam completes its expansion in the turbine. The number of reheats that are practical from a cycle efficiency and cost consideration is two.

EVOLUTION OF THE STEAM TURBINE

The first central station steam turbine in the United States was built for the Hartford Electric Light Co. in 1902. Steam conditions for this 2,000-kW unit and similar units were approximately 1.2 MPa (180 psig) and 180°C (350°F). The evolution of steam turbine power generation in the United States is summarized in Figure 2. Plotted against time are the maximum inlet steam pressure and temperature, along with plant thermal efficiency, and maximum shaft output in megawatts. The steady increase in steam turbine inlet pressure and temperature achieved an increase in plant thermal performance. From 1910 to 1920 steam turbine generators were manufactured in the 30- to 70-MW range. By 1945, the median unit sold in the United States was still only 100 MW. By 1967 the median unit had increased to 700 MW, with a peak of 1,300 MW for several fossil-fueled units placed in service in the 1970s. (A 1,300-MW unit can generate enough electricity to supply the residential needs of more than 4 million people.) During the first fifty years of the twentieth century, inlet steam pressure and temperature increased at an average rate per year of 0.3 MPa (43 psi) and 7°C (13°F), respectively. Until the early 1920s, throttle pressures were 1.4–2.1 MPa (200–300 psi), and throttle temperatures did not exceed 600°F (315°C). Above 450°F (230°C), cast steel replaced cast iron for turbine casings, valves, and so on.

Figure 3 shows the thermal performance evolution of the steam cycle as a function of material development and cycle improvements, starting in 1915. By the early 1920s, regenerative feedheating was well estab-

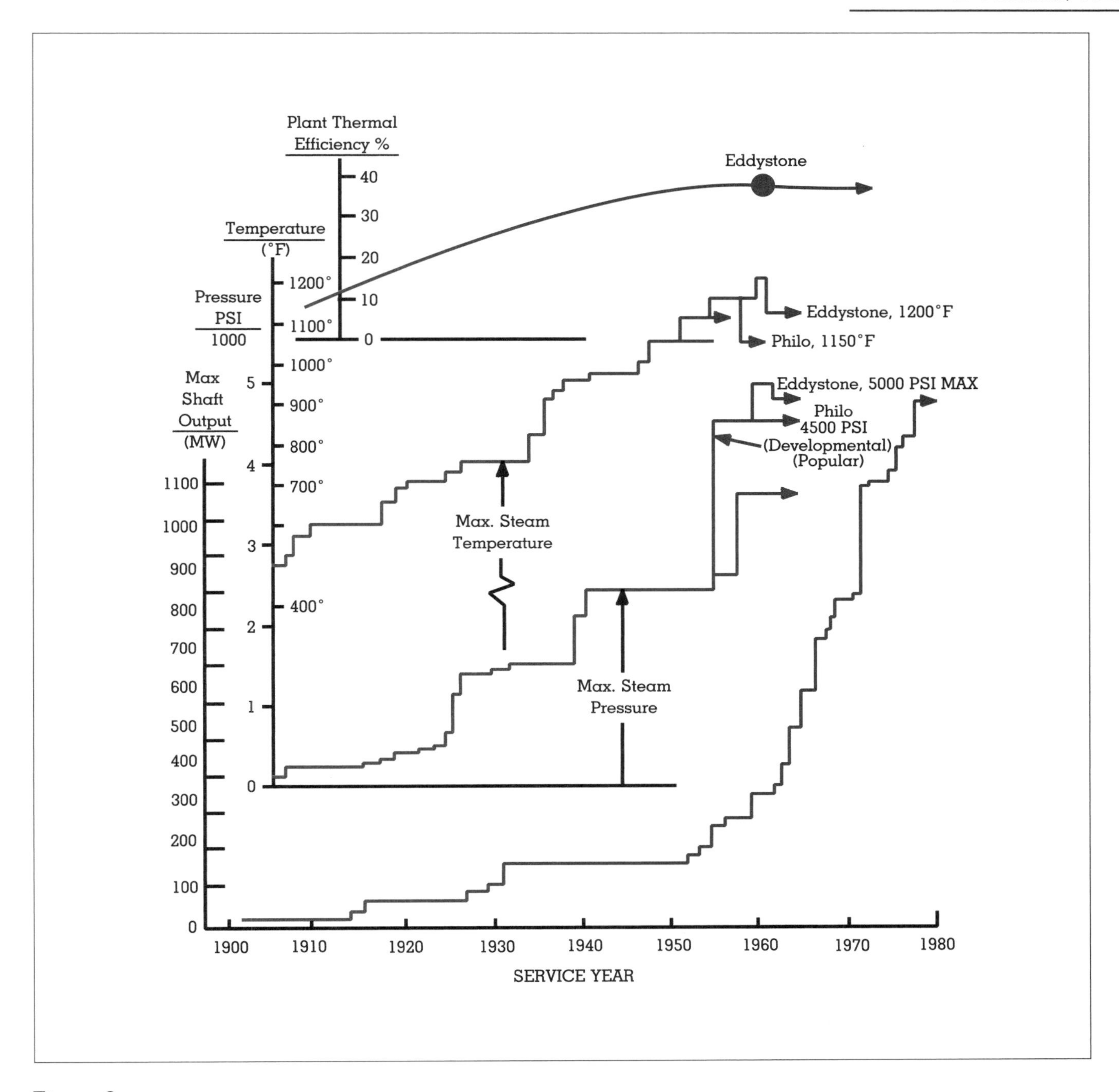

Figure 2.
The evolution of steam turbine power generation in the United States.

lished. Reheat cycles came into use in the mid-1920s. At the throttle temperature of 370°C (700°F) that was current when the pioneer 4.1- and 8.2-MPa (600- and 1,200-psi) units went into service, reheat was essential to avoid excessive moisture in the final turbine stages. As temperatures rose above 430°C (800°F) molybdenum proved effective. Using carbon-moly steels, designers pushed temperatures beyond 480°C (900°F) by the late 1930s. As a result of rising throttle temperatures, reheat fell out of use. By the late 1940s, reheat was reintroduced to improve plant efficiency, and second reheats appeared by the early 1950s.

Over the years, exhaust area was a major limitation on size. The earliest answer was the double-flow

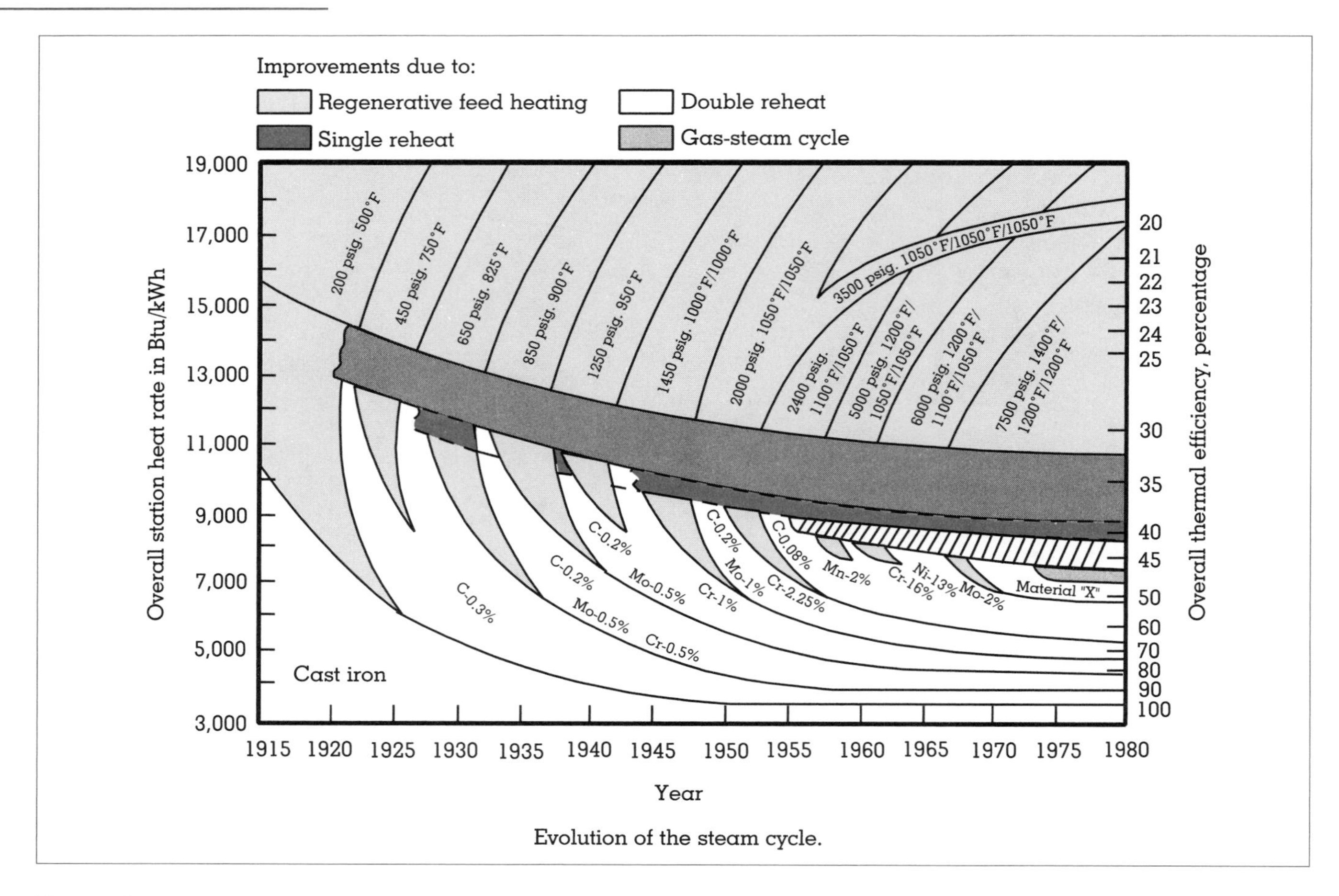

Evolution of the steam cycle.

Figure 3.
Evolution of the Steam Cycle in the United States.

single-casing machine. Cross-compounding, introduced in the 1950s, represented a big step forward. Now the speed of the LP unit could be reduced. Thus, the last-stage diameter could be greater, yielding more exhaust annulus area.

Continued advances in metallurgy allowed inlet steam conditions to be increased, as illustrated in Figure 3. In 1959 this progress culminated with the Eddystone 1 unit of the Philadelphia Electric Co. and Combustion Engineering. With initial steam conditions of 34 MPa and 650°C (5,000 psi and 1,200°F) and two reheats of 1,050°F (570°C), Eddystone 1 had the highest steam conditions and efficiency (40%) of any electric plant in the world. The generating capacity (325 MW) was equal to the largest commercially available unit at the time. Eddystone 1 has operated for many years with throttle steam conditions of approximately 4,700 psi and 1,130°F (32 MPa, 610°C), and achieved average annual heat rates comparable to today's best units.

Figure 3, initially published in 1954, summarizes the evolution of the steam cycle from the year 1915 through the steam conditions of Eddystone 1 (average station heat rates were used) and a projection of where the industry might be by the year 1980. The relationship of operating steam pressures and temperatures to available materials in this figure indicates how the increase in pressure and temperature is dependent upon metallurgical development. The magnitudes of heat rate gains resulting from the application of various kinds of steam cycles also are shown.

Since the early 1960s, advanced steam conditions have not been pursued. In the 1960s and early 1970s there was little motivation to continue lowering heat rates of fossil-fired plants due to the expected increase in nuclear power generation for base-load application and the availability of relatively inexpensive fossil fuel. Therefore the metallurgical development required to provide material "X" for advanced steam conditions was never undertaken.

Raising inlet pressure and temperature increases the cycle's available energy and thus the ideal efficiency. However, pressure increases reduce the blade heights of the initial stages and decrease ideal efficiency, offsetting some of the ideal improvement, unless unit rating is increased commensurately. Based on potential heat rate improvement, there is no reason to raise the turbine steam conditions above 48.2 MPa, 760°/760°/593°C (7,000 psi, 1,400°/1,400°/1,100°F).

NUCLEAR POWER APPLICATIONS

The first steam turbine generator for nuclear power application was placed in service at Duquesne Light Co.'s Shippingport Station in 1957. Initially rated at 60 MW, it had a maximum turbine capability rating of 100 MW. This was the first of a series of 1,800-rpm units, which were developed from a base design and operating experience with fossil-fuel machines dating back to the 1930s. Inlet steam conditions at maximum load were 3.8 MPa (545 psi), with a 600,000 kg (1.3 million-lb.) per-hour flow. This single-case machine had a 1-m (40-in.)-long last-row blade. Second-generation nuclear turbines introduced reheat at the crossover zones for improved thermal performance. Nuclear turbines, ranging in size from 400 to 1,350 MW, have used multiple LP exhausts.

Since moisture is a major concern for turbines designed for nuclear operation, a number of erosion-control techniques were used in the LP turbine. For example, adequate axial spacing between the stationary and the rotating blades minimizes blade erosion. Moisture is removed at all extraction points in the moisture region.

Nuclear turbines designed for use with a boiling-water reactor will be radioactive. Radioactivity could build up in the turbine because of the accumulation of corrosion products. A fuel rod rupture could result in highly radioactive materials entering the turbine. Therefore internal wire-drawing-type leakage paths, ordinarily unimportant in steam turbine design, must be eliminated as much as possible. Where it is impossible to eliminate, the surfaces forming the leakage paths should be faced with erosion-resistant materials deposited by welding.

FUTURE ROLE FOR STEAM TURBINE POWER GENERATION

In 2000, the power generation cycle of choice is the combined cycle, which integrates gas and steam cycles. The fuel of choice for new combined-cycle power generation is generally natural gas. Steam turbines designed to support the large gas turbine land-based power generation industry are in the range of 25 to 160 MW. Steam conditions are in the range of 12.4 MPa (1,800 psi) and 16.6 MPa (2,400 psi) and 538°C (1,000°F) and 565°C (1,050°F).

A Volt Curtis steam turbine, by GE from the designs of Charles G. Curtis, was installed for the Twin City Rapid Transit Company in Minneapolis and photographed in around 1900. (Corbis-Bettmann)

Large natural gas-fired combined cycles can reach a cycle efficiency of 58 percent, higher than a typical steam turbine power plant efficiency of 36 to 38 percent. During the 1990s, experience with optimizing advanced supercritical steam turbine cycles led to thermal cycle efficiencies in the 40 to 48 percent range for units rated at 400 to 900 MW that are existing or planned to be located in Japan, Denmark, Germany, and China.

In Denmark, two seawater-cooled 400 MW units, operating at a steam inlet pressure of 28.5 MPa (4,135 psi), have an efficiency of about 48 percent. Chubu Electric, in Japan, has been operating two 700 MW units since 1989 with inlet steam conditions of 31 MPa (4,500 psi). The efficiency gains of these two units is about 5 percent more than that of previous conventional plants of comparable size at 45 percent.

With an estimated 400 years of coal available for future power generation, coal-powered steam turbines are expected to continue to dominate global electricity fuel markets.

In the United States, coal had a 57 percent share of the electric power fuel market in 2000, up from 46 percent in 1970. This amounts to 430,000 MW generated by steam turbines that are fueled with coal. When considering other sources of generating steam for electric power—such as nuclear reactors, gas- or oil-fired broilers, and waste heat from gas turbines—steam turbines now comprise more than 600,000 MW of capacity, or approximately 75 percent of all generating capacity in the United States.

Since 1900, manufacturers have made many step changes in the basic design of steam turbines. New technology and materials have been developed to support the industry's elevation of steam conditions, optimization of thermal cycles and unit capacity. Steam turbines will continue to be the principal prime mover for electricity generation well into the twenty-first century.

Ronald L. Bannister

See also: Parsons, Charles Algernon; Rankine, William John Macquorn; Steam Engines; Turbines, Gas; Turbines, Wind.

BIBLIOGRAPHY

Artusa, F. A. (1967). "Turbine and Cycles for Nuclear Power Plant Applications." *Proceedings, American Power Conference* 29:280–294.

Bannister, R. L.; Silvestri, G. J., Jr.; Hizume, A.; and Fujikawa, T. (1987). "High-Temperature, Supercritical Steam Turbines." *Mechanical Engineering* 109(2):60–65.

Bennett, S. B.; and Bannister, R. L.; Silvestri, G. J., Jr.; and Parkes, J. B. (1981). "Current Day Practice in the Design and Operation of Fossil-Fired Steam Power Plants in the U.S.A." *Combustion* 52(7):17–24.

Bennett, S. B., and Bannister, R. L. (1981). "Pulverized Coal Power Plants: The Next Logical Step." *Mechanical Engineering* 103(12):18–24.

Church, E. F., Jr. (1950). *Steam Turbines*. New York: McGraw-Hill.

Keller, E. E., and Hodgkinson, F. (1936). "Development by the Westinghouse Machine Company." *Mechanical Engineering* 58(11):683–696.

Miller, E. H. (1984). "Historical Perspectives and Potential for Advanced Steam Cycles." *Electric Forum* 10(1):14–21.

Mochel, N. L. (1952). "Man, Metals, and Power." *Proceedings, American Society Testing Materials* 52:14–33.

Robinson, E. L. (1937). "Development by the General Electric Company." *Mechanical Engineering* 59(4): 239–256.

Silvestri, G. J., Jr.; Bannister, R. L.; Fujikawa, T.; and Hizume, A. (1992). "Optimization of Advanced Steam Conditions Power Plants." *Transactions of the ASME, Journal of Engineering for Gas Turbine and Power* 114:612–620.

Smith, D. (1999). "Ultra-Supercritical CHP: Getting More Competitive." *Modern Power Systems* 19(1):21–30.

Somerscales, E. F. C. (1992). "The Vertical Curtis Steam Turbine." *Newcomen Society of Engineering Technology Transactions* 63:1–52.

Zink, J. (1996). "Steam Turbines Power and Industry." *Power Engineering* 100(8):24–30.

TURBINES, WIND

Harnessing the wind to do work is not new. In 3000 B.C.E. wind propelled the sail boats of ancient peoples living along the coasts of the Mediterranean Sea. The Swiss and French used wind-powered pumps in 600, which was shortly followed by windmills used to make flour from grain. By 1086, there were 5,624 water mills south of the Trent and the Severn rivers in England. Holland alone once had over 9,000. As late as the 1930s in the United States, windmills were the primary source of electricity for rural farms all across the Midwest. Although the advent of fossil fuel technologies shifted energy production from animal, wind, and water, dependence upon these finite fossil energy sources and concern over atmospheric pollutants, including carbon dioxide, are causing a resurgence of interest in early energy sources such as wind.

Wind turbines in use at the turn of the twenty-first century primarily produce electricity. In developed countries where electric grid systems connect cities, town, and rural areas, wind farms consisting of numerous wind turbines produce electricity directly to the electric grid. In developing countries, remote villages are not connected to the electric grid. Smaller wind turbines, singly or in groups, have tremendous potential to bring electricity to these remote locations without requiring the significant investment in trans-

Egg-beater shaped vertical axis wind turbine (VAWT) at the test site at Sandia Laboratories. (U.S. Department of Energy)

mission lines that would be required for grid connection. Turbines range in size from less than one meter in diameter to more than 50 meters in diameter, with power ratings from 300 watts to more than one megawatt. This wide variation enables village systems to be sized to meet specific electrical demand in the kilowatt range, while large, grid-connected utility applications produce many megawatts of electricity with a few large wind turbines.

HOW WIND TURBINES WORK

Wind is the motion of the atmosphere, which is a fluid. As the wind approaches an airfoil-shaped object, the velocity changes as the fluid passes the object, creating a pressure gradient from one side of the object to the other. This pressure gradient creates a net force on one side of the object, causing it to move in the fluid. When wind hits the airfoil-shaped blade of a turbine, the lift force that is created causes the blade to rotate about the main shaft. The main shaft is connected to an electric generator. When the rotor spins due to forces from the wind, the generator creates electricity that can be fed directly into the electric grid or into a system of batteries.

Aerodynamic drag has also been used to capture energy from the wind. Drag mechanisms consist of flat or cup-shaped devices that turn the rotor. The wind simply pushes the device around the main shaft. Anemometers used to measure wind speed are often drag devices, as are traditional farm windmills.

Airplane propeller analysis relies upon the "axial momentum" theory, which is based on energy, momentum, and mass conservation laws. This theory has been applied to wind turbines as well. The power (P_w) of a fluid passing across an area perpendicular to the flow is

$$P_w = 1/2\ \rho\ A\ V_w^3$$

where ρ is the air density, A is the disk area perpendicular to the wind, and V_w is the wind speed passing through the disk area. For instance, if the wind speed is 10 m/s and the rotor area is 1,200 m^2, the available power is 600 kW. When the wind speed doubles to 20 m/s, the available power increases to 4,800 kW. This value represents the total power available in the wind, but the turbine cannot extract all of that power. If the turbine were able to extract all the available power, the wind speed would drop to zero downwind of the rotor.

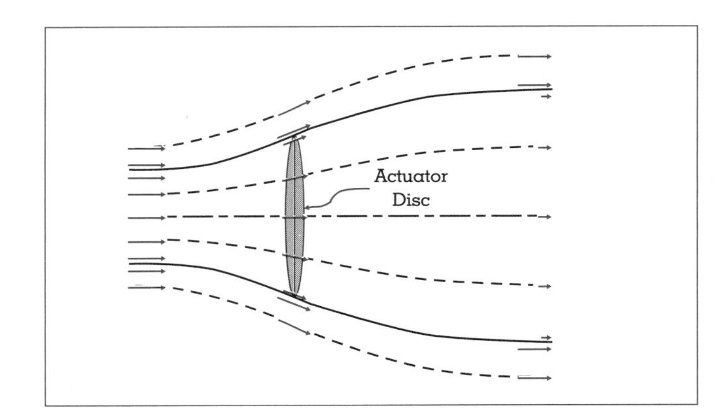

Figure 1.
Disk flow field model of an actuator.

A simple, ideal model of fluid flow through a rotor was used by both F. W. Lanchester and A. Betz (Lanchester, 1915; Betz, 1920) to study the limitation of power extracted from the wind. This "actuator disk" model is shown in Figure 1. It assumes that a perfectly frictionless fluid passes through an actuator disk, which represents the wind turbine. The fluid approaching the actuator disk slows, creating a pressure greater than atmospheric pressure. On the downwind side of the disk, the pressure drops below atmospheric pressure due to extraction of energy from the impinging fluid. As the fluid moves further downstream of the turbine, atmospheric pressure is recovered. Using axial momentum theory, Betz and Lanchester independently showed that the maximum fraction of the wind power that can be extracted is 16/27 or about 59 percent. This is known as the Lanchester/Betz limit, or, more commonly, the Betz limit, and is assumed to be an upper limit to any device that extracts kinetic energy from a fluid stream. Real wind turbines capture from 25 percent to more than 40 percent of the energy in the wind. More refined engineering analyses account for fluid friction, or viscosity, and wake rotation (Eggleston and Stoddard, 1987; Hansen and Butterfield, 1993).

TURBINE DESIGNS

There are two major types of wind turbines: horizontal-axis and vertical-axis. A wind turbine that rotates about an axis parallel to the wind is a horizontal-axis wind turbine (HAWT). Although HAWTs have not been proven clearly superior to Vertical-Axis Wind Turbines (VAWTs), they have dominated recent

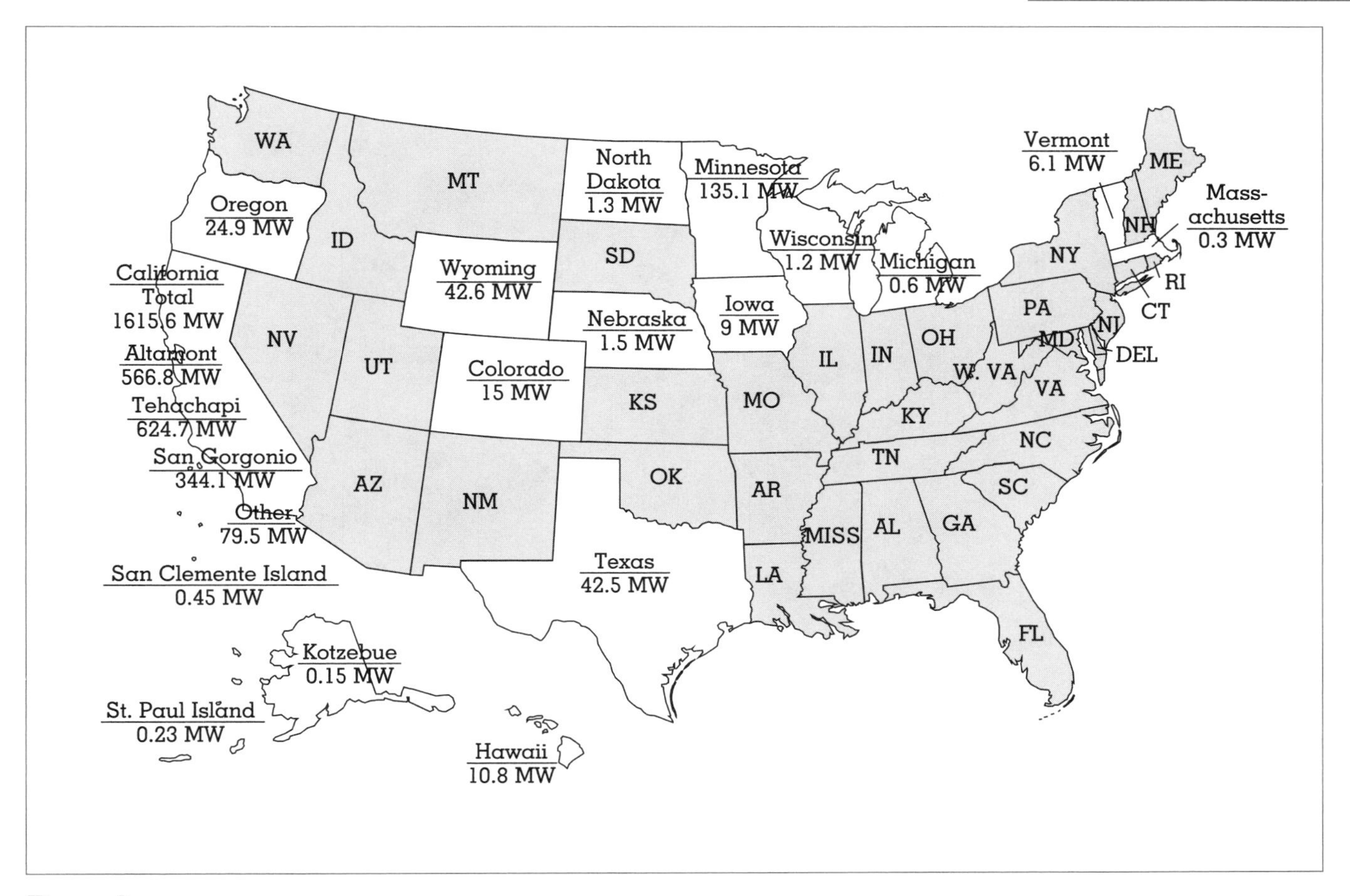

Figure 2.
U.S. capacity as of June 1999.

installations. Of all the utility-scale turbines installed today, 97 percent are HAWTs. A variety of configurations exist. HAWT rotors differ in orientation (upwind or downwind of the tower), flexibility (rigid or teetered), and number of blades (usually two or three).

Horizontal-Axis Wind Turbines (HAWTs)

An upwind turbine rotates upwind of the tower. In order to maintain the rotor's upwind position, a yaw mechanism is required to position the rotor as the wind direction shifts. In this configuration, the wind flowing toward the rotor is unobstructed over the entire rotor. Conversely, the wind flowing toward a downwind rotor is obstructed by the turbine tower over part of the rotor area. This causes fluctuating loads on each blade as it passes behind the tower, which can decrease the fatigue life of the rotor. The downwind turbine, however, aligns itself with the prevailing wind passively, eliminating the need for additional yaw drive components.

Flexibility at the rotor hub has been used to alter the load conditions on the blades. A rigid hub turbine generally has two or three blades attached to the hub. A teetered rotor, however, consists of two blades attached to the hub, which forms a hinge with the main shaft of the turbine. When the wind speed above the main shaft is higher than that below the main shaft, the rotor moves slightly downwind over the top half of the rotational cycle to reduce the loads on the blades. Although the cyclic loads are reduced by teetering the rotor, when the wind causes the rotor to exceed its maximum flexibility range hitting the teeter stops, large, transient loads can be introduced.

Although multiple-blade turbines are effective, two- and three-blade rotors are most cost-effective. Two-blade rotors commonly use teetering hinges, and all three-blade rotors are mounted on rigid hubs. Rotors up to 15 m in diameter are economically fea-

Pacific Gas & Electric engineers at verticle axis wind turbines at Altamont Pass. (Corbis-Bettmann)

sible and simplified with three blades. For large rotors exceeding 30 m in diameter, the blade weight is directly related to turbine costs. Thus, reducing the number of blades from three to two results in lower system costs with respect to the potential power available in the wind. For mid-range turbines, 15 m to 30 m in diameter, the trade off between power production and reduced cost as a result of reduced blade weight is more difficult to determine.

Most turbines are designed to rotate at a constant speed over a specific range of wind speed conditions. The generators in these turbines produce electricity compatible with the established grid system into which electricity is fed. Operating the turbine at variable rotor speeds increases the range of wind speeds over which the turbine operates. The amount of energy produced annually is increased as well. However, sophisticated power electronics is required to convert the electricity to the grid standard frequency.

Vertical-Axis Wind Turbines

A vertical-axis wind turbine (VAWT) rotates about an axis perpendicular to the wind. The design resembling an eggbeater was patented by D. G. M. Darrieus, a French inventor, in the 1920s. Because the axis of rotation is perpendicular to the ground, components such as the generator and gearbox are closer to the ground, making servicing these turbines fairly easy. Also, the turbine is not dependent upon its position relative to the wind direction. Since the blades cannot be pitched, these turbines are not self-starting. The greatest disadvantage of VAWTs is the short machine lifetime. The curved blades are susceptible to a variety of vibration problems, that lead to fatigue damage.

A modern VAWT that relies upon aerodynamic drag is known as a Savonius wind turbine. Sigurd Savonius, a Finnish inventor, developed this design in 1924. Two S-shaped panels are arranged to cup the wind and direct it between the two blades. This recirculation improves the performance of this drag device, but at best only 30 percent of the power available in the wind can be extracted.

UTILITY-SCALE APPLICATIONS

Utility-scale wind turbines range in size from 25 m to more than 50 m in diameter, with power ratings from 250 kW to 750 kW. Modern, electricity-producing wind turbines are often placed in large areas called wind farms or wind power plants. Wind farms consist of numerous, sometimes hundreds, of turbines regularly spaced on ridges or other unobstructed areas. The spacing varies depending upon the size of the turbine. Until recently California had the largest wind farms in the world. In the late 1990s, installations with capacities of 100–200 MW in the midwestern United States. Figure 2 shows the producing wind power plants in the U.S. resulting in a total capacity of 2.471 MW in 2000. This represents a small fraction of the 750 GW generating capacity in the U.S. Globally, wind power exceeded 10 GW of installed capacity in 1999. Much of the worldwide growth in wind energy use is in Europe, particularly Denmark, Germany, and Spain. To reduce pollution, many European countries subsidize wind generated electricity.

Future wind turbines will exploit the accomplishments of many years of U.S. government-supported research and development, primarily at the national laboratories, over the past twenty-five years. During the 1980s, a series of airfoils specifically designed for wind turbine applications were shown to increase annual power production by 30 percent. Detailed studies of the aerodynamics of wind turbines operating in the three-dimensional environment have led to improved models that are used for turbine design. These modeling capabilities allow turbine designers to test new concepts on paper before building prototype machines.

Several avenues for improving power production and cutting costs are being pursued by turbine designers. Building taller towers places the turbine in higher wind regimes, which increases the potential power production. In addition to taller towers, larger rotor diameters will improve power production. Increasing blade flexibility will reduce loads that reduce the fatigue life of blades, but sophisticated dynamic models are required to make such designs. Sophisticated control algorithms will monitor the turbines in order to accommodate extreme load conditions.

SMALL-SCALE APPLICATIONS

Because one-third of the world's population does not have access to electricity, many countries lacking grid systems in remote, rural areas are exploring various methods of providing citizens with access to electricity without costly grid extensions. Wind turbines are an intermittent source of electricity because the wind resource is not constant. For this reason, some form of energy storage or additional energy source is required to produce electricity on demand. Systems designed to supply entire villages or single homes are used throughout the world.

Battery systems are the most common form of energy storage. In some developing countries, battery-charging stations have been built using wind turbines. These sites simply charge batteries that people rent. The discharged batteries are exchanged for charged batteries at regular intervals. Other systems are designed for storage batteries to provide electricity during times when the wind is not blowing.

Wind turbine systems are often combined with other energy sources such as photovoltaic panels or diesel generators. Many remote areas currently rely upon diesel generators for electricity. Transportation costs limit the amount of diesel fuel that can be supplied, and diesel fuel storage poses environmental risks. By combining the generators with wind turbines, diesel fuel use is reduced. Wind and solar resources often complement each other. It is common in many areas for wind resources to be strongest in seasons when the solar resource is diminished, and vice versa. Systems that combine energy sources are called hybrid systems.

Although electricity production is the primary use of wind turbines, other applications still exist. Water pumping, desalination, and ice-making are applications that wind turbines serve. Wind turbine rotors for small-scale applications generally range in size from less than 1 m diameter to 15 m diameter, with power ratings from 300 W to 50 kW.

SITING WIND TURBINES

To obtain the best productivity from a wind turbine, it must be sited adequately. Whether establishing a wind farm with a hundred turbines or a single turbine for a home, documenting the wind conditions at a given site is a necessary step. Several databases of wind conditions have been established over the years (Elliot et al., 1987). Improvements in geographic information systems have enabled mapping wind conditions for very specific regions. Figure 3 shows the wind resource throughout the United States. Class 1 is the least energetic, while Class 7 is the most energetic. Wind turbines are generally placed in regions of Classes 3-7. This map indicates that a large part of the country has wind conditions that are conducive to wind turbine applications. These advanced mapping techniques can be used to compare one ridge to another in order to place the turbines in the most productive regions.

In addition to obtaining adequate wind resources, site selection sites for wind turbines must also consider avian populations. Several studies have been performed to determine the impact that turbines have on bird populations, with inconclusive results (Sinclair and Morrison, 1997). However, siting turbines to avoid nesting and migration patterns appears to reduce the impact that turbines have on bird mortality.

Other considerations in siting wind turbines are visual impact and noise, particularly in densely populated areas (National Wind Coordinating Committee Siting Subcommittee, 1998). Due to Europe's high population density, European wind turbine manufacturers are actively examining the potential of placing wind turbines in offshore wind farms.

ECONOMICS

Deregulation of the electric utility industry presents many uncertainties for future generation installations. Although the outcome of deregulation is unknown, ownership of the generation facilities and transmission services will most likely be distributed among distinct companies. In the face of such uncertainties, it is difficult to predict which issues regarding renewable energy will dominate. However, the benefits of integrating the utility mix with wind energy and the determination of the cost of wind energy, are issues that will be relevant to most deregulation scenarios.

Wind energy economics focuses on the fuel-saving aspects of this renewable resource, but capacity benefits and pollution reduction are important considerations as well. The capital costs are significant, but there is no annual fuel cost as is associated with fossil fuel technologies. Thus, wind energy has been used to displace fossil fuel consumption through load-matching and peak-shaving techniques. In other words, when a utility requires additional energy at peak times or peak weather conditions, wind energy is used to meet those specific needs. In addition to fuel savings, wind energy has been shown to provide capacity benefits (Billinton and Chen 1997). Studies have shown that although wind is an intermittent source, wind power plants actually produce consistent, reliable power that can be predicted. This capacity can be 15 percent to 40 percent below the installed capacity. Last, the emission-free nature of wind turbines could be exploited in a carbon-rading scenario addressing global climate change.

Two methods of determining the cost of wind energy are the Fixed Charge Rate (FCR) and the Levelized Cost of Energy (LCOE). An FCR is the rate at which revenue must be collected annually from customers in order to pay the capital cost of investment. While this incorporates the actual cost of the wind turbines, this method is not useful for comparing wind energy to other generation sources. The LCOE is used for comparison of a variety of generation technologies that may vary significantly in size and operating and investment time periods. This metric incorporates the total life cycle cost, the present value of operation and maintenance costs, and the present value of depreciation on an annual basis.

Subsidies, in the form of financing sources, and tax structures significantly impact the levelized cost of energy. The Renewable Energy Production Incentive (REPI), enacted in 1992, at $0.015/kilowatt-hour (kWh), applies only to public utilities. This tax incentive is renewed annually by Congress, making its longevity uncertain. Private owners of wind power plants are eligible for the federal Production Tax Credit (PTC), which is also $0.015/kWh. A project generally qualifies for this tax credit during its first ten years of operation. This credit is also subject to Congressional approval. Ownership by investor-owned utilities (IOUs), and internal versus external project financing, also affect the LCOE. Cost differences that vary with ownership, financing options, and tax structures can be as great as $0.02/kWh (Wiser and Kahn, 1996).

The annual LCOE for wind turbines has decreased dramatically since 1980. At that time, the LCOE was $0.35/kWh. In 1998, wind power plant projects were bid from $0.03kWh to $0.06/kWh. These numbers still exceed similar figures of merit for fossil fuel generation facilities in general. However, wind energy can become competitive when new generation capacity is required and fossil fuel costs are high, when other incentives encourage the use of clean energy sources. For example, a wind power plant installed in Minnesota was mandated in order to offset the need for storing nuclear plant wastes.

Incentives that currently encourage the use of clean energy sources may or may not survive the deregulation process. Green pricing has become a popular method for utilities to add clean energy sources, with consumers volunteering to pay the extra cost associated with renewable energy technologies. Some proposed restructuring scenarios include a Renewable Portfolio Standard (RPS), which mandates that a percentage of any utility's energy mix be comprised of renewable energy sources. Last, distributed generation systems may receive favorable status in deregulation. Small generation systems currently are not financially competitive in many areas, but through deregulation individuals or cooperatives may be able to install small systems economically.

Wind energy was the fastest growing energy technology from 1995 to 1999. This translates to an annual market value of over $1.5 billion. Supportive policies in some European countries, improved technology, and the dramatic drop in the cost of wind energy, have contributed to the growth of wind energy.

M. Maureen Hand

See also: Aerodynamics; Climate Effects; Kinetic Energy; National Energy Laboratories; Propellers; Subsidies and Energy Costs.

BIBLIOGRAPHY

Bergey, K. H. (1979). "The Lanchester-Betz Limit." *Journal of Energy* 3:382–384.

Betz, A. (1920). "Das Maximum der theoretisch möglichen Ausnützung des Windes durch Windmotoren." *Zeitschrift für das gesamte Turbinenwesen*, Heft 26.

Betz, J. A. (1966). *Introduction to the Theory of Flow Machines*. New York: Pergamon.

Billinton, R., and Chen, H. (1997). "Determination of Load Carrying Capacity Benefits of Wind Energy Conversion Systems." Proceedings of the Probabilistic Methods Applied to Power Systems 5th International Conference; September 21–25, 1997. Vancouver, BC, Canada.

DOE Wind Energy Program Home Page. U.S. Department of Energy, energy Efficiency and Renewable Energy Network. July 17, 2000. <http://www.eren.doe.gov/wind/>.

Eggleston, D. M., and Stoddard, F. S. (1987). *Wind Turbine Engineering Design*. New York: Van Nostrand Reinhold Company.

Eldridge, F. R. (1980). *Wind Machines*, 2nd ed. New York: Van Nostrand Reinhold.

Elliot D. L.; Holladay, C. G.; Barchet, W. R.; Foote, H. P.; and Sandusky, W. F. (1987). *Wind Energy Resource Atlas of the United States*, DOE/CH 10093–4, Golden, CO: Solar Energy Research Institute <http://rredc.nrel.gov/wind/pubs/atlas/>.

Gipe, P. (1993). *Wind Power for Home and Business*. Post Mills, VT: Chelsea Green Publishing Company.

Hansen, A. C., and Butterfield, C. P. (1993). "Aerodynamics of Horizontal-Axis Wind Turbines." *Annual Review of Fluid Mechanics* 25:115–49.

International Energy Agency. (1998). *1998 Wind Energy Annual Report*. NREL/BR-500-26150. Golden, CO: National Renewable Energy Laboratory.

Johnson, G. L. (1995). *Wind Energy Systems*. Englewood Cliffs, NJ: Prentice-Hall.

Lanchester, F. W. (1915). "Contributions to the Theory of Propulsion and the Screw Propeller," *Transactions of the Institution of Naval Architects* 57:98–116.

National Wind Coordinating Committee. July 17, 2000. <http://www.nationalwind.org>.

National Wind Coordinating Committee Siting Subcommittee. (1998). *Permitting of Wind Energy Facilities*. Washington, DC: National Wind Coordinating Committee.

National Wind Technology Center. National Renewable Energy Laboratory. July 17, 2000. <http://www.nrel.gov/wind/>.

Short, W.; Packey, D.; and Holt, T. (1995). *A Manual for the Economic Evaluation of Energy Efficiency and Renewable Energy Technologies* NREL/TP-462-5173. Golden, CO: National Renewable Energy Laboratory.

Sinclair, K. C., and Morrison, M. L. (1997). "Overview of the U.S. Department of Energy/National Renewable Energy Laboratory Avian Research Program." *Windpower 97 Proceedings, June 15–18, 1997, Austin, TX*. Washington, DC: American Wind Energy Association.

Spera, D. A., ed. (1994). *Wind Turbine Technology*. New York: ASME Press.

Wiser, R., and Kahn, E. (1996). *Alternative Windpower Ownership Structures: Financing Terms and Project Costs* LBNL-38921. Berkeley, CA: Lawrence Berkeley Laboratory.

U

UNITS OF ENERGY

The joule, symbol J, is the unit of energy in the science community. It is not widely used outside the science community in the United States, but it is elsewhere. For example, in the United States, the energy content of a food product is likely expressed in food calories. This energy could be expressed in joules, and that is being done in many parts of the world. A Diet Coca Cola in Australia is labeled Low Joule rather than Low Cal as in the United States. To obtain a feeling for the size of a joule, consider the following statistics:

- the sound energy that is in a whisper is about 0.01 joules,
- the kinetic energy of a 1,000-kilogram car traveling 25 meters per second (55 miles per hour) is about 300,000 joules,
- the energy from burning a barrel of oil is 6,000,000 joules,
- the annual energy use in the United States is about 100 billion billion joules, and
- the daily energy input to the Earth from the sun is about 10,000 billion billion joules.

Even if the use of units of joules were universal, the numbers involved in energy discussions would be large and cumbersome. Therefore, it is customary to use powers of ten notation and prefixes for powers of ten. For example, the energy from burning a barrel of oil is 6×10^6 joules, which can be expressed as 6 megajoules or, symbolically, 6 MJ. Table 1 summarizes the powers of ten, prefixes, and symbols usually encountered in energy considerations.

Numerical Unit	*Power of Ten*	*Prefix*	*Symbol*
million trillion	10^{18}	exa	E
trillion	10^{12}	tera	T
billion	10^{9}	giga	G
million	10^{6}	mega	M
thousand	10^{3}	kilo	k
hundredth	10^{-2}	centi	c
thousandth	10^{-3}	milli	m
millionth	10^{-6}	micro	μ

Table 1.
Symbols of Some Numerical Units

The gasoline tank on an automobile holds about 15 gallons, and the automobile can travel about 300 miles before the tank is empty. Even though the gasoline was purchased for its energy content, the driver probably did not know that the energy content of 15 gallons of gasoline is 2×10^9 J = 2 GJ. For reasons like this, there exists a variety of energy units and energy equivalents. The unit and equivalent depends on the type of energy commodity in question. A homeowner pays an electric utility for electric energy, and the

Special Unit	Study area of main use	Symbol	Equivalent in joules	Other units
Kilowatt hour	electricity	kWh	3,600,000	3413 Btu
Calorie	heat	cal	4.186	
Kilocalorie (food calorie)	heat	kcal	4,186	1000 cal
British Thermal unit	heat	Btu	1,055	252 cal
Electron volt	atoms, molecules	eV	1.60×10^{-19}	
Kilo-electron Volt	X-rays	keV	1.60×10^{-16}	1000 eV
Mega-electronVolt	nuclei, nuclear radiations	MeV	1.60×10^{-13}	1000 keV
Quadrillion	energy reserves	quad	1.055×10^{21}	10^{15} Btu
Quintillion	energy reserves	Q	1.055×10^{21}	10^{18} Btu

Energy equivalents
1 gallon of gasoline = 126,000 Btu
1 cubic foot of natural gas = 1030 Btu
1 pound of bituminous coal = 13,100 Btu
1 42-gallon barrel of oil = 5,800,000 Btu
1 therm = 100,000 Btu
Variations of these energy equivalents will appear in the literature. The values listed here are typical.

Table 2.
Some Energy Units and Their Equivalents

unit is likely a kilowatt-hour (kWh). A politician interested in imported oil will likely talk in terms of barrels of oil. Table 2 summarizes some special energy units and their equivalents.

Joseph Priest

BIBLIOGRAPHY

Hobson, A. (1995). *Physics: Concepts and Connections.* Englewood Cliffs, NJ: Prentice-Hall.

Serway, R. A. (1998). *Principles of Physics.* Fort Worth, TX: Saunders College Publishing.

UTILITY PLANNING

Electricity is a commodity that has captured the attention of Americans, from its creation to the present and onward to projections of the future. From an object of luxury to a mainstream "staff of life" and forward into the sophisticated world of tomorrow, electricity has become a critical necessity. The criteria have become: electricity on demand, reliably delivered, in sufficient quantities, and at the right price. At the heart of meeting this challenge is the utility planner.

Generating and delivering electricity requires a complicated infrastructure, significant funding, competitive markets, and adequate regulation. Broken into its simplest terms, utility planning is beginning with the most optimum operation of existing electrical facilities and foreseeing the future well enough to provide additional facility to "guarantee" an adequate supply of a reliable product (electricity) at an acceptable cost. Generation provides the product, transmission is the vehicle to deliver the product over long distances, and distribution is the functional method of providing the product to the individual customer.

Originally all facilities, financing, and decisions related to building an electrical system resided within a utility. In this simplistic approach, projections of load growth, established construction and planning criteria, and coordination of facilities remained within the single entity. In today's competitive unregulated marketplace, the planners, designers, generators, and transmission providers are within a complicated mix of organizational units; thereby providing a complicated network of system participants. The challenge for utility planning today is to coordinate all of the inputs for constructing and maintaining an electrical system that combines the economics efficiencies necessary for survival and the maintenance of a reliable network where generated power can be delivered to an ultimate customer on demand, in sufficient quantities, both today and in years to come.

HISTORY

The history of the interconnected high-voltage electric power system has roots in the early decades of the twentieth century. Moving power from generating plants remote from load centers began as utilities and customers realized that costs associated with many small generating entities placed strategically at concentrated load points would soon escalate to unacceptable values. Likewise, the many advantages of the new power source would be restricted to cities and other high-density areas. The efficiencies of larger generating units utilizing fuels of choice and availability soon became the economic choice for resources thus developed the need to "push" power across long distances.

There were many examples of early interconnected electric transmission systems extending across state territories and beyond state boundaries. Statements from a speech by Samuel Insull at Purdue University in 1924 indicated that Minneapolis, St. Paul, St. Louis, Louisville, and Cincinnati soon would be interconnected and that extension of these systems to Pittsburgh and across the Allegheny Mountains to the Atlantic Seaboard could be easily conceivable in a few years. This was but one example across the United States. To quote Insull, "It makes electric service available in new places for new purposes, so that aggregate demand for service is spread over more hours of day and night, and thus opens the way to utmost economy in production and distribution."

The interconnected power system grew and the load demand increased, over time, at varying rates and reached an average growth of some 10 percent, or a doubling every ten years. Utilities financed, constructed, and controlled the vast majority of the generation and transmission facilities in what was classified a "monopolistic society." In this environment, while the processes were complicated, the data necessary to plan these extensive projects were readily available.

One example of the complexity of the process was in load forecasting technology. By the 1960s and 1970s, utilities had developed processes to replicate the extensive electric power system for study purposes to include generation resources, transmission networks and individual points of load service to customers. The most complicated of these were the load forecast and load distribution techniques. No matter how fast technology developed, certain parameters of forecasting created significant challenges. Companies forecasted load as a composite for their territory, primarily using historical trends. Often, these forecasts were economically based, thus having boundaries driven by nontechnical issues. Load growth varied within areas of individual companies. New and increased industrial loads, which made up significant portions of the forecasts, diluted the forecast and often appeared on the system later (or earlier) than planned.

In the 1980s, as the complexity of forecasting became more challenging, utilities chose to augment their forecasting methodologies through the use of consultants who possessed sophisticated databases that included topography, diversity, trends, population growth and other intricate variables and impactors. These forecasts enabled a finer tuning of the process and in many cases a more efficient and economical forecast.

No matter how accurate the forecast, other planning techniques, while state of the art, introduced trending, estimating, and engineering judgment in the facility planning cycle. One such issue involved the time and expense to plan a project, budget the project, and place the project in service when required. Because of constraints, most planning studies considered only the estimated annual peak load for a company. This peak could occur in the summer or winter, depending on the region of the country. Obviously other variations on the annual load curve occurred, such as off-peaks (spring and fall seasons) and shoulder peaks (related to time of day). Normally these variations were trended or estimated or ignored in the interest of time, and an ultimate plan was developed. In the very early years of planning, utility planners assumed that planning for the peak load covered all other times of the year. History recorded it to be a valid theory.

In later years, studies involving variations of the peak load were initiated to answer specific questions. Generation maintenance schedules, which were initially developed to accommodate only peak load periods, were revised to protect extended off-peak periods of large energy usage. The components of generation and load include real power for load service and reactive power for voltage support. The effects of various load levels to voltage profiles were tested to plan economic generation reactive parameters and adequate capacitor levels, and to define optimum capacitor locations. In addition, off-peak voltage profiles were sensitive to generator outages and maintenance. Power transfer capabilities between entities had been routinely studied at peak load. As peak periods began to lengthen and as generation locations were more widely distributed, it became necessary to consider the impacts to transfers at load levels less than peak. The transfer capability issues have become of more specific importance with the move to greater open access and competition.

Transmission parameters for existing systems were translated into appropriate input data for calculation systems available at the time. During early periods of the twentieth century, calculations for expansion of the transmission network or grid were actually made by laborious longhand manual monotonous iterations. Fortunately, the areas of study were small or abbreviated sections not affecting other areas. In the 1940s, network analog calculators were introduced for use in modeling the power system. These calculators were large "black boxes" composed of resistors, capacitors, meters, controls, dials, plotting boards, etc., and often covered an entire room, usually at an area university.

The analog calculator employed a quantitative methodology. The engineer calculated the impedance (parameters) of a transmission line and physically dialed in the valve between points A and B. The calculator also included metering windows where various information could be readily observed. With this method an experienced planner could actually sense a "feel" of the system as the dials and data were manipulated. Likewise, the analog computer became a significant training tool for utility engineers.

By the late 1960s, digital computers began to replace the "black boxes." Despite forebodings of utility planning engineers regarding training of engineers and planning, adaptations to the new technology became commonplace.

The digital computer began as a mammoth blue or black giant mechanism hidden in some corporate information resource office. The planner's view of data processing was keypunching a "deck" of cards and dropping them into an open "hopper" to mysteriously disappear into the computer. Similarly, once the magic of the calculations was accomplished, white punched computer paper arrived from an output device with guaranteed solutions to difficult questions.

The process improved. Computers became smaller, and today it is rare for a planning engineer to be void of computer power available at his desk many times greater than the early corporate computer could muster. Computer tools of the future can only be envisioned as technology advances.

Generation data to complement the planning study were plentiful. As long as the utilities owned the physical facilities and were in control of future facility expansion, data were generally a nonissue. The parameters of generating plants were dialed into the calculations or computers, similar to transmission. Obviously, alternate locations for new plants could be evaluated, and oftentimes studies were finalized some ten years ahead of construction, with only limited analysis required as the in-service date approached.

Planning studies to identify the magnitude of generation required on a particular system for a specific time period were based on parameters such as loss of load probability (LOLP) or loss of load energy (LOLE). These studies modeled the system load curve throughout the year for projected years into the future and included existing generation resources. The purpose of the studies was to identify the probability statistics where the system could not serve its load requirement or to identify an amount of unserved energy. These data would be compared to specific criteria, and future generation requirements would be developed. The ability to obtain emergency or specific contracted generation resources from neighboring systems would normally be factored into the data. In addition, in many cases generation reserve sharing agreements were factored into the final determination of capacity required. Reserve sharing agreements were contractual arrangements, usually within power pools (groups of systems) or areas where each system involved factored in the pro rata share of neighboring generation available to that system. Both concepts of utilizing off-system resources as an integral part of a system's total resources involved risk, which had to be coordinated with economic advantages and reliability levels.

Given that there were numerous alternatives for the resources, including demand-side management, interruptible loads, and off-system purchases, many case studies were necessary. Each plan for service to future loads had to be tested for conformance with system and regional reliability criteria. The plans were further examined for cost, including losses, and examined for flexibility. Any acceptable plans had to

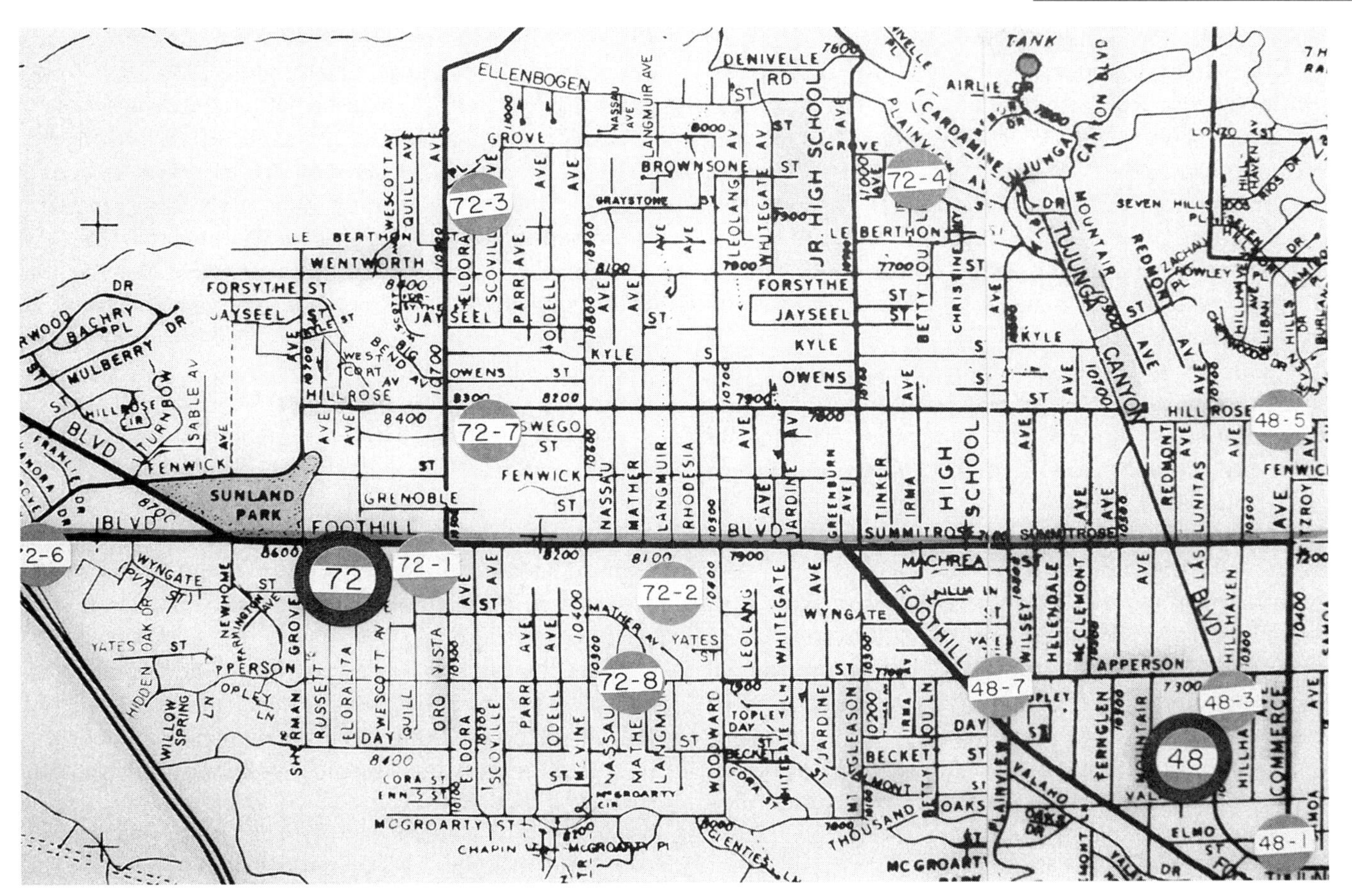

A map at Los Angeles's energy troubleshooting facility in 1987. Circuit 72–4, the source of an earlier blackout, is marked in the upper right. (Corbis-Bettmann)

be considered for ease of modification, should future load patterns differ from present projections.

Generation was normally modeled in system studies using economic dispatch schedules developed from individual unit fuel cost and heat-rate data. To serve a particular load level, the units would be "stacked" (added to the system) in order of priority based on cost and performance. Additional capacity options were available from off-system purchases or reserve sharing arrangements with neighboring systems. A system's ability to import power was a strong indicator of its territorial reserve requirements.

Prior to the 1990s, generation lead time was generally the critical factor in project planning. As smaller units, low run-time peaking plants, environmental issues, and more diversified fuel availability appeared, the lead time for transmission construction began to be of more concern than the generation construction period. In addition, line citing, environmental impact, and condemnation issues contributed to longer lead times for transmission projects.

The planning processes described herein were traditional in nature, with little variation from year to year. Growth was steady, equipment technology advancements were available for the conditions expected, and while utility facilities were capital-intensive, a regulated guaranteed return on investment resulted in adequate financing capital. The utility planner had a vision of the future with acceptable accuracy.

TRANSITION

Many events signaled the beginning of a lengthy transition period for the utility industry in general, and more specifically, for the facility planning activity. The timeline in many instances was blurred, with some of the blurring attributable to a reluctance to change.

Perhaps the Public Utilities Regulatory Policies Act (PURPA) of 1978 was the initial formal driving force for change. The introduction of nonutility generators (NUGS) with qualified facilities and almost guaranteed sales opportunities was the first major nonutility resource activity. Whether the cause was the actual construction of these facilities or the resource atmosphere it stirred, transition was on the way. Independent power producers (IPPs) soon followed. These projects were introduced into the competitive resource market on an equal footing with utility-sponsored plants and could not be ignored by utility management.

In parallel with the new approaches to generating capacity additions, the utilities, with encouragement from regulators, introduced incentives during the 1980s for reducing load demand. Since the system peak hour load provided the inertia for capacity requirement definition, "shaving" of the peak became the focus of these incentives.

Enticements to involve the retail customer base, as well as the industrial sector, in the solutions desired became popular. Interruption of large commercial load, a provision by contract, had long been used to offset capacity shortages. Interest in this method mounted. In addition, retail customers were provided enticements to "cut load," either by manual participation or by automatic devices. These enticements will like continue into the future.

Further involvement by the typical industrial or commercial utility customer, both large and small, was stimulated by time-of-day price incentives. Encouragement was provided in the form of reduced rates if use of electricity was shifted from peak periods of the day to off-peak or shoulder-peak periods. Even the residential customer was invited to participate in load shifting with price incentives or rewards. A popular example of the day was encouraging the household laundry activity to be moved to late-night hours. This suggestion was met with varying enthusiasm.

Initially there were attempts to quantify and project magnitudes of power shifted and relief provided to the capacity resource requirements. Ultimately the shifting of load became embedded in the subsequent forecast and became harder to identify.

The impact of these tactics to augment capacity resources was successful to some degree. Currently, the values identified are marginal; however, much of the impact is unidentified and embedded in the load growth.

Industrial customers, early a driving force in the industry, began to react not only to local utility incentives but also to more competitive pricing opportunities in other systems and exerted significant pressure on regulators. Forces in Washington, D.C., following the industry trends, finally acted in 1992 with open access of the transmission system to the wholesale market, and more extensive competition resulted. Possibly this one act, more than any other, provided the stimulus for a major shift in the industry through deregulation, restructure, and more specifically, in the planning process of the bulk electric system.

With the opening of the transmission network to all resource .suppliers, many marketing entities entered the game in the mid-1990s. Many of these marketers are subsidiaries of already established gas suppliers. Some have been created solely for the electric industry. Still others have been formed from the utilities themselves. All of these entities are competition-motivated. Facility planning and reliability issues, while important to their business, are left to other organizations.

The planning and construction of generation and transmission facilities by utilities came to a virtual halt. With no guaranteed market, no control over the use of their transmission networks, no assurance of stranded investment recovery, and no assurance of federal remedial treatment, the economic structure of utility planning and construction of generation facilities basically stopped in place.

The transition for utility planners has been difficult. Generation and transmission margins have deteriorated. Shortages have been noted. The number of transactions within and across systems has escalated to levels that not only test the reliability of a transmission network not designed for this level of activity but that also have challenged the ability to operate such a system with existing technology. Prices have reacted. Emergency disturbances have occurred. Basic questions regarding authority, planning, responsibility, and finances have been raised.

Several alternatives have been offered by the Federal Energy Regulatory Commission (FERC). Regional transmission groups (RTGs) were suggested. While the basic premise of these wide-area coordinating and planning organizations was sound, many of the questions involving economics and responsibility were left unanswered.

Systems of independent system operators (ISOs) were introduced as another alternative. Some have been implemented and many have been evaluated. Adoption of these systems has been slow, the same basic reasons as those affecting RTGs. The regional transmission organization (RTO) is the most current alternative being proposed and initial plans will be provided to FERC during 2000. All of these organizational arrangements are devised to segregate ownership, planning, construction, and operation in an appropriate manner to produce a nondiscriminatory, totally competitive marketplace for electricity. All of this activity will likely result in the complete deregulation of the industry, wholesale and retail. Restructuring of the electric business on all fronts is required to make deregulation complete.

The North American Electric Reliability Council (NERC), established in 1968, and its associated ten regional councils have been reliability monitors of the electric bulk power system. Their emphasis on compliance with reliability standards through peer pressure was effective in the past. NERC and the councils, with oversight from FERC, are engaged in a complete restructuring of the reliability organizations. This move will ultimately involve a move from peer pressure to mandatory compliance.

In the meantime, the utility planner struggles with how to approach the planning process in this changing environment. Perhaps these issues are more critical in the transition, since so many of the traditional parameters of planning have disappeared, and few new parameters have taken their place.

Future-generation location, timing, and ultimate customer designations are generally unknown. Transmission construction has become more of a patchwork activity. Power transactions across the various systems are at an all-time high, a condition the network was not designed to handle. System operators struggle with daily and weekly operational planning because of the volume of transactions. Many of these transactions are not identified beyond buyer and seller. The marketing of power continues to follow the traditional contractual-path power flow route from seller (source) to buyer (load), as opposed to the actual system flow path. The contract path is a predetermined contractual route from the source system to the load system, through any interconnected intermediate system or systems. The contract assumes that all or a majority of the power will flow along this path. Only systems outlined in the contract are involved or compensated. In reality the power flows over the path of least resistance, as determined by the parameters (resistance) of the transmission lines involved. While there are economic considerations involved between the two alternatives, this is an issue of significant magnitude to the system planners and operators, who must know the path of actual power flow to properly address facility planning and operating security. In addition, many operating problems arise from the lack of automated software, a requirement for the volume of daily business being handled.

One alternative to planning in this environment of unknowns is called "scenario" planning. Utility planners in 2000 continue to estimate resource requirements for five to ten years into the future. These resources could be constructed indigenous to their systems or external to their systems, or purchased from off-system. By considering multiple alternatives for generation sources, the planner can simulate power transfers from within and outside each system. The results of these scenario analyses can be used to estimate where critical transmission might be constructed to be most effective for wide-area power transfer. Similarly, analyzing multiple transfers across a system can provide further justification for a new transmission path.

Other new planning processes are being considered to aid the transition. These include variations of probability analysis, optimum planning tools, and short-lead-time projects. None of these addresses all of the constraints discussed above. This should not, however, be construed as an impossible task.

As new generation is announced from all market segments, and as physical locations are determined, the planning picture will slowly evolve as well. It is also assumed that as retail access is introduced during the first decade of the twenty-first century, it will be controlled to some extent by contract terms long enough to provide practical development of demand forecasts. It is further assumed that future legislation and regulation also will assist in defining many aspects of the planning process. While the transition may be lengthy, the utility planner may be immersed in a full competitive environment without realizing that the transition has been completed.

FUTURE

The future of utility planning is uncertain. Good engineering judgment and technological advancements, however, will prevail as future system requirements are defined.

Major wholesale load shifts, unidentified transactions, dynamic scheduling (services provided by systems remote from the load served), unknown generation locations and timing, and retail wheeling will contribute to difficulty in planning for the future. The issues of planning, however, are likely to be easier to solve than the political issues that have a major impact on the health of the electrical infrastructure. The threats are many. An environment created by new deregulation legislation that fails to consider or impacts critical electrical phenomena and proper division of responsibility between the state and federal domains will be difficult to plan for. Environmental restrictions that curtail or impact generation resources or transmission availability will be costly to overcome and may lead to undesirable shortages. Retail customer access to systems remote from the host system will present future projection issues and change obligation-to-serve regulations. While stranded investment recovery will continue to be an economic issue, it may impact future planning in the restriction of alternative choices. Finally, mandatory compliance with reliability standards, while necessary to maintain a reliable system in a competitive marketplace, can become a constraint to good planning practices if the program becomes more bureaucratic than functional.

Based on history and the inherent unknowns related to planning, it is likely that full retail access and customer choice will present the utility planner with the most difficulty in future planning of the system. Since load demand is the principal driver of the facility developmental process, it has been necessary in the past to have a high degree of probability in the forecast. Without significant improvement in planning techniques for more accurate forecasts, planning for a changing (fluctuating) load demand in a given service area will be a significant challenge.

The various restructures of the industry are all planned to address these major issues. Wide-area planning, while undefined, may solve certain issues resulting from the unknown parameters. Divestiture of the industry into generation, transmission, and distribution companies will strengthen emphasis on the facilities involved. Many of these new structures will move the industry toward full competition; however, coordination of their activities could become more difficult.

Entry into a true competitive market suggests that the market will resolve all issues created by this move. Supply and demand will likely prevail. Full retail wheeling will introduce issues into utility planning never before addressed. Planning strategy for the future is difficult to envision based on history and current transition difficulties. It is assumed, however, that the need will create the solutions.

New technology, in both long-range planning and operational planning, will aid the entrance into a full competitive market. New tools and ideas will offset increased business and downsizing of manpower created by a more competitive environment. Each significant transition in the past was met with creative solutions. Utility planning expertise and experience will be major factors in the twenty-first-century electric industry. The goal, as always, is to continue the planning, construction, and operation of an economic and reliable electric power system.

James N. Maughn

See also: Capital Investment Decisions; Economically Efficient Energy Choices; Economic Growth and Energy Consumption; Electric Power, System Protection, Control, and Monitoring of; Energy Management Control Systems; Government Intervention in Energy Markets; Regulation and Rates for Electricity; Risk Assessment and Management; Subsidies and Energy Costs; Supply and Demand and Energy Prices.

BIBLIOGRAPHY

Arrillaga, J.; Arnold, C. P.; and Harker, B. J. (1983). *Computer Modelling of Electrical Power Systems.* New York: John Wiley & Sons.

Edison Electric Institute. (1991). *Transmission Access and Wheeling.* Washington, DC: Author.

Fox-Penner, P. S. (1990). *Electric Power Transmission and Wheeling: A Technical Primer.* Washington, DC: Edison Electric Institute.

Knable, A. H. (1982). *Electrical Power Systems Engineering: Problems and Solutions.* Malabar, FL: Robert E. Krieger.

McDonald, F. (1962). *Insull.* Chicago: University of Chicago Press.

Rosenthal, B. (1991). *Transmission Access and Wheeling.* Washington, DC: Edison Electric Institute, Power Supply Division.

Spiewak, S. A., and Weiss, L. (1997). *Cogeneration and Small Power Production Manual,* 5th ed. Lilburn, GA: Fairmont Press.

Stevenson, W. D., Jr. (1982). *Elements of Power System Analysis,* 4th ed. New York: McGraw-Hill.

Wang, X., and McDonald, J. R. (1994). *Modern Power System Planning.* New York: McGraw-Hill.

VENTILATION

See: Building Design, Commercial; Building Design, Residental

VOLTA, ALESSANDRO (1745–1827)

Alessandro Volta, now known as the inventor of the electric battery and eponym of the volt, the unit of electrical potential, was a prominent figure in late eighteenth-century science. A younger son from a family of the lesser nobility, he was born in 1745 in the commercial town of Como, in Northern Italy, at that time part of Austrian Lombardy. He received an irregular education and did not attend university. As a sometime pupil of the Jesuits, however, he developed a lifelong interest in natural philosophy, combining it with a commitment to Enlightenment culture and the notion of "useful knowledge," then fashionable among the educated classes and the public administrators of Lombardy. Having chosen the science of electricity and chemistry as his fields of expertise, at twenty-four he published a treatise "on the attractive force of the electric fire." At thirty he embarked on a career as a civil servant and a teacher in the recently reformed educational institutions of Lombardy. Appointed professor of experimental physics at the University of Pavia in 1778, he held the position until he retired in 1820. He traveled extensively, sharing his enthusiasm for natural philosophy with colleagues in Switzerland, the German states, Austria, Britain, the Low Countries, and France where in 1782 he met Benjamin Franklin.

Regarded in some circles as "the Newton of electricity," by the 1780s Volta had won a European fame as an "electrician," a chemist specializing in the chemistry of airs (especially "the inflammable air found in marshes," i.e., methane, his discovery), and the brilliant inventor of intriguing machines. Volta's machines prior to the battery included the electrophorus, or the "perpetual bearer of [static] electricity," a eudiometer, the electric pistol, the "condensatore" (a device that made weak electricity detectable), and a straw electrometer. Volta's contributions to the science of electricity included the notions of tension, capacity, and actuation (an ancestor of electrostatic induction). Thanks to painstaking measurements taken throughout his life, Volta managed to combine these notions into simple, quantitative laws that offered effective guidance through the intricacies of eighteenth-century investigations into electricity. In 1794, the Royal Society of London awarded Volta the Copley Medal for his work on Galvanism. In 1801, in the wake of his discovery of the battery, Napoleon publicly rewarded him. Volta died in 1827 in Como.

The Voltaic battery conceived and built toward the end of 1799 in Como, was the first device to produce a steady flow of electricity, or electric current. The instrument enabled other natural philosophers, notably Humphry Davy, to develop electrochemistry as a new branch of science, and still others, notably Hans Christian Oersted and Georg Simon Ohm, to explore electromagnetism. Because of these later developments—showing that chemical, electri-

Alessandro Volta. (Corbis Corporation)

cal and magnetic phenomena could be converted into each other—after the early 1840s the battery was a frequent topic for reflections on what was subsequently known as energy conversion, and energy conservation.

Volta saw the battery in a different light. He had conceived it as a demonstration device to show his contact theory of electricity at work. He had developed this theory to refute Galvani's notion of a special electricity intrinsic to animals. Volta claimed that the mere contact between different conductors (especially metals) was able "to set the electric fluid in motion." He also claimed that the electric fluid—one of the several, imponderable fluids then found in physics—was the same in organic and inorganic bodies. He built the battery after reading a paper by William Nicholson, who suggested imitating the electric organs of the torpedo fish by means of an apparatus combining many electrophoruses together. Having discarded his own electrophorus as unable to perform as Nicholson expected, Volta tried instead with pairs of discs of two different metals (like silver and zinc) that he knew could give weak signs of electricity. When he managed to pile up several such pairs, always in the same order and inserting a wet cardboard disc between each metallic pair, he found that an electric current was produced at the two ends of the pile, and that the power of the current increased with the number of pairs making the pile.

In the first circulated description of the battery—two letters addressed to the Royal Society of London on March 20 and April 1, 1800—Volta mentioned no chemical phenomena associated with the new apparatus. It was William Nicholson and Anthony Carlisle who, having had access to Volta's letters prior to publication, first observed the "decomposition of water" (electrolysis) while experimenting with the battery in London in May 1800. Nicholson, in particular, emphasized the chemical phenomena accompanying the operations of the battery. After that, a struggle between Volta's contact interpretation and the chemical interpretation of the battery developed; a struggle that Wilhelm Ostwald still regarded as unsettled in 1895. The struggle had obvious if complex implications for reflections on energy conversion and conservation; the more so because people like Nicholson already perceived the battery as the herald of a new family of machines, and wondered how the "intensity of action" of these machines could be measured.

Viewing things from the perspective of his physical theory of contact electricity, Volta was intrigued by the apparently endless power of the battery to keep the electric fluid in motion without the mechanical actions needed to operate the classical, friction, electrostatic machine, and the electrophorus. He called his battery alternately the "artificial electric organ," in homage to the torpedo fish that had supplied the idea, and the "electromotive apparatus," alluding to the "perpetual motion" (his words) of the electric fluid achieved by the machine. To explain that motion Volta relied, rather than on the concepts of energy available around 1800, on his own notion of electric tension. He occasionally defined tension as "the effort each point of an electrified body makes to get rid of its electricity"; but above all he confidently and consistently measured it with the electrometer.

Giuliano Pancaldi

BIBLIOGRAPHY

Heilbron, J. L. (1981). "Volta, Alessandro." In *Dictionary of Scientific Biography,* edited by C. C. Gillispie, Vol. 14, pp. 69–82. New York: Charles Scribner's Sons.

Pancaldi, G. (1990). "Electricity and Life. Volta's Path to the Battery." *Historical Studies in the Physical and Biological Sciences*, Vol. 21. Berkeley: University of California Press.

VOLTAGE

See: Units of Energy

WASTE HEAT RECOVERY

See: Cogeneration Technologies

WASTE-TO-ENERGY TECHNOLOGY

Solid waste disposal has been an issue since the dawn of humanity. In the earliest times, land filling, or, more simply, land application, was the most likely scenario for disposal, as anything unusable or nonessential for immediate survival was discarded and left by the trail as the hunter-gatherers moved to follow life-sustaining herd migrations. As humans became more "civilized" and established permanent residences and villages and obtained an existence through means of agriculture, trade, etc., the accumulation of wastes increased due to the localized populations and increased permanence of those societies. Waste disposal practices included both landfill and combustion, as anything of fuel value was most likely burned for heat or cooking purposes. In all likelihood, landfill accounted for a minimal fraction of the waste in those early times, until the advent of the industrial revolution opened the way for greater leisure time and the economy to create and consume greater quantities of nonessentials, which, in turn, created a greater flow of disposable waste from the thriving communities.

Solid waste disposal has always consisted of two methods, burning or discarding. Requirements of communal living conditions and a greater understanding of the health and sanitary implications of haphazard waste disposal created the need to concentrate solid wastes into a landfill and bury the material. Convenience and availability of seemingly unlimited space favored land filling as the universal means of waste disposal until midway through the twentieth century. As populations increased, however, capacities in existing landfills were rapidly used up and sites for new "garbage" dumps were being pushed farther and farther from the population centers, leading to increased hauling and operating costs for the newer facilities. Since the early 1990s, a new environmental regulations continue to increase both the complexity and the costs for new landfill facilities in the United States and the other more developed nations worldwide.

Where any pressures arose concerning the siting or operating costs of land filling, consideration has been immediate for the option of combustion of the same waste stream. The earliest systems were designed to incinerate the incoming waste. With no energy recovery capabilities and only basic wet scrubber gas cleanup technology, this practice reduced the quantity of waste ultimately routed to the landfill by seventy-five to eighty-five percent. These early incinerators were typically small, only fifty to one-hundred tons per day capacity, mass burn, starved air, multistaged incinerators. During operation, the waste would be introduced into the furnace onto the first of a series of ram-activated cascading grates. In this primary zone, the fresh fuel contacted some combustion air and a stream of recycled gases from the exhaust header. As the waste began to volatilize and burn, the ram feeders on the grate pushed the pile out on the grate where it fell to the next step. In the process, the material was stirred and agitated to present a fresh combustion surface to the air stream. Likewise, the material from this second step was ram-fed inward until it "cascaded" onto the next grate. This process continued until the remaining material, primarily unburnable material plus some uncombusted carbon, fell into the water-cooled ash trough and was removed from the furnace in an ash conveyor. The combustion and volatile gases driven off from the fuel pile were mixed with combustion air in the primary chamber directly above the grate. Further combustion and quench occured in the secondary and tertiary chambers where additional ambient air and recycled flue gases were mixed into the combustion exhaust stream. All of the gases from the tertiary chamber were conveyed into the wet scrubber where particulate was removed.

In the early 1970s, the surge in energy costs, spurred by the oil embargo, created a demand for energy recovery in these waste disposal facilities. At that time, while still in its infancy stage, waste-to-energy was poised to take one of two paths to implementation—mass burn (MB) or process fuel, typically called refuse derived fuel (RDF). Like the predecessors, mass-burn technology opted to burn the waste virtually as it is received, eliminating preprocessing of the material prior to burning. The mass-burn units are typically very large furnaces, most often field erected, wherein the waste is literally plucked from a receiving floor with a grappling hook and ram-fed into the furnace onto a cascading grate design. Following the success of other European designs beginning in the 1950s, these systems were typically of the waterwall furnace design, as opposed to a more "conventional" refractory furnace. One advantage of this was the direct removal of much of the furnace heat by the waterwalls, minimizing the need for significant volumes of excess combustion air or recycled flue gas to maintain acceptable operating temperatures in the furnace. These plants have also evolved into much larger capacity facilities. They were already field erected, so up-sizing the capacity did not represent any further economic disadvantage, and have been built to capacities as great as 2,500 tons per day.

Because the mass burn systems are designed to handle the fuel as it is received, implying the obvious variations in quality, sizing, and handling therein, the combustion systems must be extremely robust and conservative in design with regard to the quality of fuel being introduced. In addition, the ash removal system must be capable of handling the size and capacity of material coming out the furnace as is fed into the furnace on the front end. While this universal acceptability has proven to be one of the most important features of the mass burn success, it has also created two major liabilities in the performance of the furnace. First, extreme variations within the fuel pile on the grate can creat significant temperature variations across the grate which, in turn, can cause high temperature damage to the equipment or generate greater emission from products of incomplete combustion. Second, unpredictable fuel composition can create surges in ash quantities, resulting in slagging and fouling of the boiler surfaces and increased emission levels of unburned hydrocarbons, acid gases, and dioxins/furans in the outlet gases. In spite of these drawbacks, the mass burn technology has captured a majority of the waste-to-energy market in the United States in the last three decades of the twentieth century. Part of that success came as a direct result of the early failures of RDF in the industry.

Simplified in process terms, RDF involves processing the incoming municipal solid waste (MSW) stream to remove a substantial portion of the noncombustible components, namely aluminum, ferrous, glass and dirt. Various sources list these components in the range as follows:

Aluminum	2%
Ferrous	6-11%
Glass	11-12%
Dirt/ grit	2-20%

A review of this composition would indicate the noncombustibles in the raw MSW range from twenty to forty percent. By removing as much of these fractions as possible from the fuel stream, the quality of fuel presented to the combustor is improved as well as the contaminants from the combustor being reduced.

Unfortunately for the RDF industry, the first attempts at implementing an RDF processing system met with disappointment and failure. With no European technology to draw from, RDF processing evolved from experience and inspiration gained from the U.S. applications. In the earliest processes, the design called for all of the incoming waste to be shred-

The Bridgeport RESCO power plant mixes garbage to facilitate the drying process, which improves the efficiency of energy generation by incineration. (Corbis Corporation)

ded as it entered the process. Conceptually, this was a good idea, providing a means to get all of the material sized to a maximum particle size, thereby enabling further separation based upon particle sizing and density; however, the actual results were less favorable. First, shredding all material as it was introduced into the process included numerous explosive items such as propane bottles, gas cans, etc. The damage caused to the equipment in these early processes was significant but the loss of life and limb from some of the early accidents was even more devastating. Aside from the catastrophic failures due to explosions, this initial shredding also contaminated the waste mixture and made further separation difficult. Shards of glass and shredded metal became imbedded in the paper or biomass fraction of the material in the shredder and were carried through the process in that fraction, thereby reducing the removal efficiencies of those noncombustible fractions. Although many of the earliest RDF processes are still in operation, numerous modifications have been made to improve their performance.

With such a rocky beginning, the RDF option soon fell out of favor and yielded the market to the mass burn technology. The complexity, added costs of material handling, and the poor operating history of the RDF processes proved to be sufficient negative factors for any significant consideration of RDF fired waste-to-energy facilities for the next ten to fifteen years. That situation did not turn around until the push toward more waste recycling activities in the late 1980s and early 1990s. With that new impetus, plus many years of experience in operating RDF processing systems, the new generation of automated waste processing plants gained favor from a political/social base as well as from a more proven technical design basis. In numerous instances, RDF systems were promoted strictly to comply with mandated recycling directives. These material recovery facilities, or MRFs, accomplished essentially the same function as the earlier RDF processes in separating combustible from noncombustible materials, but were now done in the name of "recycling" with the recovery of ferrous and aluminum metals, glass, newsprint and corrugated paper, and plastics being the primary objective. Although oftentimes reduced yields of RDF were achieved because of the higher removal of recyclable materials, the quality of the fuel stream was strongly enhanced by this approach. With a national mandate in the United States on recycling, the cost and need for preprocessing of the waste became a burden to be borne by all new waste management plans. The advantages of mass burn technology had just been eliminated via the legislated mandate for higher recycling.

With more proven methods for RDF processing being demonstrated, the increase in RDF combustion technology has followed. Some of the facilities burning RDF utilize similar grate and furnace technology as the mass burn, but others, most notably fluidized bed or circulating fluidized bed combustion, offer a new and enhanced means of combusting the waste. Fluidized bed technology refers to the concept of burning a fuel in a combustion zone comprised of small particles of sand suspended in a stream of upward flowing air. The air velocity provides a buoyancy force to lift and suspend the individual sand particles such that they are able to float throughout the mixture and move freely within the furnace. The sand bed displays the characteristics of a pot of boiling water, hence the term, fluid bed. In a circulating fluid bed, the air velocities are actually increased to the terminal velocity of the sand particle, thereby carrying the sand up and out of the furnace. The sand is mechanically separated from the air/gas stream at the outlet of the furnace, and the sand is recirculated back into the furnace at the bottom of the bed. Hence, the term circulating, or recirculated, fluid bed.

The advantages of fluid bed combustion over the more traditional technology arise from the increased turbulence provided by the bed particle action. This fluidization increases the interaction of the fuel particles with the combustion air and creates a very accelerated combustion environment for the incoming fuel. Additionally, the sand, initially heated to an ignition temperature for the incoming fuel, provides a means of heating and drying the new fuel introduced into the furnace and igniting that fuel as it heats up. The thermal stability provided by the hot sand in the bed plus the turbulence of the fluidization causes the fuel particles to be consumed in a very short period, typically a few minutes or less. Complete combustion is achieved by the uniform temperature zone within the furnace plus the enhanced intermixing of fuel and combustion air. Historically, fluidized bed combustion systems have been able to achieve better operating flexibility, lower emission levels, and improved combustion and boiler efficiencies.

All methods of combustion of the waste pose certain "problems." The fuel, or waste, stream is nonhomogeneous, so the likelihood of variances in operation and performance are typical. With mass burn systems, the inventory of fuel within the furnace is maintained fairly high, for up to thirty minutes, and the "management" of the fuel feeding system is responsible for selecting a feed blend that maintains some measure of constancy throughout the process. With RDF systems, the processing of the fuel and sizing/ shredding has enhanced the fuel homogeneity significantly, but variations in quality, content, etc., can still be expected. The fuel is known to contain measurable levels of sulfur, chlorine and other contaminants that can generate acid gases in the combustion products. These acids can, and do, attack the boiler surfaces and other equipment in the gas train, causing corrosion and necessitating continuous maintenance. Most boilers establish steam conditions of 650 psi and 750°F (398.9°C) as the maximum design condition in order to minimize the corrosion losses in the boiler. Recent improvements in metallurgy and furnace designs have enabled these design limits to be pushed to 850 psi and 825°F (440.6°C) or higher.

As a fuel, municipal solid waste (MSW) does not compare too favorably with more traditional solid fuels, such as coal. MSW averages somewhere around 4500 Btu/lb, versus coal at 10,500–13,000 Btu/lb. However, given the current U.S. population of 250 million and the annual generation of waste per person of fifteen hundred pounds, the potential energy content in the annual waste generated in the U.S. alone is comparable to nearly seventy million tons of coal and has the potential to generate over 13,000 MW of electrical power. As of a published report in 1993, 128 facilities were actually in operation, with an additional forty-two planned or under construction. Of the existing plants, ninety-two produced electricity to some degree. The remaining thirty-six produced only steam or hot water. The number of plants in the various sizes range from twenty plants less than one-hundred tons per day to twenty plants greater than two-thousand tons per day. Roughly forty plants are in the capacity of one-hundred to five-hundred tons per day and about twenty-five plants are sized for each of the five-hundred to one-thousand tons per day and the one-thousand to two-thousand tons per day capacity. Most of the smaller plants were installed in the early period of waste to energy development. Most of the facilities installed in the last ten years represent sizes from five-hundred to one-thousand tons per day. As of 1993 the capacity of waste consumed in these facilities was approximately 103,000 tons per day, representing approximately 18 percent of the projected annual waste generation in the United States.

Although waste-to-energy probably will never supply more than one percent of U.S. electricity (U.S. electricity generation in 1997 was 3,533 billion kilowatts), it is still a very useful renewable energy source. A state-of-the-art RDF cogeneration plant (hot water and electricity) burns much cleaner than the majority of existing coal-fired plants. At the same time, it is an attractive option in communities that desire inexpensive electricity and predictable waste disposal costs ten to fifteen years into the future, and that lack the real estate for additional landfill space. The major obstacle facing new waste-to-energy facilities is the considerable resistance contingency of the community who voice the "not-in-my-backyard" objection to the smell, noise and emissions. New facilities are likely to be located in rural areas, or as supplemental energy suppliers for industrial complexes, or on or near existing landfill operations.

The potential economic benefit exists for many regions to elect to become "host" facilities for disposal of wastes shipped in from greater metropolitan areas, such as New York City, which is closing its Freshkill landfill and already ships thousands of tons to Ohio, Georgia, and elsewhere for disposal, mostly in landfills. As the cost for energy (oil and natural gas) increases, the energy value of the waste will escalate. Its value as an alternate fuel will compensate for some of the underdesirable attributes and create a greater demand for future waste-to-energy facilities.

Michael L. Murphy

See also: Efficiency of Energy Use; Environmental Economics; and Environmental Problems and Energy Use

BIBLIOGRAPHY

Electric Power Research Institute. (1991). *Proceedings: 1989 Conference on Municipal Solid Waste as a Utility Fuel*, ed. E. Hughes. Palo Alto, CA: Author.

International Conference on Municipal Waste Combustion. (1989). *Vol. 1, Conference Proceedings*. Hollywood, FL: Author.

North American Waste-to-Energy Conference. (1997). *Proceedings of Fifth Annual North American Waste-to-Energy Conference*. Research Triangle Park, NC: Author.

Power Magazine Conference and Waste-to-Energy Report. (November 1986). *Energy from Wastes: Defining Public-and-Private Sector Markets*. Washington, DC: Author.

Solid Waste and Power (May/June 1993). *"WTE in North America: Managing 110,000 Tons Per Day."* Kansas City, MO: HCL Publications.

Meridian Corporation. (1986). *Waste-to-Energy: A Primer for Utility Decision-Makers*. Golden, CO: Western Area Power Administration.

WATER HEATING

Heating water consumes over 3.5 quadrillion Btus per year of primary energy which makes it the third largest energy end use for the residential and commercial sector in the United States, after space conditioning and lighting. Although indoor domestic hot water is nearly universal in the United States, this is a relatively recent historical development.

One of the earliest recorded uses of heated water was the Roman public baths. Starting in the first century C.E., many Roman cities had public baths with pools of cold, warm and hot water. The water was heated with wood or charcoal, although some of the baths incorporated passive solar features for space heating.

Prior to World War II, only about half of households in the United States had complete indoor plumbing facilities, including hot and cold water. Before indoor plumbing became widespread, water was heated on stoves in pots and pans, a difficult process that meant a lot less hot water was used.

The first water heaters used heat from a wood or coal stove to heat water in an uninsulated storage tank. Eventually, separate heating sources were

Water Heater	Installed Cost	Energy Factor	Annual Energy Cost	Life/ Reliabilty	Market Share 1995	Best Application
Gas						
Standard	$200-$600	.56	~$150	9±6 Years	47%	If Gas Available
High Efficiency	$1600+	.74	~$110	Data Unavailable	<1%	Heavy Hot Water Use Space Heating
Instantaneous	$700+	.64	~$125	Additional Maintenance	<1%	Limited Space, Low Hot Water Use
Electric						
Standard	$200-$600	.86	~$320	12±7 Years	53%	Cheap Electricity, Gas Not Available
Heat Pump	$1500+	1.7+	~$175	Poor Record To Date	<0.1%	High Hot Water Use Costly Electricity Gas Not Available
Instantaneous	$200-$400	.98		Data Unavailable		Small Intermittent Uses

Table 1.
Summary of Typical Water Heater Costs and Efficiencies

added. Either coal or wood heaters provided by a side-arm coil outside the tank. The hot water moved into the tank by convection.

The first instantaneous water heaters which started appearing in the 1890s when gas or liquid fuels started becoming available, were un-pressurized. The first models had no automatic controls and very limited safety features. Some early models were more efficient than standard modern gas-fired water heaters.

Early in the twentieth century over 150 different manufacturers were making storage tank, instantaneous, and solar water heaters. Then as indoor plumbing became more common and as people started using showers more and baths less, storage type water heaters became more popular. Accurate hot water temperatures were not possible in early instantaneous water heaters. This was not much of a problem with baths where the water temperature is adjusted before getting in the water, but it is a much bigger concern when using showers.

Solar water heating became commercially available in Southern California in the 1890s. Early models consisted of four large cylindrical tanks of heavy galvanized iron mounted horizontally in a wooden box under a glass cover.

In 1909 one company began selling solar water heaters with separate collectors and insulated storage tanks. The collectors were made of copper tubing soldered to a copper plate in a glass covered box. The hot water was transferred to the storage tank by thermosyphon action, so the insulated tank had to be installed above the collector, typically in the attic or on the roof. It was often connected to an auxiliary heater on the stove or furnace to supplement solar heat. Well insulated storage tanks meant hot water was available the next morning.

Solar water heating declined in Southern California in the 1920s due to the development of natural gas, but it continued in Florida where natural gas was very expensive. In 1941 more than half Miami's population had solar water heaters, and more than 80 percent of new homes built then were equipped with solar water heaters. By the end of the 1950s in Florida, solar water heating was displaced by electricity as the price dropped and the storage tanks of solar water heaters failed because of galvanic corrosion from connecting steel tanks to copper collectors.

Corrosion of the tank is the major factor limiting water heater lifetimes. The A. O. Smith company invented the glass-lined tank in 1939, where an enamel coating is baked onto the inside surface of the tank at high temperatures. The technology has subsequently been adopted by other manufacturers, although other linings are used such as cement or plastic.

Total sales of water heaters have climbed steadily over the past several decades with sharp increases corresponding to times of high home construction rates.

In response to the oil crisis in 1970s, and temporary curtailment of new gas connections, the percent of sales of electric water heaters grew to almost match gas-fired water heater sales. Since 1987, however, the percent of water heaters sales that are gas-fired has exceeded electric water heater sales by 5 to 10 percent.

Tax credits were provided for solar water heating systems in response to the energy crisis. After tax credits expired, the solar water heater business declined drastically. The only segment of solar water heater market that has continued to have growth is the pool heater.

Outside of the United States and Canada, storage water heaters are common in Australia, Southern Africa, Latin America, and Britain. Instantaneous water heaters are much more common particularly in Asia and Europe. In countries that were once part of the British Empire, unpressurized storage tanks that relied on gravity were common, however most have changed over to pressurized storage tanks.

STANDARD STORAGE TANK WATER HEATER CONSTRUCTION

A typical storage water heater consists of a steel tank that is lined with ceramic coating fused to the inside of the tank at high temperature during manufacturing. The tank is typically about sixteen inches in diameter and about four to five feet tall. The top of the tank is domed upward and the bottom of the tank is also domed upward in a concave manner. The outside of the tank is insulated with a polyurethane foam insulation that is squirted into the gap between the tank and a thinner sheet metal jacket. The polyurethane is made of two different components that react and harden when mixed. Included in the mixture is a blowing agent that causes the polyurethane to expand in a foam-like manner. Prior to about 1980, water heaters were insulated with fiberglass insulation. The foam insulation process was developed to allow automation and increased manufacturing speed and reduced costs. A side benefit was improved insulating ability leading to a slight increase in efficiency.

In response to concerns about ozone depletion, the original blowing agent, CFC-22 was phased out. In 1994 and 1995 the manufacturers switched to HCFC-141b, a substance with lower ozone depletion potential. HCFC-141b is much more difficult to use, and manufacturers had some difficulty learning to use the new blowing agent effectively. New formulations of polyurethane and much tighter tolerances on the manufacturing processes, such as temperatures and pressures, were required.

Electric water heaters typically use two 4,500 watt heating elements. One element is located in the lower part of the tank and provides the bulk of the energy. The other element is located near the top of the tank and is used to quickly heat a small amount of hot water after a large draw empties the tank of hot water. The elements are each controlled by separate thermostats and are interlocked so only one can come on at a time. The thermostats on electric water heaters are snap type devices that are installed directly on the outside of the tank, but inside the jacket. They are located a few inches above the element which they control.

Gas-fired water heaters use the same general method of construction, except that the elements are replaced with a burner beneath the tank. The combustion products from the burner are vented through a flue made out of the same thickness steel as the tank, that goes up through the center of the tank. To increase heat transfer from the hot flue gases to the inner wall of the flue, a baffle is inserted down the flue. This baffle is a twisted strip of sheet metal with folds and tabs on it. The folds and tabs are designed to increase the turbulence of the hot combustion products as they pass through the tank. The burner is normally a circular burner made out of sheet metal or occasionally cast iron. Nearly all gas-fired water heaters use a standing pilot for ignition purposes. The pilot is a safety pilot: If it goes out, the small amount of electricity generated by a thermocouple in the pilot flame will no longer hold open an solenoid valve, and gas to the main burner will be turned off. The thermostat on a gas-fired water heater is an iron rod within a copper tube that is inserted into the water near the bottom of the tank. As the copper tube is heated it expands and allows a click disc to close a gas valve and shut off the flow of gas to the burner. The pilot continues to burn even when the burner is not needed.

In the North American market, water heaters are almost always made with the cold water inlet and hot water outlet lines coming out of the top of the tank. The hot water outlet opens right into the top of the tank and so draws off the hottest water. The hot water has risen to the top of the tank because of its lower density. The cold water on the inlet side is directed to the bottom of the tank by a plastic dip-tube. In some models the dip-tube is curved or bent at the end to increase the turbulence at the bottom of the tank. This is to keep any sediment from settling on the bottom of the tank. As sediment— usually calcium carbonate or lime—precipitated out of the water by the increased temperature builds up, it will increase the thermal stress on the bottom of a gas-fired water heater and increase the likelihood of tank failure. On electric water heaters the sediment builds up on the surface of the elements, especially if the elements are high-density elements. Low-density elements spread the same amount of power over a larger surface of the element so the temperatures are not as high and lime doesn't build up as quickly. If the lower elements get completely buried in the sediment, the element will likely overheat and burn out.

Another required safety feature on water heaters is a temperature and pressure relief valve. This is mounted near or on top of the tank. The temperature and pressure valves are designed to vent water if the temperature or pressure in the tank becomes too high. If there is no relief valve and the controls fail to limit water temperature, there is a danger of a powerful explosion from superheated, pressurized water flashing to steam as the water heater bursts.

All glass-lined steel tanks have at least one anode. This is a long metal rod made of magnesium, aluminum or zinc, metals that are more reactive than steel. The glass lining is never perfect, and any pinholes or defects around welds would allow water to touch steel. The anode protects the steel in the tank from corrosion by "sacrificing" itself and corroding first. This prolongs the life of the tank and protects against galvanic corrosion when two dissimilar metals are in contact in water.

Storage type water heater tanks also have a drain valve at the bottom so the tank can be drained for maintenance, or in case it ever needs to be removed. These drain valves are often plastic, although higher quality models use a brass drain valve.

EFFICIENCY STANDARDS

Under the National Appliance Energy Conservation Act of 1987, residential water heaters in the United States are required to meet a minimum energy efficiency. This is measured as energy factor. The minimum allowed energy factor depends on the fuel type and decreases with increasing rated volume. The minimum energy factor for fifty gallon electric water heaters is 0.86, for forty gallon gas-fired water heaters it is 0.54.

Average efficiency of new gas-fired water heaters has increased from an estimated 47 percent in the mid-1970s to about 56 percent in 1999. Over the same period the efficiency of electric water heaters has risen from about 75 percent to 86 percent. Revised efficiency standards were expected to be adopted during 2000.

Energy factor is a measure of average service efficiency at a specified condition and hot water draw pattern. It includes the effects of both standby losses and recovery efficiency of the water heater. Currently, water heaters are shipped from the manufacturers with thermostats set at 120°F (48.9°C) to reduce the risk of scalding. This is a drop from values of 140°F (60°C) that were reported from the early 1970s.

Efficient models of water heaters have thicker insulation, up to three inches thick, on some of the most efficient electric water heaters. Another means to increase efficiency is installing heat traps, or anti–convection devices, on the inlet and outlet pipes. Standard heat traps consist of short pipe nipple containing a small plastic ball. On the inlet side the ball is lighter than water and floats up to seal the inlet pipe. On the outlet side the ball is heavier than water and sinks against the seal. This prevents the heated water, which is slightly buoyant, from setting up convection loops within the pipes.

Gas-fired water heaters are also made more efficient by a variety of designs that increase the recovery efficiency. These can be better flue baffles; multiple, smaller-diameter flues; submerged combustion chambers; and improved combustion chamber geometry. All of these methods increase the heat transfer from the flame and flue gases to the water in the tank. Because natural draft systems rely on the buoyancy of combustion products, there is a limit to the recovery efficiency. If too much heat is removed from the flue gases, the water heater won't vent properly. Another problem, if the flue gases are too cool, is that the water vapor in the combustion products will condense in the venting system. This will lead to corrosion in the chimney and possible safety problems.

One design that is used for installation convenience, but can also allow higher efficiency, is an induced draft fan that pulls in extra air to cool the flue products to temperatures that can be safely vented through plastic pipe. Because of the fan, long horizontal venting runs are possible with this system.

Flue dampers that block the flue when the burner is not firing increase the efficiency of gas-fired water heaters. These can operate electrically or thermally. Because gas-fired water heaters lose so much heat up the flue during standby periods, this can provide significant savings. These are readily available on larger water heaters used in commercial settings but haven't been applied in the residential market because of their cost.

The most effecient gas-fired water heaters are condensing models. These are designed to be so efficient that the water vapor in the combustion gases condenses out and must be drained off. The recovery efficiencies of these models can be as high as 94 percent. As of 1999, four of the major manufacturers offered condensing water heaters in the small commercial sizes (over 100 kBtu/hr rated input). To reach this efficiency level, most used a pre-mix power burner with a central flue that starts up the center of the tank, but instead of going out the top as in standard water heaters, the flue (or flues) turn and spiral down inside the tank and exit out the side at the bottom. Because of the corrosive nature of the condensate, the flues are specially protected, either by glass coating on the gas side as well as the water side or by being made out of stainless steel alloys. These models can cost as much as $1,000 to $1,500 more than standard water heaters. The best applications are where heavy hot water loads can justify the extra cost.

On standard electric resistance water heaters, not much can be done to improve efficiency other than adding more insulation and heat traps. Heat pump water heaters extract heat from air and add it to the water in the tank. Energy factors as high as 2.60 have been achieved, meaning about two-thirds less electricity is used compared to a typical electric resistance water heater. This technology is mature in space heating and household refrigerator applications (where heat is pumped out of the refrigerator into the surrounding room); however, despite several attempts by different manufacturers, heat pump water heaters have never gotten established in the market. The main problem seems to be much higher equipment costs than simple electric resistance water heaters and a poor reliability record on the part of some models.

A small fraction of homes use the same appliance to provide both space heating and water heating. This has been done for over a century with tankless coils in cast-iron boilers. Water heating can be provided by modern boilers supplying heat to an indirect water heater that uses the water from the hydronic system to heat water for domestic uses in a storage tank. Since the early 1980s a technology has been developed that uses hot water from the water heater to heat a forced air system via fan coils. Ground-source heat pumps that provide space conditioning while extracting or storing heat from the ground are often also equipped with the capability to provide water heating.

Perhaps the ultimate combined appliances, which are currently under development, will be microturbines or fuel cell power generators used as small-scale co-generation systems. These would supply not only electricity, but space and water heating as well.

James D. Lutz

See also: Heat and Heating.

BIBLIOGRAPHY

American Society of Heating, Refrigerating and Air-Conditioning Engineers, Inc. (1995). "Service Water Heating." In *1995 Handbook - HVAC Applications (I-P)*. Atlanta, GA: Author.

Barbour, C. E.; Dieckmann, J. T.; and Nowicki, B. J. (1996). *Market Disposition of High-Efficiency Water Heating Equipment, Final Report*, edited by U. S. D.o.E. Office of Buiding Equipment. Washington DC: Arthur D. Little, Inc.

Butti, K., and Perlin, J. (1980). *A Golden Thread: 2500 Years of Solar Architecture and Technology.* New York: Van Nostrand Reinhold Company.
Energy Information Administration. (1998). *Annual Energy Outlook 1999 with Projections to 2020.* Washington, DC: U.S. Department of Energy.
Gas Appliance Manufacturers Association. (1998). *Consumers' Directory of Certified Efficiency Ratings for Residential Heating and Water Heating Equipment.* Arlington, VA: GAMA.
Gas Appliance Manufacturers Association. (various years). *GAMA Statistical Release.* Arlington, VA: Author.
Greenberg, J.; Reeder, B.; and Silberstein, S. (1985). *A Review of Energy Use Factors for Selected Household Appliances*, prepared for Office of Building Energy Research and Development, Building Equipment Division, U.S. Department of Energy. Gaithersburg, MD: National Bureau of Standards.
Smith, K. (1993). *The Gas Fitter's Guide to Domestic Hot Water.* Lewiston, NY: KLS Training Corporation.
U.S. Census Bureau. (1999). "Census of Housing, Housing: Then And Now, 50 Years of Decennial Censuses." <http://www.census.gov/hhes/www/housing/census/histcensushsg.html>.
U.S. Department of Energy. (1998.) "Energy Conservation Program for Consumer Products." In *Title 10, Code of Federal Regulations, part 430.* Washington, DC: U.S. Government Printing Office.
Weingarten, L., and Weingarten, S. (1992). *The Water Heater Handbook: A Hands-On Guide to Water Heaters.* Monterey, CA: Elemental Enterprises.
Weingarten, L., and Weingarten, S. (1998). "The History of Domestic Water Heating." *PM Engineer* 4(8):74–77.

WATT

See: Units of Energy

WATT, JAMES (1736–1819)

Born fourth of five children to a merchant family in coastal Greenock, Scotland, on January 19, 1736, Watt was an ingenious modelmaker as a youth, being exposed to all the ship chandler's tackle and instruments, and was fascinated with natural philosopy, as science was called. His grandfather Watt was a very successful professor of mathematics in this remote locale. His mother's death when he was seventeen set his determination to pursue instrument making.

After nine years of some difficulty but considerable learning opportunity in London and at the University of Glasgow, while he was repairing a poorly built Newcomen engine for the university, he had the flash of inventive intuition that a separate condenser would greatly improve its efficiency. For this patented improvement, and its eventual technical and economic success, he is popularly credited with the invention of the steam engine. In 1765, while at the University of Glasgow, he married his cousin Margaret Miller.

At the age of thirty-three, Watt wrote, "Of all things in life, there is nothing more foolish than inventing." Just widowed with six children, he had put away his love of science, engineering and fine technical instrumentation to feed his family by contracting to survey for the Caledonian canal. Bankruptcy of his financial partner in steam development, John Roebuck, had derailed his efforts to improve the steam engine.

For some years previous, the successful Birmingham manufacturer Matthew Boulton had enjoyed an increasing correspondence in natural philosophy with such notables as Benjamin Franklin, Erasmus Darwin and Watt's partner, Roebuck, regarding practical uses of fuel and steam. Boulton's discussions with Watt in 1768 led him to discontinue his own developments, since Watt was clearly ahead technically and had obtained basic patents. With the bankruptcy of Roebuck, Boulton bought out Roebuck's interest, and Boulton's investment continued to fund Watt's demonstrations and developments. By then however, Watt's patent had nearly expired. With the leverage of Boulton's contacts in London, a Bill of Parliament extended the patent 24 years until 1800. The success of Watt's engine applications in the next few years brought Boulton and Watt together as partners, Boulton having two-thirds interest and managing the business, and Watt having one-third interest and managing the engineering and erection of engines.

While traveling on business, Watt met a Miss MacGregor who was connected with the University of Glasgow, and they married in 1776. Their two children did not live into adulthood.

The business acumen of Boulton provided profound industrial leverage to the technical creativity of Watt. The success of the steam engine for the rest of

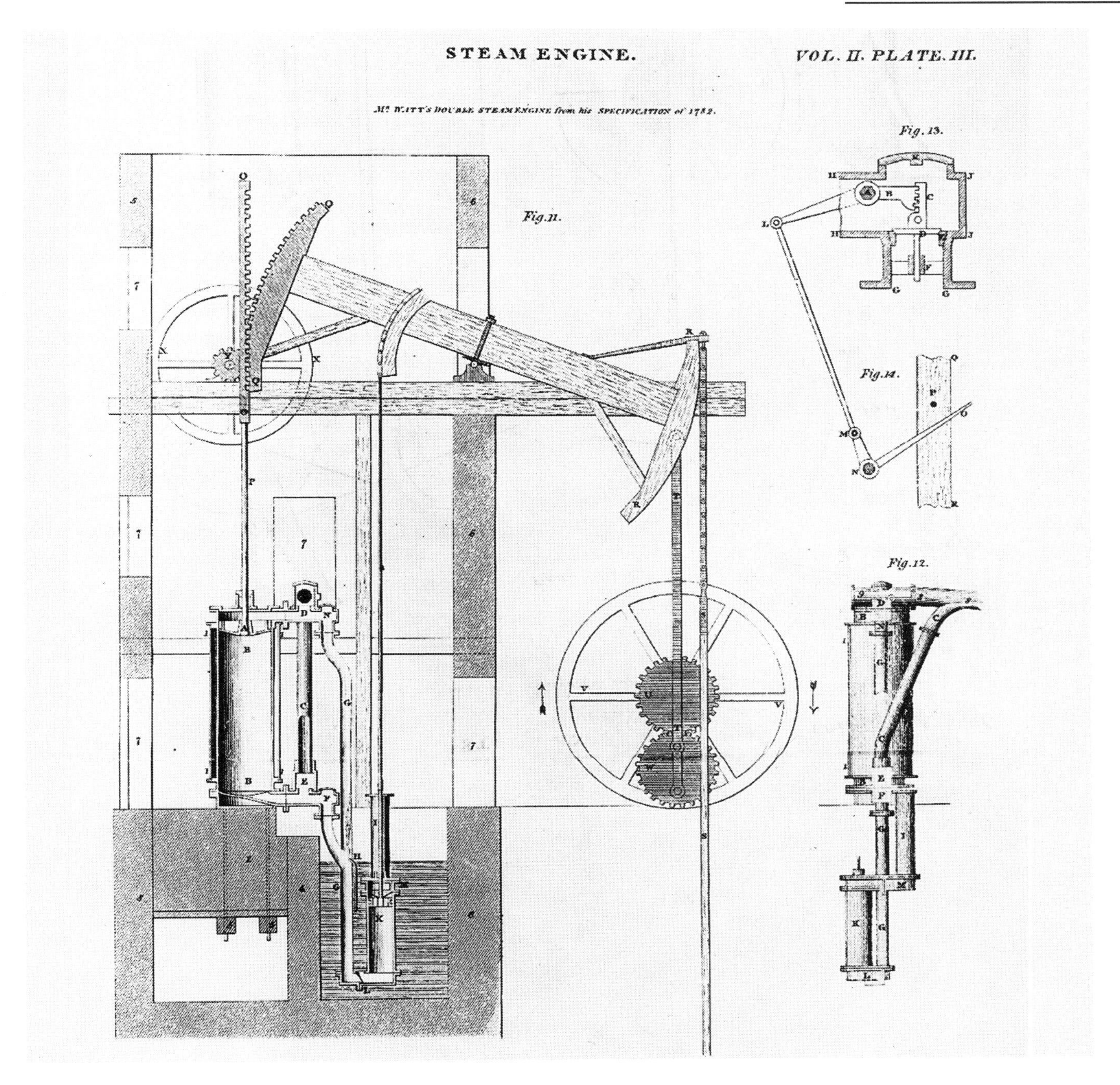

Plan of a low-pressure steam engine designed by James Watt. (Library of Congress)

the century was substantially controlled by this partnership. The partners initially obtained annual royalties based on the horsepower of their engines, then set the royalty at one-third the cost of fuel saved by the use of their engines over the Newcomen engine.

The engines were initially used in pumping and winding in the mines, then were extended to iron-rolling, spinning and weaving mills, and virtually any factory use of rotative power.

Retiring from their business just as the patents began to expire in 1800, the firm of Boulton & Watt was turned over to the junior Boulton and Watt as its industrial preeminence slipped away and royalties dwindled. Watt's friends and family aged and passed, and he settled into twilight development of three-dimensional replicating equipment for sculpture. His workshop from his home at Heathfield Hall, near Birmingham, is preserved at the Science Museum in South Kensington. The most famous engineer of his time, Watt died at home on August 19, 1819 at 82, and was buried alongside his partner Boulton in his country churchyard. A monument was erected in his memory in Westminster Abbey. Contemporary eulogies and biographies show Watt as a fine person with a brilliant mind, little business sense, and a tendency toward melancholy. It is clear that he valued Boulton as a friend as much as a business partner.

James Watt recognized the steam engine to be a heat engine, not a pressure engine, demanding the conservation and complete use of heat energy. His long acquaintance with scientists at the University of Glasgow, and his study of new information on heat and energy, gave him the intellectual tools to make these great steps, which he backed up with technical analysis. While many of the principles in Watt's 1769 patent had been previously described academically or constructed in other mechanical examples, Watt combined these principles and patented them just at the time when they could be technically applied to advantage. Coincidently, the quality of heavy machine work necessary for their success was at last becoming practical on a commercial basis. Previously, fine machine work was confined to technical, medical and other instruments.

To paraphrase the claims of Watt's 1769 patent: Keep the engine cylinder hot; condense the steam separately and keep the condenser cold; air in the steam system is defective to performance and must be removed by pumping as necessary; use the expansive energy in high-pressure steam without using a condenser; re-use the steam by compressing and reheating to avoid loss of the latent heat of evaporation. Additional patent claims were the concept of a rotary expansive engine and using materials other than cold water to seal and lubricate the engine to avoid loss of heat energy.

To more fully exploit the key patent claim for the separate condenser, the firm avoided the use of high-pressure steam since it did not absolutely require a condenser. The patent claim for expansive use of steam without a condenser was used only as ammunition to restrict development in that field, although high-pressure steam replaced low-pressure steam during Watt's lifetime, and noncondensing systems became the norm for railway locomotives and steam road vehicles.

Many of Watt's further innovations for the steam engine added to its utility, remaining in use long after patent protection expired, while some were merely stopgaps to avoid using the patents of other inventors. These include the double-acting engine in which the piston is powered in both directions with the rod passing through a seal, a parallel motion using links to guide the piston rod through a relatively straight path (Watt linkage), and the sun-and-planet gear method of driving a rotary shaft from a connecting rod, avoiding another's patent for the crank.

Watt also applied the flyball governor to the steam engine and invented the steam engine indicator, which gives a graphical representation of cylinder pressure versus displacement. In other fields, Watt invented the copying press for hand-written letters, which was in universal use for over a century, and a cloth drying machine using copper cylinders internally heated by steam.

Karl A. Petersen

See also: Trevithick, Richard.

BIBLIOGRAPHY

Dickinson, H. W. (1936). *James Watt, Craftsman & Engineer.* Cambridge: The University Press.

Robinson, E., and Musson, A. E. (1969). *James Watt and the Steam Revolution: A Documentary History.* London: Adams & Dart.

Thurston, R. H. (1878). *A History of the Growth of the Steam-Engine.* New York: D. Appleton and Company.

WAVES

Waves which are involved in many aspects of life, are disturbances that propagate through a medium with a definite speed. A wave of light conveys information to the eyes, a wave of sound brings music to the ears, a water wave rolling onto a beach can topple the swimmer, and electromagnetic waves cook food (microwave oven), and carry television reception. It takes energy to create the wave disturbance, and how the wave travels through the medium (elastic or damped) is quite variable.

There are two important categories of waves, mechanical and electromagnetic. Mechanical waves require a material medium. Water waves in a pond require water, sound waves in a room require air, a stadium "wave" must have people, and a wave produced by plucking a guitar string needs a string. Mechanical waves are produced when the medium is disturbed. A stone dropped in a pond disturbs the water, a spoken word results from a pressure disturbance, the guitar player disturbs the string by plucking it, and the sports fan disturbs the audience by standing up. The disturbance moves through the medium with a speed that is determined by the properties of the medium. In a stringed musical instrument, for example, the mass of the string and how tightly it is stretched determine the speed. That is why there are different size strings and a mechanism for tightening them.

A stone dropped in a pond pushes the water downward, which is countered by elastic forces in the water that tend to restore the water to its initial condition. The movement of the water is up and down, but the crest of the wave produced moves along the surface of the water. This type of wave is said to be transverse because the displacement of the water is perpendicular to the direction the wave moves. When the oscillations of the wave die out, there has been no net movement of water; the pond is just as it was before the stone was dropped. Yet the wave has energy associated with it. A person has only to get in the path of a water wave crashing onto a beach to know that energy is involved. The stadium wave is a transverse wave, as is a wave in a guitar string.

When air in a room is disturbed by a person speaking the molecules of the air have movements that are along the path of the wave. If you were to draw a line from the speaker's mouth to your ear, the movement of the molecules would be along this line. This type of wave, called an acoustical wave, is said to be longitudinal. The pleasant sounds of music are produced by acoustical waves. On the other hand, destruction by a bomb blast also is caused by acoustical waves. Instead of oscillating up and down, molecules in the acoustical (or compression) wave bunch together as the wave passes. It is not a transverse wave.

Imagine something floating in a pan of water with both the float and the water motionless (Figure 1A). If you were to continually bob the float up and down, you would see a continuous train of crests and troughs moving away from the float (Figure 1B). The separation of adjacent crests is called the wavelength, and given the lower case Greek lambda (λ) for a symbol. The time it takes one of the crests to travel a distance equal to the wavelength is called the period, and given the symbol T. The speed that the crest moves is just the distance traveled divided by the time so that $v = 1/T$. The period measured in seconds is just the time for one cycle of the oscillation. The reciprocal of the period is the number of cycles per second and is called the frequency, symbol f. The speed of the wave can also be written $v = f\lambda$. This simple relationship is a very important feature of all types of waves.

When waves encounter something in their path, they may bounce off (reflect) somewhat like a ball bouncing off a wall. A sound echo is produced by sound waves reflecting from something, a building, for example. We see our image in a mirror because of reflection of light. Reflection as well as other phenomena involving light led to the notion that light is a wave. It is decidedly different from a sound wave because there is no material medium required to sustain its motion. Light traveling to Earth from a distant star encounters virtually no matter until it reaches Earth's atmosphere. Modeling light as a wave has many virtues and uses.

Light is an example of electromagnetic waves. Electromagnetic waves propagate electric and magnetic fields. All electromagnetic waves travel with the same speed in a vacuum. This speed is given the symbol c and has a value 3.00×10^8 m/s (300,000,000 m/s). The general equation $v = f\lambda$ becomes $c = f\lambda$ for electromagnetic waves. Because c is the same for all electromagnetic waves in vacuum, $f\lambda$ is constant. Therefore, when the frequency of an electromagnetic wave increases, the wavelength must decrease. The range of wavelengths of electromagnetic waves is staggering: about 10^{-14} m (roughly the diameter of the nucleus of an atom) to about 1,000,000 m (approximately the distance from New York to Chicago). Because $f\lambda$ is constant, the frequency varies from a high of about 10^{22} Hz to a low of about 10^3 Hz.

Electromagnetic radiation is classified according to ranges of wavelength. Although the classifications are not precise, they are useful. For example, visible light to which our eyes are sensitive has wavelengths between about 4×10^{-7} m and 7×10^{-7} m. Microwaves have much longer wavelengths, about 0.001 m to 0.3 m.

Energy streams to Earth from the sun by electromagnetic waves. Photosynthesis depends on solar energy, and humans and animals rely on photosynthesis for food. In this sense, solar energy is not a luxury, it is essential.

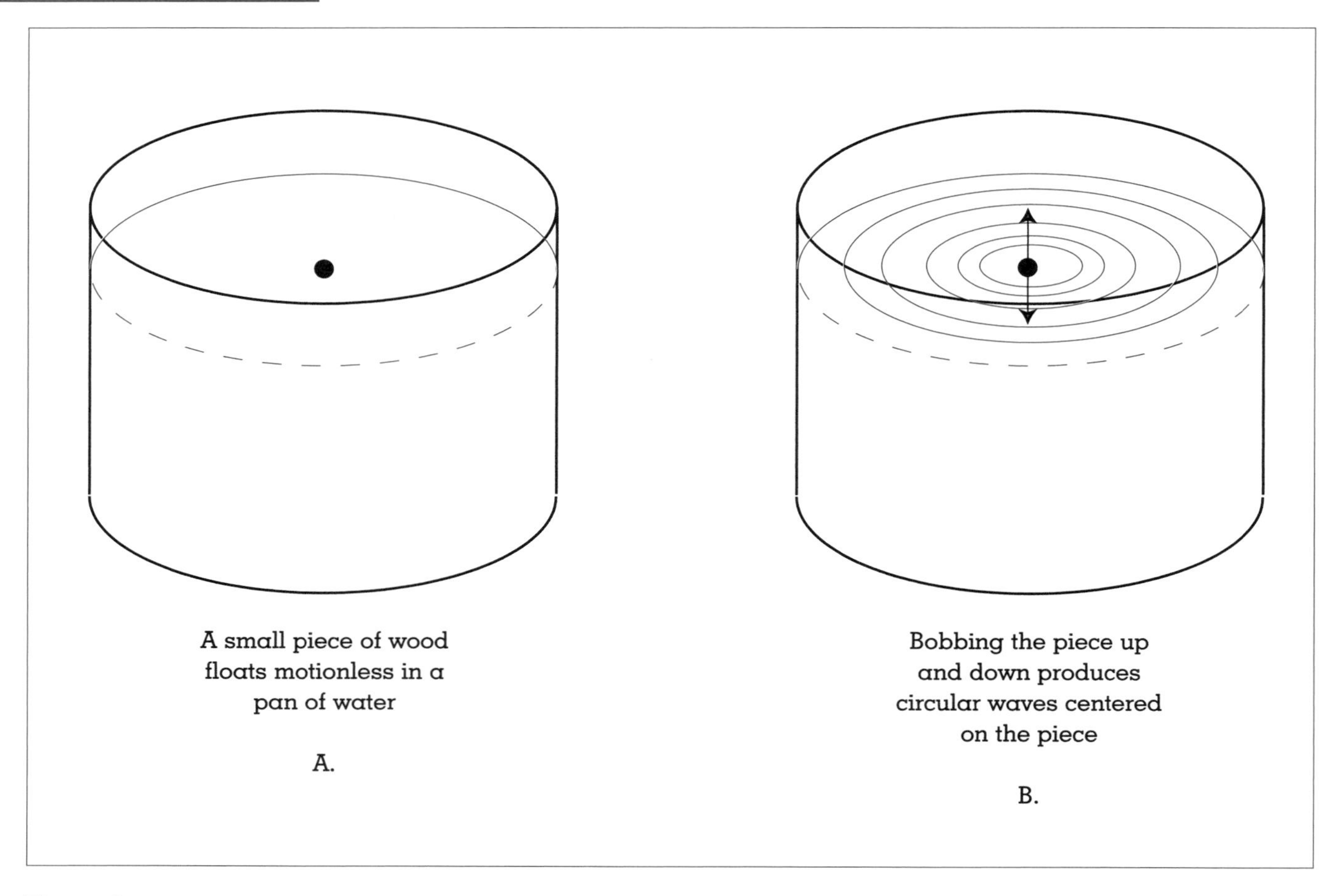

Figure 1.
Illustration of wave action inside a cylinder.

All objects, including the sun and incandescent light bulbs, emit electromagnetic radiation. The radiant energy and the type of radiation depend on the temperature. When an electric toaster is switched on, the heating element begins to glow and one can feel the radiant energy increase as the temperature increases. Visible radiation is produced by both the sun and a light bulb because the temperature of both is roughly 6,000 K. When the bulb cools to room temperature, about 300 K, the visible radiation disappears for all practical purposes. Nevertheless, the bulb at 300 K is still radiating, but the radiation is predominantly infrared that our eyes do not sense.

Radiation from the sun includes significant ultraviolet and infrared radiation in addition to visible radiation. Contributions of each type to the radiation that reaches Earth's surface are reduced significantly through absorption in Earth's atmosphere. The rate at which solar radiation falls on a square meter of Earth's solid surface is called the solar insolation. Solar insolation depends on the time of day, the day of the year, and where the square meter is located on Earth's surface. During an eight-hour period, the solar insolation in midwestern United States is roughly 600 W/m^2. If the 600 watts striking a 1 m^2 surface could be collected and converted to electricity, there would be enough electricity to operate six typical 100-watt light bulbs.

Solar energy can be used in many ways: heating buildings and houses, providing heat for producing steam for a turbogenerator in an electric power plant, producing electricity with photovoltaic cells, and so on. But since solar energy is not highly concentrated, it takes a considerable area of collection to even heat

Fiber optic cables, such as this one being spliced, allow for more information to be passed in smaller "packages." (Corbis Corporation)

a small home, which makes it uneconomical and impractical in many situations.

One of the most interesting and fascinating methods of electronic communication uses electromagnetic waves of light. The light is literally piped around in tiny bundles of transparent strings called optical fibers. Because the frequency of the light is thousands of times greater than the frequency of radio and television waves, a much higher concentration of information is permitted. The telephone industry makes great use of transmission of information via optical fibers. Using this technology allows information to be transmitted between North America and Europe via optical fiber cables strung beneath the surface of the Atlantic Ocean.

The back-and-forth motion of a child's swing is an example of an oscillating system, oscillator, for short. Set into motion, these systems oscillate at a frequency determined by properties of the oscillator. For example, the length of a child's swing and the acceleration due to gravity (g) determines the frequency of the back-and-forth motion of a child's swing. Periodically feeding energy into such a system at the same rate as the natural frequency gradually increases the energy of the oscillator. This is the phenomenon of resonance. Children achieve resonance with a swing by "pumping" at the resonant frequency of the swing. Molecules are microscopic oscillators that can be "pumped" by electromagnetic waves. If the frequencies match, the molecules absorb energy. Water molecules have a resonant frequency that falls in the microwave region of the electromagnetic spectrum. Water-based materials exposed to microwaves of the proper frequency absorb energy and heat up. This principle is exploited in a microwave cooker. When food on a plastic plate is placed in a microwave oven. the food warms because of its water content, and the plate does not warm because it is essentially devoid of water.

Ozone is a molecule formed from binding together three oxygen atoms. Gaseous ozone readily absorbs selected frequencies of ultraviolet radiation because ozone molecules have a resonant frequency in the ultraviolet region of the electromagnetic spectrum. A natural concentration of ozone exists twenty to thirty kilometers above Earth's surface. Ozone molecules absorb nearly all ultraviolet radiation from the sun having a wavelength less than about 0.3 micrometers. Accordingly, the natural ozone layer protects humans from harmful effects such as skin cancer. Human-made products migrating into the ozone layer can react with ozone molecules and convert them to other forms that do not absorb ultraviolet radiation. Chlorinated fluorocarbons, CFCs for short, once commonly used in refrigerators and aerosol spray cans, were found to be depleting the ozone, and in 1978 were banned from use in aerosol sprays. Banning the use of CFCs has not totally solved the ozone depletion problem.

Carbon dioxide is a molecule formed from binding together two oxygen atoms and one carbon atom. Gaseous carbon dioxide readily absorbs selected frequencies of infrared radiation because carbon dioxide molecules have a resonant frequency in the infrared region of the electromagnetic spectrum. Coal, natural gas, and oil all have carbon as a component, and carbon dioxide is produced when they are burned. It is a fact that the concentration of carbon dioxide in the atmosphere has been increasing since the Industrial Revolution. The carbon dioxide has little effect on solar radiation penetrating Earth's atmosphere. Earth absorbs the solar energy and, like any object, emits electromagnetic radiation. The temperature of Earth is more than 5,000 K lower than the sun, and Earth's radiation is mostly infrared. Carbon dioxide can absorb energy from the infrared radiation, resulting in a temperature increase of Earth. This is the so-called "greenhouse effect."

Coping with the greenhouse effect is a very difficult sociopolitical problem. A greenhouse effect existed on Earth long before the Industrial Revolution. Had it not, Earth's surface would be much colder than it is now. The introduction of gases absorbing infrared radiation only enhances the greenhouse effect. Carbon dioxide is not the only gas of importance; water vapor and methane, for example, are also of concern. The industrial world is dependent on coal, oil, and natural gas, so it is not easy to quit burning them to stop producing carbon dioxide.

The radiation one sees from a neon advertising sign or a laser is not produced by accelerated electric charges or by thermal effects. It comes from atoms or molecules that have absorbed energy and given this energy up as photons. A photon is a unit of electromagnetic energy having energy equal to Planck's constant (6.63×10^{-34} joule·seconds) times the frequency of the radiation ($E = hf$). The frequency of the radiation is determined by the magnitude of energy associated with atoms and molecules. Radiation from atoms and molecules is usually in the infrared, visible, and ultraviolet regions of the electromagnetic spectrum. Electromagnetic radiation also results from energy transitions in the nucleus of an atom. However, the nuclear energy scale is roughly a million times larger, making the frequency of the radiation correspondingly larger. This radiation is called gamma radiation.

All electromagnetic radiation has energy to some extent. Absorption of electromagnetic energy can be put to good use, as in photosynthesis or a microwave oven. On the other hand, electromagnetic energy can do damage. Skin cancer can result from absorption of ultraviolet radiation from the sun. This is why young and old are encouraged to stay out of the summer sun between the hours of 10 A.M. and 2 P.M. Gamma radiation is more energetic and more penetrating than ultraviolet, and the damage it can do is not confined to the surface of the skin. This is why some are concerned about the gamma radiation produced by radioactive nuclei in the spent fuel of a nuclear reactor. Ironically, gamma radiation is used to treat certain types of cancer.

Joseph Priest

See also: Atmosphere; Climatic Effects; Communications and Energy.

BIBLIOGRAPHY

Hobson, A. (1995). *Physics: Concepts and Connections.* Englewood Cliffs, NJ: Prentice-Hall.

Serway, R. A. (1998). *Principles of Physics,* 2nd ed. Fort Worth, TX: Saunders College Publishing.

WHEATSTONE, CHARLES (1802–1875)

Charles Wheatstone was born on February 6, 1802, at the Manor House, Barnwood, near Gloucester, England, and died in Paris, France, on October 19, 1875. His father, William, was in business as a shoe maker in Gloucester but he also taught music and several members of the family made musical instruments. On the death of an uncle, Charles and his older brother inherited the uncle's musical instrument manufacturing business in London. Although a competent businessman, he much preferred to give his attention to understanding the basic physics of music and musical instruments. He enjoyed the delicate work of designing and manufacturing musical instruments, and he devised new ones, in particular the highly successful concertina, patented in 1827.

Wheatstone's research into the physics of music addressed two practical questions. What is the difference between the sound vibrations of two notes having the same pitch but different timbre such as the same note played on a flute and a violin? How is sound transmitted through the bridge-piece from the string to the sound board of a violin or piano, and how far can sound be transmitted? These studies brought him into contact with Faraday, who gave several lectures at the Royal Institution in London about his discoveries. They also led to his appointment as Professor of Experimental Philosophy in the new King's College London in 1834.

Before 1831, the usual way of producing an electric current was by chemical means in the electric battery. Each cell of a battery had two different metals, or one metal and one carbon, separated by an acidic liquid. All electrical research in the first third of the nineteenth century made use of such batteries, and many combinations of materials were expired.

In 1831 Michael Faraday showed that an electromotive force is produced when a wire is moved through the lines of force of a magnet. If the wire is part of a complete circuit, a current flows. In the following years, several inventors made magneto-electric generators ("magnetos") in which coils of wire were rotated close to the poles of a fixed magnet, or a magnet was rotated close to a coil of wire. Most such machines were turned by hand and generated electricity without the consumption of expensive chemicals. They offered the prospect of larger machines driven by steam power that could generate electricity in large quantities.

A fundamental limit to the current that could be generated was imposed by the strength of the available permanent magnets. Some inventors sought to overcome this by using pairs of generators, the first being a magneto as just described, and the second being a similar machine except that electromagnets were used in place of the permanent magnets of the magneto. Current produced by the first machine was passed through the coils of the second, which then generated a larger current than could be produced by a magneto alone. Wheatstone was not primarily interested in producing large currents, but he used magnetos in some of his telegraphs.

In February 1837 Wheatstone was introduced to William Fothergill Cooke, an enthusiastic but unscientific inventor and entrepreneur who saw the commercial possibilities of a telegraph, but whose own instruments would not operate through long lengths of wire. The two went into partnership and established the first few working telegraphs in England. Their very first connected Paddington and West Drayton stations on the new Great Western Railway, using Wheatstone's five-needle system. The system also required five wires, which made it costly. Cooke preferred a telegraph with a single wire and a single needle, using a code such as Morse. Sadly, the partners quarrelled and parted. Cooke, with his Electric Telegraph Company, concentrated on the main commercial telegraph business. Wheatstone concentrated on instrument design and on user-friendly telegraphs for private use where letters were indicated directly on a dial, rather than being sent in code.

Wheatstone thought it might be possible to develop a communication system in which the vibrations of speech were transmitted many miles through solid rods or stretched wires, but he could not find a suitable material. Having become interested in the idea of communication at a distance, he took up the idea of the electric telegraph, and soon had an arrangement working in the College with which he could send signals through several miles of wire and deflect the needle of a galvanometer at the far end. He designed an arrangement in which any two of five galvanometer needles in a line could be deflected to point out a letter on a board.

The direct-reading instruments required a series of pulses to be transmitted along the line, each pulse causing the needle to advance one position. Some of Wheatstone's telegraphs used batteries and a switching system to produce the drive pulses, but magnetos naturally produced a series of pulses, and Wheatstone wanted a magneto-electric machine as his telegraph transmitter. In the course of this work he studied the action of the magneto and made some of the earliest measurements of the waveform of the current it produced. A machine that was not dependent on permanent magnets was undoubtedly an attraction for his purposes. He described his self-excited generator in a paper to the Royal Society on February 14, 1867. At that time there was widespread interest in the possibility of more powerful generators. Wheatstone's demonstration machine was driven by two large handles, possibly turned by two of his students. William Siemens described a similar machine at the same meeting. Siemens' title was "On the Conversion of Dynamical into Electrical Force without the aid of Permanent Magnets." Wheatstone approached the matter differently. His title was "On the Augmentation of the Power of a Magnet by the reaction thereon of Currents induced by the Magnet itself."

Wheatstone's self-excited generator was an important step in the development of the electric generator. Although it proved to be the last of Wheatstone's inventions in that field, the remainder of his life was devoted to further improvements in the telegraph.

Brian Bowers

BIBLIOGRAPHY

Bowers, B. (1975). *Sir Charles Wheatstone FRS.* London: HMSO for the Science Museum.

Wheatstone, C. (1867). "On the Augmentation of the Power of a Magnet by the reaction thereon of Currents induced by the Magnet itself." *Proceedings of the Royal Society*, February.

WIND ENERGY

See: Atmosphere; Turbines, Wind

WINDOWS

HISTORICAL BACKGROUND

In primitive homes, the only essential opening was a door for entry and exit. The first window was probably a hole to vent smoke from cooking and heating fires. That first opening offered the same advantages and disadvantages of windows that please and plague building occupants today: it allowed indoor pollutants to escape and daylight to enter; at the same time, it allowed heat to escape in cold weather and to enter in warm weather, creating discomfort for occupants.

Shutters were eventually added. When open, they permitted light and air to enter, along with undesirable visitors, such as intruders, rain, insects, and dust. Closed shutters provided darkness as well as protection. The next addition was translucent materials such as oiled paper or animal skin, framed into window openings.

Transparent glass was first used in Roman times, allowing daylight to enter unaccompanied by unwelcome elements, and giving building occupants a view of the outdoors. The largest surviving piece of Roman glass, three feet by four feet, came from a bath in Pompeii.

Venice became the world's glassmaking center in medieval times. Small, flat panes were produced by cutting and rolling flat hot, blown glass. This technique was used for the buildings constructed by early European settlers in America.

The next advancement in glassmaking was called sheet glass. In this method, a long sheet of glass was developed by dipping a long rod into molten glass and drawing it upward. The glass from this vertical process was then fire-polished.

During the 1600s, a plate glass casting method developed in France produced larger, higher-quality pieces of glass than previously possible. It was the first horizontal process, where glass was flattened through rollers, then ground to polish. The first uses of glass on a grand scale resulted, including the Hall of Mirrors at Versailles and the architectural feature still known as the French door.

Nineteenth-century innovations made larger, stronger, higher-quality pieces of glass more widely available. During the 1950s the float glass technique

Marble walls and crystal lamps flank windows that let light into the Hall of Mirrors in the Chateau de Versailles. (Corbis Corporation)

was developed, in which glass is produced floating on a tank of molten tin. This method produces extremely flat, uniform surfaces with few visual distortions. The desired thickness of glass can be achieved by varying the flow of molten glass from the tin bath. This process is still commonly used for residential windows in the United States.

Prior to 1965, single-pane (single-glazing) windows were standard in U.S. buildings. In cold weather, second panes of glass—storm windows—could be installed for added insulation.

WINDOWS FROM 1970 TO THE PRESENT

The energy crisis of 1973 drew attention to windows' impact on the heating and air conditioning of buildings. Windows were responsible for approximately 5 percent of total energy loss in the United States at the time, or the equivalent of all the energy provided by the Alaska pipeline. Research priorities, which had previously focused on sizing heating, ventilation, and air conditioning (HVAC) systems to meet the heating and cooling energy loads created by windows, turned to improving window energy efficiency. Double-glazed windows—two panes of glass sealed in a frame with air space in between—had been developed prior to the 1960s, but the seals on which their insulating value depended were not always reliable, and they were not widely used. As the quality of sealants improved, the market penetration of those double-pane products increased.

Window Energy Basics

To understand the energy efficiency of windows, it is necessary to define some solar, thermal, and optical properties. All heat transfer through a window is performed through conduction, convection, or radiation. Conduction is transfer of heat through solids by molecular interaction; an example of thermal heat transfer by conduction is heat being lost from a warm room, through the frame of a window. Convection is transfer of heat by fluid—in this case, air—that moves as a result of temperature differences; convection currents involving room air that comes in contact with cold glass are commonly called drafts. Drafts also can be caused by infiltration, described below. Radiation is the physical phenomenon of emitting heat; the Sun and all other objects radiate heat. Radiation can occur in winter when a fire radiates its heat to the objects and surfaces of a room to warm it. Radiation is also at work in the summer when the Sun's energy penetrates a window and heats up the room inside.

To measure the efficiency of a whole window, special testing takes into account all heat transfer from conduction, convection, and radiation. Certain values are used to represent the thermal and solar efficiency of high-performance windows by measuring reduced thermal heat loss (measured by the U-factor), solar heat loss and gain (measured by solar heat gain coefficient, or SHGC), and air infiltration (measured by air leakage).

U-factor. This is the total heat-transfer coefficient of a window system (window plus frame). It represents the amount of heat that moves through the window glass and frame. U-factors are expressed in watts per square meter per degree of temperature difference (Centigrade) or BTUs per hour per square foot per degree of temperature difference (Fahrenheit). In climates with heating requirements, low U-factors are recommended. U-factor is generally expressed as a fraction between zero and one, but in cases of very inefficient products the number can be greater than one.

Solar heat gain coefficient (SHGC). This is a measure of how much of the Sun's energy that is transmitted through the window becomes heat. It includes both the heat that is transmitted through the window and the heat that is absorbed by the window and then reradiated into a room. SHGC is also expressed as a fraction between zero and one. The SHGC can be determined for any angle at which sunlight hits the glass, but the most commonly used reference is for "normal incidence" —sunlight striking the glass at a ninety-degree angle. In climates with cooling requirements, low SHGC numbers are recommended. Because low SHGC numbers also improve the thermal performance of a window (i.e., U-factor), they are also recommended in heating climates, depending on the architectural design of the building.

Visible transmittance (VT). This is a window system's transmittance across the visible portion of the solar spectrum; it is a measure of the visible light that comes through window glass without being reflected or absorbed. VT is expressed as a fraction between zero and one. It is desirable to preserve visible transmittance—building occupants' access to daylight and views of the outdoors—while reducing U-factors, SHGC, and transmission of UV radiation. In all climates, a high VT is recommended.

Infiltration rate. This is the rate at which air moves through a window system as a result of pressure differences between the inside and the outside. Infiltration rates are expressed as airflow rates—cubic meters per hour or cubic feet per minute per square foot of window. An infiltration rate below 0.3 cfm/sq ft is recommended in all applications.

Technological Advances in Window Energy Efficiency

Reducing unwanted heat loss. Technological advances in window performance have come in the components of windows (glass, frames, spacers, sealants, warm edge design) and in the way those components fit together to make the whole window more efficient. A major goal in window energy efficiency starting in the late 1970s has been to design products with lower U-factors to achieve the benefits described above. Efforts to produce highly insulating window products in the 1970s and early 1980s focused on existing technologies: double-glazed windows led to triple- and quadruple-glazed windows, which lowered U-factors by increasing the number of still air spaces in the window system. Each layer reduced U-factors by approximately 1 hr-ft^2-F/Btu but also increased the window's weight and cost of fabrication as well as significantly lowering visible transmittance, which diminished the quality of the light and view.

In the mid-1970s it became clear that most of the heat transfer through a typical double-glazed window was through thermal radiation exchange from one layer of the window to another. Although glass blocks direct transmittance of long-wave infrared radiation in sunlight, it absorbs the radiation and reemits it at a high rate. To solve this problem, the first transparent, optically clear, low-emissivity (low-e) coating was developed in 1983 to reduce radiative heat transfer without compromising visible transmittance.

Radiant heat transfer had historically been the biggest heat transfer mechanism for windows. Low-e materials were developed and have historically been used to control for heat transfer. An example of the popularity of using metals to reflect heat to control for radiant heat transfer is the thermos bottle. Applying that new technology to windows, and getting materials that normally would affect transparency of the product to remain visually neutral, was a huge advance to the industry.

Low-e coatings are microscopically thin, virtually invisible metal or metallic oxide layers deposited on a window or glazing surface. Low-e coatings are virtually transparent to visible light. At the time of its invention, technology had not been developed to apply this product, a thin polyester film with a silver-based coating, directly to large pieces of glass, so it was applied to a substrate that was then suspended

An estate of houses installed with solar panels and other energy-saving devices at Shenley Lodge in Milton Keynes, England. (Corbis Corporation)

between two glazing layers. This technique had the advantage of creating another insulating air space in the double-glazed window unit, cutting window heat loss rates by up to 50 percent, but it also increased manufacturing costs.

Two processes, referred to as sputtered and pyrolitic, were developed to produce large volumes of quality, low-e coated glass. Pyrolitic coatings are incorporated into float glass production and tend to be more durable. Sputtered systems use a stand-alone vacuum deposition process to produce coatings that are have lower emissivities but that are softer and need more protection than pyrolitic coatings.

Reducing heat transfer with gas fill. Conduction and convection cause heat transfer across the air spaces in multilayer windows. Although air is a relatively good insulator, other gases that have lower thermal conductivity can be sealed into the cavities between panes. These gases include argon, carbon dioxide, and krypton. Developments in sealants and quality control during the 1980s reduced gas leakage in some window units to less than 1 percent per year.

By the late 1980s, many manufacturers offered double-glazed units with low-e coatings and inexpensive gas fills such as argon. Even greater efficiency could be achieved with extra-low-conductivity gases such as krypton filling small gaps between panes. Three-layer units with two gas-filled gaps and low-e coatings to reduce radiative heat transfer were designed at Lawrence Berkeley National Laboratory (LBNL), a leader in U.S. energy-efficient windows research. Products using these concepts appeared on the market in 1991.

Reducing heat transfer through warm-edge design. As new efficient glass technologies emerged, the relative fraction of heat lost through window frames decreased. During the 1990s, existing metal-framed windows were increasingly replaced by lower-conductivity vinyl. Pultruded fiberglass also became an energy-efficient framing option. The high-conductivity aluminum spacers typically used to hold panes in place in multilayer windows were replaced with more energy-efficient materials: metal-reinforced butyl, stainless steel, thermally broken aluminum or steel, and silicone foam.

Reducing infiltration. Traditional "brute force" approaches to weather stripping were redesigned to reduce infiltration with less material that contacted a larger surface area. New materials such as thermoplastic elastomers offered additional potential to control for infiltration. But perhaps the most important way to control infiltration is proper installation.

Controlling unwanted solar heat gain. The first generation of low-e coatings, described above, was designed for colder climates, where controlling heat loss is important. The next generation, introduced during the 1980s, selectively blocked the UV and shortwave infrared or "heat" radiation parts of the solar spectrum while admitting as much daylight as possible. These new spectrally selective coatings were useful in warmer areas, where reducing solar heat gains in summer and interior heat loss in winter is important. Where building energy codes are restrictive about the percentage of glazing allowed or the total energy performance of the unit, use of low-solar gain coatings can permit architects to increase total window area compared to what would be allowed in a building design with less efficient glazing.

TODAY'S WINDOWS

Today, nearly 90 percent of all residential windows sold in the United States have two or more layers of glazing. Approximately one-third of all windows sold for residential buildings and 20 percent of windows sold for commercial buildings in 1996 had low-emissivity coatings and insulating frames appropriate for the local climate. Besides the energy savings, energy-efficient windows increase thermal comfort, increase daylight and view, decrease condensation, decrease utility bills, and offer protection from UV fading.

Consumers buy homes with windows for the aesthetics, daylighting, and improved view of windows, yet they don't want to sacrifice comfort, that condition of mind that expresses satisfaction with the environment. Energy efficient windows transfer less energy, so less indoor heat is lost in cold weather and less unwanted heat from the outside enters in warm weather. Heating and cooling energy costs are therefore reduced, and occupants are more comfortable because the window's surface temperature is closer to the ambient indoor temperature. Windows without energy-efficiency features have very warm surfaces in warm weather and cold surfaces in cold weather, so occupants near a window feel uncomfortable.

During the heating season, high performance windows create warmer interior glass surfaces, reducing frost and condensation. The development of low-conductance spacers to reduce heat transfer near the edge of insulated glazing, and improved seals have significantly reduced condensation under all conditions.

Fading is probably one of the least understood consequences of visible and ultraviolet (UV) light transmitted through windows. Fabrics fade at different rates; UV radiation can fade organic materials such as carpet, artwork, paints, and wood differently than it fades synthetic materials. Energy efficient windows that reduce transmission of the UV spectrum not blocked by ordinary glass can reduce the potential for fading damage. Although most fading damage occurs from UV, visible light also has the potential to cause some damage.

People prefer windows in buildings to add light, open up space, and create aesthetically pleasing environments in which to live and work. Windows are also used for natural ventilation in some climates. Energy efficient windows allow increased daylight to enter a building without compromising the comfort of a room by allowing in unwanted heat or cold, particularly infiltration that can create drafts. Several studies also suggest daylighting can increase productivity, health and welfare.

Homeowners can realize substantial reductions in energy bills by selecting energy efficient windows. It is possible to save up to 30 percent on the heating and cooling portion of one's utility bill by selecting high performance windows. Moreover, building owners can rely on smaller heating and cooling equipment to meet smaller loads, and power suppliers that can meet customer demand with less generating capacity.

Research has shown that if all new windows sold in warmer U.S. regions had low-solar gain coatings, cooling energy use could be reduced by 25 percent in 2010; if residential windows in all parts of the United States had these coatings, heating energy use would

Buildings in downtown Detroit with sunlight reflecting off the myriad windowpanes. (Field Mark Publications)

decline by 19 percent. The total projected energy savings would be $1.2 billion per year. However, the market share for windows purchased with low-e coatings in the late 1990s is less than 20 percent in the southern United States, compared to 50 percent or more in the coldest states.

Choosing Efficient Windows

Window recommendations for a particular installation depend on local climate variations and the intent of the occupant and designer. If reducing energy costs is the primary goal, local utility rates must be taken into account along with which space-conditioning loads (heating or cooling) dominate. If comfort is the primary goal, occupant preferences must be addressed along with local site characteristics such as nearby shade trees that may help keep the building cool.

To achieve maximum energy performance, an energy-efficient window's characteristics must compensate for local climate conditions. In residential buildings in cold climates, the majority of building conditioning energy goes to heating, so efficient windows must keep heat indoors. In some cases, the solar gains that enter through a window may exceed the energy lost through the window, capturing passive solar gains and retaining captured heat. By contrast, solar heat gain is a liability in a residential building in a warm climate where cooling accounts for the majority of space conditioning energy use. Passive solar design in cooling climates necessarily employs different strategies: maximizing shading, using light roof colors and, in some cases, allowing natural ventilation.

Commercial buildings, such as offices with large windows, present a different challenge from residences. In almost all climates, commercial buildings suffer from excess solar heat gains, which produce large cooling loads. These buildings are best served by windows that filter out unnecessary solar radiation (and complementary technologies that redirect daylight to spaces away from windows to offset electrical lighting needs, see below).

A number of efforts assist U.S. consumers and homeowners in choosing windows. The National Fenestration Rating Council (NFRC), a non-profit organization created in 1989 by members of the window industry, code officials, energy officials, utilities and architects to devise a technically sound, impartial way to compare windows, has established a rating, labeling, and certification system. All NFRC ratings are for the whole window, that is, window plus frame. In the past, ratings were based on glass performance only, which sometimes over-rated the performance of a real installed window production. Laboratory tests and computer simulations determine a window's basic thermal and optical properties, which appear in NFRC certification documents for most windows sold in the U.S. Air infiltration and ratings of annual heating and cooling energy performance may be added sometime in the early 2000s. Work is under way to rate condensation resistance and long-term energy performance.

The U.S. Department of Energy (DOE) and the Environmental Protection Agency (EPA), have established the Energy Star program for energy-efficient windows to certify which products meet minimum energy performance criteria. Because window efficiency depends on the local climate, Energy Star window certifications are based on three U.S. climate zones: northern (heating energy dominated), southern (cooling energy dominated), and central (roughly equal heating and cooling).

The Efficient Windows Collaborative (EWC), managed by the Alliance to Save Energy, is a coalition of window and window component manufacturers, research organizations, non-profits, and federal and state government agencies interested in expanding the market for high-efficiency window products. Established in 1997, the EWC educates builders, remodeling contractors, home energy raters, utilities, and consumers about efficient window products.

WINDOWS OF THE FUTURE

Future energy-efficient window technologies will likely emphasize further control of heat loss and solar gain and integration with other building components. New technologies also will likely find ways to harness the power of thermal and solar heat transfer rather than just controlling it. Harnessing the power of the Sun, wind, and elements to make windows power-generation sources rather than energy losers will occur in the future. Designers, planners, and window industry visionaries foresee windows as dynamic, integrated, controllable "appliances in the walls." Windows of the future may very well be automated home systems that serve many purposes. The technologies listed below are only a few in a list of hundreds that may improve efficiency, performance, and visibility.

Future Window Technologies for Controlling Heat Loss

The gas fills that reduce the U-factors of low-e, insulated windows require very low pressure—that is, a vacuum or evacuated window. Research is focusing on improving seals, which are critical in vacuum windows, and methods of trapping the gas fill, which would otherwise diffuse through the seals. Spacers that keep the glass layers from collapsing under the required low pressure are significant sources of heat transfer; reducing their thermal bridging effect is under study.

Another promising means to reduce conductive, convective, and radiative heat transfer through windows is to fill the space between panes with a transparent insulating material. Silica aerogel is under development for this purpose. Because the silica particles are much smaller than the wavelengths of visible light, aerogel is transparent to the human eye. Its thermal conductivity is lower than that of air, and it is opaque to most longwave infrared radiation, so radiative heat losses through an aerogel-insulated window are minimal. It is much easier to vacuum-seal a window with aerogel insulation than with gas insulation. Aerogel is also strong enough to balance external atmospheric pressure on the window, so no spacers are needed. Testing and evaluation to confirm the performance of silica aerogel is under way.

Future Window Technologies for Controlling Solar Heat Gain

Solar heat gain through windows can add to or detract from energy efficiency, depending on a building's type and use, the local climate, the season, and even the time of day. For some applications, SHGCs should be maximized; for others, SHGCs should be minimized.

For residential buildings in climates where heating energy use dominates, it is important to maximize solar gains through windows. Solar heat transmission rates can be decreased by minimizing impurities in the glass (e.g., using low-iron glass). Low-e coatings may also be designed to be "smart" about maximizing or minimizing gains as a result of inside environmental conditions.

Controlling solar heat gains is equally important because

1. Residences are increasingly being built with air conditioners; in northern climates, cooling costs may be similar to heating costs even though the annual demand is heating-dominated, because electricity used for cooling is generally much more expensive than gas for heating.
2. Almost all commercial buildings are cooling-dominated, and solar gains through windows are major contributors to cooling loads.
3. Cooling loads are most significant on hot summer afternoons when electric utilities typically have annual peak loads. Lower solar heat gains lead to reduced peaks, which reduce peak demand charges and power plant capacity requirements.

Solar heat gain needs to be controlled in three ways: by intensity, by variations in time, and by spatial distribution

Controlling solar heat gain intensity. As described above, low-solar gain low-e glazing blocks out parts of the solar spectrum. Approximately half of incident solar energy is in the solar infrared; this solar radiation will not light interior spaces, but it is responsible for most of the heat gain, so it should be the first to be rejected by a selective glazing.

Controlling solar heat gain variations in time. Optical coatings that control solar heat gain on a time-dependent basis may be advantageous in applications such as commercial buildings where the benefits of solar gain and daylight vary daily or hourly, and in residential buildings where the desirability of solar gains varies seasonally. Glazings with variable properties, typically called chromogenic (switchable) glazings or "smart" windows, can be more effective and efficient than traditional shading devices.

There are several different types of switchable glazings. All are activated by a physical phenomenon (light, heat, or electricity), and, once activated, they switch, either incrementally or completely, to a different state. The state in which they permit the highest solar transmittance (typically the inactivated state) is called the bleached state; the state with the lowest solar transmittance is called the colored state.

Three switching mechanisms have been identified: photochromic (light-sensitive), thermochromic (heat-sensitive), and electrochromic (electrically activated). Photochromic materials alter their optical properties when exposed to light, typically UV, and revert to their original state in the absence of light. Probably the most familiar and widespread consumer use of photochromics is in sunglasses. Photochromic glass cannot yet be produced using the float glass process, so it is prohibitively expensive for windows. Current research is on developing plastic photochromic layers that are durable enough for use in windows. Photochromic glazing may be appropriate to control daylight transmission but is insensitive to solar heat gain.

Thermochromic materials change optical properties when subjected to temperature change. Their primary commercial uses to date are in inks, paints, security devices, and temperature indicators. Research in window thermochromics is focusing on durability, specification of switching or activation temperatures and ranges, and achievement of sufficient optical clarity in the colored state. Thermochromics can respond well to thermal effects but may not allow desirable daylight transmittance and may not distinguish between high ambient temperatures and incident solar radiation.

Electrochromic glazings are complex, multilayer coatings whose optical properties vary continuously from a highly transmissive (bleached) state to a low-transmitting (colored) state. A small electrical potential difference, typically one to five volts, triggers the change in state. The longer the potential difference is applied, the more the electrochromic layer will switch until it reaches its maximum or minimum. Electrochromics can be tied into building energy management control systems that optimize trade-offs between lighting energy and cooling consumption. Electrochromics appear to have greater potential for energy savings and user comfort than photochromics or thermochromics.

Controlling solar heat gain distribution. Commercial buildings often have a surplus of daylight near exterior windows and a lack of daylight deep in their interiors. Skylights can help on the top floor; strategies are being explored to redistribute daylight from side windows to the center of the building on lower floors.

Various technologies are being studied to maximize daylight, direct daylight and reject heat gain, and avoid visual glare. Two types of technology are being studied: holographic and angular selective glazing. Light shelves and complex light pipes also can redirect daylight. Light shelves often are simple horizontal elements on a building's exterior. Light pipes are rectangular "pipes" extending into a building from the exterior. Variations on these kinds of architectural features are being studied.

Transparent or translucent insulating materials (TIMs) can provide light or solar gains without view. TIMs typically have thermal properties similar to conventional opaque insulation and are thicker than conventional insulating glass units, providing significant resistance to heat transfer.

TIMS often have the same thermal conductance as a wall, so they are appropriate to admit daylight in areas where insulated walls would otherwise be installed. In Europe, where there are many uninsulated masonry buildings, TIMs with solar control mechanisms could be excellent retrofits. TIMs' optical properties are different from those of conventional insulating glass units; solar radiation often is scattered as is passes through a TIM. As a result, measuring the optical properties of TIMs can be challenging. Laminated and safety glazings were originally designed to help structurally support the building. They also may help to control daylight and reduce solar heat gain.

Integration of Windows with Other Building Components

Windows often are thought of as independent components of a building. However, to maximize windows' energy efficiency, they must be integrated with the rest of a building system.

Projects around the country are beginning to demonstrate that the energy consumption of new houses can be reduced by as much as 50 percent with no behavioral changes and little or no impact on the cost of construction. This is possible because cost savings in one component can then be reinvested to improve energy performance and product quality. New techniques for improving the building envelope, such as upgrading to high-performance windows, often enable builders to install smaller, less expensive heating and cooling systems than would be needed with traditional windows. Equipment cost savings in heating, ventilation, and air conditioning (HVAC) offset the added envelope improvement costs.

At a demonstration project in Las Vegas, low-solar gain low-e windows were a key technology used to reduce measured cooling energy by 20 percent and heating energy by 50 percent in new homes. After improvements to the construction, including energy-efficient windows with an SHGC of 0.4 and a U-factor of 0.35, the HVAC was sized at three tons, with a resulting cost savings of $750 to the builder. Energy-efficient windows accounted for half a ton to one ton of those savings.

Specific integrated residential applications. An Integrated Window System (IWS) is a window-wall panel with a number of energy-efficient features:

- movable interior or exterior insulating night shades recessed into the wall cavity;
- seasonally deployed solar shading, that can be interior, exterior, or in-between;
- optional overhangs;
- insulated box beam headers;
- larger glazed area;
- elimination of structural window form elements.

IWSs would most likely be factory-built and integrated into conventional construction, replacing likely framing components. IWS prototypes are being designed.

Commercial applications. Exterior building design, glazing, lighting systems, and heating, ventilation, and air-conditioning equipment are typically specified and installed separately in commercial buildings. However, to realize the energy-saving potential of advanced products such as electrochromics or light shelves, it will require concurrent advances in sensor, software, and hardware of energy management control systems.

Other Integrated Energy Issues

In addition to the examples cited above, many manufacturers and researchers in the industry are trying to identify methods of reducing embodied energy costs. Glassmaking is an energy-intensive industry; vinyl, aluminum, and fiberglass extrusion also is energy-intensive. Making wood windows out of an ever-shrinking supply of forestry is becoming difficult. Researchers hope to identify whether there is some advantage to recycling whole window products at the end of their useful life and whether that process is more energy- and resource-efficient than beginning again with virgin materials.

Alicia Ward

See also: Air Conditioning; Building Design, Commercial; Building Design, Energy Codes and Building Design, Residential; Climatic Effects; Cool Communities; Efficiency of Energy Use; Energy Management Control Systems; Insulation.

BIBLIOGRAPHY

Arasteh, D. (1996). "Advances in Window Technology: 1973–1993." In *Advances in Solar Energy*, ed. K. W. Boöer. Boulder, CO: American Solar Energy Society.

Arasteh D. and Selkowitz, S. (1989). "A Superwindow Field Demonstration Program in Northwest Montana." *ASHRAE Transactions* 95(1).

Arasteh, D.; Selkowitz, S.; and Hartmann, J. (1985). "Detailed Thermal Performance Data on Conventional and Highly Insulating Window Systems." *ASHRAE Transactions* 91(1).

ASHRAE. (1997). *ASHRAE Handbook: Fundamentals*. Atlanta: Author.

ASHRAE. (1981). *Thermal Environmental Conditions for Human Occupancy*. Atlanta: Author.

ASTM. (1988). *Standard Test Method for Solar Absorptance, Reflectance, and Transmittance of Materials Using Integrating Spheres*. Philadelphia: Author.

ASTM. (1992). *Standard Test Method for Determining the Rate of Air Leakage Through Exterior Windows, Curtain Walls, and Doors Under Specified Pressure Differences Across the Specimen*. Philadelphia: Author.

Beck, F. A., and Arasteh, D. (1992). "Improving the Thermal Performance of Vinyl-Framed Windows." Proceedings of the ASHRAE/DOE/BTECC Conference on the Thermal Performance of the Exterior Envelopes of Buildings V, Clearwater Beach, FL.

Carmody, J.; Selkowitz, S.; and Heschong, L. (1996). *Residential Windows*. New York: W. W. Norton.

Carpenter, S., and McGowan, A. (1993). "Effect of Framing Systems on the Thermal Performance of Windows." *ASHRAE Transactions* 99(1).

Chahroudi, D. (1991). *Weather Panel Architecture*. Albuquerque NM: Suntek.

Collins, R. E.; Fischer-Cripps, A. C.; and Tang, J. Z. (1992). "Transparent Evacuated Insulation." *Solar Energy* 49(5): 333–350.

Ducker Research Company. (1997). *Study to Quantify and Profile the U.S. Market for Residential and Light Commercial Windows and the Technology for High-performance Windows*. Berkeley, CA: Lawrence Berkeley National Laboratory.

Efficient Windows Collaborative. <http\\:www.efficientwindows.org>.

Elsherbiny, S. M.; Raithby, G. D.; and Hollands, K. B. T. (1982). "Heat Transfer by Natural Convection Across Vertical and Inclined Air Layers." *ASME Transactions* 204:96.

Finlayson, E. U.; Arasteh, D.; Huizenga, C.; Rubin, M. D.; and Reilly, M. S. (1993). WINDOWS 4.0: *Documentation of Calculation Procedures*. Berkeley, CA: Lawrence Berkeley National Laboratory.

Frost, K.; Arasteh, D.; and Eto, J. (1993). *Savings from Energy-Efficient Windows: Current and Future Savings from New Fenestration Technologies in the Residential Market*. Berkeley, CA: Lawrence Berkeley National Laboratory.

Frost, K.; Eto, J.; Arasteh, D.; and Yasdanian, M. (1996). *The National Energy Requirements of Residential Windows in the U.S.: Today and Tomorrow*. Berkeley, CA: Lawrence Berkeley National Laboratory.

Geller, H., and Thorne, J. (1999). *U.S. Department of Energy's Office of Building Technologies: Successful Initiative of the 1990s*. Washington, DC: American Council for an Energy-Efficient Economy.

Heschong, L. (1999). *Daylighting in Schools: An Investigation into the Relationship between Daylighting and Human Performance*. Fair Oak, CA: Heschong Mahone Group.

Hunt, A. J. (1992). "Optical and Thermal Properties of Silica Aerogel." In *Chemical Processing of Advanced Materials*, ed. L. I. Hench and J. K. West. New York: John Wiley & Sons.

Johnson, R.; Connell, D.; Selkowitz, S.; and Arasteh, D. (1985). "Advanced Optical Materials for Daylighting in Office Buildings." Tenth National Passive Solar Conference Proceedings, American Solar Energy Society, Inc.

Kim, J. J.; Papamichael, K.; Selkowitz, S.; Spitzglas, M.; and Modest, M. (1986). "Determining Daylight Illuminance in Rooms Having Complex Fenestration Systems." *International Daylighting Conference Proceedings* 2:219.

Klems, J. (1983). "Methods of Estimating Air Infiltration through Window." *Energy and Buildings* 5:243–252.

Lampert, C. M., and Ma, Y. (1992). *Fenestration 2000: Advanced Glazings Materials Study*. St. Helens, UK: Pilkington Glass.

Lee, E. S., and Selkowitz, S. E. (1993). "Integrated Design Yields Smart Envelope and Lighting Systems." *Consulting-Specifying Engineer* 64.

Papamichael, K.; Beltraán, L.; Furler, R.; Lee, E. S.; Selkowitz, S.; and Rubin, M. (1993). *The Energy Performance of Prototype Holographic Glazings*. Berkeley, CA: Lawrence Berkeley National Laboratory.

Papamichael, K.; Klems, J.; and Selkowitz, S. (1988). "Determination and Application of Bidirectional Solar-Optical Properties of Fenestration Systems." Proceedings of the 13th National Passive Solar Conference, Massachusetts Institute of Technology.

Reilly, M. S.; Arasteh, D.; and Rubin, M. (1990). "The Effects of Infrared Absorbing Gasses on Window Heat

Transfer: A Comparison of Theory and Experiment." *Solar Energy Materials* 20:277–288.

Reilly, S.; Arasteh, D.; and Selkowitz, S. (1991). "Thermal and Optical Analysis of Switchable Window Glazings." *Solar Energy Materials* 22:1–14.

Selkowitz, S. E. (1981). "Thermal Performance of Insulating Window Systems." *ASHRAE Transactions* 85(2).

Selkowitz, S. E., and Lampert, C. M. (1989). "Application of Large-Area Chromogenics to Architectural Glazings." In *Large Area Chromogenics: Materials and Devices for Transmittance Control,* ed. C. M. Lampert and C. G. Granqvist. Bellingham, WA: Engineering Press.

Shakerin, S. (1987). "Wind-Related Heat Transfer Coefficient for Flat-Plate Solar Collectors." *Journal of Solar Energy Engineering* 109:108–110.

Sullivan, R.; Chin, B.; Arasteh, D.; and Selkowitz, S. (1991). "RESFEN: A Residential Fenestration Performance Design Tool." *ASHRAE Transactions* 98(1).

Sullivan, R., and Selkowitz, S. (1987). "Residential Heating and Cooling Energy Cost Implications Associated with Window Type." *ASHRAE Transactions* 93(1):1525–1539.

Warner, J.; Reilly, S.; Selkowitz, S.; and Arasteh, D. (1992). "Utility and Economic Benefits of Electrochromic Smart Windows." Proceedings of the ACEEE 1992 Summer Study on Energy Efficiency, Asilomar Conference Center, Pacific Grove, CA.

Weidt, J. L.; Weidt, J.; and Selkowitz, S. (1981). "Field Air Leakage of Newly Installed Residential Windows." *ASHRAE Transactions* 85(1):149–159.

Windows and Daylighting Group. (1992). *WINDOWS 4.0, A PC Program for Analyzing Window Thermal Performance.* Berkeley, CA: Lawrence Berkeley Laboratory.

Wright, J. (1993). *VISION 3.* Waterloo, Ontario: University of Waterloo.

Yazdanian, M., and Klems, J. (1994). "Measurement of the Exterior Convective Film Coefficient for Windows in Low-Rise Buildings." *ASHRAE Transactions* 100.

WOOD

See: Biofuels

WORK AND ENERGY

Work, in general, is an activity in which one exerts strength or faculties to do or perform something. In physics, work is defined as a force on an object multiplied by the displacement of the object. Measuring force in newtons (N) and displacement in meters (m) yields newtons times meters (Nm) as the units of work. To honor James Prescott Joule, who did pioneering work in the science of thermodynamics, a newton-meter is called a joule (J). A person pushing on a box with a force of 200 N (about 50 pounds) and moving the box 1 m would do 200 J of work. Energy is capacity, or ability, for doing work. Work and energy are both measured in joules.

After being outside on a cold day, a person's hand warms when shaking hands with someone who has spent time in a warm room. Heat flows from the warm hand to the cool hand to produce the warming. Rather than shaking hands, the person could briskly rub his hands together. In this case, warming is a result of doing work, not due to a flow of heat. Heat is energy in transit and, accordingly, is measured in joules. It takes 4,200 joules of heat to raise the temperature of 1 kilogram of water 1 degree Celsius. The same temperature rise would occur if 4,200 joules of mechanical work were done on the water using a food mixer, for example.

The concepts of heat and work are very important for understanding the roles of automobiles, airplanes, trucks, trains, and electric power plants that are essential for a modern industrial society. Examining the technology, one always sees the involvement of an engine of some sort. Most automobiles employ a gasoline engine, trucks and trains use diesel engines, and an electric power plant uses a steam turbine to drive an electric generator. There are significant differences in the size and complexity of the engines involved, but they all convert heat to useful work.

The word kinetic implies motion. An object in motion has a capacity for doing work and has kinetic energy. Forces acting on an object do work when the object moves and the kinetic energy changes. When pulling on a sled to start it moving, force and displacement are in the same direction. Work done by this force is taken to be positive. While the sled is moving, frictional forces on the runners pull backwards. The force and displacement are in opposite directions and the work is taken to be negative. The net amount of work is the algebraic sum of the work due to all forces acting on an object. For example, if pulling on the sled produced 100 J of work and friction produced –50 J the net amount of work is 100 J – 50 J = +50 J.

A very important principle called the work-energy principle reads as follows: net amount of work is equal to the change in kinetic energy. Emphasis is placed on net and change. If the net work is positive, this means the net force and displacement have the same direction, and the kinetic energy increases. Increasing the kinetic energy of a car requires positive work. Conversely, decreasing the kinetic energy requires negative work. When a braking system brings a car to rest, forces must act in a direction opposite to the direction the car is moving. If a car travels down a highway at constant speed, its kinetic energy does not change. This means the net work on the car is zero. Air always pushes against the car and does negative work. To keep the speed from changing, forces that propel the car forward must do positive work of equal magnitude. The faster the car travels, the greater the negative work done by the air. As a result, forces driving the car forward must do more work. This means more consumption of gasoline. This is the basic reason why the fuel economy of a car is sensitive to speed.

Joseph Priest

See also: Joule, James Prescott.

BIBLIOGRAPHY

Hobson, A. (1995). *Physics: Concepts and Connections.* Englewood Cliffs, NJ: Prentice-Hall.

Serway, R. A. (1998). *Principles of Physics*, 2nd ed. Fort Worth, TX: Saunders College Publishing.

YUCCA MOUNTAIN

See: Nuclear Waste

ZERO-WATER DISCHARGE

See: Emission Controls, Power Plant

Z-PINCH

See: Nuclear Fission

ENERGY TIMELINE

ENERGY SOURCE, PROCESSING, AND STORAGE EVENTS

B.C.E.

2600 Construction of Lake Moeris in Egypt, a reservoir created by a dam 27 miles long.

2589 Construction of Great Pyramid of Khufu begins and lasts until 2566 B.C.E.; innovative use of available power for transportation and construction.

1500 Water clocks being used by Egyptians.

850 Natural gas utilized in China.

400 An oil well is completed on an island in the Ionian Sea and the oil used in lamps.

180 Revolving mill invented; it is powered by slaves and asses.

65 Earliest known reference to the use of a windmill is made by Antipater of Thessalonica.

C.E.

600 Arabs develop a windmill in which the paddlewheel revolves in the vertical plane.

1000 Burmese successfully drill oil wells.

1003 Wells are drilled in China for natural gas, which flows through bamboo pipes to be used perhaps in porcelain manufacture.

1200 Coal being mined in Europe.

1269 Frenchman Pèlerin de Maricourt writes a treatise on magnetism that includes the earliest description of the compass in the Western world.

1430 Turret windmill invented.

1600 England suffers from timber and fuel wood scarcity.

1619 Coke first used instead of charcoal in a blast furnace.

1635 John Winthrop, Jr., opens America's first chemical plant in Boston. They produce saltpeter (used in gunpowder) and alum (used in tanning).

1640 Oil well completed in Italy; kerosine from the oil later used for lighting

1670 First distillation of gas from coal.

1678 Dutch mathematician Christiaan Huygens first states his wave theory of light, published in *Traité de la lumière* in 1690.

1682 Law of gravitation announced by English mathematician and physicist Isaac Newton; five years later his *Philosophiae Naturalis Principia Mathematica* is published, setting forth the laws of motion as well as gravitation.

1694 Oil produced in England by retorting oil shale and cannel coal.

1793 British physicist Benjamin Thompson (Count Rumford) shows that work is convertible into heat and vice versa.

1799 Production of coal gas.

1800 Italian physicist Alessandro Volta demonstrates the galvanic cell, also known as the voltaic cell.

1802 Gas lighting.

1819 English chemist and physician John Kidd obtains naphthalene from coal tar, pointing the way toward the use of coal as a source of many important chemicals.

1830 American physicist Joseph Henry discovers the principle of electromagnetic induction. English chemist and physicist Michael Faraday independently discovered the same principle a year later but is the first to publish his findings.

1850s The first petroleum refinery consisting of a one-barrel still is built in Pittsburgh by Samuel Kier.

1850 Scottish chemist James Young starts to produce "coal oil" (kerosine) from coal.

1853 Kerosene is extracted from petroleum.

1854 The Pennsylvania Rock Oil Company becomes the first oil company in the United States.

1855 Chemist Benjamin Silliman, of New Haven, Connecticut, obtains valuable products by distilling petroleum. They include tar, gasoline, and various solvents.

1856 The first synthetic dye is developed by William H. Perkin (English); he accidentally creates a mauve dye from the impure aniline in coal tar.

1857 Oil is discovered in Romania and Ontario.

1859 The first commercially successful U.S. oil well is drilled by E. L. Drake near Titusville, Pennsylvania. This 70-foot well launches the petroleum industry.

1867 Swedish philanthropist Alfred Nobel patents dynamite.

1886 There are forty to fifty water-powered electric plants reported to be online or under construction.

1888 The first dry cell battery, consisting of a moistened cathode and a swollen starch or plaster of paris separator, is invented.

1895 The Edison dry cell nickel/cadmium battery and the Jungers nickel/iron cell are developed; work on these endeavors lasts until 1905.

1900 American chemist Charles Palmer makes a breakthrough in devising a thermal process to produce gasoline from crude petroleum.

1900 First off-shore wells, fixed to piers, are drilled in the Caspian Sea.

1901 Oil drilling begins in Persia.

1905 German-American physicist Albert Einstein formulates the Law of Mass-Energy Equivalence ($E=mc^2$) and the Photon Theory of Light.

1907 Atlantic Refining Company introduces the tower still refinery, in which petroleum is separated in a continuous process rather than in batches.

1913 The world's first geothermal power station begins operation in Italy.

1918 Shell introduces the first drill with diamond tooth edges.

1921 Federal Energy Regulatory Commission statistics show hydropower generation at 3,700 MW (cf. 1992 figure of 91,600 MW).

1918 Ethyl gasoline is developed by General Motors Laboratories in the United States.

1926 Du Pont and Commercial Solvents begin synthetic methanol production in the United States.

1931 Oil drilling becomes more accurate because of the gyroscopic clinograph that stabilizes the drill.

1932 After American physicist Ernest Orlando Lawrence invented the cyclotron three years earlier, the E. O. Lawrence Cyclotron becomes operational. It helps scientists discover what an atom is composed of, how it behaves, and how its energy can be tapped.

1933 Construction of Grand Coulee Dam begins. Originally built to meet irrigation needs, it has more electric generating capacity than any dam in North America by 1975.

1935 From 1920 to 1935, 6.5 million windmills have been erected in the United States to pump water, run sawmills, or generate electricity.

1936 The Hoover Dam is completed.

1936 The Houdry Process is used in the catalytic cracking of petroleum.

1937 Westinghouse constructs its "Atom Smasher" in Forest Hills, Pennsylvania. The five million volt van de Graaff generator represents the first large-scale program in nuclear physics established in industry, makes possible precise measurements of nuclear reactions, and provides valuable research experience for the company's pioneering work in nuclear power.

1939 Enrico Fermi (Italian-American), Otto Hahn (German), F. Strassman, Lise Meitner (Austrian), and Otto Frisch (Austrian) discover and describe nuclear fission.

1940 Standard Oil Co. (Indiana) develops catalytic reforming to produce higher octane gasoline.

1941 American nuclear chemist Glenn Seaborg's team of experimenters isolates plutonium, which proves to be a better fuel for nuclear reactors than uranium because of its greater energy yield.

1942 The Manhattan Project at the University of Chicago Laboratory, headed by Enrico Fermi (Italian-American), creates the first self-sustaining nuclear chain reaction.

1942 Natural gas liquified for first time in Cleveland, Ohio.

1944 Seismic profiling for oil deposits begins in the Gulf of Mexico.

1947 Offshore oil wells drilled off coast of Louisiana.

1951 An electricity-producing nuclear breeder reactor commissioned by the U.S. Atomic Energy Commission.

1952 The first hydrogen (fusion) bomb to be tested by the United States is exploded at Bikini Atoll.

1954 First submersible drilling unit, "Mr. Charlie," is used.

1954 First solar cell developed by Bell Telephone Laboratories researchers.

1959 Liquified natural gas is shipped via cryogenic tanker from Lake Charles, Louisiana, to London.

1960s First large-scale U.S. nuclear power plants go on line.

1960 Geysers near San Francisco begin supplying geothermal electric power.

1976 Clinch River Breeder Reactor Project, the first large-scale demonstration breeder reactor is constructed near Oak Ridge, Tennessee. The project died for lack of support, however.

1982 The Tokamak Fusion Test Reactor at the Princeton Plasma Physics Laboratory produces fusion.

1992 Federal Energy Regulatory Commission statistics show hydropower generation at 91,600 MW (cf. 1921 figure of 3,700 MW).

TRANSPORTATION AND AGRICULTURE

B.C.E.

8000 Settled agriculture occurs in the Near East and other centers of human habitation.

7500 Dugout canoes are used in northwestern Europe. Reed boats are developed in Mesopotamia and Egypt.

7000 Agriculture starts in Mexico.

3500 The earliest illustration of a sail dates from this period and was found near Luxor, Egypt. The sail is fixed to a single mast and there is a shelter aft.

3200 The first wheeled vehicles are believed to have appeared in Sumer (now Iraq).

2700 Spoked wheel appears; traction plow already developed.

2400 First canal for ships built at Elephante in Egypt.

2000 Horse-drawn vehicles are used.

312 Roman road building prowess is exemplified in the construction of the Appian Way.

C.E.

400 Paddlewheel propulsion invented in China.

500 Invention of the modern horse collar in China helps produce agriculture surpluses.

700 Lateen sailing vessels are established in the Mediterranean Sea, increasing directional sailing ability.

730 Stern post rudder invented for sailing vessels.

1701 English agriculturist Jethro Tull invents a seed drill that sows seed in neat rows, saving seed and making it easier to minimize weeds.

1705 English inventor Thomas Newcomen builds his atmospheric steam engine.

1769 French engineer Nicholas-Joseph Cugnot builds his steam road-carriage.

1786 Scottish inventor Andrew Meikle builds his first thresher.

1783 First manned flight via balloon.

1787 American John Fitch launches the first U.S. steamboat.

1804 English inventor Richard Trevithick's steam railway locomotive.

1805 Twin-screw propeller developed by American John Stephens.

1814 Railway locomotive by Englishman George Stephenson.

1816 The forerunner of the bicycle is patented by German inventor Karl D. Sauerbronn.

1819 Atlantic first crossed by a steam-powered vessel, the *Savannah.*

1825 Stockton and Darlington Railway completed, first public steam powered railway.

1827 Steam automobile invented by Hancock.

1829 First horsedrawn carriage "omnibus," which carries eighteen passengers, introduced by G. Shillibear in London.

1830 Liverpool and Manchester railroad.

1831 American inventor Cyrus McCormick develops the harvester.

1834 Electric streetcar invented by Thomas Davenport (American).

1844 American inventor Charles Goodyear patents "vulcanizing" of rubber.

1847 German-English electrical engineer William Siemens creates the regenerative steam engine.

1852 American inventor Elisha G. Otis develops the first "safety" elevator; it incorporates a brake that prevents elevators from falling even if the main cable is completely cut.

1852 The first nonrigid, powered, manned airship is flown by its builder, French engineer Henri Giffard; this marks the beginning of the practical airship.

1860 French inventor Jean-Joseph-Étienne Lenoir builds the first practical internal-combustion engine, fueled by illuminating gas.

1865 Samuel Calthrop (American) creates a streamlined locomotive.

1867 Gas engine built by German engineers Nikolaus August Otto and Eugen Langen.

1868 American engineer George Westinghouse introduces the air brake. The new power braking system uses compressed air as the operating medium.

1869 The Transcontinental Railroad is completed as the Golden Spike is driven in at Promontory Point, Utah.

1875 Internal combustion engine invented by Siegfried Marcus (Austrian).

1877 German aeronautical engineer Otto Lilienthal invents his first glider.

1877 American meatpacker Gustavus Franklin Swift invents the refrigerated railcar.

1883 Elevated electric railroad opens in Chicago.

1884 English engineer Charles Algernon Parsons invents his compound steam turbine.

1885 The gasoline automobile is developed by German engineer Karl Friedrich Benz. Before this, gasoline was an unwanted fraction of petroleum that caused many house fires because of its tendency to explode when placed in kerosene lamps.

1886 The first modern oil tanker, *Gluckauf,* is built for Germany by England.

1887 Scottish inventor John Boyd Dunlop creates an air-inflated rubber tire.

1888 Richmond Union Passenger Railway electric street railway system designed by American electrical engineer Frank Julian Sprague.

1889 Carl Gustaf de Laval (Swedish) improves the steam engine by devising a small, high-speed turbine in which jets of steam hit a single set of blades set on a rim of a wheel.

1889 Petroleum-fueled agricultural tractor developed in the United States.

1900 Worldwide railways total 470,000 miles of track.

1901 First merchant vessel, *King Edward,* driven by steam turbines in Scotland.

1902 Steam superheaters dramatically improve the performance of railway engines.

1903 American inventors Orville and Wilbur Wright fly the first powered aircraft at Kitty Hawk, North Carolina.

1903 The Ford Motor Company is founded.

1905 Long Island Railroad is the first to abandon steam completely in favor of electrification.

1911 American industrialist and electrical engineer Elmer A. Sperry creates a gyrocompass.

1924 Constant speed propeller brings better efficiency to aircraft.

1926 American physicist R. H. Goddard builds and launches the first liquid-fuel rocket.

1927 First electrically powered, automatically controlled pipeline is built in California.

1930 English inventor Frank Whittle patents the basic design of the turbojet engine.

1939 Russian-American aeronautical engineer Igor Sikorsky makes the first successful tethered helicopter flight.

1941 The Whittle jet engine is flown from Britain to the United States and provides the model for the first practical American jet engines that will be built by General Electric.

1947 Diesel-electric locomotive is built.

1952 First regular jet air passenger service begins; it goes from London, England, to Johannesburg, South Africa.

1954 U.S. Navy launchs the first nuclear-powered submarine, *U.S.S. Nautilus,* the first use of nuclear propulsion. It could cruise 62,500 miles before refueling.

1956 German engineer Felix Wankel develops the prototype for his Wankel engine, a rotary-piston engine.

1956 English engineer Christopher Cockerell designs the hovercraft.

1959 *NS Savannah* launches; it is the first nuclear-powered merchant ship.

1961 *USS Enterprise,* a nuclear powered aircraft carrier and the world's largest ship, is launched. It operates at speeds of 20 knots for distances of up to 400,000 miles.

1966 Electronic fuel injection is developed; it eventually replaces the carburetor.

1968 Supersonic transport plane.

1972 Ford invents the sodium sulfur battery.

1977 Hydrogen gas used to power vehicles in two California experiments.

LIGHT, HEAT, ELECTRICITY, AND COMMUNICATION

B.C.E.

c. 500000 Peking man uses fire for heat, light, and cooking.

c. 17000 Stone oil lamps enable Paleolithic artists to paint and engrave the walls of caves.

c. 2400 Greek civilization designs buildings to take advantage of solar heating.

c. 8000 Oil lamps are used in Mesopotamia.

c. 2500 Glass-making occurs in Egypt and Mesopotamia.

c. 600 Greek philosopher Thales observes that amber, a fossilized type of tree sap, attracts bits of paper and certain materials, like straw, when rubbed. This is the first mention of static electricity.

285 The Lighthouse at Alexandria is constructed; a mirror projects the light of fire for thirty miles.

1 Roman engineers develop a vertical watermill known as a Vitruvian mill.

C.E.

551 Jerome Cardan, an Italian mathematician, determines that while amber attracts light objects, a magnetic black stone attracts only iron. This is the first in a series of discoveries that links electricity to magnetism.

1340 Blast furnaces are developed in Europe.

1600 English physician William Gilbert discovers that materials like glass, sulfur, and diamonds behave just like amber. He calls these materials electrics, which means amber in Latin.

1646 Walter Charlton coins the word "electricity" to explain the attraction between these substances.

1672 German physicist Otto von Guericke molds a large sphere out of sulfur. Holding a piece of wool against this spinning sphere produces a large spark. This is the first generator to use friction to create electricity.

1729 Englishman Steven Gray first discovers that metals are conductors and non-metals are non-conductors.

1745 Dutch mathematician and physicist Pieter van Musschenbroek invents the Leyden Jar, which stores an electric charge.

1747 American statesman and inventor Benjamin Franklin deduces the existence of positive and negative electric charges.

1752 Benjamin Franklin flies a kite during a thunderstorm with a key dangling on the end of a wire. A silk string collects a charge from the thunderclouds which is conducted into a Leyden Jar. Thus, he makes the connection between lightning and electricity. This experiment leads to his invention of the lightning rod.

1762 Oil street lamps used in New York City.

1767 The city of Philadelphia lights streets with whale-oil lamps.

1790 First steam-heated factory.

1792 Coal-gas lighting invented by William Murdock in Cornwall.

1794 Italian physicist Allessandro Volta creates the first continuous electrical current by making a battery out of silver and zinc strips placed in salty water. Prior to this discovery all man-made electrical sources came from static.

1795 British physicist Benjamin Thompson (Count Rumford) invents the Rumford stove, a close-toped range that economizes fuel.

1800 The first commercial battery is manufactured. Scientists realize that if chemical changes can create electricity, then electricity can create chemical change.

1803 First factory illuminated by coal-gas lighting, in James Watt's foundry (Scotland).

1803 Italian cities of Genoa and Parma are lighted by kerosine from an oil well in Modena.

1813 London Bridge is lighted by gas.

1819 Danish physicist Hans Christian Oersted creates a magnet with electrical current, establishing the connection between electricity and magnetism.

1826 German physicist George Simon Ohm publishes *Die galvanische kette, mathematisch bearbeitet,* in which he describes his discovery that the voltage across an electrical conductor is proportional to the electrical current, and that the current is inversely proportional to the resistance of the conductor. His formulation becomes known as Ohm's Law.

1828 English chemist and physicist Michael Faraday discovers electromagnetic induction.

1829 American physicist Joseph Henry develops a coil magnet that grows stronger as more wire is wound around an iron core. He succeeds in lifting more than a ton of metal.

1830 Thermostat is invented.

1831 English chemist and physicist Michael Faraday creates the first electrical generator by using a magnet and a spinning copper plate to produce a current. Using a steam engine to keep the copper plate spinning within the magnetic field, electrical current is produced.

1831 American physicist Joseph Henry, by reversing Faraday's discovery, passes an electrical current through a magnetic field to turn a copper wheel, creating the first electric motor. For the first time in history, electrical energy can be used to power machines to do work that was formerly done by humans and animals.

1834 The first practical liquid refrigerating machine is patented by American inventor Jacob Perkins.

1835 The electric automobile is created by American inventor Thomas Davenport.

1838 American inventors Samuel F. B. Morse and Alfred Vail first demonstrate practical telegraphy.

1844 Samuel F. B. Morse builds the first electric telegraph. By transmitting short or long signals along a wire, messages can be sent anywhere. The Morse code makes it possible to send messages long distances at the speed of light.

1845 Safety matches are developed.

1849 American engineer James B. Francis invents the hydraulic turbine.

1851 American inventor John Gorrie patents an expansion cycle refrigerating machine.

1852 Heat pump is invented.

1868 The Siemens brothers (German) design the regenerative gas furnace.

1864 Scottish physicist James Clerk Maxwell publishes his theory of light and electricity.

1866 The first transatlantic cable is laid, creating a permanent electrical communications link between the old world and the new.

1868 French chemist Georges Leclanche invents the zinc-carbon battery, a precursor of the dry cell and the modern portable battery.

1875 A building in France is iluminated by electricity.

1876 American inventor and educator Alexander Graham Bell develops the telephone by converting electrical impulses into sound.

1878 American inventor Charles F. Brush invents the arc lamp.

1879 The streets of Cleveland are lit by carbon-arc lamps.

1879 Thomas Edison invents the incandescent lamp.

1879 English physicist William Crookes develops the cathode ray tube.

1882 October. The pioneering firm of United States Electric Illuminating Company starts up South Carolina's first central station for incandescent electric lighting one month after Thomas Edison opened his central station on New York City's Pearl Street. In the following years, U.S. Electric becomes one of Edison's main competitors.

1885 American electrical engineer William Stanley invents the transformer.

1886 Austrian chemist Carl Auer (Freiherr von Welsbach) invents the Welsbach mantle, tripling the output of kerosene lamps and gas burners.

1884 American engineer Lester A. Pelton develops the hydraulic turbine.

1884 First large-scale use of natural gas in Pittsburgh.

1887 Italian-American physicist Nikola Tesla invents a motor that produces alternating current. This discovery changes the way electricity is transmitted over long distances.

1889 The first commercial, long-distance transmission of electricity takes place when a direct-current line provides power from Willamette Falls for street lights in Portland, Oregon.

1893 German physicists Julius Elster and Hans F. Geitel invent the first photoelectric cell as a result of studying the photoelectric effect.

1895 The first hydroelectric generator at Niagara Falls, New York, produces alternating current from a Nikola Tesla design.

1900 A hydroelectric plant is built at Niagara Falls.

1901 The first reception of transatlantic radio signals.

1904 Gas is used for the first time for heating and hot water in London, England.

1904 English electrical engineer John Ambrose Fleming patents the first electron tube, which he calls a diode vacuum tube.

1904 American electrical engineer Ernst Alexanderson creates a high-frequency alternator; it allows reliable transoceanic radiotelegraph communication.

1909 Shoshone Transmission Line, power generated by the Shoshone Hydroelectric Generating Station to Denver.

1910 French chemist and physicist Georges Claude invents the first neon light.

1911 English engineer Charles Algernon Parsons improves the turbo-alternator for generating electricity in power stations.

1913 Austrian inventor Victor Kaplan patents his turbine; the invention enables hydroelectric power stations to be more consistently efficient.

1920 Martin Hochstadter introduces a three-core power cable that does not become deformed or burn by the high voltage electricity.

1930 First domestic gas water heater to work efficiently is developed.

1934 Power from the Boulder (Colorado) dam is transmitted 270 miles to Los Angeles, California.

1937 First commercial convective heater equipped with an electric fan.

1937 The five million-volt van de Graaff generator, Westinghouse "Atom Smasher, 1937," represents the first large-scale program in nuclear physics established by industry.

1939 The Massachusetts Institute of Technology builds a solar house.

1939 American physicist John Vincent Atanasoff collaborates with Clifford E. Berry to design the first digital electronic computer.

1940 The first gas-powered turbine to generate electricity is developed in Switzerland.

1943 Electric cables are filled with pressurized gas for insulation in England.

1945 Perry L. Spencer (United States) patents the first microwave oven.

1948 Bell Telephone Laboratories first demonstrates the transistor, a non-vacuum device that will eventually replace the conventional electron tubes.

1952 Four thousand die in London, England, from smog during air-inversion event.

1953 Four hundred die in New York City from smog event.

1954 Americans D. M. Chapin, C. S. Fuller, and G. L Pearson develop the silicon photovoltaic cell.

1955 Narinder Kapany invents optical fiber in Germany.
1958 American Gordon Gould develops the laser.
1959 Jack Kilby and Robert Noyce invent the integrated circuit.
1962 American Nick Holonyak, Jr., invents the light-emitting diode.
1963 Solar furnace capable of generating temperatures greater than 9,300°F becomes operational in Japan for scientific research.
1964 Generating electricity without turbines via magnetohydrodynamics (MHD) could theoretically double the output from nuclear power plants.
1964 George Heilmeier develops the liquid-crystal display.
1966 First superconducting motor.
1971 Ted Hoff invents the microprocessor.
1986 April. A severe nuclear accident occurs at Chernobyl in the former Soviet Union.
1986 High Temperature Superconductors developed by J. Georg Bednorz and Karl A. Muller.

MECHANICAL ENGINEERING

B.C.E.
1550 Levers used for well sweep in Egypt as well as India.
1500 Simple pulley employed in Egypt.
900 Rotary bucket pump invented.
700 First mention of the pulley; chain of pots used to raise water.
600 Compound pulley crane first used in Greece.
300 Tooth wheels and gears developed.
180 Quern or revolving mill is invented; turned by slaves or asses.
150 Force pump appears.
100 Undershot water-wheel first designed.
27 Vitruvian waterwheel created; first known instance of the transmission of power through gearing.

C.E.
1–100 Aeolipile, earliest recognized steam-powered mechanism, built by Hero of Alexandria.
200–300 Barbagal water-mills developed in Provence, France.
600–700 Windmills appear on Iranian plateau.
1240 Water-powered saw and jack invented by Honnecourt in France.
1335 Mechanical clock erected in the tower of Milan.
1472 In the next fifty years, Leonardo Da Vinci constructs the following devices: centrifugal pump, dredge for canal building, breech-loading cannon, rifled firearms, universal joint, rope-and-belt drive, link chains, bevel gears, spiral gears, parachute, and propeller.
1530 Foot-driven spinning wheel invented.
1698 Steam pump created by Thomas Savery (English).
1698 Denys Papin's steam engine, France.
1705 Reciprocating steam engine developed by Thomas Newcomen (English).
1720s Thomas Newcomen's steam engine comes into general use.
1769 James Watt receives main patent on condensing steam engine.
1769 Richard Arkwright invents his spinning water frame to spin yarn and silk; later to be used in North America as an industrial mill.
1772 Ball-bearings developed.
1772 James Watt's double-acting steam engine first used in Britain.
1782 Double-acting steam engine patented by James Watt.
1785 Screw propeller invented invented by Joseph Bramah.
1801 Richard Trevithick demonstrates first successful steam-powered road vehicle in Cornwall, England.
1816 Robert Stirling patents a forerunner of the Stirling engine, tauted as "A New Type of Hot Air Engine with Economiser."
1827 Benoit Fourneyron develops a water turbine in France.
1877 Invention of a hydraulic elevator with jugger mechanism.
1884 First practical steam turbine patented by Charles Algernon Parsons.

WARFARE AND SPACE

B.C.E.
30000 Bow and arrow first used.

1000 Assyrians use battering rams mounted on wheeled fighting towers.

1000 Crossbow invented in China.

400 Catapult and mechanical crossbow (ballista) invented at Syracuse and used against Carthage.

400 Catapult first used in China.

215 Archimedes' catapult used against the Romans.

C.E.

950 Gunpowder invented in China.

1000 The trebuchet, a missile thrower of great force operated by a hundred or more men, appears in China.

1118 Cannon used by Moors.

1232 Chinese use rockets against the Tartars and invent hot-air balloons.

1259 First cannons used in China.

1405 Portable firearms appear.

1490s First portable siege guns used in a military campaign in France.

1515 Wheel-lock gun invented in Germany.

1525 Rifled musket designed.

1835 American Samuel Colt invents the revolver.

1840s Guncotton (nitrocellulose) and nitroglycerine developed.

1850s First ocean-going steam-powered naval vessels.

1860s Railroads provide important support for armies in the U.S. Civil War and the Wars of German Unification.

1864 Self-propelled torpedo invented by Robert Whitehead (England).

1867 Alfred Nobel invents dynamite.

1880s Electric power introduced on warships, which allows significant improvement in their capabilities.

1904–1905 Japan defeats the Russian army; first use of telephone and telegraph to coordinate troops and supplies.

1914 Tank invented by E. D. Swinton (England).

1926 Liquid-fuel rocket developed by R. H. Goddard (United States).

1939 August 2. Albert Einstein writes to U.S. President Franklin Roosevelt to warn him of the possibility of Nazi Germany developing atomic weapons.

1944 Ballistic missile developed by Wernher von Braun (Germany).

1945 July 16. First atomic bomb test at Alamogordo, New Mexico, under the code name "Manhattan Project."

1945 August. United States drops atomic bombs on Hiroshima and Nagasaki, Japan, during World War II.

1949 The Soviet Union tests its first atomic bomb, which launches the "arms race" with the United States.

1952 United States tests the first hydrogen bomb.

1955 World's first nuclear submarine, *Nautilus,* is tested (United States).

1957 First artificial satellite, Sputnik, in orbit (Soviet Union).

1958 Neutron bomb developed in the United States

1965 First nuclear reactor in space is launched.

1983 U.S. President Ronald Reagan announces the Strategic Defense Initiative, intended to shield the United States from nuclear attack.

1991 January 15. United States and allied countries launch Operation Desert Storm against Iraq, a military operation characterized by some as an "energy war."

1992 May. The nuclear arms race ends—for the first time since 1945, the United States builds no nuclear weapons.

BUSINESS, GOVERNMENT, AND CULTURE

C.E.

529 St. Benedict founds Monte Cassino; the Benedictine Order, which is to become a very powerful force in Western Christianity, adopts manual labor as a virtuous action.

716 St. Boniface, an English Benedictine, visits Germany and establishes abbeys and country estates as centers of industry and material progress.

1066 With the Norman conquest of England, industry moves from the abbeys and country estates to the towns because of energy—water mills and water transportation.

1095 The First Crusade begins (ending in 1099); the Crusades, of which there were ten—if one counts the tragic Children's Crusade—reflected a growing energy surplus (as did the universities) in western Europe, not yet channeled into national military establishments.

1150 Cistercians introduce use of city garbage and sewer water as fertilizer near Milan (Italy); both Benedictines and Cistercians drain swamps and lakes in Germany, France, the Low Countries, and Italy.

1273 Coal smoke in London, England, provokes complaints from the gentry.

1306 King Edward I of England makes it a capital offense to burn coal in London.

1345 Division of hours and minutes into sixties (Germany); without linear time, the Industrial Revolution would not have been possible.

1798 T. R. Malthus publishes *An Essay on the Principle of Population,* as pessimistic in its conclusions as the Marquis de Condorcet's work was optimistic

1863 The British government passes the "Alkali Works Act" in an attempt to control environmentally harmful emissions.

1870 John D. Rockefeller founds the Standard Oil Company.

1882 Standard Oil Trust is formed and buys controlling interest in a number of oil companies.

1901 Oil found at Spindletop, Texas, which leads to the formation of Gulf, Texaco, and Sun oil companies.

1907 First drive-in gas station opens in St. Louis, Missouri.

1908 Oil is discovered in Persia; Anglo-Persian (later British Petroleum) is formed.

1911 U.S. Supreme Court breaks Standard Oil monopoly into thirty-four companies.

1920 The U.S. Congress passes the Federal Water Power Act of 1920, which authorizes the first Federal Power Commission (later Federal Energy Regulatory Commission). It has authority to issue licenses for hydroelectric projects that are best adapted to the comprehensive development of a waterway.

1930 Huge oil discovery in East Texas.

1935 Federal Water Power Act becomes part of the Federal Power Act to regulate interstate commerce in electricity.

1937 The Bonneville Project Act creates the Bonneville Power Administration (BPA), which markets electricity generated at Federal hydro projects to the northwestern United States; it also owns the nation's largest network of long-distance, high-voltage transmission lines needed to bring hydropower to market.

1938 Oil discovered in Kuwait and Saudi Arabia.

1941 The United States, Britain, and the Netherlands put an oil embargo on Japan after it takes over Indonesia.

1952 Smog identified for the first time in Los Angeles, California, from the combination of a large number of automobiles, bright sunlight, and frequently stagnant air.

1953 December 8. U.S. President Dwight Eisenhower delivers his "Atoms for Peace" speech before the United Nations.

1954 The Atomic Energy Act of 1954 permits and encourages the participation of private industry in the development and use of nuclear energy.

1955 Atomic Energy Commission announces the Power Demonstration Reactor Program, a cooperative effort with private industry in constructing experimental power reactors.

1956 Celilo Village, a traditional Indian tribal fishing ground, is flooded by the Dalles Dam. The sovereign rights of tribes lead to several court cases and agreements that affect use of rivers and hydropower in the United States.

1957 The United Nations establishes the International Atomic Energy Agency in Vienna, Austria, to promote the peaceful use of nuclear energy.

1960 The Organization of Petroleum Exporting Countries (OPEC) founded in Baghdad, Iraq.

1961 The United States and Canada sign the Columbia River Treaty. Under the treaty, Canada builds two storage dams and one dam for generation, resulting in greater power and flood control benefits at U.S. facilities downstream.

1964 The Pacific Northwest Coordination Agreement is signed; it seeks to meet the region's electricity needs most efficiently by operating the diverse generating resources as a coordinated system, as if they were owned by a single utility.

1965 First major blackout occurs in northeast United States and Canada.

1966 The Public Power Council is formed to give a voice to publicly owned utilities in the Northwest. PPC represents and advocates the common legal and technical interests of the Northwest's consumer owned utilities.

1967 The Pacific Northwest-Pacific Southwest Intertie creates the only direct way to move electricity between the Northwest and California. Billions of dollars are saved by the Northwest trading some spring and summer surplus power for fall and winter power from California.

1967 The Air Quality Act becomes law in the United States.

1968 Oil is discovered on Alaska's North Slope.

1969 Santa Barbara, California, oil spill.

1970 Production of crude oil in the United States reaches an all-time peak.

1970 Clean Air Act goes into effect in the United States.

1973 October 6. The Yom Kippur War breaks out in the Middle East.

1973 October 17. The Organization of Arab Petroleum Exporting Countries declares an oil embargo; prices rise to nearly $12 a barrel from $3.

1975 December. U.S. President Gerald Ford signs the Energy Policy and Conservation Act, extending price controls into 1979, mandating automobile fuel economy standards, and authorizing creation of a strategic petroleum reserve.

1977 June 3. The U.S. Department of Energy is created by the consolidation of the Federal Energy Administration, the Energy Research and Development Administration, and the Atomic Energy Commission.

1978 U.S. Congress passes Public Utility Regulatory Policies Act (PURPA). This law requires utilities to purchase electricity from qualified independent power producers. Portions of the act helped stimulate growth of small scale hydro plants as a means of meeting the nation's energy needs.

1978 November. U.S. President Jimmy Carter signs the National Energy Conservation and Policy Act, which promotes conservation activities, requires development of standard measures of energy efficiency and its reporting to the public.

1978 Last new nuclear power plant ordered in the United States.

1979 March. An accident occurs at the Three Mile Island nuclear power plant in New York.

1979 June. U.S. President Jimmy Carter announces program to increase the nation's use of solar energy.

1979 July. U.S President Jimmy Carter proposes $88 billion effort to enhance production of synthetic fuels from coal and shale oil reserves.

1980 U.S. Congress passes the Pacific Northwest Electric Power Planning and Conservation Act. The Northwest Power Planning Council is formed. The Council is charged with developing a plan to meet Northwest energy needs. The act also called for the Council to develop a fish and wildlife mitigation and enhancement plan.

1980 June. U.S. President Jimmy Carter signs the Energy Security Act.

1981 July. U.S. President Ronald Reagan signs Executive Order 12287, effectively decontrolling crude oil and refined petroleum products.

1979–1981 Panic caused by Iran's revolution and the Iran-Iraq war sends oil prices as high as $34 a barrel from $13.

1983 January. U.S. President Ronald Reagan signs the Nuclear Waste Policy Act of 1982, the first comprehensive nuclear waste legislation.

1986 Congress amends the Federal Power Act, increasing environmental review of hydropower projects.

1986 Oil price collapses to $12 a barrel.

1987 December. U.S. Congress approves Yucca Mountain, Nevada, as the only repository site for high-level nuclear waste.

1988 The Northwest Power Planning Council designates 44,000 miles of Northwest streams as "protected areas" because of their importance as critical fish and wildlife habitat.

1989 New York Governor Mario Cuomo and the Long Island Power Authority announce that the already built Shoreham Nuclear Power Plant will never open.

1989 *Exxon Valdez* runs aground off the Alaska coast.

1990 August. Iraq invades Kuwait, triggering an international crisis.

1991 January 15. United States and allied countries launch Operation Desert Storm against Iraq to end its invasion of Kuwait.

1991–1995 By some estimates, fish and wildlife protection measures reduced firm electric generation by about 850 megawatts annually.

1992 June. Representatives from many nations convene at the Earth Summit in Rio de Janeiro, Brazil.

1993 April. U.S. President Bill Clinton announces that the United States will stabilize greenhouse gas emissions at 1990 levels by the year 2000.

1994 The U.S. Supreme Court rules that states have the authority under the Clean Water Act to establish minimum streamflows at hydro projects. The ruling gives states more authority in hydro licensing and relicensing decisions.

1997 December 11. The Kyoto Protocol is adopted by the United Nations Framework Convention on Climate Change.

1998 United States oil and utility industry companies spend over $100 million to influence federal government energy policy.

1998 Oil prices fall sharply; collapses in Asian economies severely curtail demand.

2000 Oil prices surge to highest levels since the mid-1970s.

INDEX

Key to codes: b=box, f=figure, t=table. Bold page numbers indicate main articles. Italic numbers indicate photos.

C

F

I

J

K

L

M

N

O

P

S

Y

Z